Industrial Drafting

Principles
Techniques
Industry Practices

by John D. Bies

Macmillan Publishing Company
New York
Collier Macmillan Publishers
London

Macmillan Publishing Company
866 Third Avenue, New York, N.Y. 10022
Collier Macmillan Canada, Inc.

Library of Congress Cataloging-in-Publication Data

Bies, John D., 1946–
Industrial drafting.

Includes index.
1. Mechanical drawing. I. Title.
T353.B465 1987 604'.2 86-31156
ISBN 0-02-510610-4

10 9 8 7 6 5 4 3 2 1

Printed in the United States of America

Contents

CHAPTER 1

Drafting Room Practices

- Technical Personnel
- Drafting Organization and Standards
- Project Procedures

The rapid development of technology, with its proliferation of new products, coupled with the implementation of automated and computer systems, has resulted in an increased demand for information presented in a clear and understandable format. The need for industrial technical and engineering drawings to translate this information is greater than ever. The future for drafters, technicians, technologists, and engineers is bright.

It may appear that the skills required for industrial drafting have changed significantly with the use of computer-aided drafting (CAD) systems; this, however, is a false assumption. Even with the aid of CAD systems, the professional drafter must still be familiar with all drafting concepts, and know how to prepare and present various kinds of drawings. CAD and all the other automated systems are only new *tools*, and are not substitutes for drafting competence.

It is important for drafters to comprehend the nature of their profession and its working environment. All drafters function as part of a team rather than as individuals. Engineering firms have set procedures that they follow for assigning responsibilities and completing projects. It is, therefore, the purpose of this chapter to review those drafting room practices that are frequently encountered in the workplace.

Technical Personnel

Since the early 1960s the engineering field has evolved significantly, especially in the nature of technical personnel and their responsibilities on the job. During the first half of the twentieth century, engineers were often assigned to drafting tables, where they rose through the ranks from preparing working drawings to becoming project leaders. An essential part of the early university preparation of engineers was a series of drafting courses.

The engineering field has changed substantially since the 1950s, turning its attention to the creation and development of new products and a more theoretical approach to technology. In fact, very few engineering programs still require their students to take drafting courses. In some situations this lack of preparation means that the engineer must receive on-the-job training, especially in smaller companies or those with an applied approach to product development. It should be noted, however, that even when engineers are not concerned with the preparation of drawings, they still must be able to read them with complete understanding and make any necessary changes.

Technicians and Technologists

With engineers having now oriented themselves to a more theoretical approach to technology, industry has assigned drafting tasks to two major groups of technically prepared professionals: technicians and technologists. Drafters themselves may be assigned such job titles as technician, technical aid, technologist, or drafter. In many organizations, the drafting function has become so completely merged with these positions that all technical personnel, in effect, are drafters.

Though the responsibilities of drafters, technicians, technologists, and engineers may overlap in many companies, there are formal distinctions among the occupations. To make matters even more confusing, it is not uncommon to find the job title "engineer" assigned to technical personnel who have no formal training in engineering but who do have the practical technical experience and competency to warrant such a title.

A technician is a person with specific technical skills and knowledge who works closely with the engineers and the engineering project staff to whom his specialized training can be of most benefit.

Technicians might specialize in mechanical drafting and design, illustration, manufacturing, machining, computer programming, and production estimating. Formal education for a technician includes the completion of a one- or two-year training program beyond the high-school level, and can lead to an Associate degree. The vast majority of technician programs require course work in drafting.

Technologists are considered to have not only an in-depth training in a particular technology, but a broader preparation and expertise than technicians. Technologists are required to apply scientific and mathematical principles to mechanical design and manufacturing. In many companies, the technologist is considered an engineer. Technologists work directly with engineers on various aspects of product design, planning, engineering, and manufacturing. Formal education for technologists requires the completion of a four-year degree program in one of two types of technologist program: engineering technology and mechanical or industrial technology. The first requires more course work in math and science, and is closely related to that of an engineer. The second technology program is less theoretical and more applied. Both programs, however, require competence in drafting procedures.

Drafters

The success of the day-to-day operation of an engineering department or firm may well depend upon the quality of its drafting staff. Industrial drafters are more likely to be generalists, compared to architectural or structural drafters. They must therefore have a sound understanding of the principles and practices of drafting.

In some industries, it is possible for drafters to specialize in particular areas of drafting, such as tool-and-die design, special machine design, auto-body design, or jig and fixture drawings. Whatever the area of specialization, the drafter's job is to translate ideas and designs into complete sets of working drawings. These drawings are then used by manufacturing engineers, tradesmen, machine operators, and service technicians for the production, maintenance, and servicing of products.

Most firms that hire beginning drafters have basic expectations. Industrial and engineering employers most commonly want employees with a solid understanding of and proficiency in basic drafting practices and procedures. With this as a foundation, the employee will then develop specialized skills needed for the job.

The formal education of drafters is varied, ranging from one-year training programs to four-year technology and engineering degrees. In most cases, drafters have had at least one full year of general drafting course work, and in some cases more advanced and/or specialized courses.

Drafting Organization and Standards

The drafter's working environment has changed dramatically over the last fifty years. The physical appearance, furniture, and equipment used in modern drafting facilities no longer resemble the clustered and crowded offices of yesteryear. Today more tools are available to the drafter, ranging from CAD systems and special printing devices to electric erasers and customized templates.

Personnel Organization

The exact nature and structure of the drafting department will vary from one firm to the next. The smaller the firm, the fewer the organizational levels, and the broader the responsibilities of each individual. Thus, drafters employed in small engineering departments or firms will be expected to produce a wide range of drawings and perform other nondrafting duties (e.g., writing specifications and technical reports). Those employed in large departments will often be required to pass through a series of positions, each with its specific duties and responsibilities.

Regardless of the training and educational experience of newly graduated drafters, most will be required to start at the bottom, or near the bottom, and work their way up to more responsible positions. The specific titles and number of job levels may vary from firm to firm, but they will generally include the following positions:

1. Tracer
2. Junior drafter
3. Detailer
4. Layout drafter
5. Checker
6. Designer

The tracer, or first-level drafter, has little or no on-the-job experience. To introduce the beginning drafter to the company's standards and procedures, and to develop drawing skills and speed, the tracer will produce plans for reproduction by tracing them onto some drafting medium.

In many small and medium-sized drafting departments, the first position will be that of junior drafter. These drafters trace drawings and make reproductions, and are expected to make minor corrections and changes on existing engineering drawings, such as dimension and part location changes.

The next level of drafter is the detailer, the backbone of the engineering department. He or she is responsible for making detailed drawings of all parts, which are then used for planning and production purposes. Details are basically the working drawings that are prepared from product designs and layouts.

Once the drafter has developed enough experience and skill as a detailer, the next step is layout work. Layout drawings are used in designing complex products with many parts and components. These drawings are used to show how the parts will be assembled and will function. Information provided in layout drawings will have less to do with individual parts than with the overall working of the product.

From layout drafter, the next promotion is usually checker. The checker reviews all drawings, including revised drawings, details, and layouts, to make sure that they are correct. In addition to the review process, checkers will also be responsible for design revisions, layout drawings, and details.

Although in some companies, the top position might be project or production engineer, designers are considered to be at the upper end of the drafting field. Designers develop new products and devices, and manage all the drafting personnel serving under them. Any drawing errors are the responsibility of the designer. Several years of experience as detailer, layout drafter, and checker often precede promotion to designer.

Drafting Standards

To ensure that a drawing can be read and interpreted by anyone, a system of standards has been developed that is used by industrial and governmental organizations. The standardization of drafting practices and procedures was made possible by the work of many professional and trade organizations and societies.

Through the American National Standards Institute (ANSI), representatives of federal and state government and industry involved with mechanical and industrial engineering and design have produced a standards handbook establishing acceptable principles of drawing presentation, dimensioning, tolerancing, surface texture, and graphic symbols. For more specialized drafting fields, such as electrical and piping drafting, other standards are also available.

The importance of these standards cannot be too strongly emphasized, for they represent drawing practices that are used throughout the industry. With a working knowledge of these practices, one can function in any drafting department.

MILITARY STANDARDS

Because ANSI standards are so widely used, one sometimes forgets that there are other standards organizations. One of the largest is the Department of Defense (DOD). All engineering drawings prepared under contract with the DOD must meet their requirements.

Originally each standard was prefaced with the letters *JAN*, which stood for Joint Army-Navy Specifications. Most JAN standards and specifications are now identified by the MIL, for military, preface. In some cases the MIL preface has been superseded by the DOD preface.

Anyone who is preparing drawings for the military should be working from MIL and DOD guidelines, which can be obtained from any regional office of a given branch of the military. A listing of all military standards and specifications can be obtained through the Superintendent of Documents, U.S. Government Printing Office, Washington, DC 20402. Examples of military standards are:

MIL-STD-12	Abbreviations for Drawings in Publications
MIL-STD-12B	Abbreviations for Use on Drawings
MIL-STD-17-2	Mechanical Symbols Used in Aeronautical and Aerospace Drawings
DOD-STD-100	Practices for Engineering Drawings
MIL-STD-1472	Human Engineering Design Criteria

AMERICAN NATIONAL STANDARDS INSTITUTE

ANSI is an organization that was formerly known as the American Standards Association (ASA) and the United States of America Standards Institute (USASI). It consists of engineering and technical organizations that have united for the promotion of standardization. (It should be noted that ANSI does not write the standards; it merely promotes their use.)

Standards and specifications promoted by ANSI will carry their initials as a preface. Other organizations who are members of ANSI and have developed specific standards for their professions use their own initials for the same set of standards. It is more common, however, to use the ANSI notation.

Examples of national organizations that belong to ANSI and have a significant influence on industrial drafting standards and practices are:

American Iron and Steel Institute (AISI)
American Society of Mechanical Engineers (ASME)
American Society for Testing and Materials (ASTM)
Society of Automotive Engineers (SAE)
Society of Manufacturing Engineers (SME)

A listing of standards and specifications available from ANSI can be obtained by writing to the American National Standards Institute, Inc., 1430 Broadway, New York, NY 10018.

ANSI coding procedures are similar to those used by the DOD. The code will be prefaced by ANSI, followed by a letter and number code and the date that the standard was accepted by ANSI, or the date of its latest revision. Some codes will have the date followed by another date with an R preface (e.g., 1964, R1980), which indicates a revision date. For example, ANSI-B6.1-1968 notes that the ANSI standard B6.1 was accepted by ANSI in 1968. Examples of ANSI standards codes of interest to drafters are:

ANSI-Y14.5M-1982	Graphic Symbols for Engineering Drawings
ANSI-Y14.5-1982	Dimensioning and Tolerancing
ANSI-Y14.35-1978	Surface Texture Drawing Symbols
ANSI-Y10.3-1968	Standard Symbols for Mechanics

ANSI-B46.1-1978	Standard Surface Texture
ANSI-Y32.2-1975	Graphic Symbols for Electrical and Electronics Diagrams
ANSI-B4.1-1967, R1979	Standard Limits and Fits

INTERNATIONAL AND NATIONAL STANDARDS

Perhaps the best-known international standards are the International Metric Standards, which give specifications for all metric units. The International System of Units, designated by the letters *SI*, was formed in 1960 and stands for the Système International d'Unités. The basic units of measure controlled by SI are:

Length—meter or metre (m)
Mass—gram (g)
Time—second (s)
Electric current—ampere (A)
Thermodynamic temperature—kelvin (K)
Amount of substance—mole (mol)
Luminous intensity—candela (cd)

Though SI units do play a role in industrial drafting (dimensioning and tolerances), this international body does not provide specific guidelines for drafting practices. These are usually set by individual countries. In turn, each country has a standards designation, or coding system, with which all standards and specifications are classified.

The first European standards organization was the International Standards Association (ISA), later replaced by the International Organization for Standardization (ISO). Membership in the ISO meant that a country would have to consider ISO guidelines and the adoption of the SI units. Thus, after becoming a member of that body, Great Britain found it necessary to consider conversion from the inch to SI metric standard to keep in step with other member nations.

Although international standards are agreed upon by member nations; member nations are not required to adopt them. For example, although the United States is a member of the ISO, it did not agree upon ISO steel flange size standards because U.S. tolerances are significantly different from European manufacturing tolerances, and the conversion process would be too costly. In fact, no agreement was possible between the American, British, and European communities for this particular standard.

Because there are differences in standards from country to country, it is important to be aware of variations in national standards designations, particularly for drafters working with firms involved in international contracts and trade. Table 1-1 gives a representative listing of countries and their standards designations.

Table 1-1 International and National Standards

Standard	Designation
International	
International Organization for Standardization	ISO
Pan American Standards Commission	COPANT
National	
Arabia	ASMO
Argentina	IRAM
Australia	AS-SAA
Austria	ONORM-ONA
Barbados	BNSI
Belgium	NBN-IBN
Bolivia	DGNT
Brazil	NB-ABNT
Bulgaria	NDS-ISMIV
Canada	CSA
Central Africa	CAS
Chile	INDITECNOR
China, People's Republic	GB
China, Taiwan	CNS
Colombia	INCONTEC
Czechoslovakia	CSN-RVH-PVUM-ON
Denmark	DS
Dominican Republic	DIGENOR
Ecuador	INEN
Egypt	EOS
Ethiopia	ES
Finland	SFS
France	AFNOR-NF
Germany	DIN, VDE, and VDI
Greece	ELOT
Guatemala	ICAITI

Table 1-1 International and National Standards

Standard	Designation
Hungary	MSZ-MSZH
India	IS-ISI
Indonesia	NI-YDNI
Iran	IAIRI
Iraq	IOS
Ireland	IA-IIRS
Israel	SI-SII
Italy	UNI
Jamaica	BJS
Japan	JIS
Korea, South	KS
Kuwait	KSS
Malaysia	MS
Mexico	DGN
Netherlands	DEN-NNI
New Zealand	SANZ
Norway	NS
Pakistan	PS-PSI
Peru	ITINTEC
Philippines	SAO-PBS
Poland	PN-PKN
Portugal	NP
Rumania	STAS-OSS
Saudi Arabia	SSA-SASO
South Africa	SABS
Spain	IRATRA-UNE
Sweden	SIS-SEN-SMS
Switzerland	SNV and VSM
Syria	SNS
Thailand	TIS
Turkey	TSE-TS
United Kingdom	BS-BSI
Uruguay	UNIT
USSR	GOST
Venezuela	COVENIN-NORVEN
Yugoslavia	JZS-JUS

International standards that are of interest to drafters include:

ISO R.128	Engineering Drawing, Principles of Presentation
ISO R.129	Engineering Drawing, Dimensioning
ISO R.406	Inscription of Linear and Angular Tolerances
ISO R.1101	Tolerances of Form and of Position
ISO R.1047	Architectural and Building Drawings: Presentation of Drawings
ISO R.1302	Method of Indicating Surface Texture on Drawings

IN-HOUSE DRAWING MANUALS

Companies with large drafting departments often provide drawing manuals that outline standard procedures and practices for the firm and the industry. These manuals may vary from a loose-leaf binder of ten to thirty pages to several volumes of five hundred pages each. Most drafting departments will have some form of in-house manual that outlines policies and procedures. In many instances, it will present sample drawings and project procedures. Standard forms of view placement, dimensioning, tolerancing, notations, and abbreviations are also provided. These manuals are not only used as a reference but sometimes also as an instructional tool for orienting new personnel. They will usually be based upon industry-wide standards.

In addition to the in-house manual, drafters are constantly working from product catalogs, which provide dimensional and design specifications for a wide range of parts used in manufacturing. These catalogs can save not only time but also money. For instance, detailed drawings are provided for parts that have to be specially made. If a standard premade part can be substituted, the detailed drawing could be eliminated, and the part would need to be seen only in the layout or assembly drawing. This reduces not only the time a drafter spends on the board, but also the time needed to make the part, which translates into reduced drawing time and expense and in reduced production time and cost.

Project Procedures

The engineer, technologist, and drafter are all part of a team, and very seldom work alone. The team is responsible for the planning, designing, and specification of products that may range from simple brackets to complex machinery. In some cases, these products must be not only functional, but also have a pleasing appearance.

It is important that the drafter be able to work well with other people, for the first step in a project is to talk with the project engineer and designer about the overall project goals, costs, and requirements. The engineers and designers will then develop rough sketches, which may or may not include dimensions and specifications. These sketches are then reviewed by small groups for further discussion and input.

Once several sketches have been accepted, they are turned over to the detail or layout drafter, who prepares a set of working drawings. From the sketches, the drafter must determine all the dimensional requirements and specifications. This information is usually obtained from industrial standards, product catalogs, and input from the in-house engineers and designers.

Once the initial working drawings have been completed, the project team reviews the drawings and reworks the concept, design, innovation, planning, and manufacturing of the product. The team's various suggestions are then implemented by the drafter.

Once all the revisions have been made and incorporated into the drawings, production procedures can begin. A prototype product is first made to work out any problems and determine if any modifications should be made in the design. Again, changes to be made are given to the drafter, who must then incorporate the modifications. This process will be repeated until a final design is accepted.

This is the usual procedure in many engineering and design organizations. In fact, much of the work done by drafters consists of revisions and changes. One can thus easily see why the drafter must not only be technically competent, but able to work well with others.

CHAPTER 2

Drawing Materials, Instruments, and Equipment

- Drawing Surfaces
- Leads, Lead Substitutes, and Inks
- Drafting Instruments and Equipment

The quality and clarity of a set of industrial drawings depends not only on the competence of the drafter, but also on the quality of the drawing materials, instruments, and equipment. Accurate drawings can only be made with precision instruments and measuring scales, while sharp, clear drawings require high-quality drawing surfaces and marking media. The drafter must therefore know how to identify and select high-quality instruments and materials, and develop good working techniques for using them.

Beginning drafters are frequently confused as to which product they should select, and from what source. The large number of manufacturers and drafting supply outlets often causes the inexperienced to purchase too many items. A relatively small number of drafting tools can do the job quite satisfactorily if care is taken in their selection. The rule of thumb when selecting drafting instruments and supplies is to purchase the highest quality that you can afford, rather than the largest quantity.

Industrial drafters should consider only reputable companies when making a purchase. If you do not know who these companies are, ask other drafters or a drafting instructor at a local educational institution to recommend them.

One other fact should be kept in mind: the most expensive and highest quality instruments will only perform as well as their owner knows how to use them. Drafting instruments should always be handled in a careful and professional manner. Learn how to properly use each drafting instrument and piece of equipment, and know which one is appropriate for which purpose. This chapter will discuss the types and uses of the various drafting materials, instruments, and equipment.

Drawing Surfaces

Drawing surfaces pertain to the medium upon which drawings are made, that is, drafting papers, films, and cloths. A variety of drawing surfaces are available to the drafter, some for general use, others primarily for particular situations and problems. The correct selection of these mediums can greatly affect the appearance and quality of a drawing.

Drafting Sheet Sizes and Formats

Drafting paper, film, and cloth can be obtained in standard sheet sizes or roll widths. Preprinted standard engineering and design formats may also be ordered by the sheet on a variety of drawing materials. To select the correct sheet or roll sizes and format, the drafter should be familiar with industrial standards.

DRAWING SHEET SIZES

There are standard sizes of paper that have been assigned specifications by both ANSI and ISO. ANSI paper sizes are typically specified in inch units (comparable metric sizes are also available through ANSI), while ISO standards are given in millimeters (mm). The basic ANSI sizes are specified by letter, while ISO standards are given in letter-number combinations. In addition to ANSI and ISO sizes, other sizes of sheets are available through supply companies.

ANSI size standards should be used for all drawings prepared in the United States. ISO sizes should be used when drawing specifica-

Table 2-1 Standard Sheet Sizes

Standard	Designation	Dimensions
ANSI	A	8.5 × 11 inches
	B	11 × 17 inches
	C	17 × 22 inches
	D	22 × 34 inches
	E	34 × 44 inches
ISO	4A0	1682 × 2378 mm
	2A0	1189 × 1682 mm
	A0	841 × 1189 mm
	A1	594 × 841 mm
	A2	420 × 594 mm
	A3	297 × 420 mm
	A4	210 × 297 mm
	A5	148 × 210 mm
	A6	105 × 148 mm
Other Sizes		12 × 18 inches
		18 × 24 inches
		24 × 36 inches

tions (contracts) require it—namely, for international work. Standard sheet sizes are presented in Table 2-1.

In addition to the individual sheets of drawing material, it is also possible to obtain paper, film, and cloth in rolls. Rolled material, however, is not available in preprinted formats. Most rolls come in lengths of 20 or 50 yards, and in standard widths of 24, 30, 34, 36, 42, and 54 inches.

FORMATS

Drawing formats are used to "frame" the drawing and provide room for written information. Usually the format will include a drawing number at the lower right-hand corner, which is sometimes repeated at the upper left-hand corner.

Most firms use preprinted drawing sheets. The drawing format will include a border within the edge of the sheet, and the information required to accompany the presentation. This will include the name of the company; drawing number and project references; notes pertaining to material, tolerances, and amended procedures; the drafter who prepared the drawing; the checker; the pertinent dates; and other necessary information.

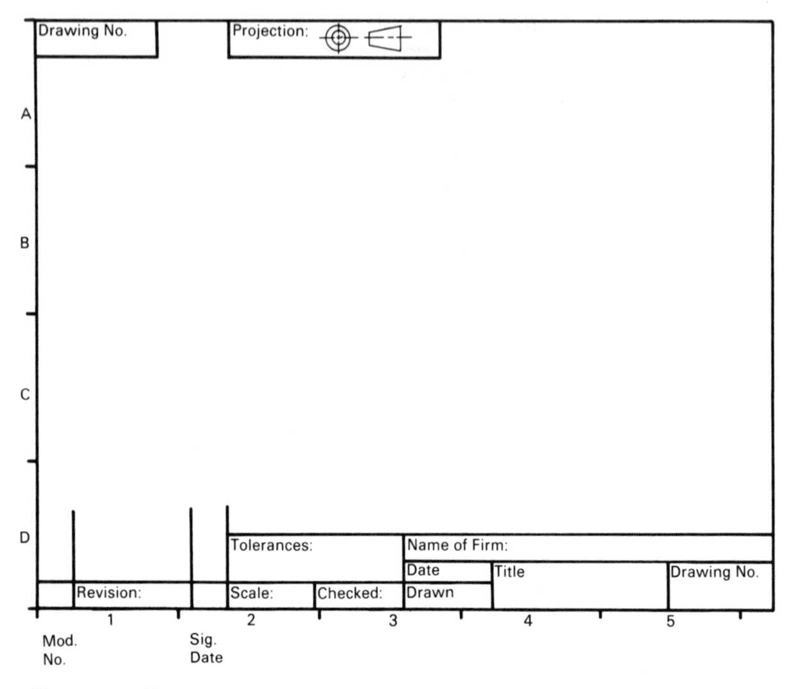

Fig. 2-1. Typical drawing format.

Figure 2-1 shows a typical preprinted layout. The only feature of this format that is not found in many U.S. formats is the reference to projection. Because the method usually followed in the United States for detail presentation is third angle projection, no reference is needed. In some countries, however (most of them European), first angle projection techniques are employed. Whenever you are preparing drawings for situations in which both first and third angle projections are used, notation should be made as to the procedure being used. (The difference between first and third angle projections will be discussed in further detail in Chapter 6.)

Drawing Surface Materials

The quality and variety of drawing surface materials have improved markedly since the turn of the century. Through modern manufacturing techniques and resins, materials are now available that are dimensionally stable, resistant to deterioration, and relatively inexpensive. Three basic types of drawing material are used by industrial drafters: paper, film, and cloth.

DRAWING PAPERS

Drawing paper is used for the vast majority of industrial drawings. Drafting departments use two basic types of drawing paper. The first is a thicker, nontranslucent paper that is often used for drawing maps, charts, and drawings for photographic reproduction. Sometimes a cream or buff paper is used for layouts and detailed drawings.

The second type of paper, vellum and tracing paper, is by far the most commonly used, and is recommended for general industrial drawings. There is usually little difference between vellum and tracing paper; in fact, the two terms are often used interchangeably. Where there is a difference, the tracing paper will be more translucent than the vellum.

Vellums and tracing papers are available as either *natural* or *prepared* papers. Prepared papers are treated with a resin and tend to be thicker and stronger, with little or no loss of transparency. Natural vellums are untreated. Though prepared papers are slightly more expensive than natural papers, because of their resistance to wear they are preferable where frequent changes are made on drawings.

DRAFTING FILMS

The newest drawing material used in drafting is film. Because of its dimensional stability, high transparency, resistance to age and heat, nonsolubility, and waterproofness, polyester film provides an excellent medium for industrial drawings. Film is used primarily when a large number of high-quality reproductions is required.

Another advantage of polyester film is that pencil lines appear on it as high-quality black lines. This, however, is due in part to the abrasiveness of the film, which causes a more rapid wear on drawing leads. A harder lead (two or three degrees harder than that typically used on paper) is therefore recommended. There is also available a plastic lead that is especially designed for film. Inks and other drawing media are also quite effective.

The major drawback in using polyester film is its cost, which is higher than that of paper. Film is therefore reserved for drawings where high quality is important.

DRAFTING CLOTH

The first drawing material ever used was cloth. Ancient drawings and writings were produced on material made of cellulose fibers that were stabilized by sizing material of glue or starch. Thanks to the excellent properties of drawing papers and films, there is little need nowadays to prepare industrial drawings on cloth. However, cloth is still sometimes used to meet contractual requirements.

Drafting cloth is produced from a high-quality linen or cotton, and sized with a starch. In some cases, the starch will be moisture resistant. Cloth will resist aging, and reproduces well. Though cloth will accept lead and pencil, ink usually works better.

Leads, Lead Substitutes, and Inks

One change in recent years is in the kind of drawing media used, namely, leads, lead substitutes, and inks. Most drawings are still prepared with leads, but owing to the improvements in modern inks, an increasing number of firms are adopting ink. Of limited and more specialized use are lead substitutes.

Leads and Lead Substitutes

Most sketching and mechanical drawings are still made with lead. The original lead medium was the pencil, followed by the mechanical lead holder, and more recently, the stick lead holder. Whatever the lead medium used, care should be taken in its selection. All leads are made of a combination of clay and graphite and come in fourteen different ratings of hardness, ranging from a very soft 6B lead to very hard 9H lead (see Table 2-2). Lead substitutes are made of a synthetic plastic that has but one grade. Lead substitutes are mainly used in situations in which lead wear is a problem, such as polyester film drawings.

The two softest leads, 6B and 5B, are much too soft for any practical industrial drafting; they are used in some forms of freehand sketching. Leads ranging between 4B and HB are useful for preparing certain types of illustration, rendering, and sketch. Most industrial drafting work uses leads between F and 4H. The very hard leads, 5H and higher, are excellent for layout, light construction work, and nonreproducing lines. The exact degree of hardness that should be used

Table 2-2 Rating of Lead Hardness

Rating	Level of Hardness
6B	
5B	
4B	Soft
3B	
2B	
B	
HB	
F	Medium
H	
2H	
3H	Hard
4H	
5H	
6H	
7H	Very Hard
8H	
9H	

by individual drafters will depend upon the amount of pressure they normally apply when drawing. Those who are "heavy-handed" would tend to select a harder lead rating, while the "lighter-handed" would choose a softer-rated lead. Probably the two most commonly used leads are 2H and 3H.

The most basic form of lead medium is the drafting pencil, which can be purchased in each of the lead hardness ratings. Pencils are wood-bonded leads that must be sharpened by first removing a section of the wood with either a knife or a mechanical sharpener. With the lead exposed, the tip can be sharpened to a conical, wedge, or bevel point. If a conical point is desired, a special pencil sharpener known as a *pointer* can be used (Fig. 2-2). The other points must be sharpened manually with either a file or a piece of sandpaper.

A mechanical lead holder is an ejector-type device that holds individual pieces of lead (Fig. 2-3). The lead can be ejected to a desired length by a clamping chuck that can be opened or closed by either pressing or turning a release mechanism at the opposite end. Once exposed, the lead can be sharpened in the same manner as pencils.

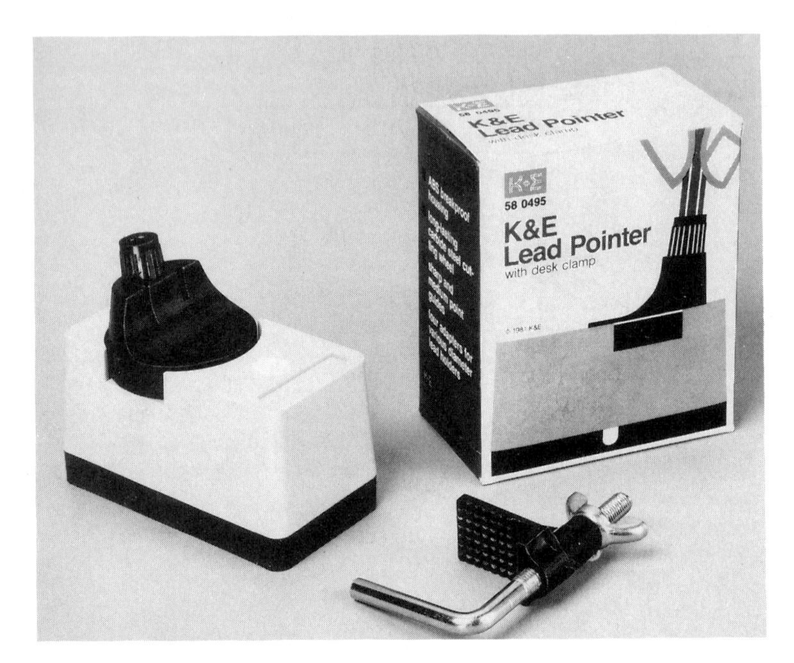

Fig. 2-2. Lead pointer. *(Courtesy of Keuffel & Esser Co., Parsippany, NJ)*

The main advantage of this holder is that it eliminates the need to remove the wood bonding, and the sharpened end can be protected by retracting the lead.

One of the problems in using pencils or mechanical lead holders is maintaining the lead point. To minimize sharpening, the pencil and holder should always be rotated as the line is being drawn.

The newest product is the stick lead holder (Fig. 2-4). Unlike other lead products, these holders are designed to hold leads of specific thicknesses or diameters (ranging from 0.3 mm to 0.9 mm). These thicknesses correspond to the various line weights used in drawings. In addition to lead thicknesses, stick leads can also be purchased in

Fig. 2-3. Mechanical lead holder. *(Courtesy of Keuffel & Esser Co., Parsippany, NJ)*

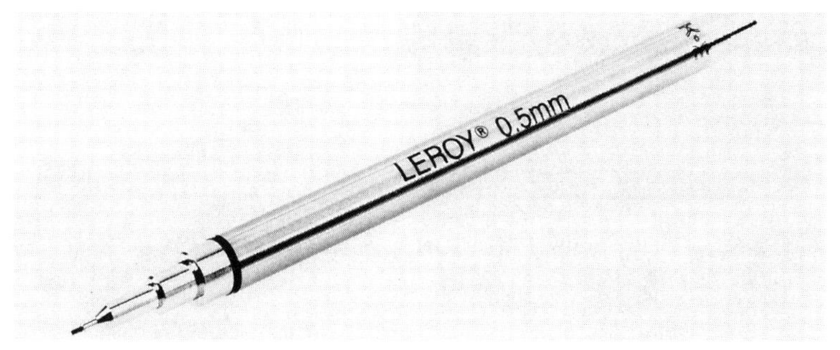

Fig. 2-4. Stick lead holder (LEROY®). *(Courtesy of Keuffel & Esser Co., Parsippany, NJ)*

the total range of hardness ratings. Since the lead maintains a constant diameter, there is no need for sharpening, which has made the stick lead holders popular among drafters.

Inks and Inking

Until the early part of the twentieth century, ink drawings were the standard of the industry. From about the 1920s on, lead media drawings began to dominate the drafting field. Within the last ten years there have been major advances in ink and inking products, so that inking is beginning to make a comeback. Some engineering firms are now preparing their final drawings in ink, and there are still specialized areas of drafting, such as copyright drawings, in which all plans must be inked.

Inking not only permits the production of a high-quality drawing, but also ensures good, clear reproductions. Drawings that must be "camera ready," or photographed for reprinting known as PMTs (Photo-Mechanical Technique), are best prepared by inking.

One of the first inks used in preparing engineering drawings was *India ink*. This substance proved difficult to work with because of its long drying time and tendency to smear and to clog drawing instruments. Such problems have been all but eliminated by modern ink that is quick drying, water or nonwater soluble, and anticlogging. It is also possible to obtain ink in a variety of colors.

Along with ink, a wide range of inking instruments and attachments is available. The inking pen is the most important of these. The first technical pen used was the *ruling pen*, which is made up

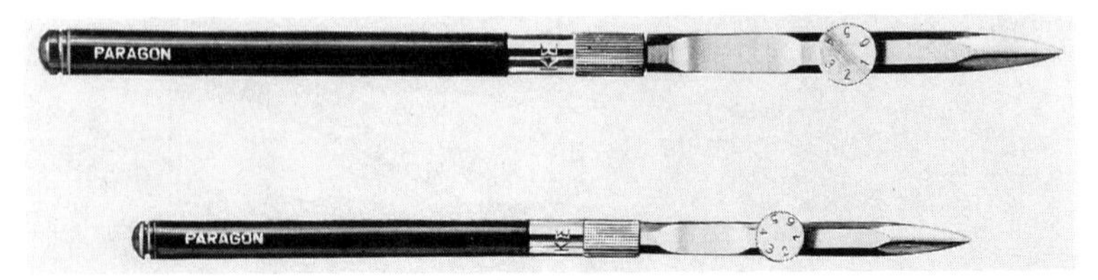

Fig. 2-5. Ruling pens (PARAGON®). *(Courtesy of Keuffel & Esser Co., Parsippany, NJ)*

of a set of nibs that can be adjusted by a small thumbscrew (Fig. 2-5). Though the ruling pen is still in use, it has all but been replaced by the technical drawing pen (Fig. 2-6). Like stick leads, technical drawing pens are available in a variety of standard widths specified in both inch and metric units. For technical illustrations and freehand drawings, technical fountain pens (Fig. 2-7) are also found in standard metric widths. Fountain pens are primarily used for artline and lettering work, while technical drawing pens are designed for mechanical line drafting. Shown in Table 2-3 are the standard widths for technical drawing and fountain pens.

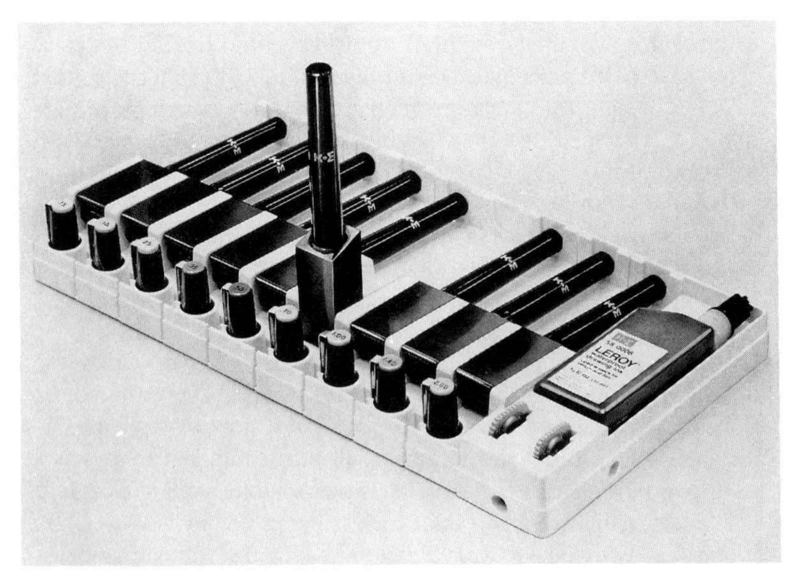

Fig. 2-6. Set of technical pens (LEROY®). *(Courtesy of Keuffel & Esser Co., Parsippany, NJ)*

Fig. 2-7. Technical fountain pen. *(Courtesy of Keuffel & Esser Co., Parsippany, NJ)*

Mechanical instruments and lettering sets are available with inking attachments. Inking compasses are designed with several types of inking adapters: nib ends, technical pen ends, and a plunger end (Fig. 2-8). Lettering sets (Fig. 2-9) are also available in various lettering styles and with different inking adapters.

Other Drawing Media

Though transfer overlays are not lead or inking devices, they are used to illustrate various symbols, numerals, and letters. These products were developed to provide a practical method for placing information onto drawings (Fig. 2-10). Overlay symbols and letters can be easily transferred from a master sheet to the drawing. First the symbol is

Table 2-3 Standard Inking Pen Widths

Type of Inking Pen	Available Line Widths	
	Sizes in Inches	*Sizes in mm*
Technical drawing pens	0.008	0.13
	0.010	0.18
	0.013	0.25
	0.017	0.35
	0.021	0.50
	0.026	0.70
	0.035	1.00
	0.043	1.40
	0.055	2.00
	0.067	
	0.083	
	0.098	
	0.125	
	0.150	
	0.200	
	0.250	

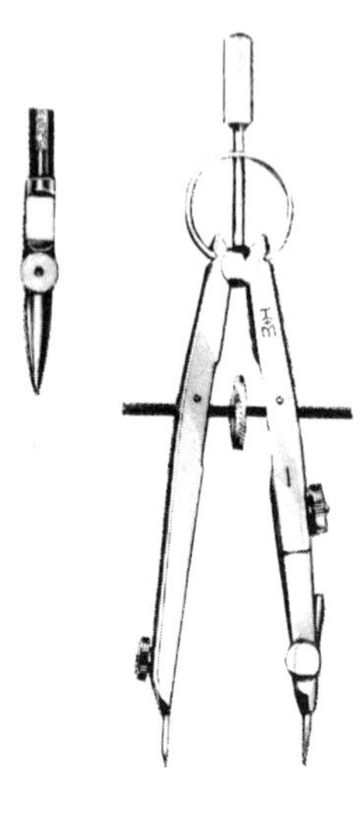

Fig. 2-8. Inking compass and inking nib attachment. *(Courtesy of Keuffel & Esser Co., Parsippany, NJ)*

placed at the proper location on the drawing and is rubbed with a smooth stylus or blunt pencil point. Then the master sheet is lifted, and the symbol remains. The major disadvantage of this procedure is that the overlay in time will become brittle and can be removed by abrasion or rubbing.

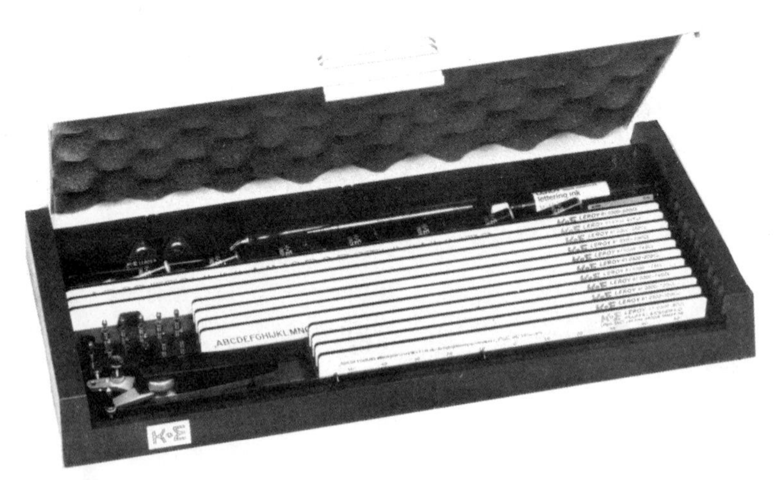

Fig. 2-9. Lettering set (LEROY®). *(Courtesy of Keuffel & Esser Co., Parsippany, NJ)*

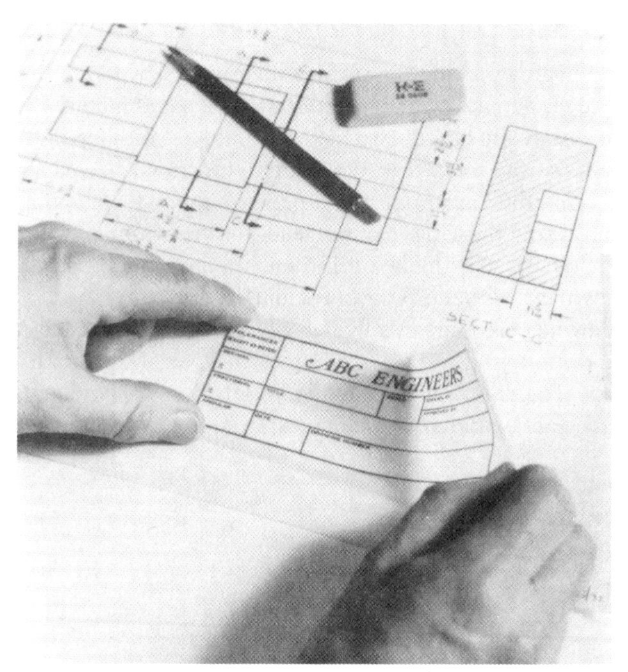

Fig. 2-10. **Transfer overlays.** *(Courtesy of Keuffel & Esser Co., Parsippany, NJ)*

When large quantities of information must be provided on a drawing, as in production specifications, typewriters are sometimes used; the drawing is inserted into a typewriter with a special carriage, or the information is typed on a transparent material and adhered to the drawing. By using a plastic ribbon, sharp, clear notations can be made.

Drafting Instruments and Equipment

Various types of drafting instrument and equipment are available to the industrial drafter. Some have general application for all kinds of drafting, while others are specifically designed for specialized purposes, such as architectural, mapping and topographic, and electrical drafting. This chapter will describe the purpose and proper uses of basic industrial drafting instruments and equipment.

Drawing Boards and Drafting Furniture

Drawing boards provide the surface upon which working drawings are prepared, and are available in different sizes and materials. When selecting a drawing board, two factors should be considered: the type of drawing surface on which plans are to be drawn (i.e. paper, film, or cloth), and the maximum size of paper required for drawings. The size of industrial drawings can range all the way from 8 × 10 inches to more than 42 × 84 inches; it is therefore important to know what size drawings are made by a particular firm.

Drawing boards are usually made of basswood or white pine (Fig. 2-11), and are found in sizes ranging from 12 × 17 inches to 23 × 31 inches. The reason why drawing boards do not come any larger is that they are considered to be portable equipment, and any larger

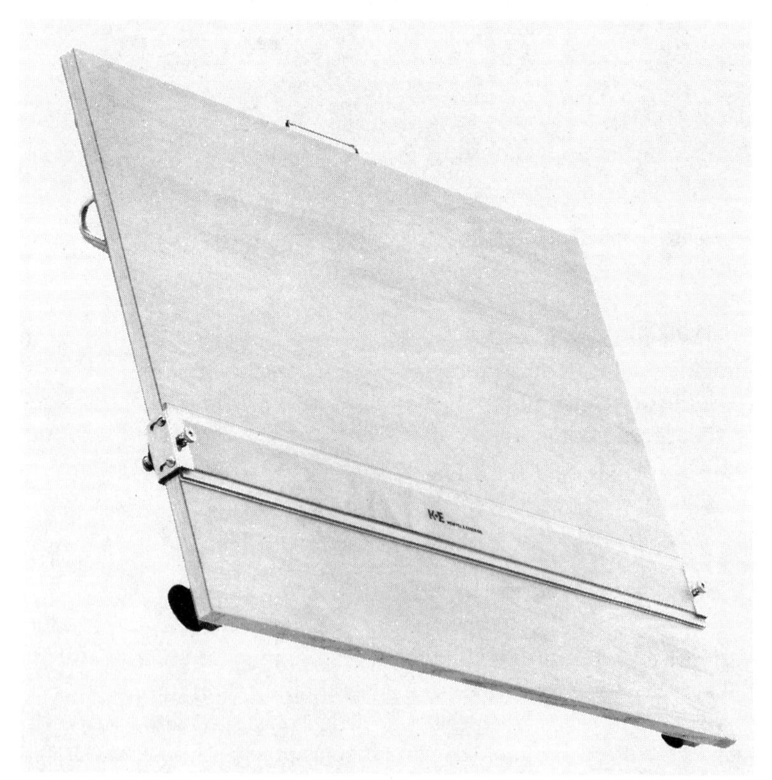

Fig. 2-11. Drawing board. *(Courtesy of Keuffel & Esser Co., Parsippany, NJ)*

size would be difficult to handle. Drawing boards are not commonly used within the industry. They have a wider use in specialized areas such as technical illustration and rendering, but they are primarily used in drafting classes and at home.

The major piece of equipment on which drawings are prepared is the drafting table. When portability is not a consideration in industry, tables are the most advantageous. They are found in a wide range of designs, and are constructed out of either wood or metal. The table has two distinct features. First, the drawing surface is so constructed that it will stand up to a variety of drawing operations. Second, the drawing surface can be adjusted to a variety of angles or pitches, and often to different heights. Such adjustments make it possible to "customize" the drawing environment for the drafter. Fig. 2-12a and b shows two drafting table designs and their use.

The most economical drafting table will usually be constructed out of wood and will provide only for a drawing surface. More elaborate and expensive drafting tables are known as *work stations* or *work centers*, and combine a drawing table and a desk. Most tables are specified by drawing surface size, which varies from 32 × 42 inches to 48 × 120 inches.

T-Squares, Parallels, and Drafting Machines

The T-square is perhaps the most characteristic symbol of drafting. (Fig. 2-13). At one time engineering and drafting students were recognizable on the campus by the T-square they always carried. Today T-squares are no longer widely used in industry. They are, however, still preferred by some drafters. They are usually used in combination with a drawing board.

T-squares can become cumbersome when used for larger drawings, and they have a tendency to loosen where the head and blade join. When this happens, the attaching screws should be tightened and their square (alignment of the head to the blade) should be checked. When loosening does occur, the drawn lines will not be parallel and all other lines drawn from the T-square will be inaccurate. If loosening persists, the T-square should be discarded.

One piece of equipment that led to the demise of the T-square is the parallel (Fig. 2-14). Parallels provide excellent guides for drawing parallel horizontal lines. They come in standard sizes ranging from 30 to 96 inches, and are made of hardwood or metal with a clear

(a) Two Types
of Drawing Tables

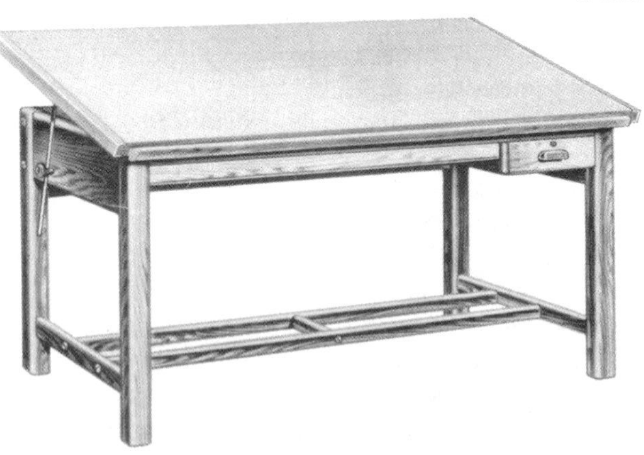

Fig. 2-12. Drawing tables. *(Courtesy of Keuffel & Esser Co., Parsippany, NJ)*

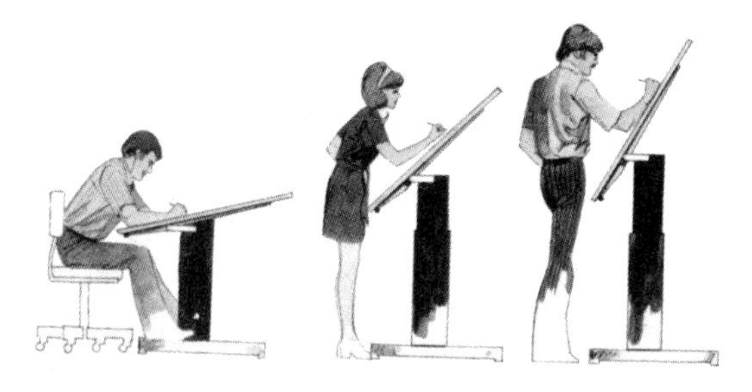

(b) Flexibility of Drawing Tables Meet the Requirements of Most Drafters.

plastic edge. They move up and down the drawing surface and are always kept horizontal by the set of guide wires they ride on. These wires pass over a set of pulleys to the corners of the drawing board. Many parallels have fitted rollers that keep them off the drawing surface and minimize the chance of damaging the drawing or getting dirt on it.

The second piece of equipment that led to the T-square's unpopularity among professionals is the drafting machine. Of the three devices, this is the newest. Drafting machines are designed to draw horizontal, vertical, and angular lines. There are two types of drafting machine: standard and track-type (Figs. 2-15 and 2-16). The standard drafting machine is the most common and is best for drawings that are drawn on E-size paper or smaller. The track-type drafting machine can be used for drawings in excess of 34 × 44 inches. The reason for this is that the arms of the machine can move a track along the entire length of the drawing surface, while the arms of the standard drafting machine are fixed.

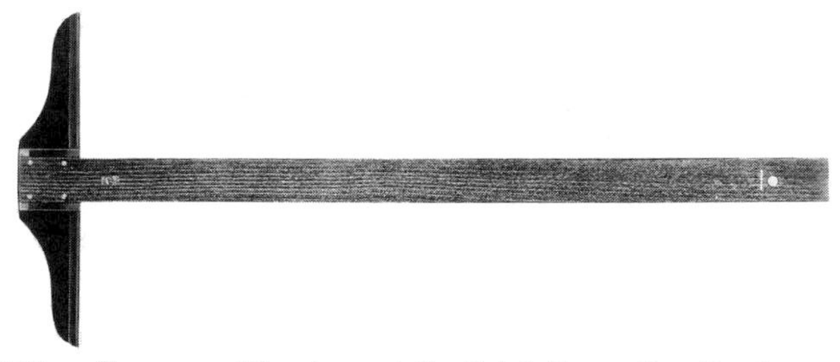

Fig. 2-13. T-square. *(Courtesy of Keuffel & Esser Co., Parsippany, NJ)*

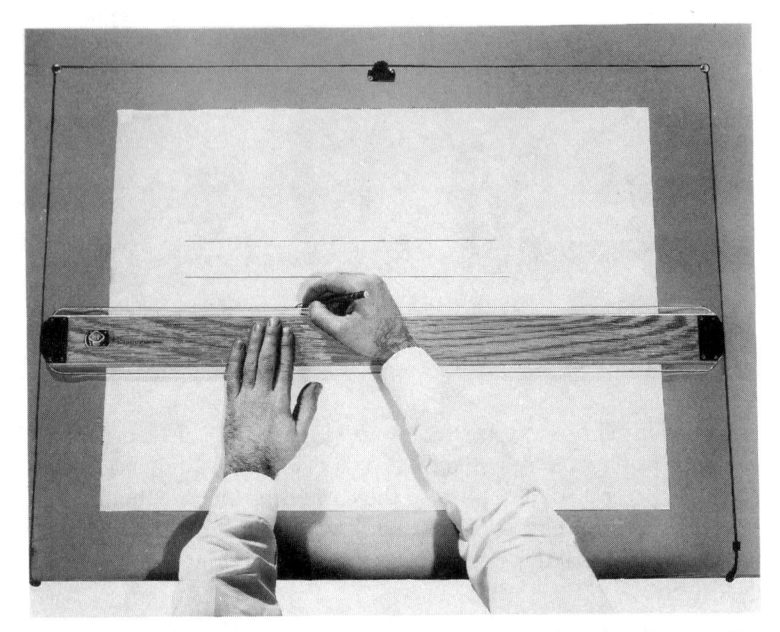

Fig. 2-14. Parallels. *(Courtesy of Keuffel & Esser Co., Parsippany, NJ)*

Fig. 2-15. Standard drafting machine. *(Courtesy of Keuffel & Esser Co., Parsippany, NJ)*

Fig. 2-16. Track-type drafting machine. *(Courtesy of Bruning)*

All drafting machines are equipped with a rotating head. Attached to the head are two blades set at a 90° angle to each other). These blades (Fig. 2-17) can be made out of transparent plastic, boxwood, or aluminum with scales. When ordering blades, different scales can be specified, such as metric, civil engineering, architectural, or mechanical scales. Standard blade lengths are 12, 18, 24, and 36 inches. When installed, the horizontal blade should always be longer than the vertical blade.

The drafting machine head is also movable, and rotates about a protractor (Fig. 2-18). It can lock in automatically at 15° increments.

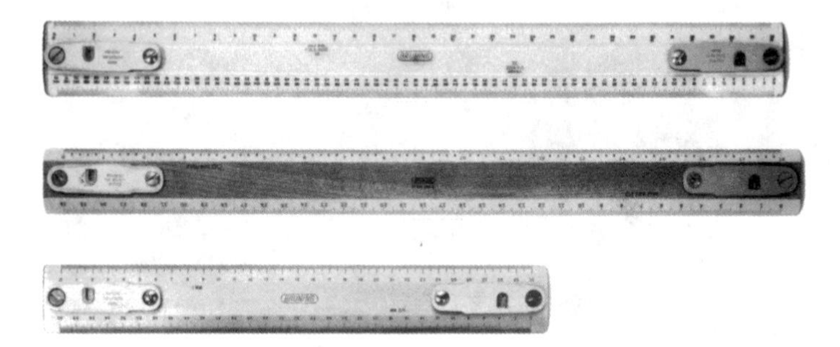

Fig. 2-17. Drafting machine blades. *(Courtesy of Keuffel & Esser Co., Parsippany, NJ)*

When a line must be drawn in at a specific angle, the head is rotated to that angle, then is locked into place by a lock lever. Where angle measurements must be held to close tolerances, drafting machine heads are available with a vernier that allows accurate angle measurement to within 0° 30′ (½°).

Triangles, Templates, and Curves

Triangles, templates, and curves help make the drafter's job easier, and his work more precise. Triangles are used as straightedges to draw vertical and angular lines, usually in 15° increments. There are three basic types of triangle: 45°, 30°–60°, and adjustable triangles (Fig. 2-19). Triangles can be purchased in standard heights at 2-inch increments,

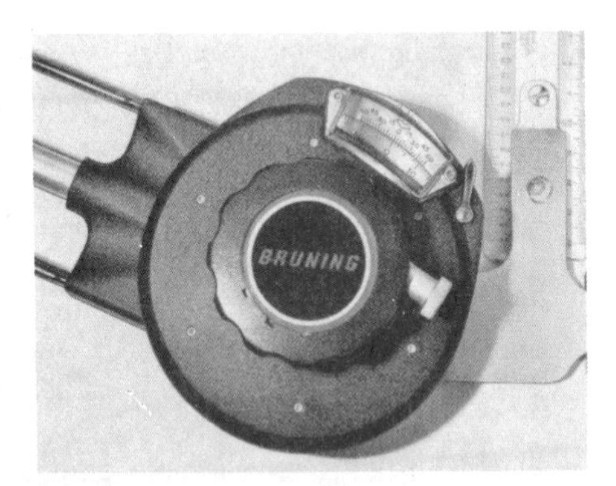

Fig. 2-18. Protractor head of a drafting machine, with vernier. *(Courtesy of Keuffel & Esser Co., Parsippany, NJ)*

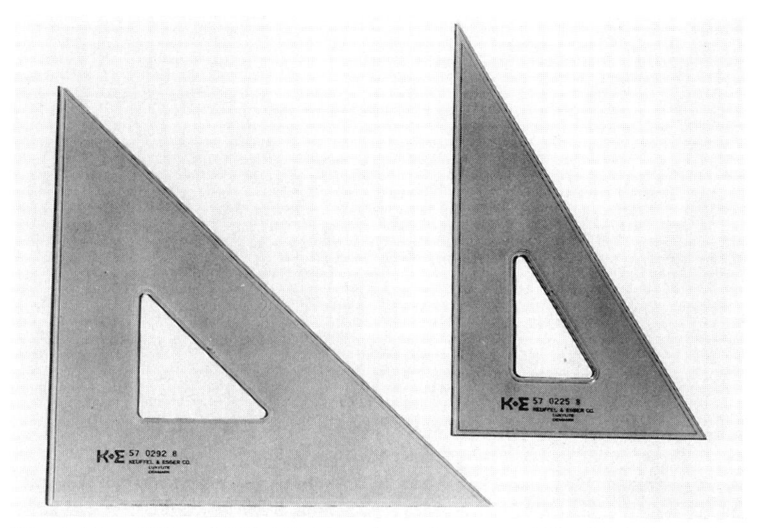

Fig. 2-19. 45° and 30°-60° triangles. *(Courtesy of Keuffel & Esser Co., Parsippany, NJ)*

ranging from 4 to 18 inches. The 45° and 30°-60° are usually purchased as a set where the 30°-60° triangle should have a height 2 inches greater than the 45° triangle.

The adjustable triangle combines the functions of the triangle and protractor. With a protractor attached to the base portion of the triangle, the adjustable hypotenuse can be set at any desired angle to within 0° 30′. Because of this flexibility, some drafters make regular use of the adjustable triangle. To obtain angles other than 30°, 60°, 45°, and 90° using the 30°-60° and 45° triangles, the drafter must use them in combination (Fig. 2-20), and even then they only provide angle measures at 15° increments. All three types of triangle are most frequently made of a clear or fluorescent colored plastic, but are also available in aluminum and wood.

Templates are time-saving drafting tools used to draw a wide variety of shapes and figures (Fig. 2-21). Some of the templates used by industrial drafters are listed in Table 2-4. Occasionally, a particular template is not available. It is then common practice for drafters to make their own customized templates out of thick polyester or acrylic film. This should be done only when many of a particular shape must be drawn; otherwise it is too time-consuming and expensive.

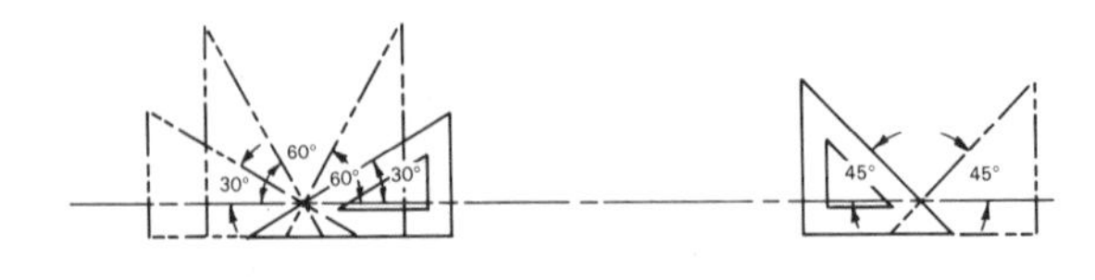

Fig. 2-20. Obtaining 15° increments using 45° and 30°-60° triangles.

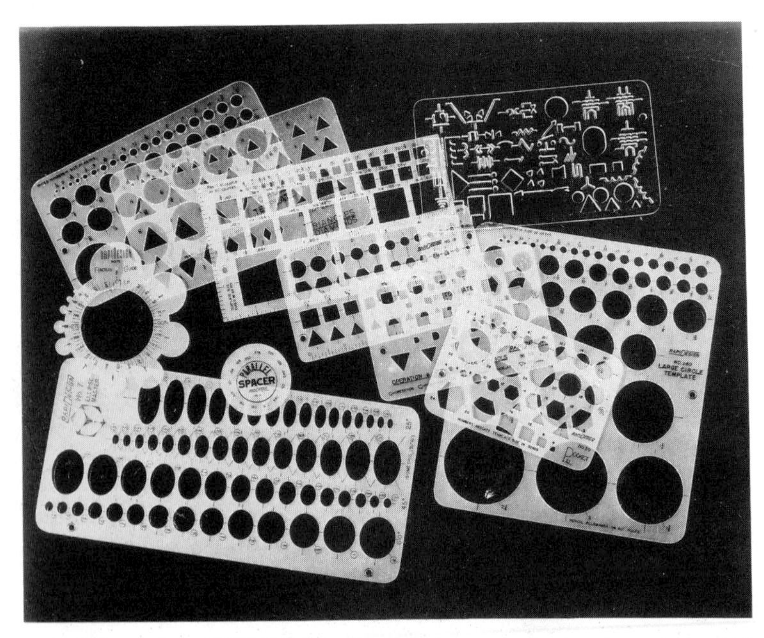

Fig. 2-21. Templates. *(Courtesy of Keuffel & Esser Co., Parsippany, NJ)*

Table 2-4 Common Templates Used in Industrial Drafting

Type	Description
General Templates	Squares
	Circles
	Triangles and diamonds
	Radius guide
	Large and extra-large circles
	Decimal circles
	Small and large ellipses
	Isometric ellipses
	10°–60° series ellipses
	Vertical and slant lettering
Metric Templates	Screw thread
	Squares
	Circles
	Triangles and diamonds
	Ellipse series
	Radius guide
	Lettering
Mechanical Templates	Screw threads
	Small machine screws
	Nuts and bolts
	Gears
	Tool design
	Tri-wing fasteners
	Isometric hex heads and nuts
	Isometric springs
Other Templates	Isometric piping
	Pipe fittings
	Instrument symbols
	Computer diagramming
	Organizational chart
	Arrowheads
	Plotting symbols
	Welding symbols
	Fluidic symbols
	Electrical/electronic symbols

Curves are used by drafters who have to draw precise curved and parallel curved lines. Three general types of curve are available: French, radius, and flexible curves. French curves, also known as irregular curves, are used for drawing curves with variable slopes, and can be purchased individually or as a complete set (Fig. 2-22). Radius or cir-

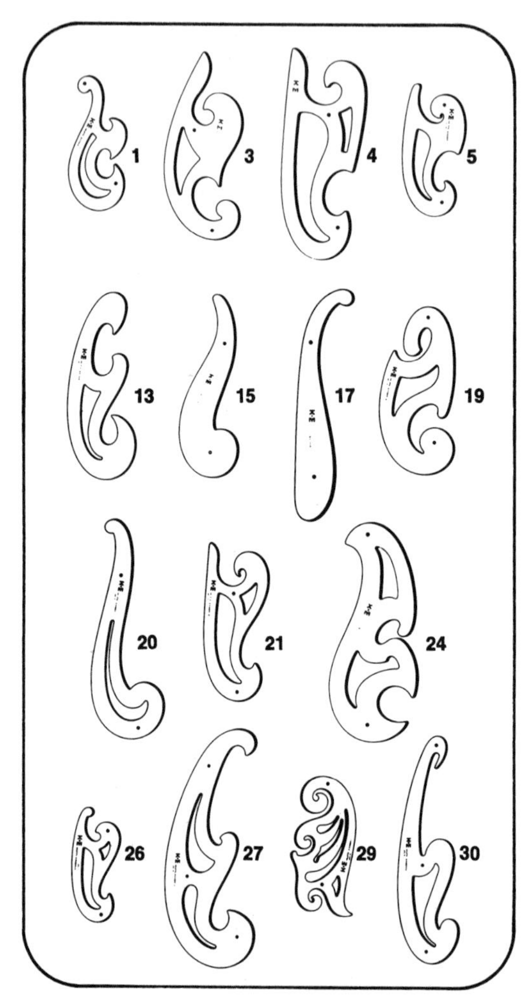

Fig. 2-22. Set of irregular curves. *(Courtesy of Keuffel & Esser Co., Parsippany, NJ)*

cular curves (Fig. 2-23) are true curves; that is, they have the same curvature on both edges, so that either edge can be used. Flexible or adjustable curves (Fig. 2-24) are designed so that their shape can be changed to meet any curved pattern. They are generally available in lengths of 12, 18, 24, and 30 inches, and can be bent to fit any contour within a 3-inch radius, with no supports.

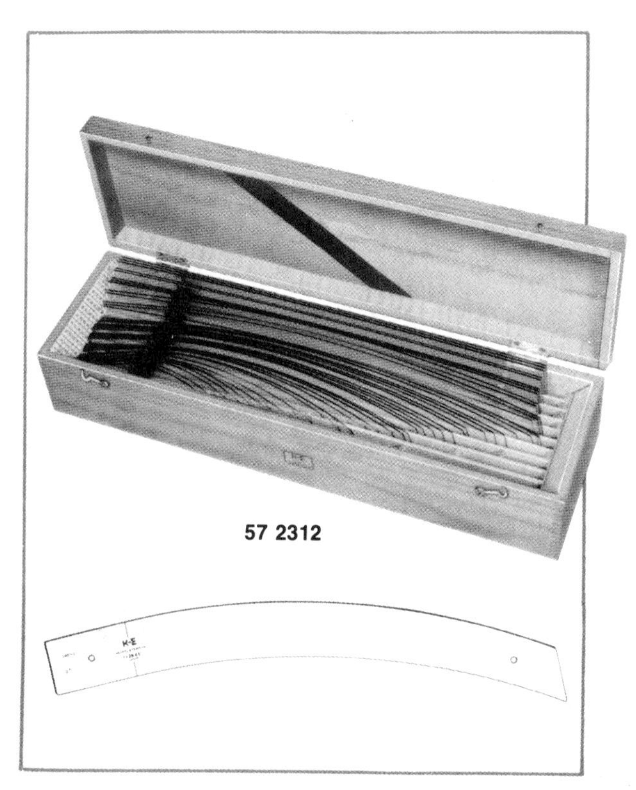

Fig. 2-23. Radius curves. *(Courtesy of Keuffel & Esser Co., Parsippany, NJ)*

Two other types of specialized curve are available. Ship curves are used in marine engineering and architecture. They are usually purchased in a set of fifty six patterns, and are manufactured to meet federal specifications. Railroad curves are designed for railroad layout work, and are available in sets of fifty five curves.

Fig. 2-24. Flexible curve. *(Courtesy of Keuffel & Esser Co., Parsippany, NJ)*

Scales

Scales are one of the most important of the drafter's tools, for they provide the means by which all measurements are made. Four basic types of scale are used by drafters: mechanical engineer's, architect's, civil engineer's, and metric scales. The mechanical engineer's scale provides measurements for drawings ranging from full size down to 1/8 size. Architect's scales are based upon the foot measurement, ranging from 3/32" = 1'-0" (1/128 size) to 1" = 1'-0" (full size). This scale is also used by industrial drafters when larger scaling is required.

The civil engineer's scale divides the inch measure into ten to eighty divisions of ten. Here the inch can represent such common civil engineering units as feet, rods, or miles. The metric scale is used for preparing full-size and scaled drawings in metric units in reduction from full size to 1/150. Table 2-5 gives a summary of the various graduations available with these scales. In addition to these, other special scales are also available through drafting supply companies.

Table 2-5 Standard Scales

Type of Scale	Graduations Available	
Mechanical Engineer's	1" = 1"	
	1/2" = 1"	
	1/4" = 1"	
	1/8" = 1"	
Architect's	12" = 1'-0"	1/2" = 1'-0"
	6" = 1'-0"	3/8" = 1'-0"
	3" = 1'-0"	1/4" = 1'-0"
	1 1/2" = 1'-0"	3/16" = 1'-0"
	1" = 1'-0"	1/8" = 1'-0"
	3/4" = 1'-0"	3/32" = 1'-0"
Civil Engineer's	10 divisions per inch	
	20 divisions per inch	
	30 divisions per inch	
	40 divisions per inch	
	50 divisions per inch	
	60 divisions per inch	
	80 divisions per inch	
Metric	1:1	1:33.3
	1:2	1:50
	1:5	1:75
	1:10	1:80
	1:20	1:100
	1:25	1:150

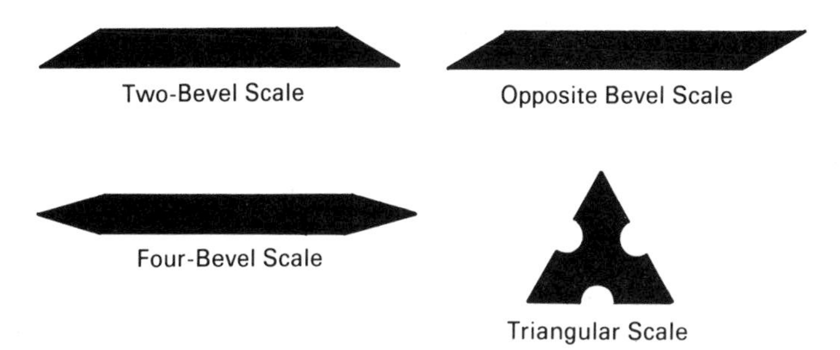

Fig. 2-25. Scale shapes.

SCALE SHAPES

The number of graduations on a scale will depend on its shape. Scales come in four different shapes, shown in Fig. 2-25 and described as follows:

1. *Two-bevel scales* are designed with a wide base so that they lie flat. These scales will include no more than four different measuring scales, all of which are printed on one side. These scales are made at 6- and 12-inch lengths.
2. *Opposite bevel scales* are easier to handle and lift from a tilted drawing surface. They have the same graduations as the two-bevel scale, but are usually available only in 12-inch lengths.
3. *Four-bevel scales* are mostly used as pocket scales; they have four faces, contain eight different scale graduations, and come in lengths of 6 and 12 inches.
4. *Triangular scales* are a combination of six faces with twelve different scale graduations. This scale is available only in a 12-inch length.

READING SCALES

Reading a standard mechanical engineer's scale is quite simple. The scale is clearly marked in inches, and the smaller graduations are labeled according to the scale used (i.e., 1/2, 1/4, 1/8, and 1/16). Hence, each scale (half size, quarter size, eighth size, and sixteenth size) is divided and read in the same manner as a full-size scale (Fig. 2-26a). The metric and civil engineer's scale (Fig. 2-26b and c) is read in the same way as the mechanical engineer's scale, the only difference being that the civil engineer's scale had more divisions per unit of measure.

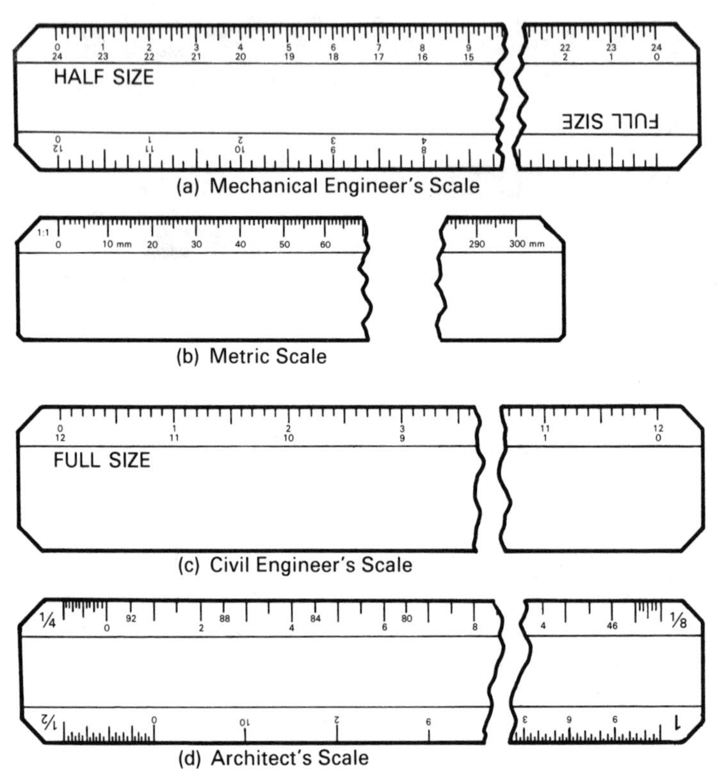

(a) Mechanical Engineer's Scale

(b) Metric Scale

(c) Civil Engineer's Scale

(d) Architect's Scale

Fig. 2-26. Reading different scales.

The architect's scale requires a little more study for accurate reading, because of the smaller divisions followed by a "0" marking, followed by progressive notations such as 2, 4, 6, 8, and so on. The smaller markings are divisions over a foot length. For example, Fig. 2-26d illustrates the commonly used scale ¼″ = 1′-0″. Each line to the right of the zero represents one inch. To the left of zero, each long line represents one foot (note that the short lines to the left of zero and the "92" foot marking represent an overlapping of the ⅛″ = 1′-0″ scale at the opposite end).

Drafting Instruments

The two basic drafting instruments used by drafters are the divider and compass (Fig. 2-27). There are, however, a number of different

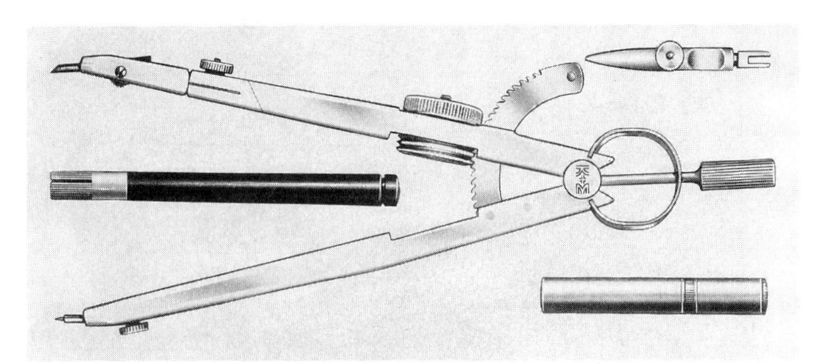

Fig. 2-27. Divider and compass. *(Courtesy of Keuffel & Esser Co., Parsippany, NJ)*

compass and divider designs available for industrial drafting purposes. Dividers are used for transferring measurements from one location to another (e.g., from an assembly to a detailed drawing, or from a scale to a working drawing), while compasses are used to draw circles and circular arcs.

The type and number of drafting instruments selected will depend upon the type of work you are doing. Instruments can be purchased individually or in complete sets (Fig. 2-28). What is important is that you choose exactly the right drawing instruments for your purpose, and buy the best quality that you can afford.

The basic design of the divider has not changed in over a thousand years. It consists of two arms with printed ends that are opened or closed by the action of a friction head. Though dividers are available with various maximum openings, the most common type can be opened to a span of approximately 8 to 9 inches.

Compasses are similar in design to the divider, except that there is a lead or ink drawing element at the end of one arm. Three types of compass are used in industry. The first is a friction-head compass, a general-purpose instrument that can be quickly set to size. Its only disadvantage is that when pressure is increased (to draw a darker line), the arms will tend to spread. Another is the bow compass, whose arms open and close by rotating a wheel attached to a threaded shaft fitted into each arm. Bow compasses provide more stability than friction-head types.

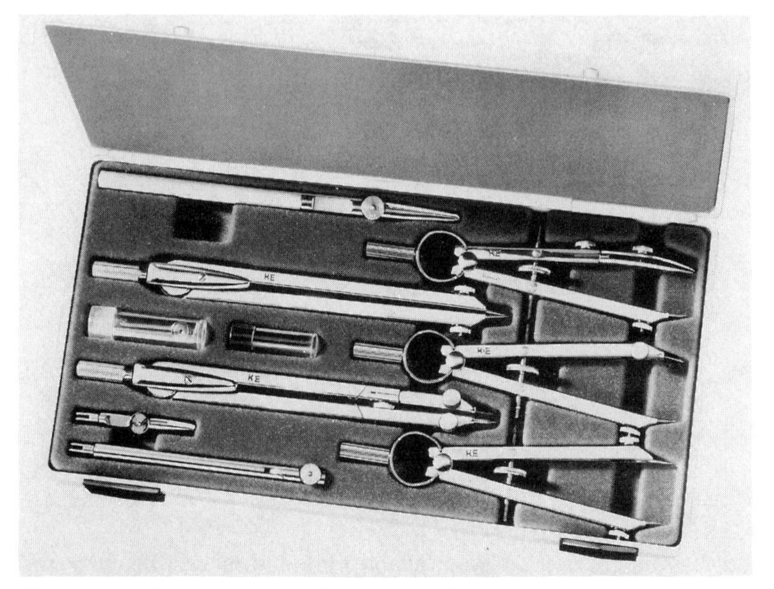

Fig. 2-28. Set of drafting instruments. *(Courtesy of Keuffel & Esser Co., Parsippany, NJ)*

A beam compass is used for drawing large-diameter circles. Beam compasses employ a central beam that functions as a track along which the pointer and drawing element ride. Most beam compasses range in expanse from 8 to 24 inches (Fig. 2-29).

SPECIAL PURPOSE DIVIDERS AND COMPASSES

There are several types of divider and compass that are designed for special purposes. The drop or drop-bow compass (Fig. 2-30) is used

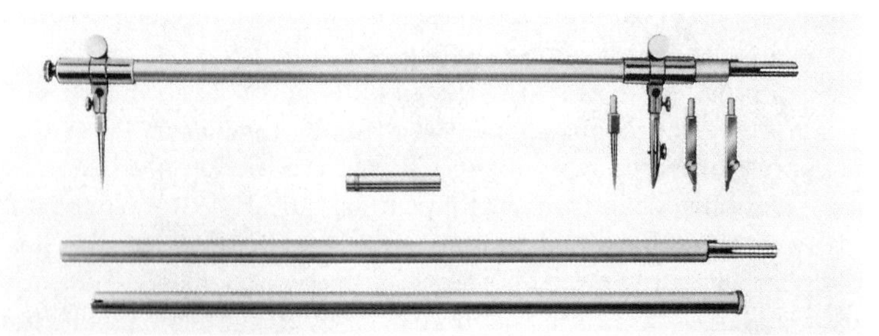

Fig. 2-29. Beam compass. *(Courtesy of Keuffel & Esser Co., Parsippany, NJ)*

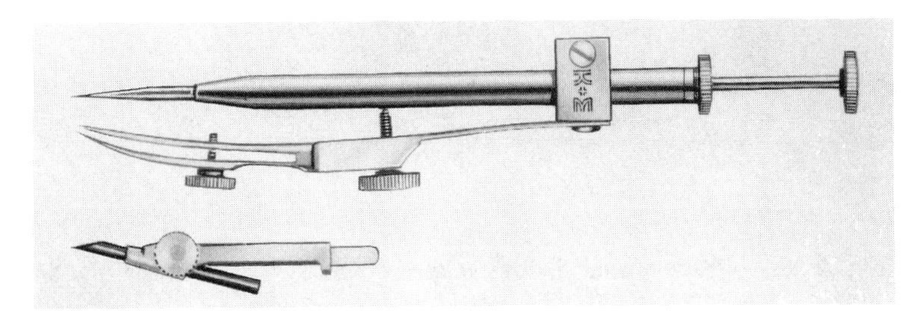

Fig. 2-30. Drop-bow compass. *(Courtesy of Keuffel & Esser Co., Parsippany, NJ)*

for drawing circles with a diameter as small as 1/64 inch (0.012 inch or 0.3 mm). Slide compasses (Fig. 2-31) are direct reading beam compasses. Along the beam are one or more scales by which circle radii can be set.

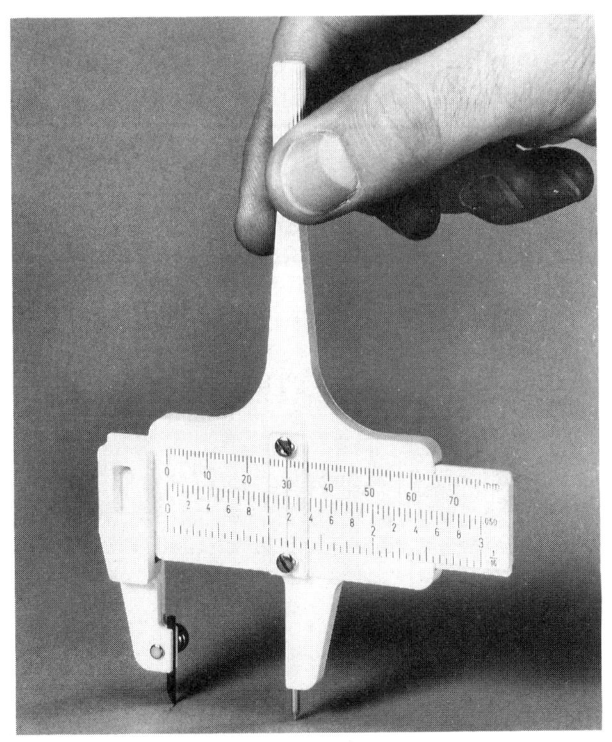

Fig. 2-31. Slide compass. *(Courtesy of Keuffel & Esser Co., Parsippany, NJ)*

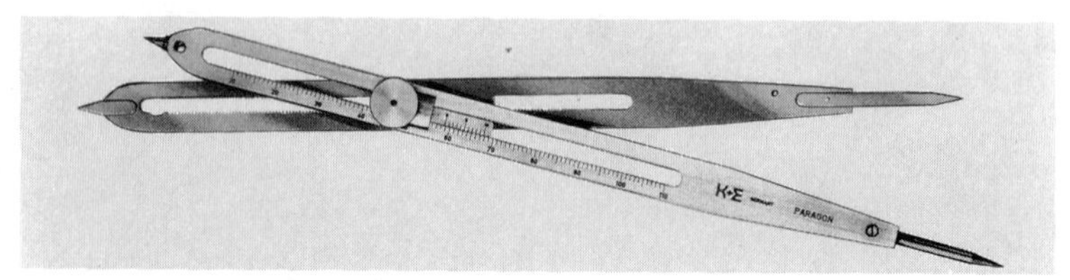

Fig. 2-32. Proportional divider. *(Courtesy of Keuffel & Esser Co., Parsippany, NJ)*

Proportional dividers (Fig. 2-32) are different from standard dividers in that their whole length is divided into 100 equal parts, which can be further divided into tenths. With this scale it is possible to set off distances at any ratio from 1:1 to 1:0.

USE OF DIVIDERS AND COMPASSES

It is important that drafters become competent in the use of dividers and compasses, since many drawing procedures require them. Without absolute control of these instruments, drawings will be inaccurate and unattractive.

Friction-head dividers should be opened with one hand by clasping the two arms between the thumb and second finger. The second and third finger should be positioned between the arms. To open the dividers, spread the second and third fingers, and to close them, squeeze the thumb and second finger (Fig. 2-33). These movements should be practiced until the dividers can be easily opened and closed to the desired spread.

To transfer a measurement with fiction-head dividers, open them to the desired position and hold the instrument at the top of the friction head with the thumb and forefinger. Carefully place one point of the divider at one of the measurements and swing the other end in the desired direction. If a series of measurements are to be "stepped off," swing the divider along a line in alternate directions (see Fig. 2-34). Friction compasses use the same opening and closing procedures.

Fig. 2-33. Adjusting a friction divider.

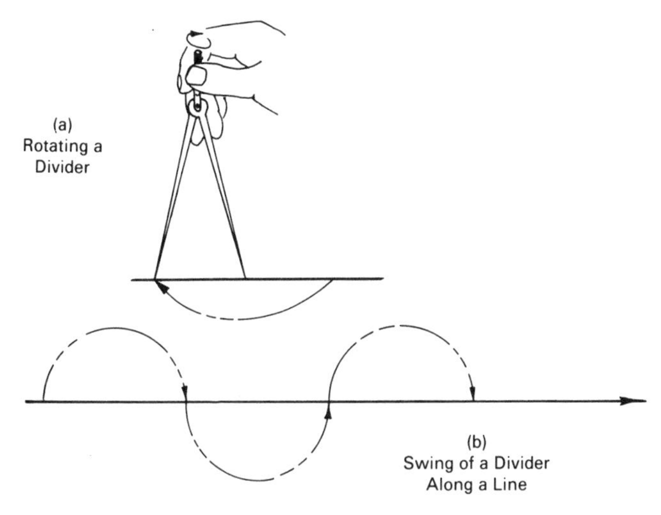

Fig. 2-34. Stepping off distances with a divider.

With a bow compass, the top should be held as with a divider, while the adjusting wheel is rotated with the other hand. When adjusting the compass, the length of the point and marker should be vertically centered (Fig. 2-35).

To draw a circle, set the compass to the desired radius. Place the point at the center of the circle. Grasp the top of the compass between thumb and forefinger, tilt the compass slightly, and draw the circle in one smooth rotation (Fig. 2-36). When a circle must be drawn with great accuracy, it is advisable to draw it in lightly first, check its diameter with a scale, and then redraw it with a heavier hand.

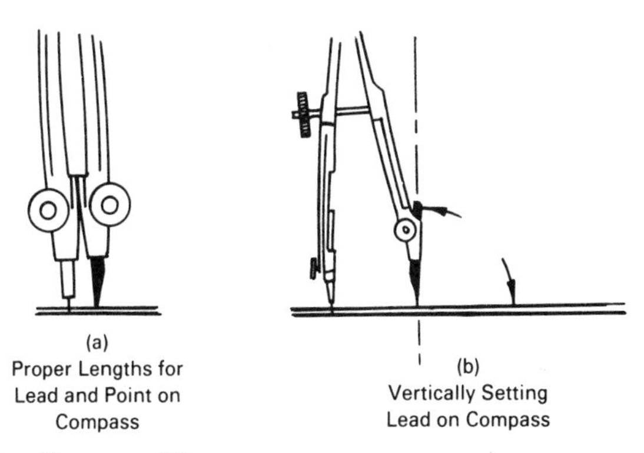

Fig. 2-35. Proper settings on compass.

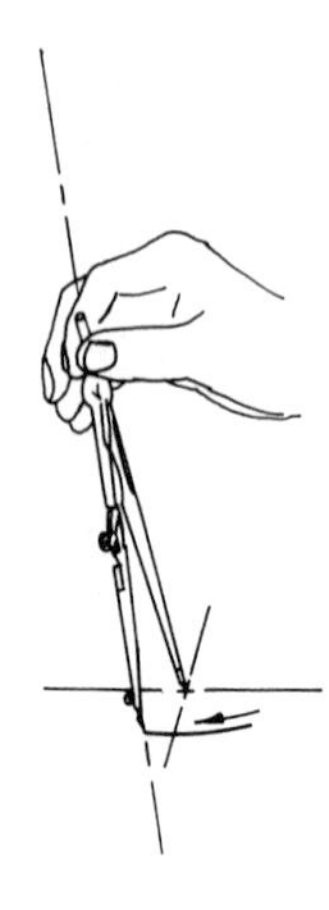

Fig. 2-36. Rotating compass by twisting the knurled handle at a slight angle.

PROPER USE OF DRAWING INSTRUMENTS

Professional drafters maintain high levels of performance by the proper use of their drawing instruments. Several points that should be kept in mind when preparing industrial drawings are:

1. The scale is a measuring instrument, and should not be used as a straightedge for drawing lines.
2. Make sure that you are working with the proper hardness of lead. Never use a dull lead point or a lead stick of the wrong diameter.
3. When sharpening your lead, be sure to do it away from the drawing surface.
4. Wash your hands before you start drawing.
5. Be sure that your triangles, scales, and other equipment and instruments are clean and well maintained.
6. Inks should not be kept open and on the drawing surface.
7. Inked lines should not be dried with a blotter; if you are in a hurry, use a fast-drying ink.
8. Before mounting any drawing material to the drawing surface, make sure that it is completely clean and free of foreign material.
9. Return all instruments to their proper storage place after use.
10. Drawings should not be folded.
11. Have on the drawing table only the instruments that you are going to use.

CHAPTER 3

Graphic Standards and Lettering

- Line Expressions
- Lettering
- Exercises

Understanding and interpreting drawings is an essential skill for all drafters. To prepare clear, understandable, and precise drawings, drafters must be able to apply the drafting standards that are in general use throughout the industry. These standards are of two basic categories: graphic and letter expressions.

Lines provide specifications about the size and shape of objects, while letters, numbers, and symbols are used to express additional information. The preparation of industrial drawings, therefore, requires that drafters use the correct graphic expressions for each situation, and be competent in the presentation of alphanumeric information. This chapter will discuss the different types of lines used in industrial drawings and professional lettering techniques.

Line Expressions

The basic element in all drawings is the drawn line. Different kinds of line are used to communicate different information and visual characteristics. The set of conventional line expressions used in drafting is sometimes referred to as the "alphabet of lines." Line expressions have been standardized by both ANSI and ISO. The two sets of standards are quite similar, with slight variations which will be noted.

Alphabet of Lines

The wide variety of drawings prepared in industry necessitates variety in the types of line used in making a set of working drawings. The alphabet of lines (Fig. 3-1) includes the different types of line accepted by the drafting profession. They are thick, medium, and thin, and may be continuous, broken, or irregular. Within every drawing, each different type of line conveys a specific kind of information to the reader of the drawing.

Eight common line groupings are used to provide different information about the object drawn. Fig. 3-2 illustrates the use and characteristic of these line symbols, which are described as follows:

1. *Visible lines* are used to represent all edges and visible outlines that are not hidden in a particular view. These lines should be drawn as thick, continuous lines. Visible lines, sometimes referred to as *object lines*, should stand out from the other lines in the drawing, so that the general shape of the object is apparent.
2. *Hidden lines* appear as short dashes of medium thickness. They are used to show those surfaces, edges, or object corners that are hidden from view. In some cases, hidden lines may be omitted if they would complicate the drawing or make it difficult to interpret. Whenever hidden lines are used, the person reading the drawing should always refer to other views for further clarification.
3. *Center lines* are used to show the axes of round or symmetrically shaped holes and solids, and reference lines for dimensions. They can also be used to represent path lines, pitch circles, axes of symmetry, and the extreme positions of movable parts. Center lines are drawn thin, with a chain pattern of long and short dashes.
4. *Dimension, extension, and leader lines* are drawn as continuous thin lines. Extension lines are extended from visible lines to indicate the limits of the feature to be dimensioned, and are usually continuations of the object line. When drawn, the extension line should not come in contact with the dimensioned feature and should extend approximately 1/8 inch or 3 mm beyond the dimension line.

   Dimension lines are drawn between extension lines. Arrowheads are placed at the end of the dimension line, and

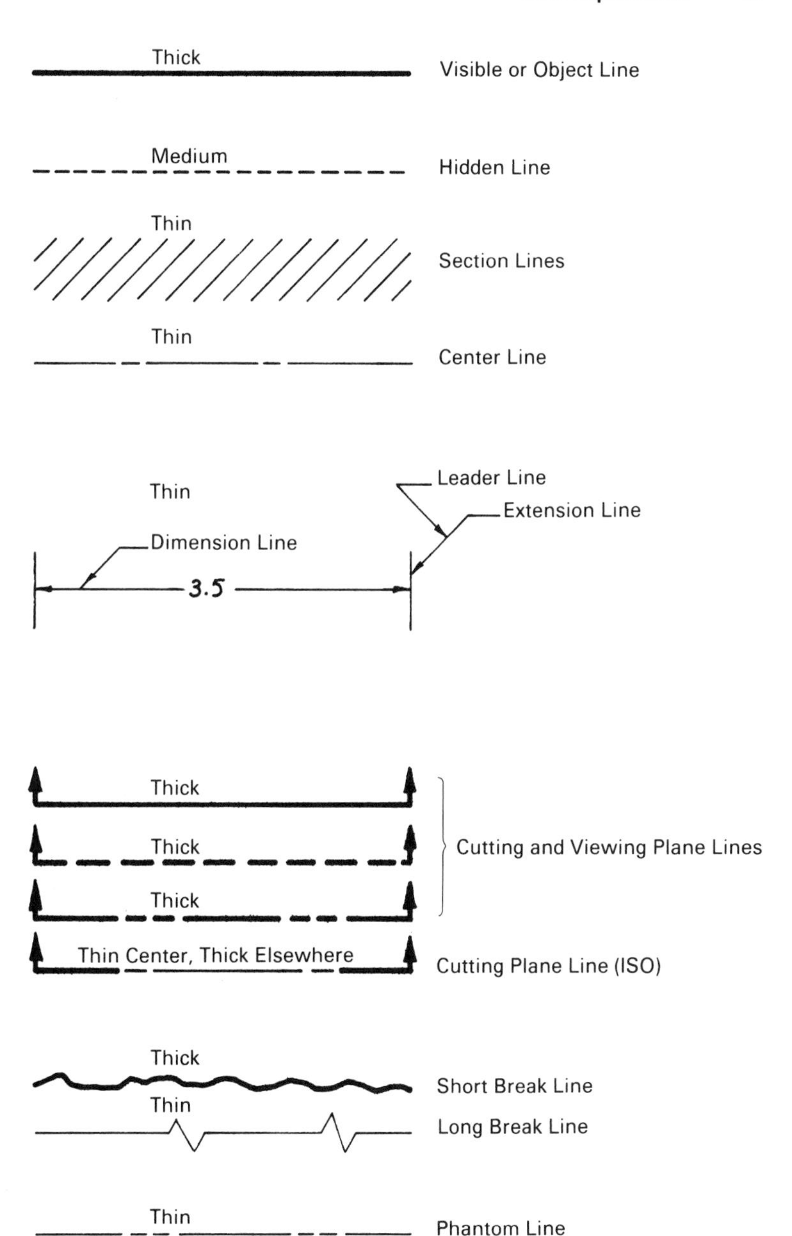
Thick
Visible or Object Line
Medium
Hidden Line
Thin
Section Lines
Thin
Center Line
Thin
Leader Line
Extension Line
Dimension Line
3.5
Thick
Thick
Thick
Cutting and Viewing Plane Lines
Thin Center, Thick Elsewhere
Cutting Plane Line (ISO)
Thick
Short Break Line
Thin
Long Break Line
Thin
Phantom Line

Fig. 3-1. Alphabet of lines.

touch the extension line. The dimension itself is usually placed in an opening along the dimension line, approximately midway between the extension lines. In some cases, the dimension is placed above the line.

The leader, or leader line, is used to make note of a part or feature. It is drawn at an angle of 60°, 45°, or 30° to the center of the feature. At the end of the leader, the arrowhead is used to touch circumferences or perimeters, while the dot is placed on surfaces.

5. *Cutting and viewing plane lines* are extra-thick lines that can be drawn either continuously or as chains. Cutting plane lines indicate the position of imaginary cutting planes. The cutting plane is found in section-view drawings where portions of an object are "cut away" to show what the interior looks like. Viewing plane lines are used to indicate the direction and height of a partial view.

   At the end of cutting and viewing plane lines will be a set of arrowheads that indicate the viewing position. Cutting plane lines drawn to ISO specifications appear as thick lines at the end and thin along the length, but all other features will be the same.
6. *Section Lines* are thin, continuous lines that are usually drawn at a 45° angle. The spacing between these lines will vary, but normally will have a 1/8 inch or 3 mm separation. Section lines are employed to show the cut surface in section views. It should be noted that a number of industries make use of special symbols to signify various materials (this will be discussed in further detail in Chapter 8).
7. *Break Lines* are used to shorten a view of a long part. There are three basic types of break line: a thick, wavy line that is primarily used for short breaks; a thin, straight line with a zigzag; and an "S" break used for cylindrical objects.
8. *Phantom Lines* are drawn as thin lines with a dash pattern of one long and two short. These lines are used to show movable parts in one location (extreme positions), the path of a moving part, finished or machined surfaces on castings, the outline of a rough casting on finished parts, the position of a part that is next to or fits with the part being drawn, or the portion of the part to be removed.

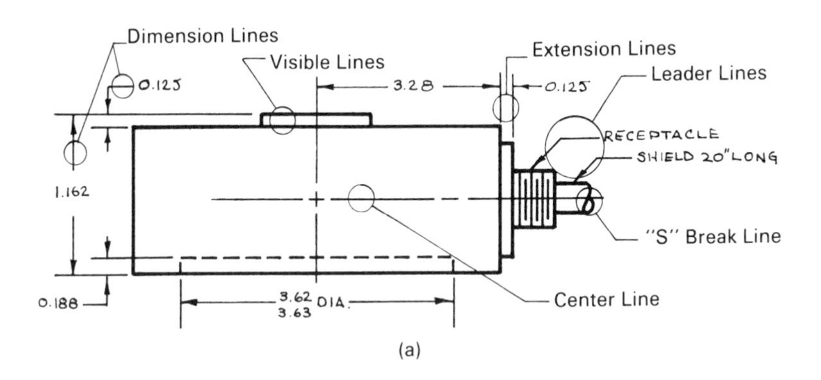

(a)

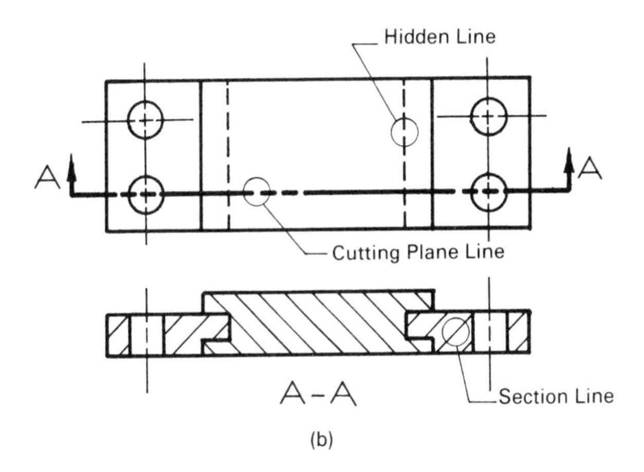

(b)

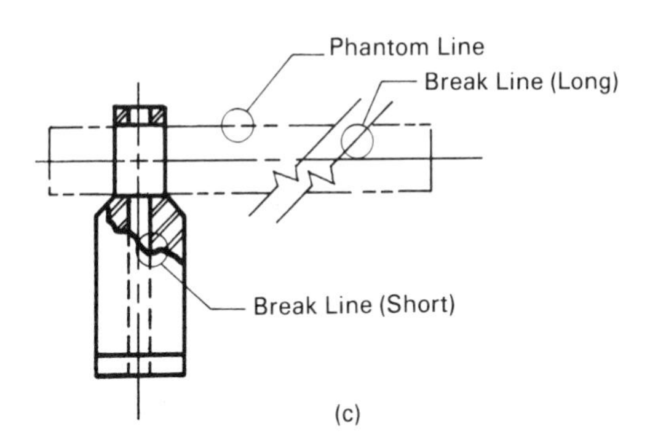

(c)

Fig. 3-2. Correct usage of line expressions.

Line Practices

All lines should be drawn bold and black, with an even density and thickness throughout their length. All lines appearing on the same drawing may be drawn in either ink or pencil. Seldom is it necessary to combine pencil and ink; where conditions warrant this, a uniform density should be maintained throughout the drawing. As a general rule of thumb, thick lines will be 0.032 inch or 0.80 mm, medium will be 0.020 inch or 0.50 mm, and thin will be 0.0125 inch or 0.30 mm. This, however, will vary according to the size of the drawing and clarity of visualization; lines can be as thin as 0.006 inch or 0.16 mm, or as thick as 0.050 inch or 1.25 mm.

To draw lines correctly and with confidence requires repeated practice. The drawing of hidden lines is difficult at first, because of the need for line and spacing consistency. Each dash of the hidden line is about 1/8 inch (3 mm) long with a 1/16 inch (1.5 mm) spacing. Drawing such lines accurately, and without measuring, requires good hand-eye coordination and skill.

When using leads, attempt to draw all lines uniform in thickness and density, which can only be accomplished by keeping the lead point sharpened (or using the appropriate stick lead) and pressing down with consistent pressure. It is good practice to rotate your lead holder or pencil as you draw your lines.

If ink is used, pay attention while drawing to the technical pen size selected and the flow of the ink. If the line appears imperfect, determine what is wrong and correct the situation immediately. Be careful not to smear the ink with drafting tools or your hand.

Lettering

Alphanumeric symbols are the word supplements on drawings. They are used to provide dimensions and notations relative to the drawn part, and complete the descriptions needed for manufacturing and servicing. Lettering can make all the difference in the overall appearance of an industrial drawing. A poorly lettered drawing will be difficult to read and interpret.

Most of the lettering placed on drawings will be freehand, done without the aid of a lettering set or machine. The ability to letter well is a basic requirement for all drafters, and can only be achieved by frequent and careful practice. Anyone who has average muscular control can learn to letter well.

Lettering should appear bold, clear, and upright. Letters should be of sufficient size to be read easily even when the drawing is reproduced at a smaller size. Capital letters should be used throughout the drawing, unless lower-case characters are specified or are more appropriate (i.e., abbreviations for units or other standards).

Lettering Styles

The most common lettering style recommended for industrial drawings is the single-stroke vertical, commercial Gothic letter (Fig. 3-3). The size of the lettering will depend upon the nature of the information to be provided. Most dimensions, notations, and specifications will use lettering 1/8 inch (3 mm) in height (A, B, and C size drawings). Lettering sizes between 3/16 and 1/4 inch (5 to 6 mm) should be used for headings, while letters up to 1/2 inch (13 mm) can be used for title blocks. For sizes other than these, check the drafting standards used by your company.

Small, condensed, or extended lettering is not recommended. The vast majority of drawings will require vertical lettering, though an inclined style can be used. *Never* use vertical and inclined lettering on the same drawings. The standard slope angle used for inclined lettering is 67.5°, and can be approximated by a slope of 2:5.

Lettering Practices

Drafters seldom letter in notations and specifications without the use of guide lines. In fact, it is highly recommended that the beginning

,ABCDEFGHIJKL
MNOP,QRSTUVWXYZ&
(%)"=;0123456789· abcd
efghijklmnopqrstuvwxyz

Fig. 3-3. Typical Gothic lettering style used in drafting.

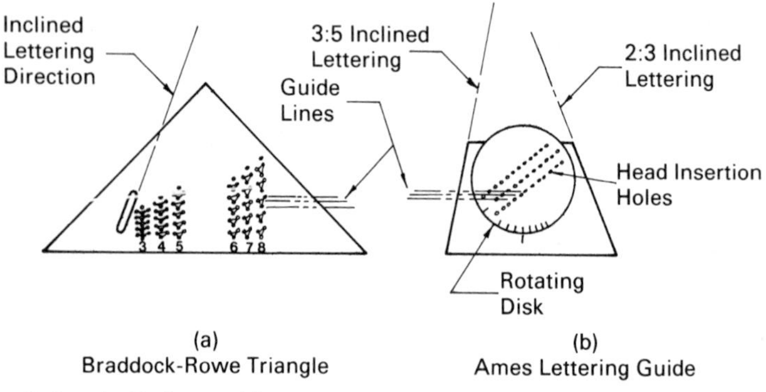

Fig. 3-4. Lettering guides.

drafter employ guide lines for all dimensioning work, even though it might take longer.

Guide lines should be drawn for both the top and bottom of the letters. The base line should be drawn first, from which the height of the letters is then marked off and drawn. Whichever letter height is selected for a drawing should be maintained throughout the drawing for all dimensioning and notations.

To speed up the process of drawing in guide lines, many drafters make use of letter guides. Two convenient devices are the Braddock-Rowe triangle and the Ames lettering instrument (Fig. 3-4). The holes along these devices are grouped for both upper- and lowercase letters. In addition, slanted edges are provided for inclined lettering.

When using lettering guides, a sharpened hard lead pencil is inserted into the proper row of holes. With the guide resting on a straightedge (e.g., parallel, drafting machine, or T-square), the guide is moved along by the pencil drawing the guide line. The Ames lettering guide has a center disk that can be adjusted to a letter height number that gives equal spaces or 2:3 and 3:5 waist-line proportion lines. Note that waist-line proportion lines are used only for lowercase lettering (see Fig. 3-5).

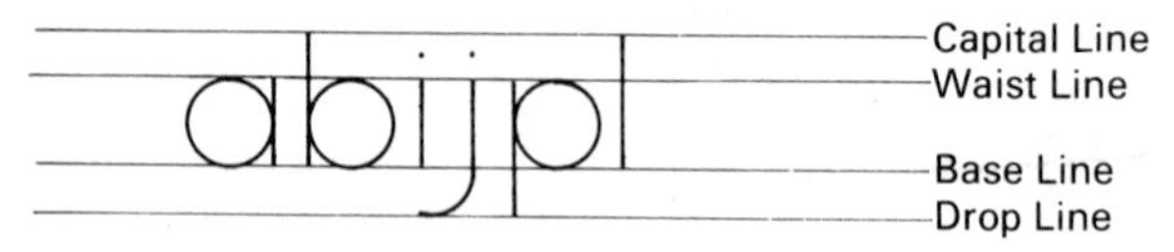

Fig. 3-5. Guide lines.

If lettering is not drawn in dark enough it will not reproduce well, and the result will be an unreadable print. To avoid this problem, the appropriate lead hardness should be selected. If a pencil or mechanical holder is used, the lead should be clean, firm, and well sharpened. When using a stick lead holder, select the appropriate diameter lead stick. If your experience in drafting and lettering is limited, first experiment on the drawing material with different hardness and sizes of lead, then make your selection. This trial and error process is useful for all drafters.

In lettering, there are three basic pencil movements: vertical strokes made entirely by finger movements; horizontal strokes, which

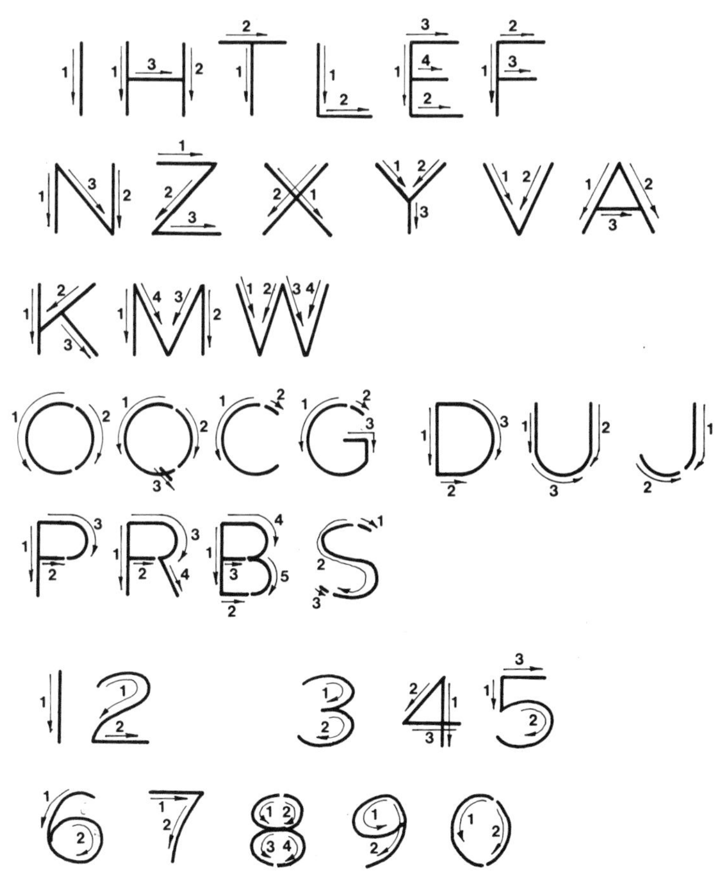

Fig. 3-6. Stroke order for Gothic lettering.

should be made by pivoting the hand at the wrist while the fingers move slightly to keep the lead at horizontal; and the circular stroke, made primarily by finger movement, with a slight wrist action.

All letters are designed to be made using a specific stroke order. Fig. 3-6 shows the letters of the alphabet and numerals, together with the recommended stroke order (given in numerical sequence—1, 2, etc.). Special note should be taken of the presentation of fractions, which are always made with a straight horizontal bar. If a 1/8-inch lettering height is used, the total height of the fraction should be 1/4 inch. Here each of the fraction numerals will be *less than 1/8 inch* because of the slight gap between the numerals and the horizontal bar.

Lettering Devices and Equipment

Sometimes freehand lettering is not sufficient for a finished drawing. Drawings that require a very high quality of uniform lettering usually make use of one or more lettering devices or equipment. Several of the more common lettering devices and equipment used by industrial drafters include:

1. *Lettering templates.* One of the lettering devices that is easiest to use is the lettering template (Fig. 3-7). Templates are available in various lettering styles and sizes, though the most commonly used will be vertical lettering. Lettering templates are available in standard sizes of 5/32, 1/8, 3/16, 1/4, 5/16, 3/8, 1/2, and 3/4 inch. Examples of typical lettering styles are vertical and slant letter, Neo-Gothic, Old English, and microfont.
2. *Mechanical lettering sets.* The mechanical lettering set employs the use of a template, positioning guide, and scriber (Fig. 3-8). The scriber is made up of three elements: a metal point that

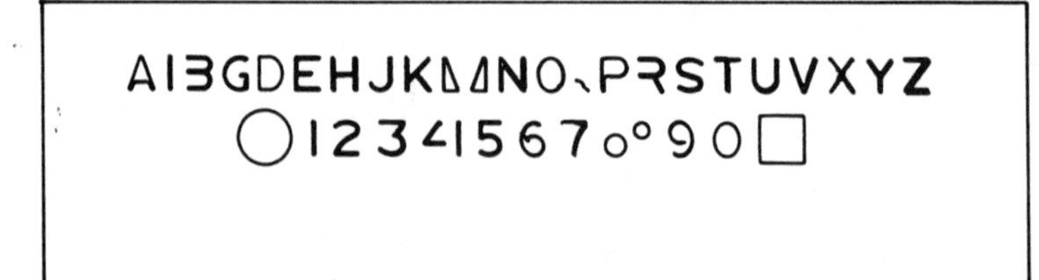

Fig. 3-7. Typical lettering template.

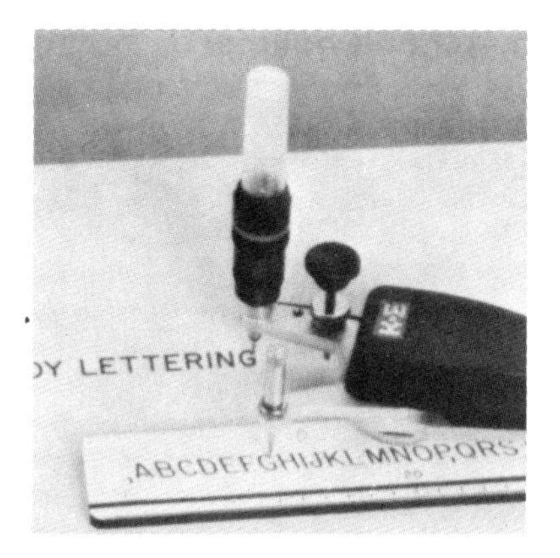

Fig. 3-8. Mechanical lettering (LEROY®). *(Courtesy of Keuffel & Esser Co., Parsippany, NJ)*

fits into the positioning guide and keeps the letters on horizontal; the template point, which fits into the lettering template and traces the letter or number; and the marking end, which can be fitted with a lead or inking device, and is used to draw the letter onto the drawing surface.

Like lettering templates, mechanical lettering sets are available in vertical or slant lettering and in sizes ranging from 0.050 to 0.500 inch. When inking is desired, lettering sets can be purchased with a technical lettering pen attachment or with an ink reservoir and plunger end.

3. *Transfer lettering.* A technique that produces an exceptionally high-quality letter is transfer lettering. Letters, numbers, and symbols are transferred directly from a master sheet or strip onto the drawing. This is accomplished by placing the master on the drawing and positioning the letter at the desired location. Then the entire letter is rubbed over with a smooth stylus or blunt point, which adheres it to the drawing surface. The master sheet is then lifted, leaving only the transferred letter (Fig. 3-9).

   Transfer lettering is available in a wide variety of lettering styles and sizes. The most common lettering styles are Gothic and Helvetica. Lettering size, however, is specified in terms of points. Table 3-1 lists common lettering sizes and their equivalencies.

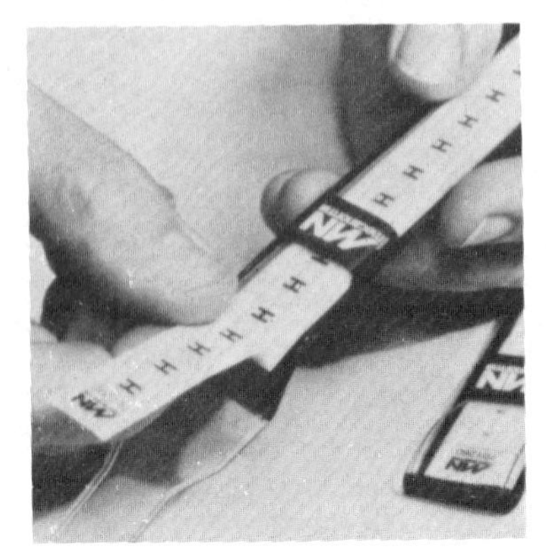

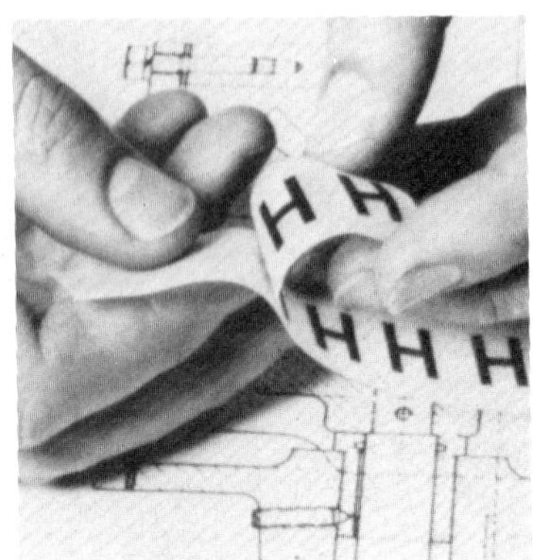

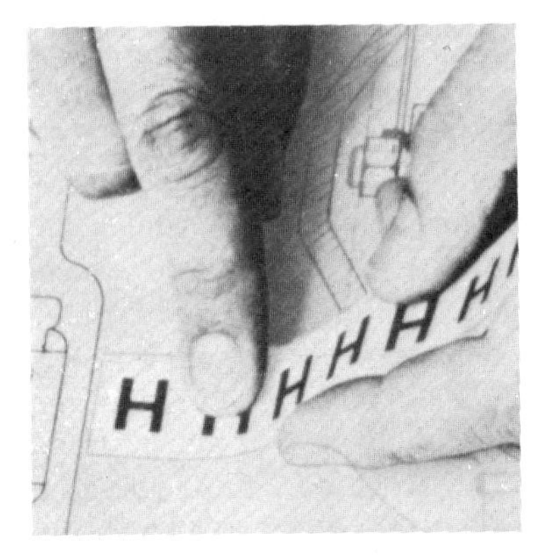

Fig. 3-9. Transfer lettering. *(Courtesy of Keuffel & Esser Co., Parsippany, NJ)*

Table 3-1 Sizes and Equivalencies for Transfer Lettering

	Equivalence	
Point Size	*Inch*	*Metric (mm)*
8	0.079	2
12	0.118	3
14	0.157	4
18	0.197	5
20	0.236	6
24	0.276	7
28	0.315	8
36	0.394	10
48	0.472	12

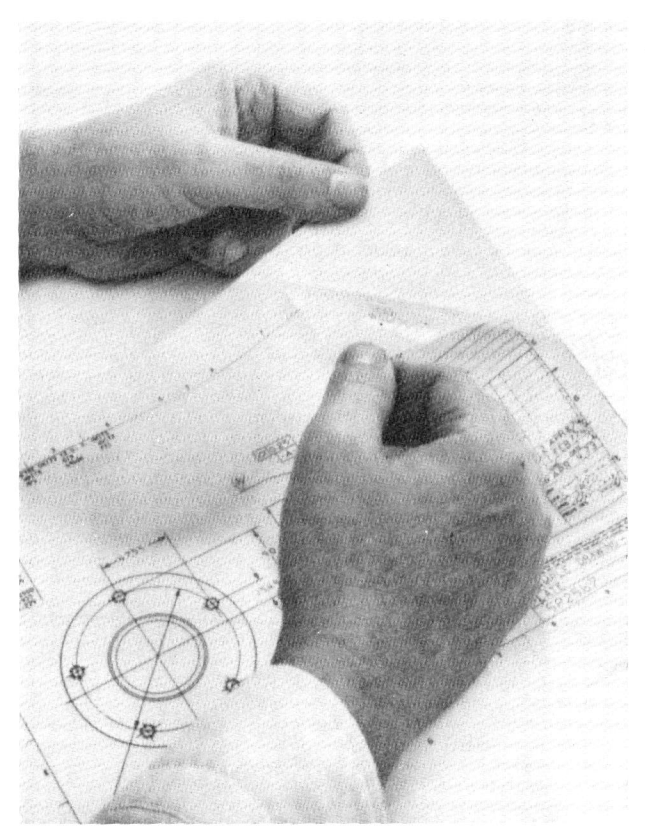

Fig. 3-10. Special transfer symbols.

Specialty symbols, such as logos and legends, can be printed by many supply companies on special order. Larger figures, such as title blocks, are printed on a clear polyester film with adhesive on the back. These overlays are then peeled off the master sheet and adhered to the drawing surface (see Fig. 3-10).

4. *Typewriters.* Occasionally it is desirable to use a typewriter, particularly when extensive specifications, guidelines, or instructions must be given. It is not usual, however, to make extensive use of the typewriter.

Exercises

3.1 Identify the graphic line expressions A through J noted in Fig. 3-11.

3.2 Study the drawing in Fig. 3-12, and identify lines A through H.

3.3 Letter the following statement in 1/8- and 5/32-inch vertical capital letters, using the appropriate lead hardness, and draw in the necessary guide lines:

+ 0.0018 VARIATION ALLOWED WHERE FRACTIONAL DIMENSIONS ARE USED LOCATING FINISHED SURFACES. TOLERANCE DIMENSIONS SPECIFY ACTUAL GAUGE SIZES.

3.4 Letter the statement in Fig. 3-13 in 1/8-inch vertical and inclined capital letters. The necessary guide lines should be drawn in with a hard pencil.

3.5 Letter the following statements in 1/4- and 1/8-inch vertical capital letters:

COUNTERSINK ALL TAPPED HOLES 90° INCLUDED ANGLE TO MAJOR DIAMETER OF THREAD.

THE ANGLE BETWEEN THE LINE OF ACTION AND THE LINE PERPENDICULAR TO THE CENTER OF THE MATING GEARS WILL BE 14° 30′ 0″ UNLESS OTHERWISE NOTED.

5/16 DRILL—3/8—16 UNC—2A

NECK 3/8 WIDE × 1/4 DEEP

S.A.E. 1050-COLD DRAWN STEEL BAR

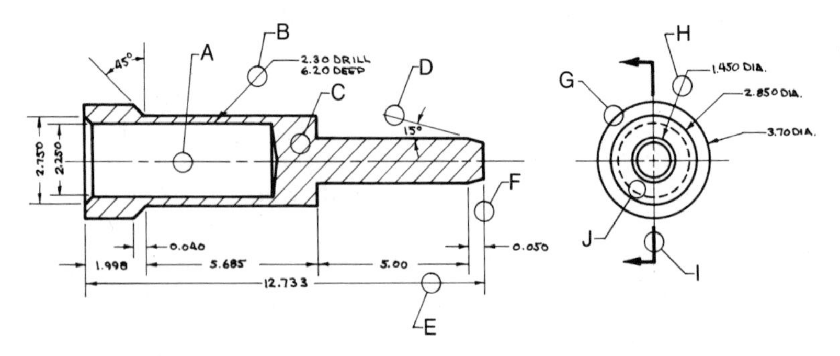

Fig. 3-11. Problem 3.1.

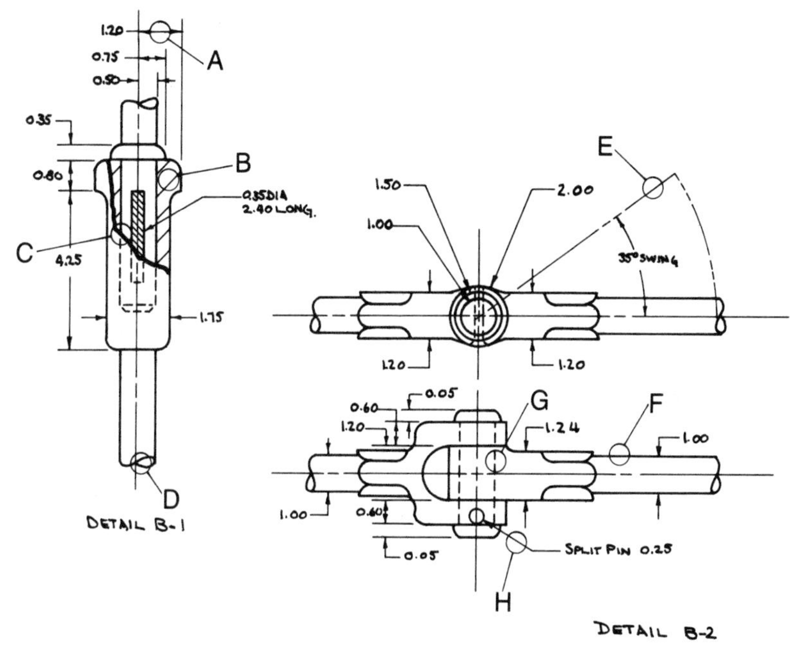

Fig. 3-12. Problem 3.2.

$\frac{5}{8}$ DRILL $\frac{3}{8}$-16UNC 1930 N/mm² 20% Ni-Ti-Al

WELD 75Ni-15 Cr-9 Fe+Mo+Ag

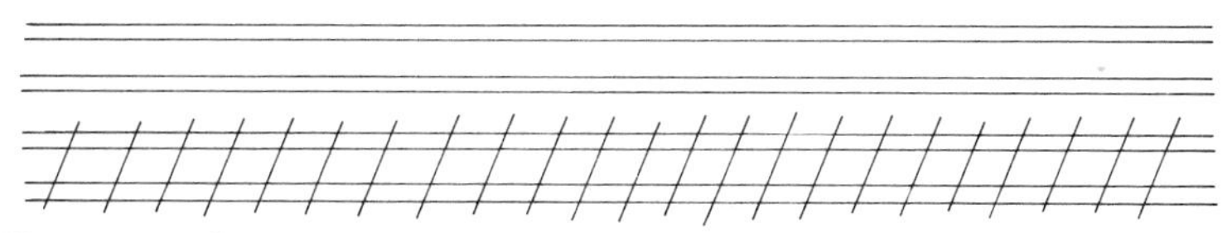

Fig. 3-13. Problem 3.4.

CHAPTER 4

Sketching and Instrument Drawing

- Technical Sketching
- Instrument Drawing
- Exercises

The preparation and production of industrial drawings is based upon the principles of freehand sketching and instrument drawing. Sketching is used for the development of designs and the modification of existing plans, while instrument drawings are initially prepared from sketches, and later changed from freehand modifications and notations. Both sketching and instrument drawing are used in the drafting profession, and drafters must be competent in both techniques. This chapter presents the basic principles and techniques employed in sketching and instrument drawings.

Technical Sketching

In engineering and drafting departments, sketching is primarily used by project engineers, designers, and chief drafters to express their ideas about the design or production of a product. Checkers also use sketching techniques as a method for noting modifications, changes, or corrections on working drawings. It is also common for beginning drafters to use sketching as a way of clarifying ideas before they begin preparing a set of working drawings.

Anyone presently employed, or seeking employment, in the drafting field should not underestimate the importance of freehand sketching. In some cases, sketching is the ultimate form of communicating information about product concepts and designs.

Sketching Materials

Sketches are made on a variety of drawing materials ranging from bond papers to drafting films and vellums. The quality of the sketch will depend, in part, upon the grade of paper and the hardness of lead used for the drawing. If the purpose of the sketch is to convey a bold, coarse effect, then a soft pencil on a rough textured paper should be used. On the other hand, if the sketch is intended to give information about intricate details, then a harder pencil on a smoother paper should be used.

The best leads for sketching will range from a soft 6B to a harder 2H, 6B leads producing a relatively wide and rough line. With harder leads (4B, B, F, and 2H), the sketched lines will become progressively lighter and narrower. Narrow lines can be produced with the softer leads, but they require continual sharpening (or the use of a narrow stick lead). As a rule of thumb, softer leads are used for bold, rough line work, and harder leads are best for smooth, light lines.

Drafters who have been improperly trained in sketching techniques will often make use of straightedges and cheap compasses to produce their sketches. Sketches prepared this way not only waste valuable time, but also result in a poor presentation because they are difficult to read and interpret.

Sketches that are made on existing drawings for the purposes of clarification, corrections, changes, or modifications do not employ any guide lines. Sometimes, however, guide lines are useful, particularly in the development of a new design. It is not uncommon to find engineers, designers, and drafters using coordinate paper, sketching paper with rulings that divide one-inch squares into 1/8-, 1/4-, or 1/10-inch squares. Coordinate paper has two major advantages over plain sketching paper: it speeds up the drawing process considerably, and it enables the drafter to produce a sketch to scale. Fig. 4-1 shows examples of two sketches, one prepared on coordinate paper and another on plain paper.

One other factor must be considered in the selection of sketching materials: whether or not the drawing is to be reproduced. Sometimes prints are made of product sketches, and are then distributed for review. In that case, it is best to use a paper with sufficient transparency

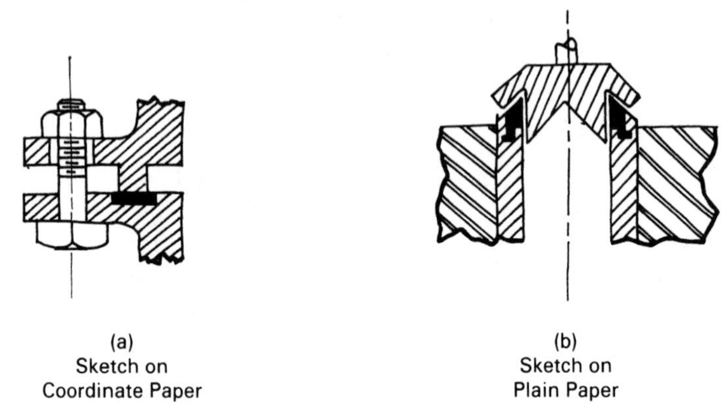

Fig. 4-1. Two types of sketching material.

(e.g., tracing or vellum paper). Most coordinate papers are printed with a nonreproducible ink that will not appear when the drawing is reproduced.

Drawing Techniques

All sketched drawings use a combination of straight and circular lines. This section will explain how straight and circular lines should be drawn, together with the proper sequencing that is used in preparing a sketch.

DRAWING STRAIGHT LINES

Once the proper type of paper and lead hardness has been selected, sketching can begin. The pencil should be held comfortably, about one to one and a half inches above the point, so as to allow for free and easy movement of the pencil or lead holder.

Vertical lines should always be sketched in a top-to-bottom movement (i.e., downward strokes), while horizontal lines are made in a left-to-right movement. For left-handed drafters, vertical lines are still drawn top to bottom, however, horizontal lines are drawn right to left. Lines that incline downward from right to left are drawn in the same manner as prescribed for vertical lines. Lines inclined downward from left to right present a unique problem. They are perhaps the most difficult to draw, and are best done by rotating the paper so that the line can be drawn as if it were horizontal. Fig. 4-2 illustrates sketching straight lines.

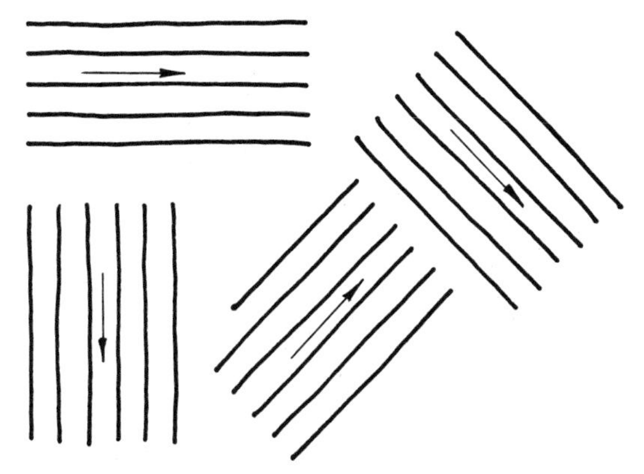

Fig. 4-2. **Sketching straight lines and direction of line movement.**

When drawing different types of line, try to maintain the appropriate line weights as specified in the alphabet of lines. The basic procedures used to draw straight lines are as follows (Fig. 4-3):

1. Mark the end points of the line. (Note: This is recommended for long lines.)
2. Before drawing, move your hand as though you were drawing along the path of the line so as to develop an appropriate motion.
3. Sketch a very light line from one point to the next. Do not attempt to draw long lines as one continuous line; longer lines should be drawn as a series of short lines.
4. As you are drawing the line, make any necessary corrections to keep the line in a true path.
5. Once the line is drawn in lightly, darken the finished line in the same manner so that it is dark, uniform, and straight.

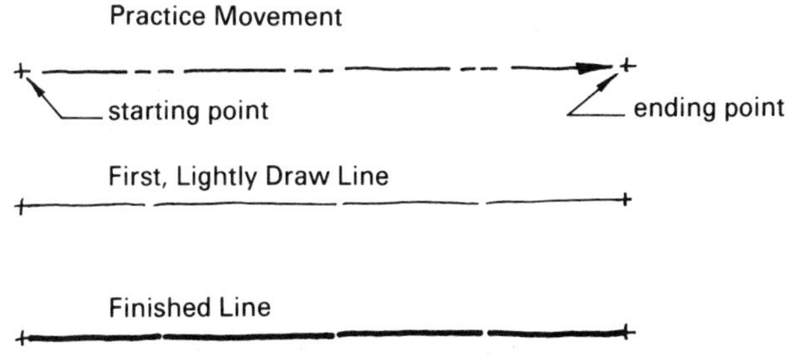

Fig. 4-3. **Procedures for drawing long, straight lines.**

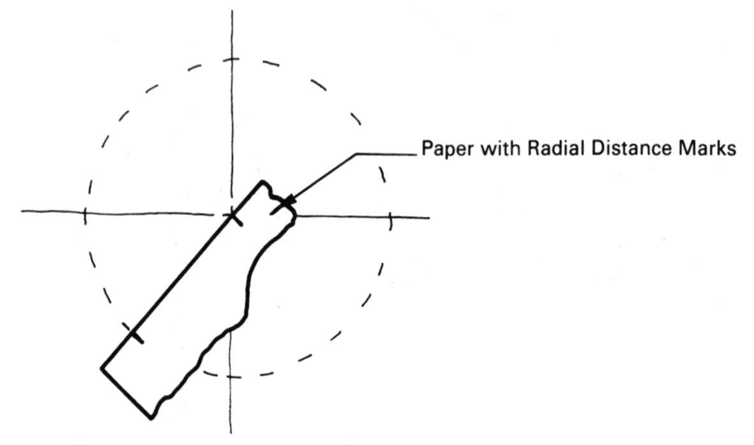

Fig. 4-4. Using radial distances to mark off a circle.

DRAWING CIRCULAR LINES

Circular lines include circles, ellipses, and curves. The best method for drawing circles is first to sketch in perpendicular center lines and mark off radial distances along their length. If additional distances are required, they can be marked off by eye or with a pieces of paper that has the radial distance marked off (Fig. 4-4).

Larger circles will require more radial distance points than smaller circles; it is recommended that additional diagonals be drawn in and marked off. Through each of these points, draw a short line perpendicular to the diagonals. These lines will serve as tangents to the circle, and provide an excellent guide for sketching. The circle can then be completed by drawing short arcs from tangent point to tangent point (Fig. 4-5).

For the beginning drafter or student, another sketching technique might prove easier. This technique makes use of a constructed square with diagonals (Fig. 4-6), and incorporates the following procedures:

1. Sketch in perpendicular center lines and mark off radial distances.
2. Draw a square about the center lines so that its sides run through the distances. The square should be drawn in very lightly.
3. Sketch in the diagonals of the square.

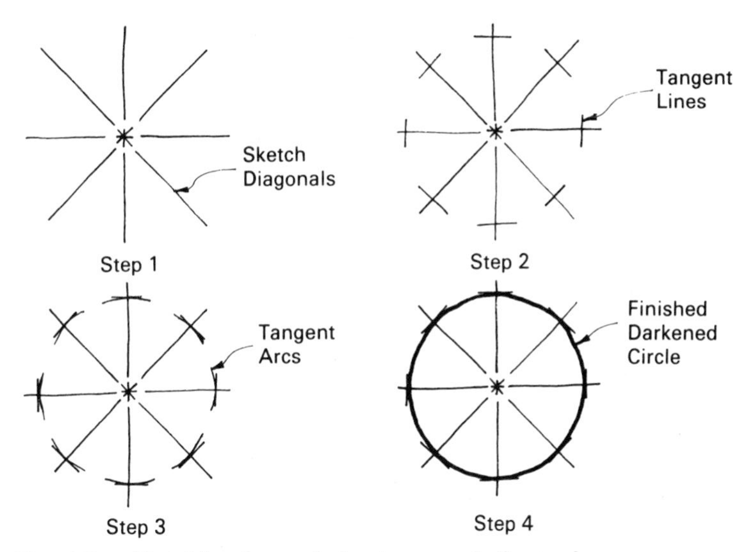

Fig. 4-5. Sketching large circles by use of diagonals.

4. Mark off radial distances along the square's diagonals.
5. Lightly sketch the circle with a series of short arcs that run through the radial distances and are tangent to the sides of the square.
6. Make any corrections needed to provide for a smooth circle, and darken it in.

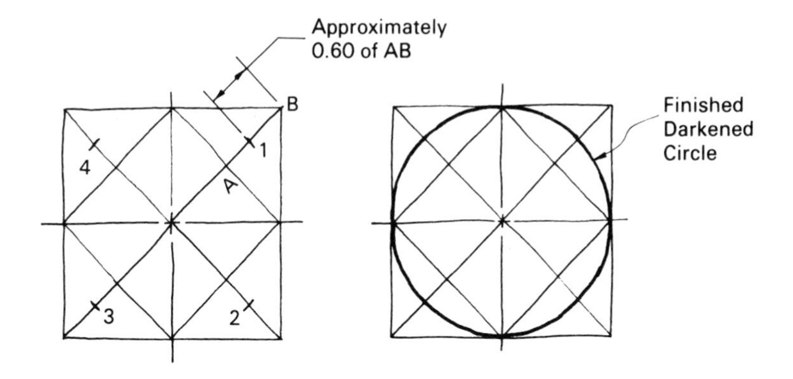

Fig. 4-6. Sketching a circle by constructed square method.

A construction square can also be used for sketching ellipses. The procedures are the same, except that a rectangle is used instead of a square.

SKETCH SEQUENCING

To make a working sketch requires systematically following all the practices and standards recognized in drafting. The following procedures (Fig. 4-7) are recommended:

1. If an existing object is to be drawn, examine it carefully, giving particular attention to detail.
2. Determine the view(s) that will best illustrate the design.
3. "Block in" the general view, using light construction lines.
4. Lightly sketch in all center and construction lines used for drawing circles, ellipses, and curves.
5. Draw in all circular lines first, followed by straight lines, with a light line.

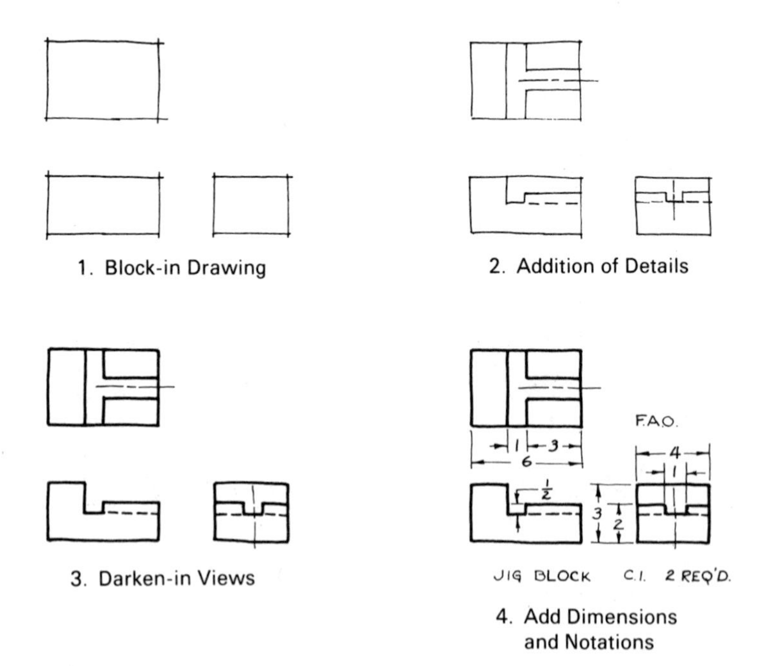

Fig. 4-7. Sketching sequence.

6. Darken in all object and hidden lines.
7. Sketch in extension and dimension lines, including arrowheads.
8. Add needed dimensions, notations, and other drawing specifications.
9. Look over drawing for omissions or incorrect information.

PROPORTIONING TECHNIQUES

A problem that sometimes arises while making a sketch is the establishment of appropriate relationships between dimensions. All width, height, and depth measurements must be appropriately proportioned to one another. Therefore, the drafter should be able to draw sketches that show the proper size relationship between any component parts, such as holes, keyways, slots, threads, springs, and gears. If proper proportioning is not shown in a drawing, its interpretation will be difficult, and even impossible.

Many sketches, especially those drawn in the design phase of a project, are not drawn to scale. The drafter must then determine the best proportions for a clear understanding of the product. For example, should the width be one-half the height, the hole one-fourth the width, or the slot three-fourths the depth?

Experienced drafters become so proficient in sketching that they can easily proportion a product's component parts by eye. Until one can accomplish this very quickly and skillfully, it may be necessary to use construction techniques. These techniques tend to increase the amount of time required to prepare the drawing, add additional construction lines that distract the reader, and detract from the neat appearance of the drawing. However, they will provide the beginning drafter with an accurate method for proportioning drawings.

The basic procedure used for proportioning a surface area is shown in Fig. 4-8. In Fig. 4-8a, a rectangular object is proportioned into one-half, one-quarter, and one-eighth sections by drawing in the diagonals of the rectangle. Fig. 4-8b shows the technique for finding a one-third proportion, and 4-8c for a one-sixth proportion.

Measuring Instruments

There are times when a sketch will serve as a working drawing, and must contain all the needed dimensions, specifications, and notations. This situation will most likely arise when a part must be replaced ow-

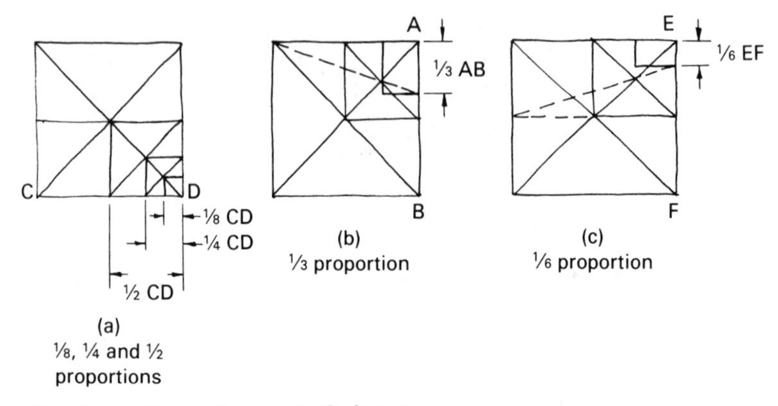

Fig. 4-8. Proportioning techniques.

ing to wear or breakage. In such cases, a proper measuring instrument must be selected so that measurements can be taken directly from the old part and placed on the sketch. General dimensions can be made with a steel rule that measures to the nearest 1/64th inch or 1 mm. If more precision is needed, a micrometer can be used for measurements as small as 0.0001 in. or 0.01 mm.

The use of a measuring instrument for a sketch drawing is generally not preferred to coordinate paper, but if coordinate papers are not used, drawing scales are employed to establish general proportions rather than to produce an accurate drawing.

Instrument Drawing

The vast majority of working drawings prepared by drafters are made with the use of drafting instruments. Instrument drawings are sometimes referred to as *mechanical drawings*, because mechanical devices are used in the drawing process. Instrument drawings require significantly more time and care than do sketches, and should be used only after most problems have been worked out in the design phase. This is not to say, however, that no changes will be required once an instrument drawing is completed.

Characteristics of Instrument Drawings

Drawings prepared with mechanical instruments are more precise and finished in appearance than are sketches. All lines will appear straight,

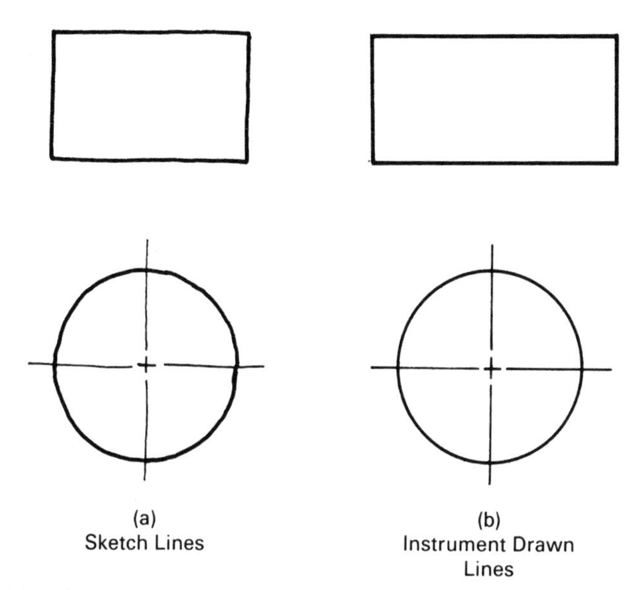

Fig. 4-9. Comparison between sketched and instrument-drawn lines.

with no waviness. Shown in Fig. 4-9 is a comparison of straight and circular lines drawn by sketching and mechanical instrument techniques.

Except in a few special cases, all instrument drawings are drawn to scale. Scale notations, usually provided in the title block of the drawing, will indicate the ratio of the drawing size to the actual size of the part. Typical scales, with their appropriate notations, used in industrial working drawings are:

2 = 1, 2:1, or double size
1 = 1, 1:1, or full size
$^1/_2$ = 1, 1:2, or half size
$^1/_4$ = 1, or 1:4
$^1/_8$ = 1, or 1:8

There are times when it is necessary to draw details to a different scale than is specified on the drawing sheet. Wherever this occurs, a special scale notation should be made under the individual detail. The scale notation in the title block should also have the additional note: "EXCEPT AS NOTED." Another practice is known as *framed structuring*. This technique consists of drawing the center lines of sections in one scale and superimposing the details, or partial details, to larger

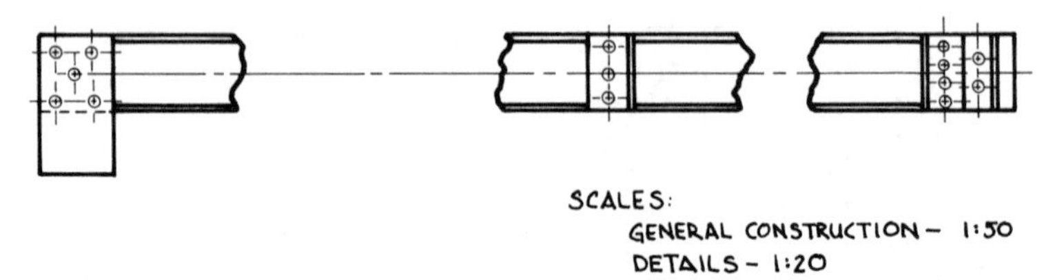

Fig. 4-10. Framed structuring drawing using two scales on single drawing.

scales at the points of intersection. This scaling technique is shown in Fig. 4-10.

Instrument Drawing Techniques

Instrument drawing, like technical sketching, requires the use of a proper technique and systematic procedure to produce quality drawings. To become a competent drafter, these techniques must become almost second nature.

Before drawing can begin, the drawing paper must be properly placed on the drawing surface and fastened. Regardless of the size of paper used, it should be located well up on the drawing surface toward the left-hand edge. (If the drafter is left-handed and using a left-handed drafting machine, the paper should be located toward the right-hand edge.) The lower edge of a plain sheet of paper, or the lower border line of a preprinted sheet, should be aligned with the working or ruling edge of the drafting machine, parallel, or T-square. Once positioned, the paper is fastened to the drawing surface at all four corners with masking tape.

Some drafters use Scotch tape, thumbtacks, or staples to fasten their drawings. These are not recommended; thumbtacks and staples will leave holes in the drawing paper, and Scotch tape tends to tear the paper when it is removed. Masking tape, on the other hand, can be easily removed without damaging the paper. Once fastened, the paper should not be removed until the drawing is completed.

DRAWING STRAIGHT LINES

Again, the lead should be properly sharpened, or the proper stick lead size selected, before drawing begins. All horizontal lines are to be drawn from left to right (right to left if you are left-handed), so that

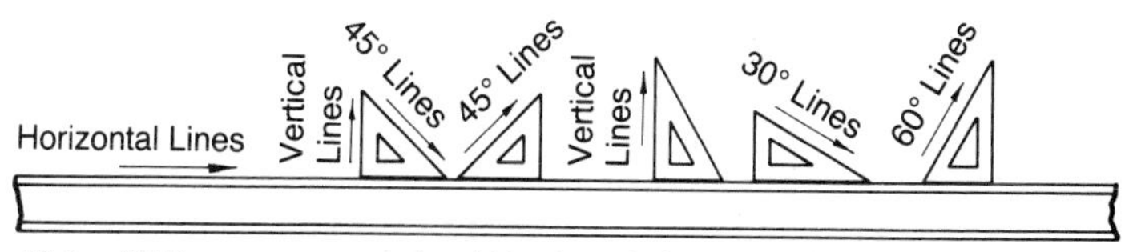

Note: All line movement should be from left to right for horizontal and inclined lines, and from bottom to top for vertical lines.

Fig. 4-11. Drawing horizontal, vertical, and inclined lines.

you are always drawing away from the head of the drafting machine or T-square. The lines should be drawn along the upper edge of the blade. As the pencil moves along the blade, take care to ensure that the blade does not move, by pressing the free hand down on it as the line is drawn.

Vertical lines will be drawn with either a triangle or the short vertical blade of the drafting machine. Unlike in sketches, mechanical lines are drawn in an *upward direction*, along the vertical side of the triangle. The horizontal leg is rested against the horizontal blade of the drafting machine, parallel, or T-square. When drawing, the palm of the left hand should rest on the blade as the fingers hold the triangle in place. The vertical leg of the triangle should be facing the left side of the drawing surface.

To draw vertical lines with a drafting machine, use the left edge of the vertical blade as the ruling edge. Place it along the desired length and hold it in position with the left hand pressing down on the horizontal blade, and draw in an upward direction.

The direction for drawing inclined lines will vary according to the angle of the incline. Angles that are greater than 45° should be drawn in an upward direction, while those less than 45° are drawn in a downward direction. 45° lines can be drawn in either an upward or downward direction. The general movement, however, should always be from left to right or right to left for left-handed drafters. Fig. 4-11 illustrates the drawing of straight lines.

DRAWING CIRCULAR LINES

The primary instrument used to draw circles and arcs is the compass. Curved lines, on the other hand, are drawn with irregular curves, because the radius of the curvature will not be constant. The basic

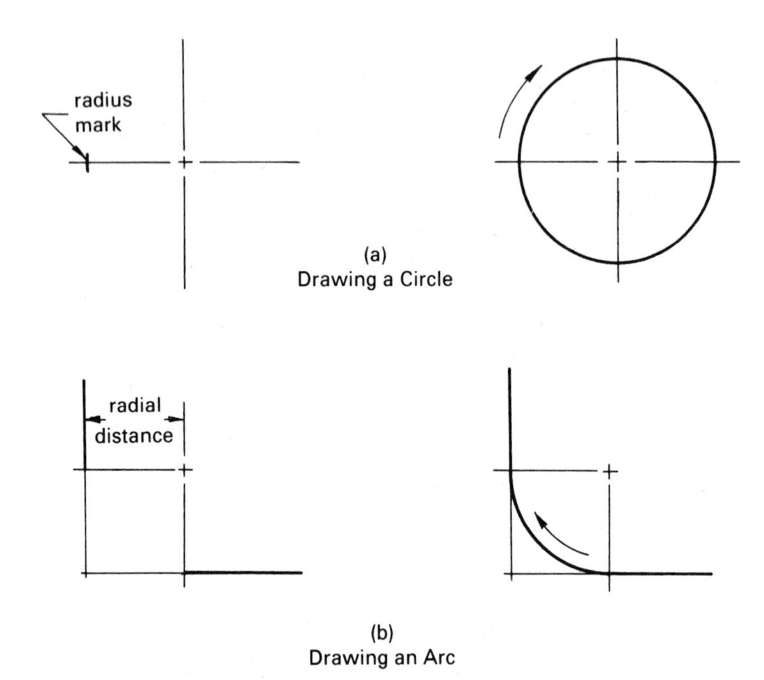

Fig. 4-12. Drawing circular lines with instruments.

procedure used to draw circles and arcs is illustrated in Fig. 4-12, and described as follows:

1. Locate all circular centers. The center radius for arcs can be constructed by the intersection of radial distances with a compass from the end locations of the arc.
2. Draw in all center lines, through the circular centers, with light construction lines.
3. Adjust the compass a distance equal to the radius of the circle or arc.
4. Lightly draw in the circle or arc.
5. Darken in all center lines, circles, and arcs.

Curved lines are made by connecting a series of points along a given path. To use an irregular curve properly, it is first necessary to have a sufficient number of points along the curved path, usually at least three points that will pass through the ruling edge of the curve at any one time (Fig. 4-13).

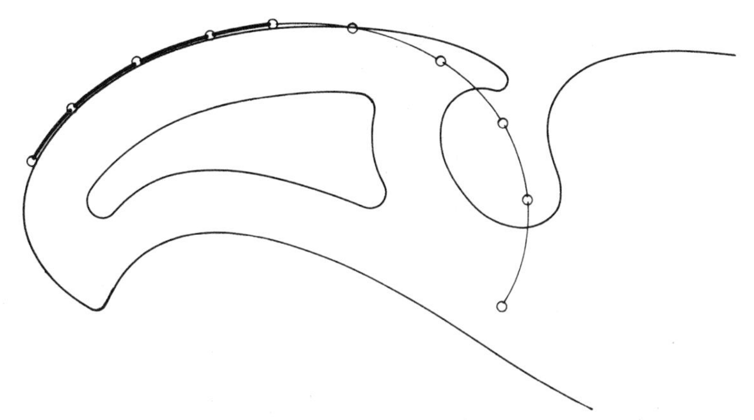

Fig. 4-13. Proper use of an irregular curve.

The irregular curve should be placed along the plotted path so that the increasing curvature of the ruling edge follows the direction of that portion of the curve. To eliminate humps and bumps, each section of the curve should be drawn so that the line drawn will stop short of the first and last points to which the irregular curve has been fitted. After drawing a section, the curve is moved so that the new position of the ruling edge will fit the section of the line previously drawn. This procedure should be followed until the last point is reached.

The recommended procedures (Fig. 4-14) for drawing irregular curves are as follows:

1. Locate as many points along the curved line as needed to use an irregular curve properly. Use at least three points.
2. Very lightly sketch in the curve freehand.
3. Begin drawing at the first point, and continue so that the line will stop just short of the last point along the ruling edge.
4. Adjust the irregular curve to pick up the drawn line, and continue through at least three more points.
5. Continue procedure until the curved line is completed.

SEQUENCING OF DRAWINGS

Most industrial drawings will require that the drafter be able to draw both straight and curved lines in combination. Since these lines are

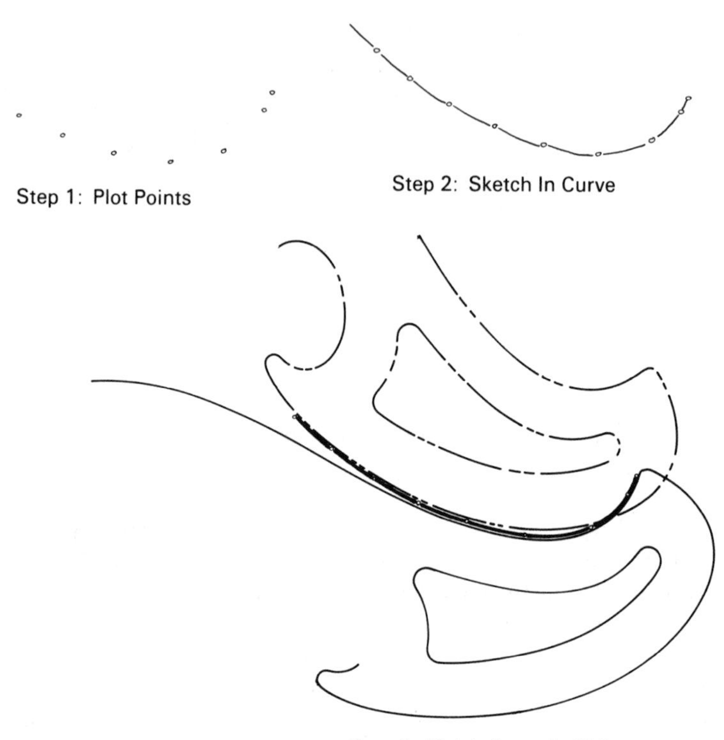
Step 1: Plot Points
Step 2: Sketch In Curve
Step 3: Finish Curve by Using
Irregular Curve

Fig. 4-14. Procedures for drawing an irregular curve.

often used in combination to form a part or view, they must blend together, with no bumps or humps along their length. The recommended sequence for preparing drawings is illustrated in Fig. 4-15, and consists of the following steps:

1. Locate and draw in all center lines lightly.
2. Identify all radial distances and draw them in along the center lines.
3. Draw all circles, followed by arcs. If curved surfaces are present, draw these in first.
4. Draw in all tangent lines.
5. Complete all remaining object lines.
6. Draw in hidden lines, and darken in all center lines.

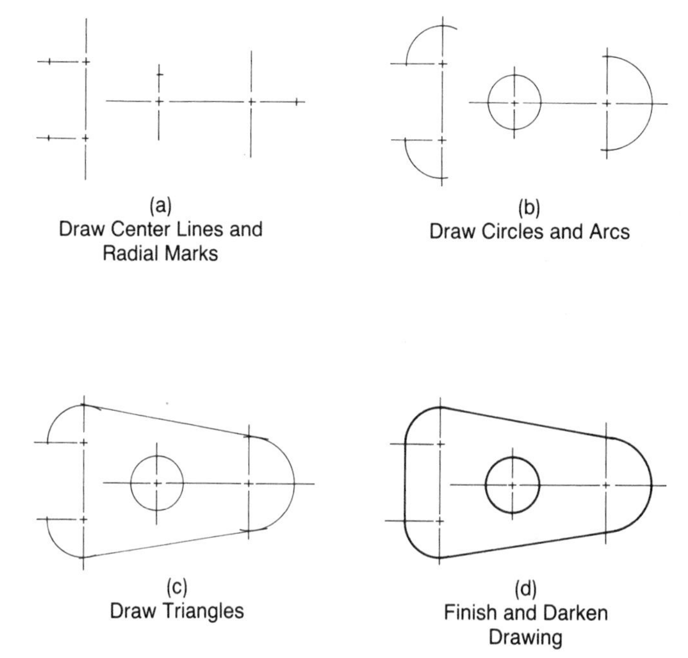
(a)
Draw Center Lines and Radial Marks
(b)
Draw Circles and Arcs
(c)
Draw Triangles
(d)
Finish and Darken Drawing

Fig. 4-15. Proper sequence for producing an instrument drawing.

7. Locate and draw all phantom lines, followed by extension and dimension lines.
8. Place all dimensions and notations on the drawing.

Inking

With the advent of improved inks and the availability of easy-to-use inking instruments, more and more working drawings are being prepared in ink. It is therefore essential that drafters not only understand proper inking procedures and techniques, but also be competent in preparing inked drawings.

INKING PEN TECHNIQUES

Technical inking pens have all but replaced the older ruling pen. In some situations, however, the ruling pen might prove helpful (e.g., touching up drawings). Ruling pens are used to ink mechanical lines.

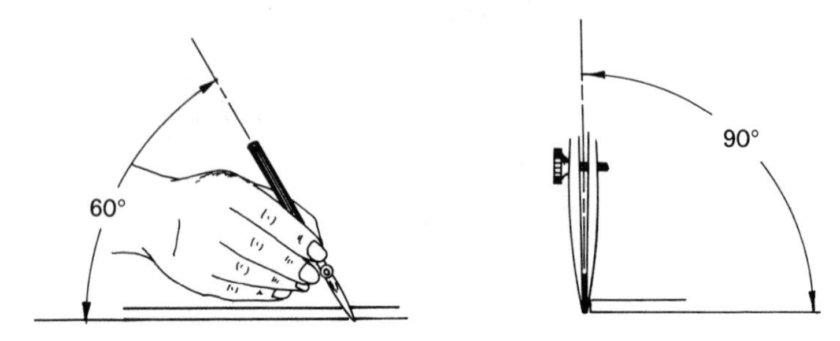

Fig. 4-16. Proper position of ruling pen.

When used, they should be held at an angle of 60° in the direction of the movement and perpendicular to the drawing paper (Fig. 4-16). Because the flow of ink is difficult to control, inking problems are likely to arise (see Fig. 4-17).

Most compass inking attachments are used in the same way as ruling pens, and proper inking procedures should be followed. They include the following steps:

1. Set the desired width of the line by regulating the adjusting screw to the side of the nibs.
2. Place ink into the nibs with a dropper device or quill of the ink bottle. Take care not to get any ink on the outside surfaces of the nibs. Also, be sure that sufficient ink is available to finish the line. As a rule, never have more than 1/4 inch of ink in the nibs at one time.

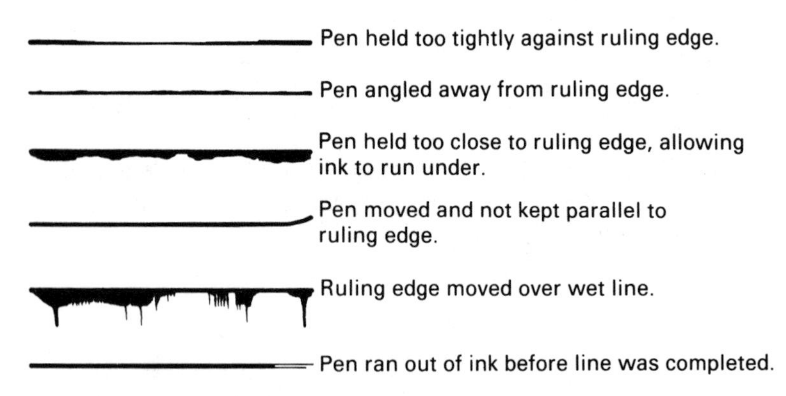

Fig. 4-17. Common errors made with ruling pens.

3. Before inking, use a test sheet to make sure that the nibs are set at the appropriate line thickness.
4. Proceed to ink in the drawing, taking care not to cause any inking errors.

Technical inking pens are much easier to use than ruling pens. Here line width is based upon the size of the pen itself. Before using, make sure that there is sufficient ink in the ink tube and that the ink flows freely through the pen point. If the point is clogged with debris or dried ink, remove the ink tube and clean the head of the pen thoroughly. If clogging is heavy, it may be necessary to disassemble the pen head completely.

All pens are equipped with a thin plunger that will move up and down the barrel of the point. If the plunger does not move when the pen is slightly shaken up and down (the plunger will make a small clicking sound), it must be cleaned. Some pens are designed with a spring mechanism attached to the plunger that inhibits the natural clogging of ink. Such pens cannot be tested for clogging by shaking the pen. Regardless of the type of pen used, it is good practice to clean the technical pen periodically, or after an extended period of nonuse.

Though technical pens are much easier to use than ruling pens, they are not error-free. When used, the pen should be held in a perpendicular position to allow for the proper flow of ink. Fig. 4-18 shows correct inking procedures and several errors that can be made with a technical pen.

INKING PROCEDURES

Inked drawings are prepared by drawing the plan in pencil first (for overinking), or as a tracing from an existing drawing. Because the ink is permanent, it would be impractical to ink a drawing in the same manner as a pencil drawing (most inks can be erased with a special eraser, but excessive erasures will damage the paper). Ink drawings therefore require more drawing time than pencil drawings, but they offer a significantly higher quality product that can be used to generate crisp, sharp prints.

Once the paper has been placed on the drawing surface for tracing or overinking, the drafter must be conscious of an inking order, which will not only make the inking process easier but will also help to minimize errors. Before inking, all radial centers and tangent points

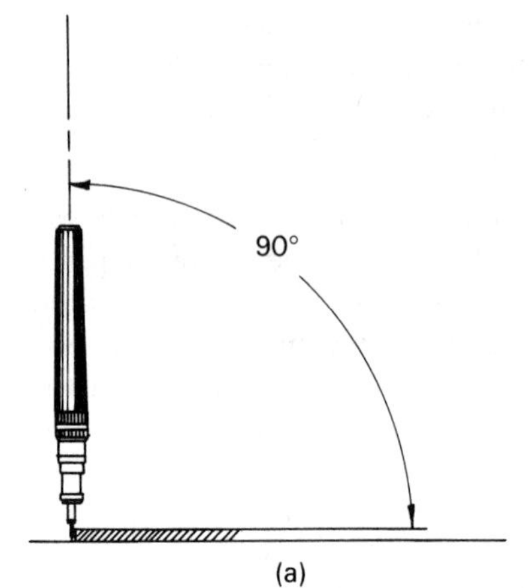

(a)
Proper Positioning of Technical Pen

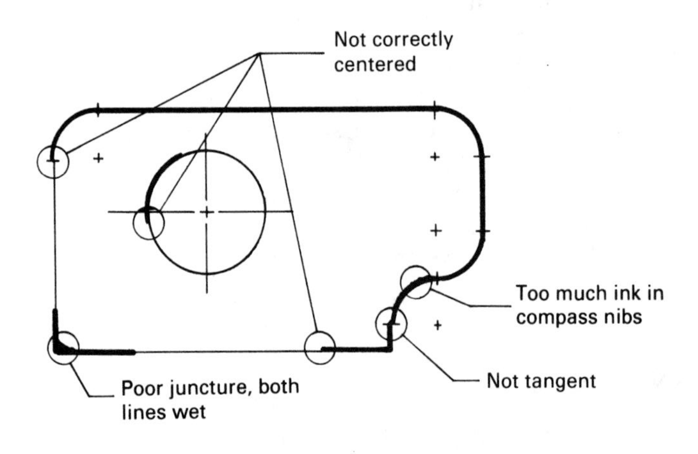

(b)
Poor Inking Practices

Fig. 4-18. Technical pen inking.

should be located and marked. Once this is accomplished, inking order should be as follows:

1. Circles and arcs, with smaller radii first, and in the order of:
   a. object lines
   b. hidden lines
   c. center and dimension lines

2. Irregular curves in the order of:
   a. object lines
   b. hidden lines
3. Straight object lines in the order of:
   a. horizontal lines, beginning at the top of the sheet and working down
   b. vertical lines, beginning at the left side and working across
   c. inclined lines, working from left to right
4. Hidden lines in the order of:
   a. horizontal lines
   b. vertical lines
   c. inclined lines
5. All other lines, such as center, extension, dimension, and phantom lines, in the order of:
   a. horizontal lines
   b. vertical lines
   c. inclined lines
   d. section lines
6. Dimensions and arrowheads
7. Notations and specifications
8. Titles
9. Border

Exercises

The following series of exercises are intended to provide experience in technical sketching and instrument drawing that will help to develop drafting skills. All drawings should be drawn lightly with a hard lead before they are darkened in. Make sure that all needed constructions are correct, and that line representations are of appropriate weight.

4.1 On a sheet each of plain and coordinate drawing paper, reproduce by freehand sketching techniques the line formations shown in Fig. 4-19.

4.2 Reproduce the line formations shown in Fig. 4-20 using freehand sketching on both plain and coordinate drawing paper.

4.3 Prepare two drawings of the object in Fig. 4-21. The first should be by freehand sketching, and the second by instrument drawing techniques. Do not draw the object to scale in your freehand sketch. Use a 1:1 scale for your instrument drawing, and include all dimensions and notations.

4.4 Copy the instrument drawing in Fig. 4-22.

4.5 Trace the drawing prepared in exercise 4.4 with ink. Be sure that you follow the correct inking procedures and sequence.

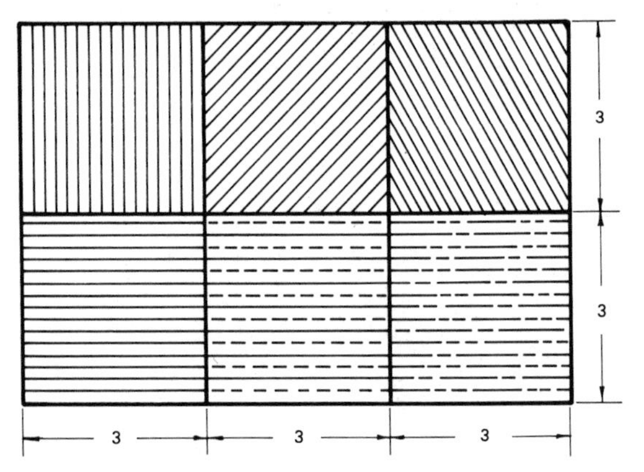

Fig. 4-19. Problem 4.1.

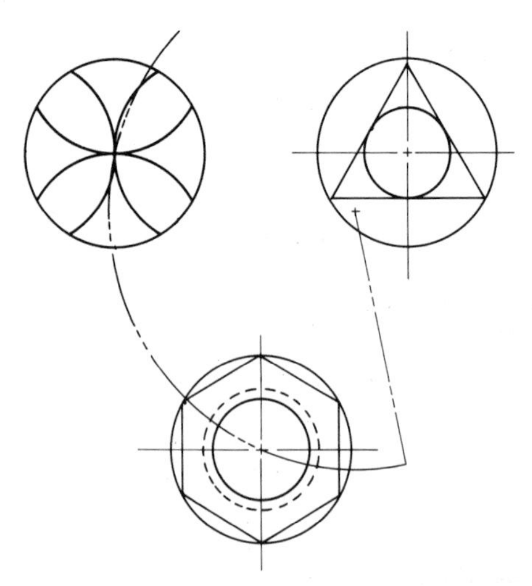

Fig. 4-20. Problem 4.2.

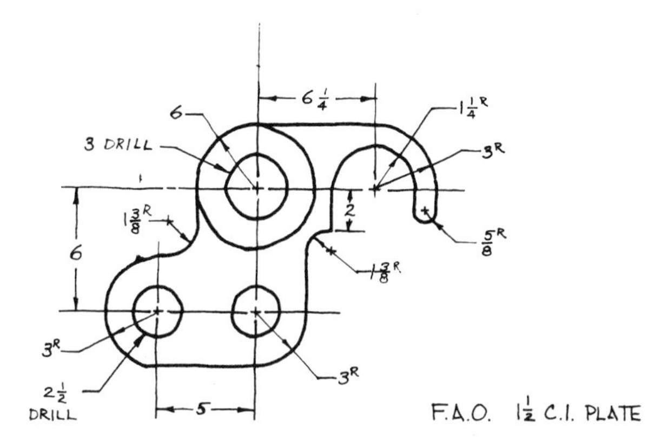

Fig. 4-21. Problem 4.3.

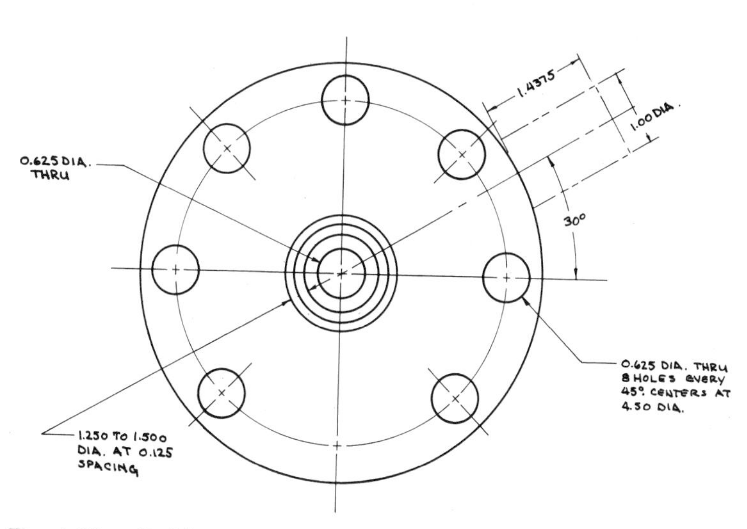

Fig. 4-22. Problems 4.5 and 4.6.

CHAPTER 5

Geometric Construction

The solutions of many drafting and design problems are based upon the graphic elements of plane geometry and trigonometry. It is not uncommon to find graphic procedures chosen over strictly mathematical calculations when solving engineering problems, for graphic methods are often both easier and quicker. It is interesting to note that all geometric and trigonometric theories, laws, postulates, and proofs are derived from drawings. The geometric constructions used by drafters, then, are applied forms of geometry and trigonometry.

All the geometric constructions presented in this chapter have their counterpart in mathematical principles. It is therefore important that the drafter be familiar not only with drafting techniques, but also with the mathematical concepts upon which they are based. The geometric construction procedures presented here, however, have been adapted for use with drafting equipment and instruments, and are sometimes slightly different from those taught in mathematics courses.

Basic Geometric and Trigonometric Concepts

Present-day geometric and trigonometric concepts were first expressed by early mathematicians as drawings. Plane and solid shapes were constructed using drawing techniques, and later explained in mathematical terms. Geometry is the phase of mathematics that deals with points, lines, angles, surfaces, and solids, while trigonometry is the study of the application of angles, triangles, and trigonometric functions.

An adequate grasp of the concepts involved in geometric construction requires a clear understanding of the basic terms used in geometry and trigonometry. This section will describe and discuss those elements of geometry and trigonometry that are directly related to drafting.

Points and Lines

A point is defined as a location in space. It has no height, width, or length—it is dimensionless. In a drawing, a point may be represented in one of three ways: as a small cross, as a perpendicular line crossing an existing line, or as the point where two lines intersect (Fig. 5-1). A point should never be represented as a dot on a drawing; this is considered poor practice.

A line may be either straight or curved, and has but one dimension: length. Mathematically, a line is defined as the path of a point moving through space. A more practical definition with application to drafting is that a line is the connecting of two points.

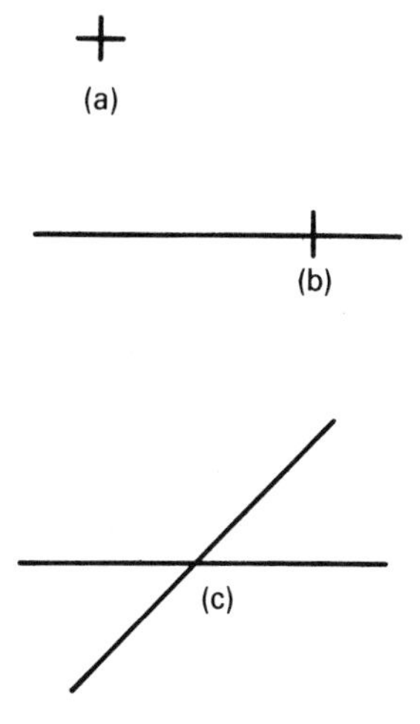

Fig. 5-1. Three representations of points.

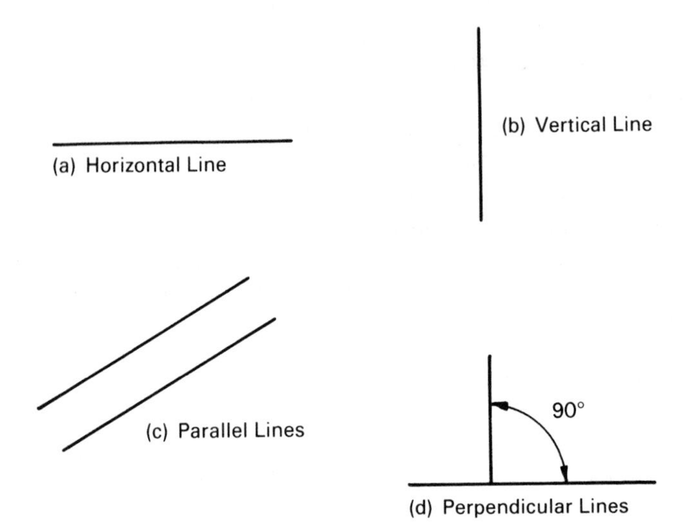

Fig. 5-2. Relationships between lines.

In drawings, there are two types of line representations. The first includes lines with known lengths that are drawn to length with recognizable end points. The second includes lines with unknown lengths that are drawn to a convenient length. Lines are perhaps most commonly described in terms of their position and relationship to other lines. Examples of these are horizontal, vertical, parallel, and perpendicular lines (Fig. 5-2).

Angles

An angle is formed by the intersection of two lines, and is measured in terms of degrees (°). The maximum size of a drawn angle is 360°, which is a circle. Each degree is divided into 60 minutes ('), which is further divided into 60 seconds ("). Thus, the angle 23 degrees, 14 minutes, 32 seconds is noted as 23° 14' 32".

Angles are classified according to their size and relationship (Fig. 5-3). In all, there are seven major classifications of angles:

1. *Complete circles* are equal to 360°.
2. *Straight lines* are equal to 180°.
3. *Right angles* are equal to 90°.
4. *Acute angles* are angles that are less than 90°.

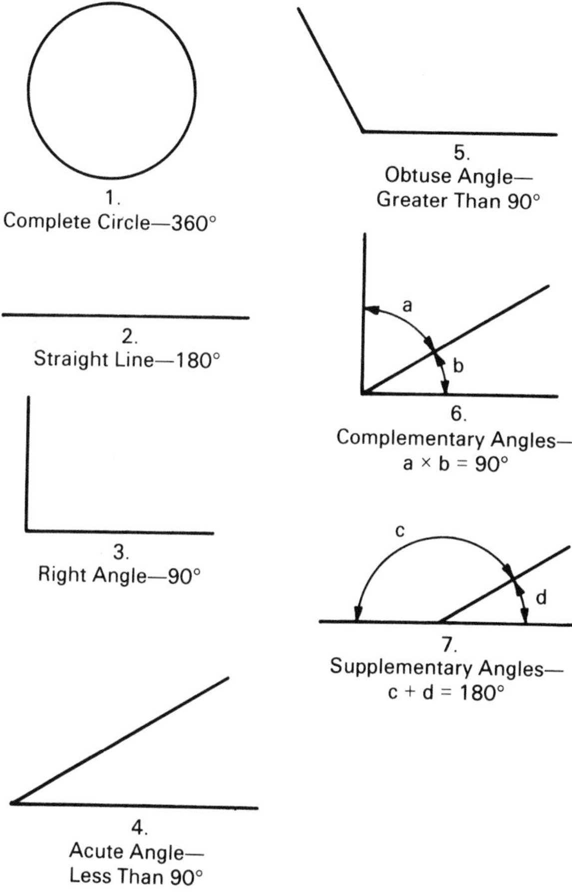

Fig. 5-3. Classification of angles.

5. *Obtuse angles* are greater than 90°.
6. *Complementary angles* are angles whose sum is equal to 90°.
7. *Supplementary angles* are angles whose sum is equal to 180°.

Triangles

A triangle is a plane figure with three sides whose interior angles will add up to 180°. The four general classifications and characteristics of triangles (Fig. 5-4a) are as follows:

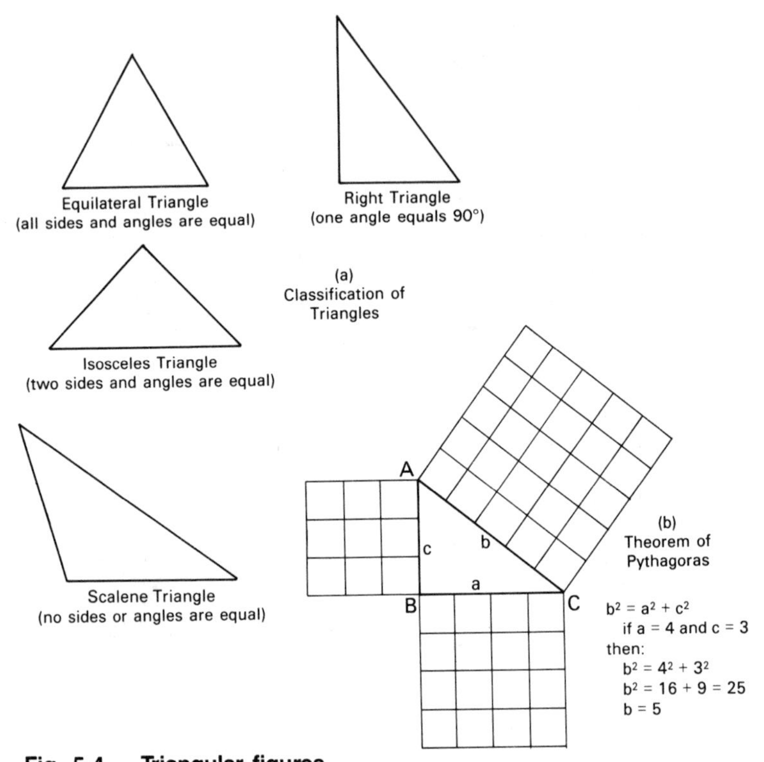

Fig. 5-4. Triangular figures.

1. *Equilateral triangles* have three equal sides and angles.
2. *Isosceles triangles* have two equal sides and angles.
3. *Scalene triangles* have no sides or angles equal to each other.
4. *Right triangles* have one angle equal to 90°.

It should be noted that all side dimensions for right triangles are governed by the *theorem of Pythagoras* (Fig. 5-4b), which states that the square of the hypotenuse is equal to the sum of the squares of the other two sides. Mathematically, this is expressed as:

$$AB^2 = BC^2 + CA^2$$

Quadrilaterals

A quadrilateral is any four-sided plane figure whose interior angles add up to 360°. Any quadrilateral having opposite sides parallel to

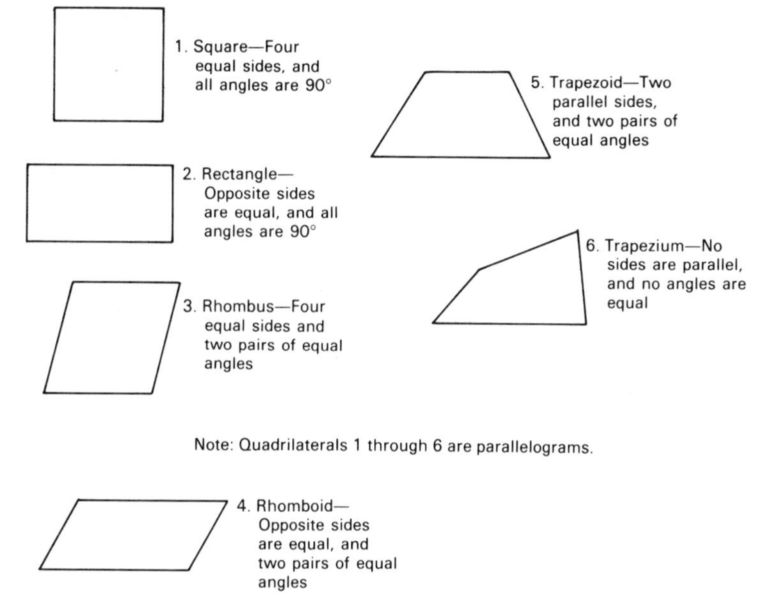
1. Square—Four equal sides, and all angles are 90°
2. Rectangle—Opposite sides are equal, and all angles are 90°
3. Rhombus—Four equal sides and two pairs of equal angles
5. Trapezoid—Two parallel sides, and two pairs of equal angles
6. Trapezium—No sides are parallel, and no angles are equal
Note: Quadrilaterals 1 through 6 are parallelograms.
4. Rhomboid—Opposite sides are equal, and two pairs of equal angles

Fig. 5-5. Types of quadrilateral.

each other is known as a *parallelogram*. Several common quadrilaterals are illustrated in Fig. 5-5, and described as follows:

1. *Squares* are parallelograms having four equal sides that are perpendicular to each other.
2. *Rectangles* are parallelograms with opposite equal sides that are at right angles to each other.
3. *Rhombuses* are parallelograms having four equal sides with two different interior angles.
4. *Rhomboids* are parallelograms having opposite equal sides with two different interior angles.
5. *Trapezoids* have two sides that are parallel, with two different interior angles.
6. *Trapeziums* have no parallel sides and four different interior angles.

Polygons

A polygon is any multisided plane figure that is bounded by straight lines. One of the most recognizable types of polygons is the *regular polygon*. Regular polygons differ from others in that the lengths of all sides are equal, and all interior angles are equal. Regular polygons may therefore be constructed within a circle (inscribed) so that each corner is touching the circumference, but does not extend beyond it; or they can be constructed about a circle (circumscribed) so that each side will be tangent to the circle's circumference, but not penetrate it.

Examples of eight common regular polygons (Fig. 5-6), along with the number of equal sides, are:

1. *Equilateral triangle:* three equal sides
2. *Square:* four equal sides
3. *Pentagon:* five equal sides
4. *Hexagon:* six equal sides
5. *Heptagon:* seven equal sides
6. *Octagon:* eight equal sides
7. *Nonagon:* nine equal sides
8. *Decagon:* ten equal sides

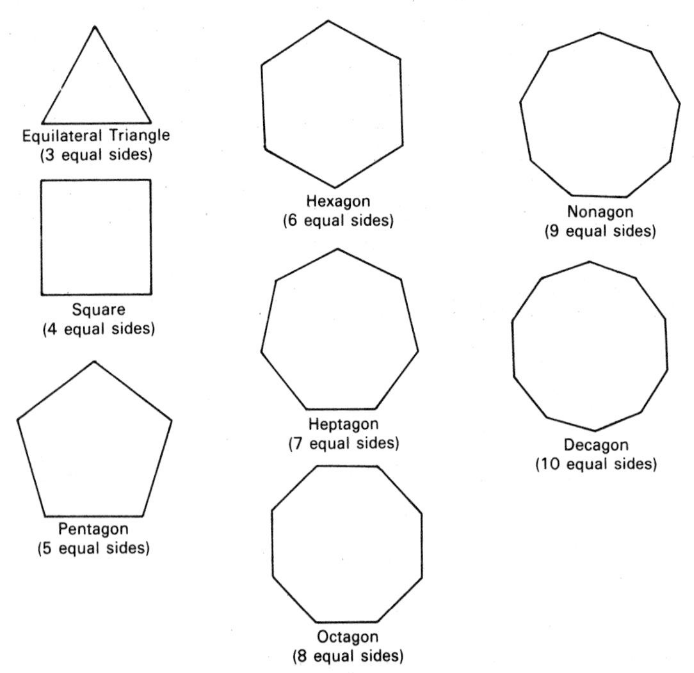

Fig. 5-6. Common regular polygons.

Circles

A circle is a closed curved plane figure in which every point on that curve is an equal distance from a center point. The circle plays a leading part in geometric construction and other drafting problems; it is therefore important for the drafter to be familiar with its various elements (Fig. 5-7). The basic elements of a circle, and their definition, include:

1. *The circumference* of a circle is the distance around the circle.
2. *An arc* is any portion of the circumference.
3. *The diameter* of a circle is the straight-line distance from one point on the circumference, through the center point of the circle, to the opposite side of the circumference.
4. *The radius* of a circle is the distance from the circle's center point to the circumference.
5. A *chord* is any line joining two points on the circumference of the circle.
6. A *quadrant* is a one-fourth, or quarter, section of a circle.
7. A *sector* is a section of a circle bounded by two radii and an arc.
8. A *segment* is that section of a circle bounded by a chord and arc.

There are two basic circle measurements that are commonly calculated in drafting problems. The first is the circumference, which is found by using the following formula:

$$C = \pi d$$

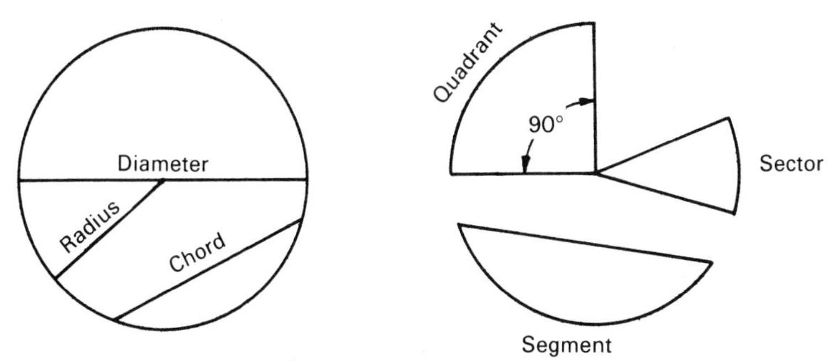

Fig. 5-7. Basic elements of a circle.

where: C = circumference
π = (pi) 3.14159 or 22/7
d = diameter

The second calculation made is for the area of a circle. The formula used for these problems is:

$$A = \pi r^2$$

where: A = area
π = (pi) 3.14159 or 22/7
r = radius

Solids

Solids are different from the one- and two-dimensional figures presented thus far. Unlike points, lines, and planes, solids are described in terms of three dimensions: length, width (or depth), and height. Solids incorporate the use of planes, lines, and points for their construction, but within a three-dimensional context. For the purposes of discussion, solid figures have been divided into six broad categories, each of which will be presented in this section.

POLYHEDRONS

Any solid figure with multiple plane surfaces is a polyhedron. When the plane surface is made of equal regular polygons, the solid will be known as a regular polyhedron. The five basic regular polyhedrons used for many engineering and design problems are shown in Fig. 5-8, and include:

1. *Tetrahedron:* four regular polygon surfaces
2. *Hexahedron:* six regular polygon surfaces
3. *Octahedron:* eight regular polygon surfaces
4. *Dodecahedron:* twelve regular polygon surfaces
5. *Icosahedron:* twenty regular polygon surfaces

PRISMS

Solids that have equal polygons as ends that are joined by parallelograms are known as prisms. Prisms are named according to the characteristic of their ends or bases. A prism with a triangular

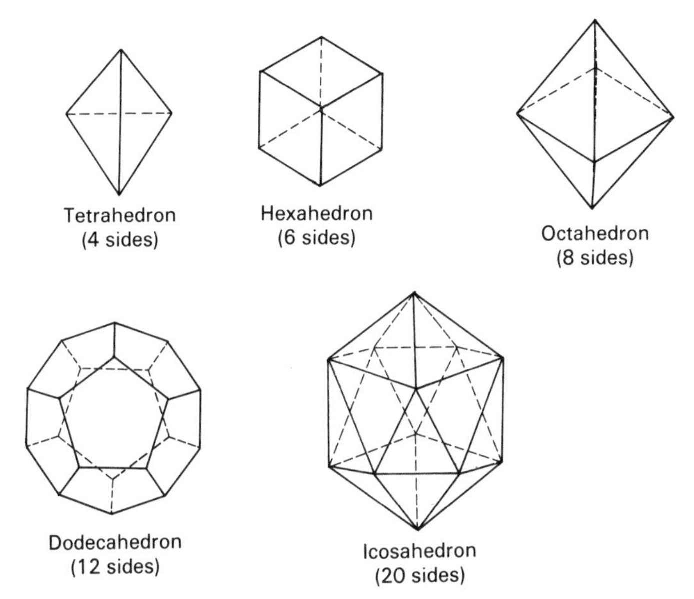

Fig. 5-8. Five regular polyhedrons.

base is a triangular prism, a rectangular prism has a rectangular base, and a pentagonal prism has a pentagonal base.

The *axis* of a prism is that line drawn from the center point of one base to the center point of the other base. A right prism is one whose axis is at 90° to both bases. Oblique prisms, on the other hand, have axes that are not perpendicular to the bases.

When the ends of a prism are parallelograms, the prism is called a *parallelepiped.* A prism that has been cut off at an angle not parallel to the other base is called a *truncated prism.* Fig. 5-9 illustrates different types of prism.

PYRAMIDS

Solids with a polygon base and triangular sides that meet at a common point (vertex) are called pyramids (Fig. 5-10). Like prisms, pyramids are named according to their base, e.g., triangular pyramids, square pyramids, and pentagonal pyramids.

The axis of a pyramid is found by drawing a line from its vertex to the center point of the base. If that axis is perpendicular to the

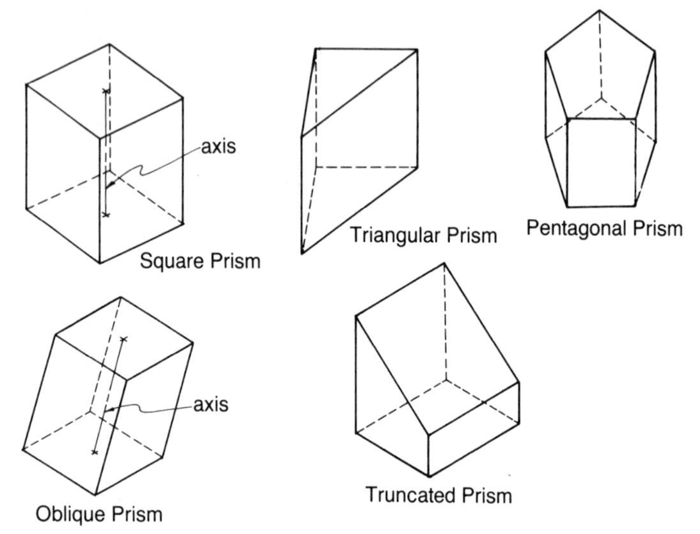

Fig. 5-9. Types of prisms.

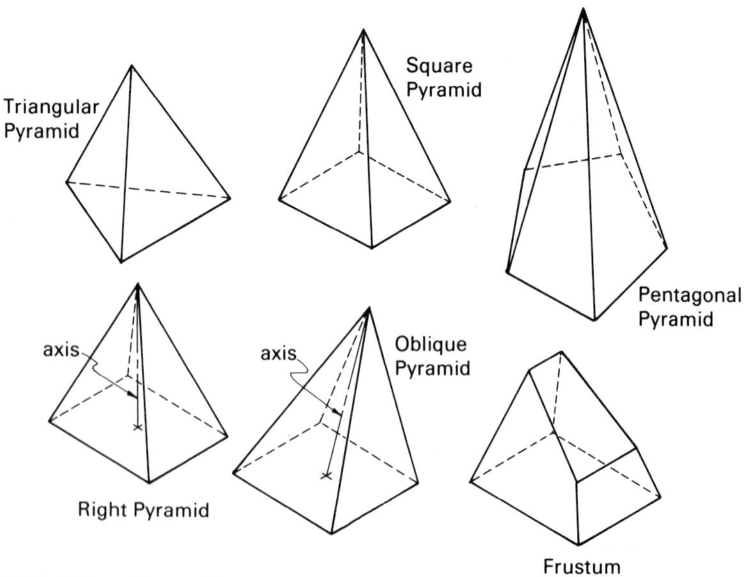

Fig. 5-10. Pyramids.

base, the solid will be a right pyramid. When the axis is not perpendicular, it will be an oblique pyramid. Truncated pyramids, also known as *frustums*, are those in which a section near the vertex has been cut off.

CYLINDERS

A solid figure that has been formed by the edge of a rectangle rotating about the parallel edge of an axis is a cylinder. Thus, the ends of a cylinder will be circles of equal size and parallel to each other. A more technical definition of a cylinder is that it is an object that is formed by a straight line known as a *generatrix*. The generatrix moves about a circle in such a way as to be always parallel to its previous position. Each position of the generatrix is called an *element*.

The axis of a cylinder is found by drawing a line from the center of one base to the center of another. As a result, the generatrix will also be parallel to the cylinder's axis. Cylinders that have axes perpendicular to their bases are known as right cylinders, while those with nonperpendicular axes are oblique. Truncated cylinders have one end of the cylinder cut at a nonparallel angle to the base. See Fig. 5-11.

CONES

Solids that have a curved base and sides that taper evenly to a point (vertex or apex) are known as cones (Fig. 5-12). Hence, any point located on the surface of a cone is in a straight-line position between the vertex and base. Cones can also be described as being formed by a generatrix where one end is fixed at the vertex and the other moves

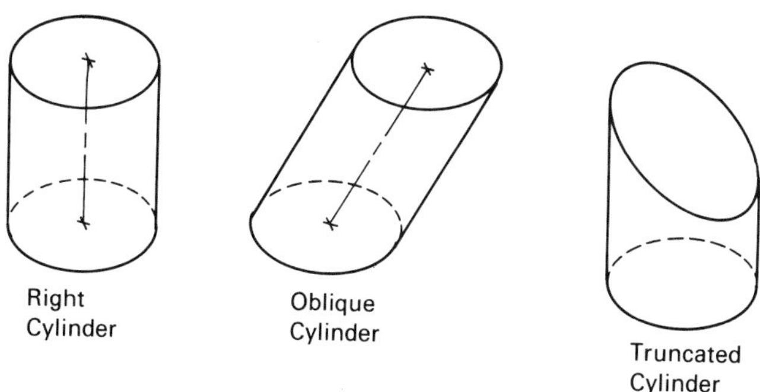

Fig. 5-11. Cylinders.

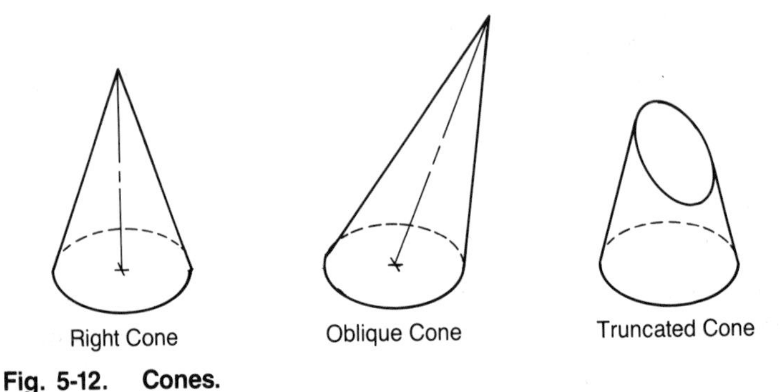

Fig. 5-12. Cones.

around a curved surface. Again, each position of the generatrix is known as an element.

The axis of a cone is drawn from its vertex to the center of the base. Right cones have axes perpendicular to the base, oblique cones have nonperpendicular axes. Cones that are cut near the vertex at an angle parallel to the base are known as frustums. Truncated cones, on the other hand, have been cut near the vertex at an angle not parallel to the base.

Cones also are used in engineering and drafting problems to produce *conic sections* (Fig. 5-13). These sections produce curves that are created by cutting planes that intersect right cones at various angles. The four basic conic sections produced are the circle, the ellipse, the parabola, and the hyperbola.

SPHERICALS

Solids that have design relationships to spheres are known as sphericals (Fig. 5-14). A sphere is a round solid figure in which every point on its surface is the same distance from its center point. The axis of a sphere passes through the center point; the points where the axis emerges on the sphere's surface are known as *poles*.

Elongations or distortions of spheres whose cross sections are all ellipses and circles are called *ellipsoids*. Ellipsoids that are elongated along the axis of the sphere are *prolate ellipsoids*, while those elongated away from the axis are known as *oblate ellipsoids*. Another spherical solid is the *torus*. Tori are doughnut-shaped solids that are produced by a circle rotated about an axis so that it will not intersect itself. This axis will therefore be eccentric to the circle.

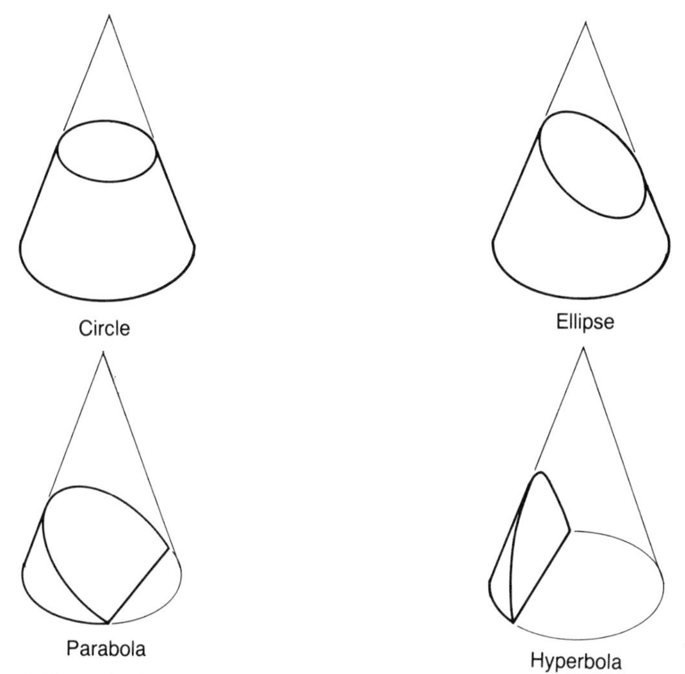

Fig. 5-13. Conic sections.

Geometric Constructions

There are a large number of geometric construction procedures available to drafters. To present all of them would be a monumental task, and beyond the scope and purpose of this book. The geometric constructions covered here will deal only with those procedures commonly used by drafters. They have been divided into the five major areas: lines and angles, triangles and other polygons, curves and circles, ellipses, and other curvature forms.

Lines and Angles

Most geometric construction procedures involving lines and angles are concerned with dividing, transferring, and creating relationships. Sometimes more than one construction procedure can be used to accomplish the same end. When that is the case, the most advantageous construction will be identified.

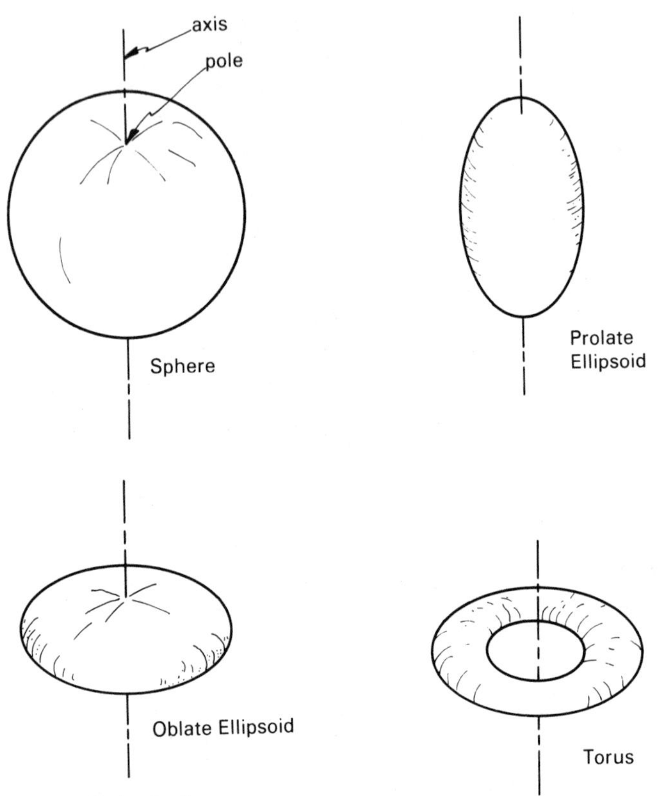

Fig. 5-14. Sphericals.

BISECTING A LINE

This construction consists of dividing a line into two equal parts, and it is frequently used in industrial drawings. Two basic techniques can be used for line bisection: the first is more closely allied to mathematical procedures, while the second makes effective use of drafting equipment. There is no major advantage, however, of one procedure over the other.

Given: Line AB

Procedure A—bisecting a line with a compass (Fig. 5-15):

1. With a compass, draw two arcs with a radius greater than one-half the length of line AB from points A and B.

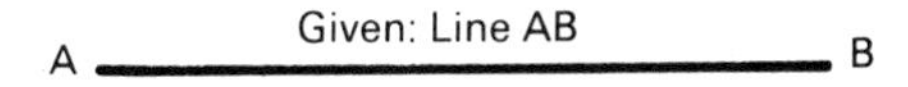

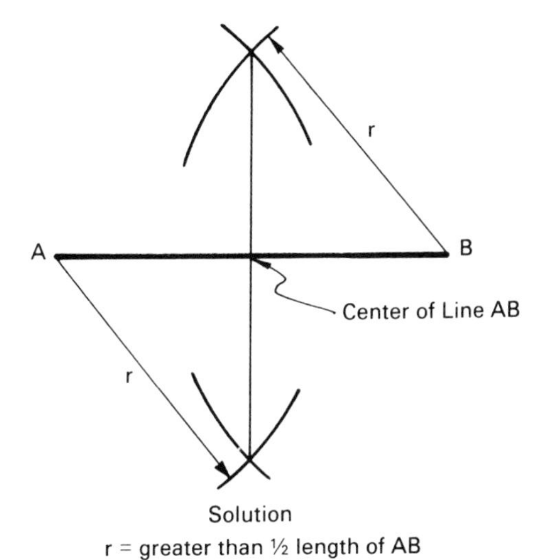

Fig. 5-15. Bisecting a line with a compass.

2. Where the arcs intersect above and below line AB, connect with a straight line.
3. The point of bisection will be found where the connecting line intersects line AB at point C.

Procedure B—bisecting a line with a triangle (Fig. 5-16):

1. With a 45° or 30°-60° triangle, draw a line from points A and B. The lines must be at the same angle from both points.
2. Where the two lines intersect, draw a perpendicular to line AB.
3. The point of bisection is where the perpendicular intersects AB at point C.

TRISECTING A LINE

Dividing a line into three equal parts is called trisection. One primary method is used for this construction technique.

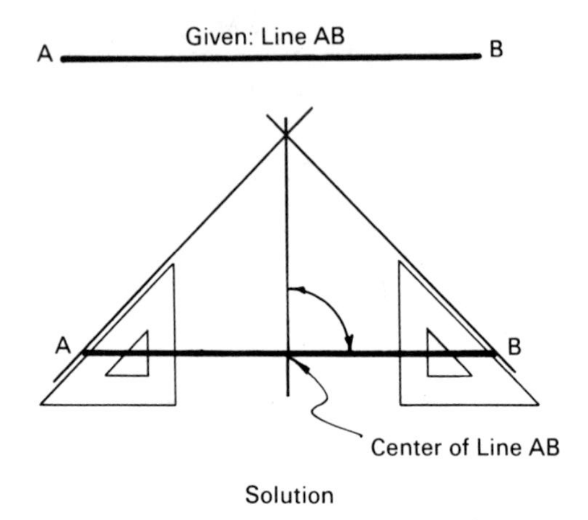

Note: Triangle can be 45°, or 30°-60°, as long as the same angle is used through A and B.

Fig. 5-16. Bisecting a line with a triangle.

Given: Line CD

Procedure—trisecting a line (Fig. 5-17):

1. Draw a line from points C and D at 30° angles until they intersect one another.
2. From the point of intersection, draw two 60° lines until they intersect line CD at points E and F.
3. The points of trisection will be where the 60° lines intersect CD at E and F.

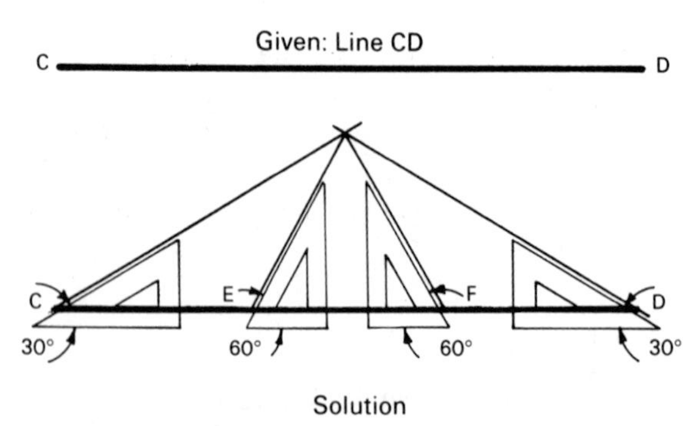

Fig. 5-17. Trisecting a line.

DIVIDING A LINE INTO A GIVEN NUMBER OF EQUAL PARTS

This geometric construction is used when a line must be divided into three or more parts. Of the two techniques described here, the first tends to be the fastest for the experienced drafter, who is able to manipulate a straightedge more quickly than a scale.

Given: Line EF to be divided into seven equal parts

Procedure A—mark-off method (Fig. 5-18):

1. Draw a line at any convenient angle to one end of line EF (point E in this example), and mark off seven equal units along its length.
2. Draw a line from the seventh point to the opposite end of line EF (point F).
3. Draw lines through the six remaining marks so that they will be parallel to line 7F.
4. The points of equal division will be where the parallel lines intersect line EF.

Procedure B—scale method (Fig. 5-19):

1. From one end of line EF (point F), draw a vertical line to a convenient length.

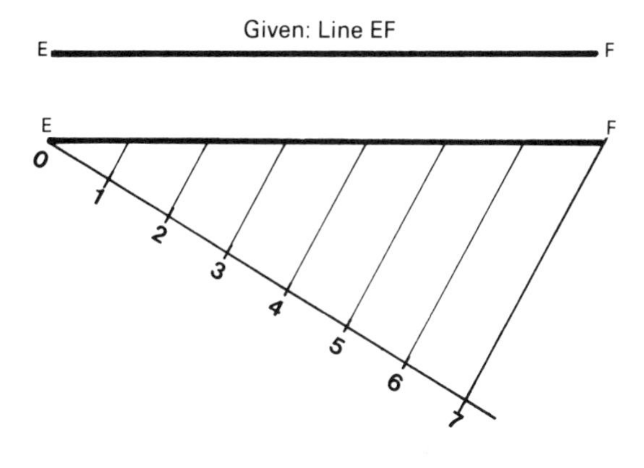

Fig. 5-18. Mark-off method for dividing a line into a given number of equal parts.

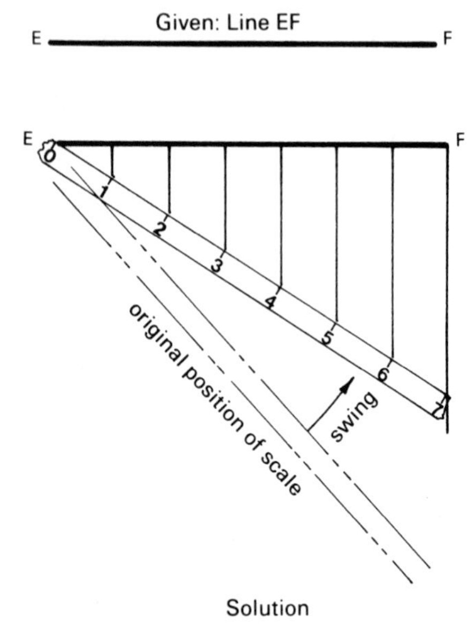

Fig. 5-19. Scale method for dividing a line into a given number of equal parts.

2. From the other end of line EF (point E), set the zero of the scale and swing it until the seventh unit touches the vertical line.
3. Mark off each unit along the scale with a dot or prick point, and draw vertical lines through those dots to line EF.
4. The points of equal division will be where the vertical lines intersect line EF.

DIVIDING A LINE INTO PROPORTIONAL PARTS

There are times when a line must be divided into proportional parts at given ratios, as in the drawing of cam displacement diagrams. Either of the two procedures described here can be used with the same degree of speed and accuracy.

Given: Line GH to be divided into four proportional parts with the ratios 1:2:3:5

Procedure A (Fig. 5-20):

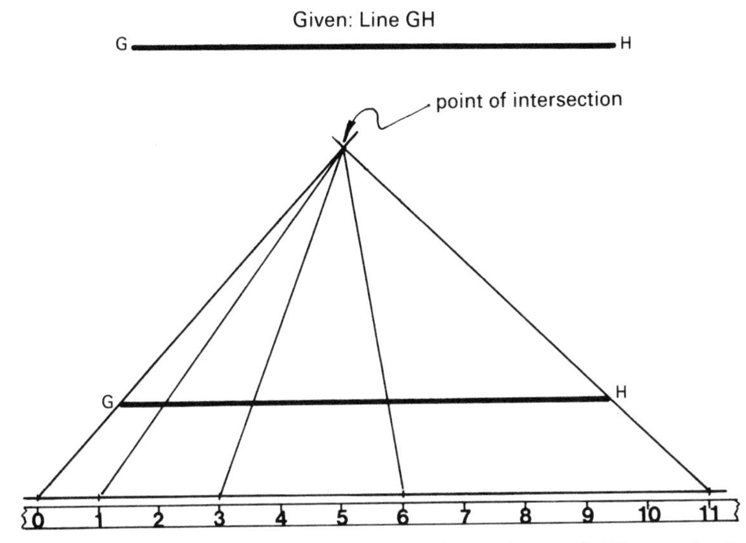

Fig. 5-20. Dividing a line into proportional parts by parallel line method.

1. Draw a line parallel to, and at a convenient distance from, line GH.
2. On the drawn line, mark off distances of 1, 2, 3, and 5 units.
3. From the zero point of the line, draw a line through point G. Draw a second line from the last unit mark through point H until it intersects the line drawn through G.
4. From the point of intersection, draw a line to the 2- and 3-unit marks.
5. The points of proportional division will be where the drawn lines intersect line GH.

Procedure B (Fig. 5-21):

1. From point G, draw a line at a convenient angle and length, and mark off distances of 1, 2, 3, and 5 units so that the zero point is at G.
2. From the 5-unit mark (the last mark), draw a line to point H.
3. Draw lines parallel to 5H through all other measured distances until they intersect line GH.
4. The points of proportional division will be where the parallel lines intersect line GH.

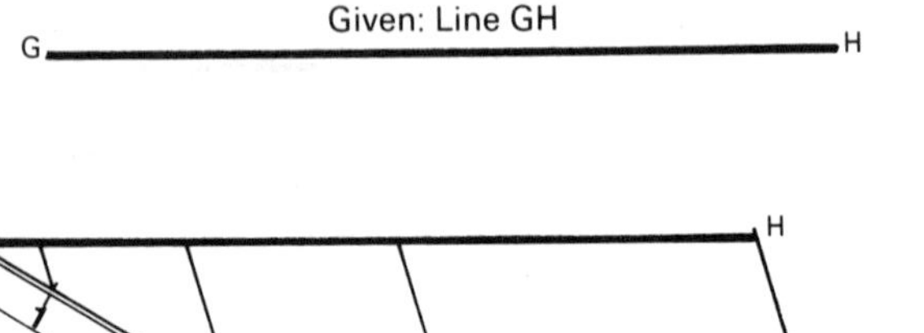

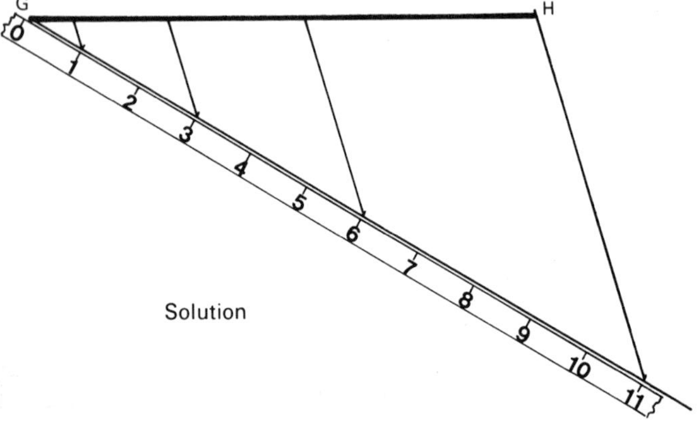

Fig. 5-21. Dividing a line into proportional parts by scale method.

DRAWING A LINE PARALLEL TO ANOTHER LINE

A construction technique that is frequently required in working drawings is the drawing of lines parallel to existing lines. The first method presented is more applicable where greater distances are required between parallel lines, while the second should be used for drawing parallel lines in close proximity.

Given: Line KL

Procedure A—tangent method (Fig. 5-22):

1. Select any two points along line KL and draw two arcs at a radius equal to the desired distance between the proposed parallel lines. If the line is curved, select random points to draw your arcs.
2. Draw a tangent to the two drawn arcs. Tangents are drawn with the aid of straightedges for straight lines and irregular curves for curved lines.
3. The parallel line will be the drawn tangent.

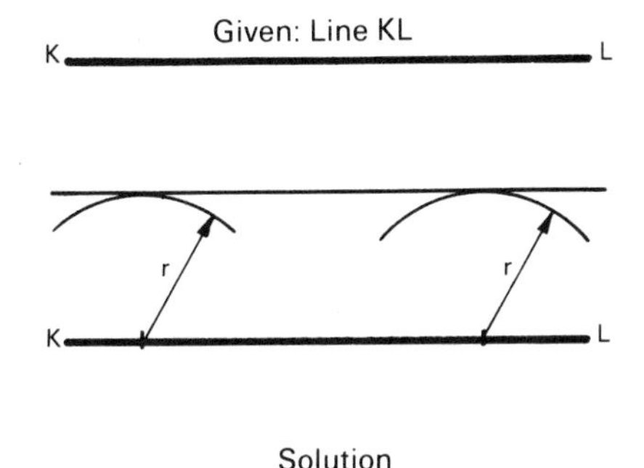

Fig. 5-22. Drawing a parallel line by tangent method.

Procedure B—straightedge method (Fig. 5-23):

1. Select any one point on line KL, and measure the desired distance that the parallel line is to be drawn.
2. Set the ruling edge of a triangle on line KL and move a second straightedge (e.g., another triangle or T-square) to one side of the triangle.
3. Slide the triangle down the straightedge until it reaches the desired distance, and draw the parallel line.

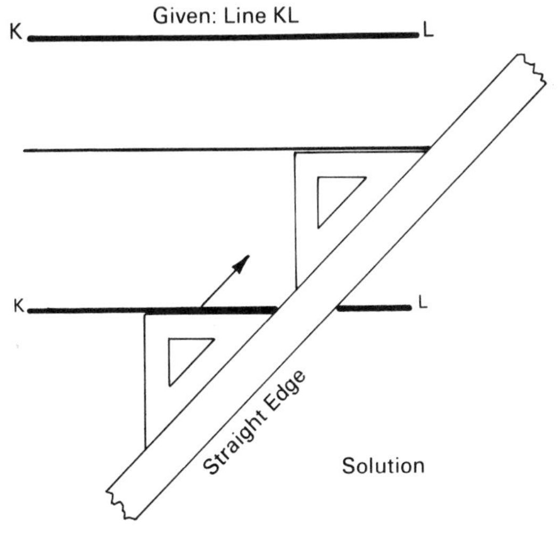

Fig. 5-23. Drawing a parallel line by straightedge method.

DRAWING A PERPENDICULAR FROM A POINT TO A LINE

There are two basic procedures that can be used to construct a perpendicular from a point to a line. The first is more commonly used in the field of mathematics, the second is more appropriate for drafting.

Given: Line MN and point P

Procedure A—arc method (Fig. 5-24):

1. From point P, draw an arc that will intersect line MN at two separate locations.
2. With the two intersection points as centers, draw two additional arcs of greater distance than the midpoint of line MN, so that they intersect at a given point.
3. Draw a straight line from point P to the intersection of the two arcs.
4. The perpendicular from point P to line MN will be the intersecting line drawn from P to the intersecting arcs.

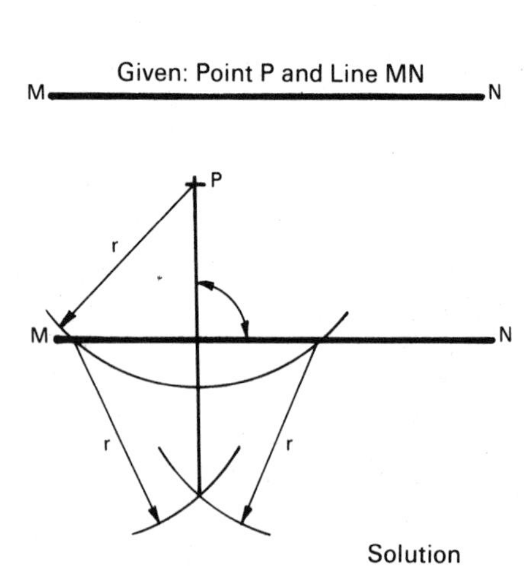

Fig. 5-24. Arc method for drawing a perpendicular from a point to a line.

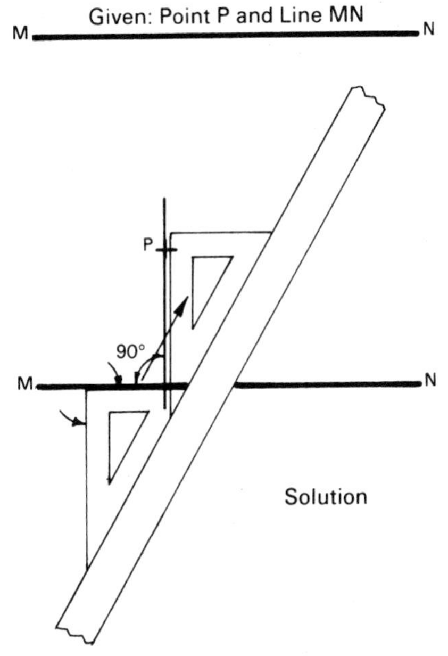

Fig. 5-25. Straightedge method for drawing a perpendicular from a point to a line.

Procedure B—triangle method (Fig. 5-25):

1. Move a triangle on a supporting straightedge, until one leg of the perpendicular corner lines up with line MN.
2. Slide the triangle until the other leg of the perpendicular corner reaches point P, and draw a line to MN.
3. The perpendicular from point P to MN will be the vertical line drawn from P.

BISECTING AN ANGLE

There is one primary method used for dividing an angle into two equal parts. This technique is used in both mathematics and drafting.

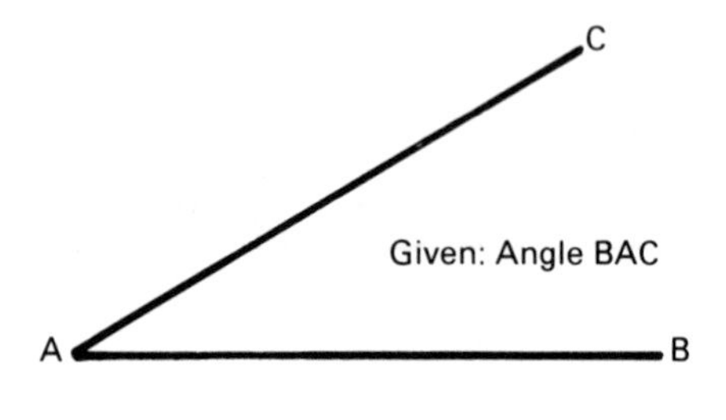

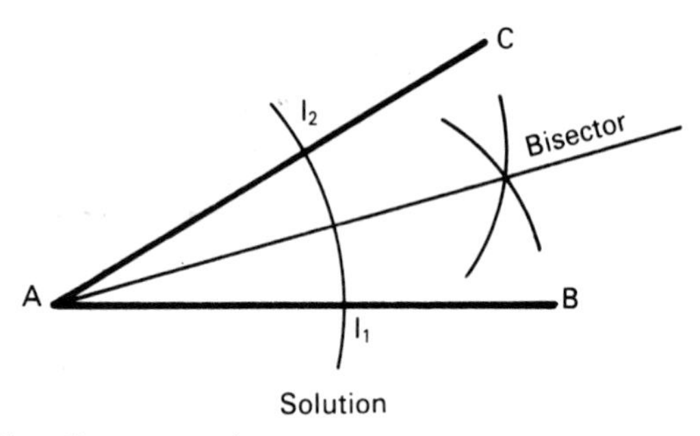

Fig. 5-26. Bisecting an angle.

Given: Angle BAC

Procedure—bisecting an angle (Fig. 5-26):

1. Using A as the center point, strike an arc so that it will intersect side AB and AC of the angle.
2. From the two points of intersection, I_1 and I_2, strike two more arcs at such a radius so that they will intersect one another.
3. The line of angle bisection will be drawn from point A to the point of intersection of the two arcs.

RECONSTRUCTION OF AN ANGLE

When an angle must be transferred from one drawing to another, it will be necessary to reconstruct it using drafting techniques. There is one primary method used for this process.

Given: Angle EDF

Procedure—reconstructing an angle (Fig. 5-27):

1. Using D as the center point, strike an arc so that it will intersect side DE and DF.
2. Draw one side of the angle you are reconstructing (D′E′), and strike the same radius arc found in EDF, with D′ as the center point.
3. Measure the distance, by compass or divider, between the arc intersections on EDF, and transfer to the arc drawn across D′E′.
4. The reconstructed angle will be completed when a line is drawn from D′ to the measured distance of the arc, thus establishing side D′F′.

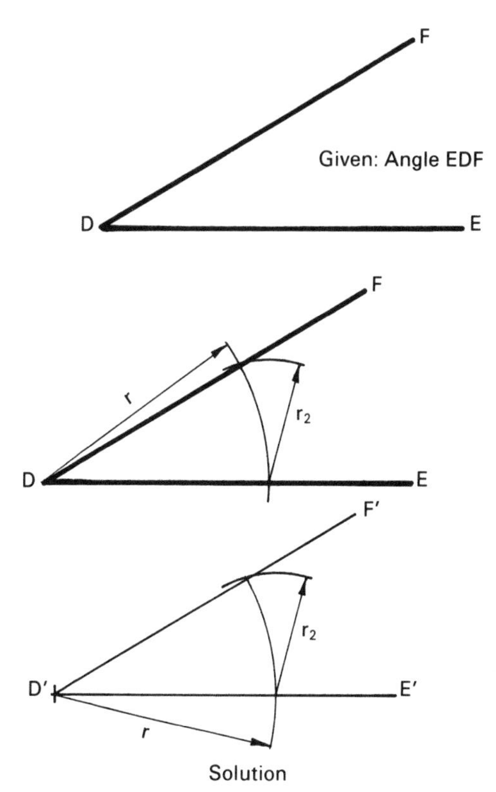

Fig. 5-27. Reconstructing an angle.

LAYING OUT AN ANGLE

When an angle must be laid out to a specific angle, a triangle or protractor may be used. There are times, however, when greater accuracy is required. To meet these close tolerances, three construction procedures are available to the drafter.

Given: Angle specification of 32° 20′

Procedure A—tangent method (Fig. 5-28):

1. Draw one side of angle HGJ from point G a total of 10 units. (Note: the length of the side can be any distance; the 10-unit measure was selected for ease of computation.)
2. Using a trigonometric function table or scientific calculator, we find that the tangent of 32° 20′ is 0.6330. Multiply the tangent of the angle by the length of GH, which will equal 6.330.
3. From H, draw a perpendicular to the length of 6.330 units.
4. The angle 32° 20′ will be established when a line is drawn from G to the end of the perpendicular, producing angle HGJ.

Procedure B—sine method (Fig. 5-29):

1. Draw one side, GH, of angle HGJ to a length of 10 units.
2. Using a trigonometric function table or scientific calculator, we note that the sine of 32° 20′ is 0.5348. Multiply the sine of the angle by the length of GH, which will equal 5.348.

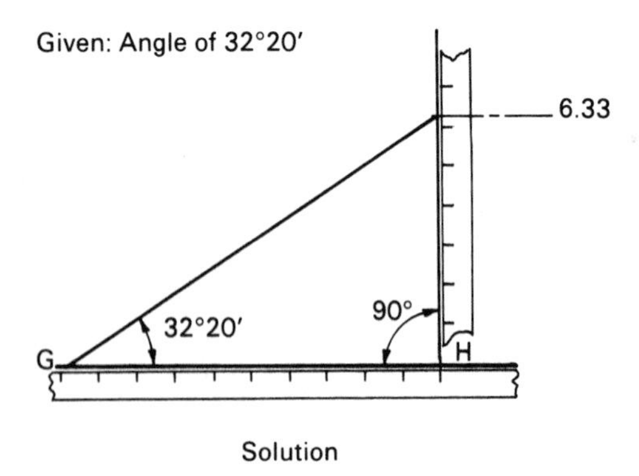

Fig. 5-28. Tangent method for laying out an angle.

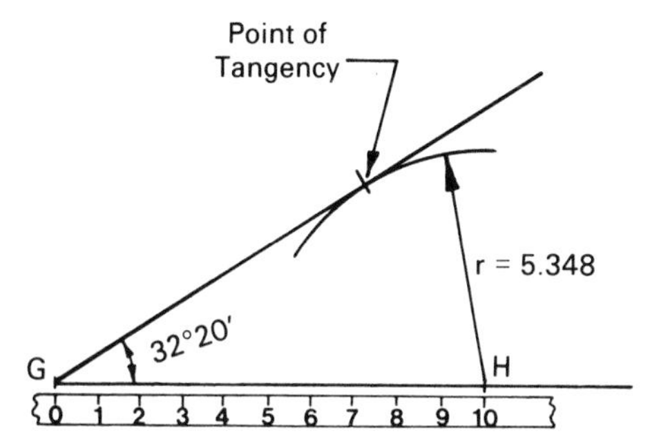

Fig. 5-29. Sine method for laying out an angle.

3. Strike an arc from H with a radius of 5.348 units.
4. The angle of 32° 20′ will be established when a tangent line is drawn from G to the arc, producing angle HGJ.

Procedure C—chord method (Fig. 5-30):

1. Draw one side, GH, of angle HGJ, and strike an arc with a radius of 10 units.
2. Using a math table (usually found in a table of segments of circles), we find that the chord of 32° 20′ is 0.5569. Multiply the chord of the angle by the length of GH, which will equal 5.569.

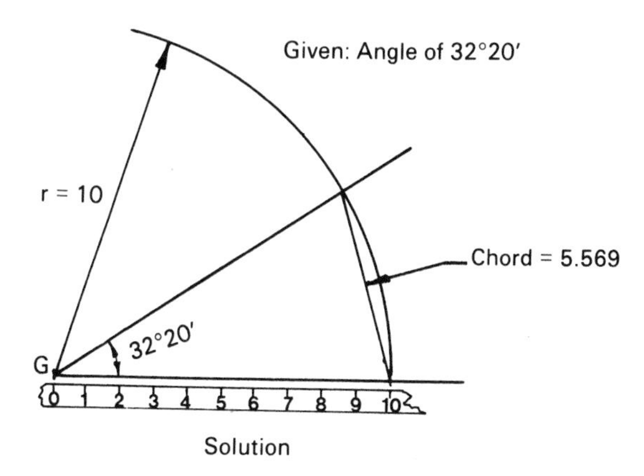

Fig. 5-30. Chord method for laying out an angle.

3. Where the arc intersects side GH, measure the chord distance of 5.569 and mark it on the arc.
4. The angle of 32° 20′ will be established when a line is drawn from G through the mark on the arc, producing angle HGJ.

Triangles and Other Polygons

Designers and drafters are frequently called upon to employ and draw triangular and polygonal shapes in product design plans. Though some of these figures can be drawn with the aid of templates, frequently many figures must be constructed, for not all figures are of standard size. The two primary methods used for the geometric construction of triangles and other polygons are development and transfer techniques. Many of the specific procedures presented here can easily be categorized as one or the other of these techniques.

DEVELOPING AN EQUILATERAL TRIANGLE

As defined, an equilateral triangle has three equal sides and interior angles. Since the interior angles of all triangles total 180°, each angle of an equilateral will be equal to 60°. Of the two construction methods presented, the second is the most commonly used by drafters.

Given: Side AB of an equilateral triangle

Procedure A—arc method (Fig. 5-31):

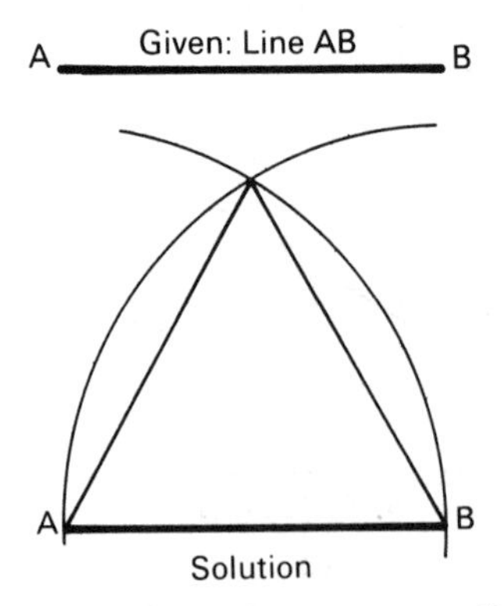

Fig. 5-31. Arc method for developing an equilateral triangle.

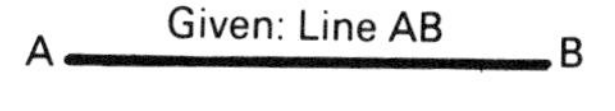

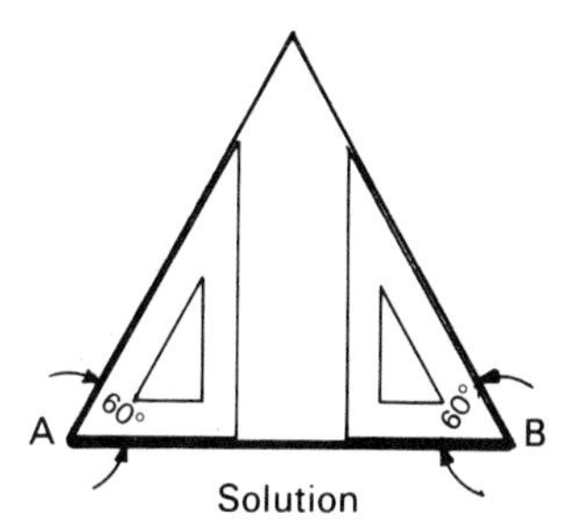

Fig. 5-32. Triangle method for developing an equilateral triangle.

1. From points A and B, strike two arcs with a radius equal to AB so that they will intersect at a common point C.
2. The equilateral triangle will be completed when a straight line is drawn from points A and B to C.

Procedure B—triangle method (Fig. 5-32):

1. Using a 30°-60° triangle, draw a 60° line from A and B until they intersect at point C.
2. The drawn lines ABC will produce the equilateral triangle.

CONSTRUCTING A TRIANGLE GIVEN THREE SIDES

When none of the angles of a triangle are known, it is still possible to construct a triangle if the length of each side is given. One primary method is used for this geometric construction.

Given: Sides DE, EF, and FD of triangle DEF

Procedure (Fig. 5-33):

1. Draw one side, DE, of the triangle in an appropriate location.
2. From D, strike an arc with a radius equal to FD. Strike a second arc from E with a radius equal to EF.
3. The intersection of the two arcs will be point F.
4. The constructed triangle will be completed when a straight line has been drawn between the three points.

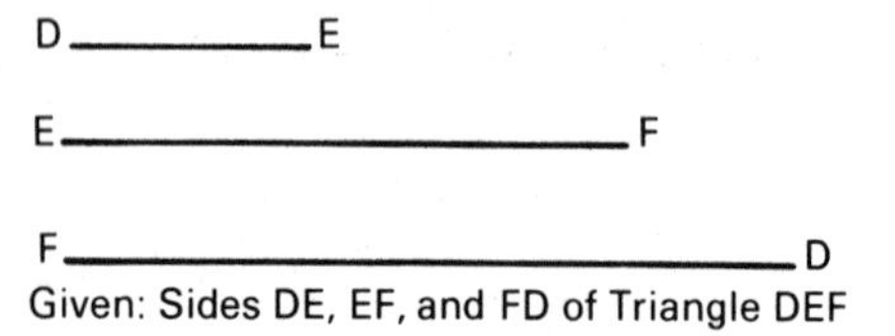

Given: Sides DE, EF, and FD of Triangle DEF

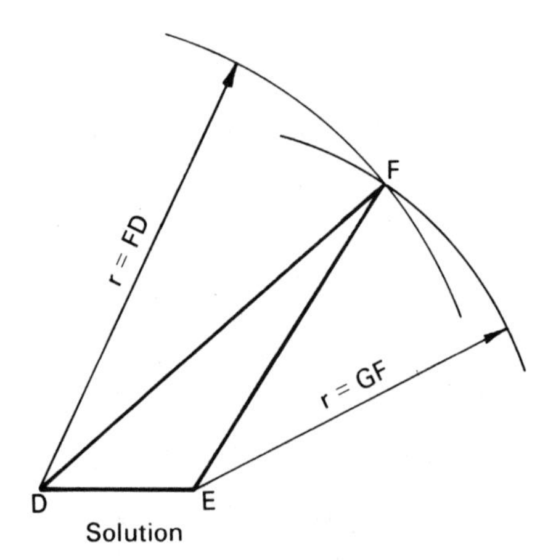

Solution

Fig. 5-33. Constructing a triangle given three sides.

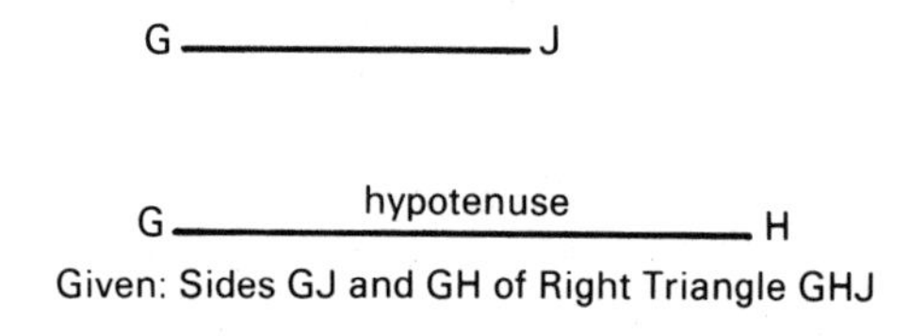

Given: Sides GJ and GH of Right Triangle GHJ

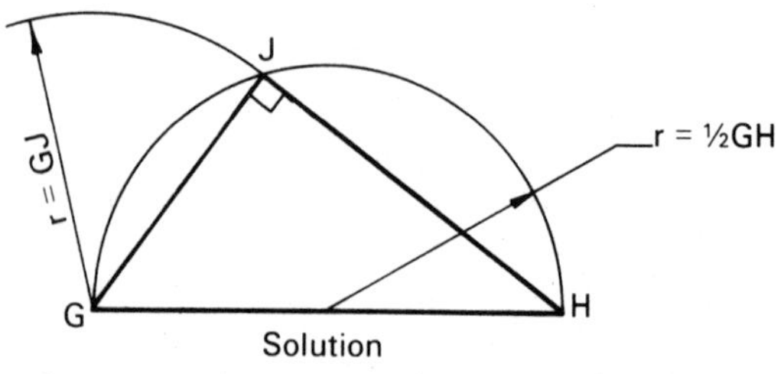

Solution

Fig. 5-34. Constructing a right triangle given one side and the hypotenuse.

CONSTRUCTING A RIGHT TRIANGLE GIVEN ONE SIDE AND THE HYPOTENUSE

The construction procedure used for this problem is based upon the principles of right-angle trigonometry, and is often used by drafters and engineers in the solution of graphic problems.

Given: Hypotenuse GH and side GJ

Procedure (Fig. 5-34):

1. Draw a semicircle, with side GH serving as the diameter.
2. From point G, strike an arc whose radius is equal to distance GJ, so that it will intersect the semicircle at point J.
3. The right triangle will be constructed when a straight line is drawn from G and H to J.

DEVELOPING A SQUARE GIVEN ONE SIDE

The geometric construction procedure used to draw a square is based upon the definition of this plane figure: namely, that it is an equal-sided parallelogram each of whose interior angles is 90°.

Given: Side KL of square KLMN

Procedure (Fig. 5-35):

1. Draw perpendicular lines from points K and L.

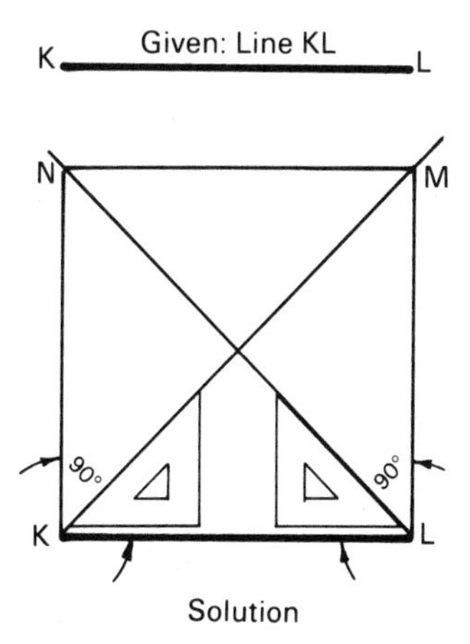

Fig. 5-35. Developing a square given one side.

2. Using a 45° triangle, draw a 45°-angled line from points K and L until they intersect the perpendiculars, locating points M and N.
3. The constructed square will be completed when M and N are connected by a straight line.

DEVELOPING A SQUARE GIVEN THE CIRCUMFERENCE OF A CIRCLE

Two construction procedures can be used for this problem. In the first, the square is inscribed within the circle; it should be used when the diameter of the circle is equal to the length of the square's diagonal. The second construction technique circumscribes the square about the circle, and should be used when the circle's diameter is the same as the square's sides.

Given: Circle O

Procedure A—inscribed method (Fig. 5-36):

1. Through O, draw two perpendicular center lines until they intersect the circle at points A, B, C, and D.
2. The constructed square will be completed when A, B, C, and D are connected by straight lines.

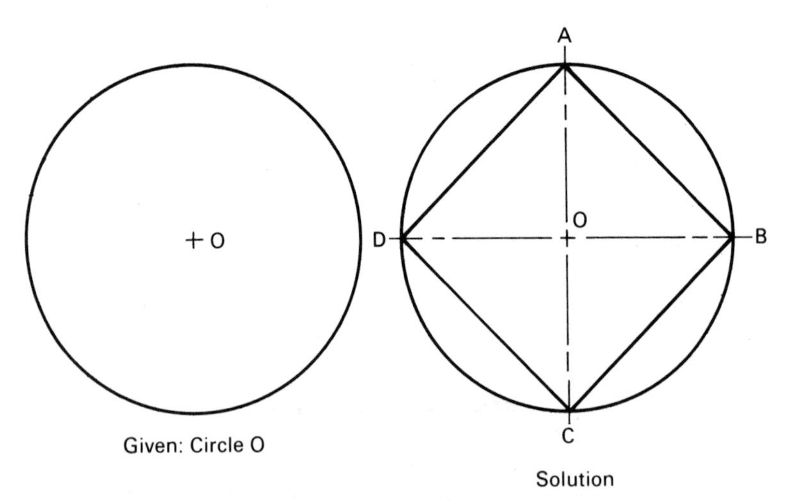

Fig. 5-36. Inscribed method for developing a square.

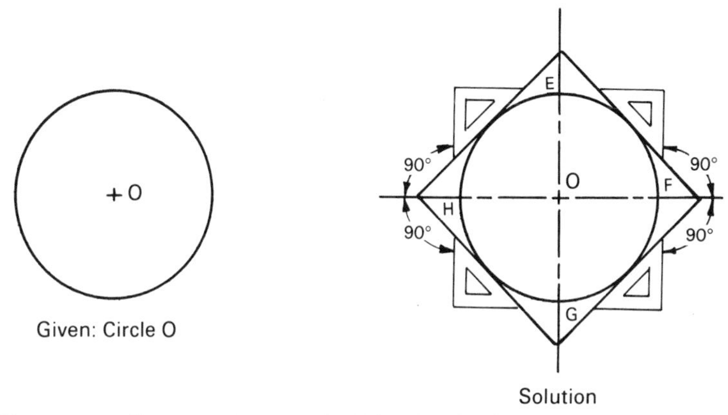

Fig. 5-37. Circumscribed method for developing a square.

Procedure B—circumscribed method (Fig. 5-37):

1. Through O, draw two perpendicular center lines until they intersect the circle at points E, F, G, and H.
2. Using a 45° triangle, draw four tangents to the circle so that they are 45° to the center lines.
3. The constructed square will be completed when the tangents are drawn to one another.

CONSTRUCTING A REGULAR PENTAGON

The first procedure is used when accuracy is important (since angle size can be accurately set by a vernier head drafting machine). The second procedure should be used when speed is required and accuracy is not critical. The last technique is the least used, since it requires a series of complex construction procedures. Of the three, however, the first is perhaps the most frequently used by drafters.

Given: Circle P

Procedure A—angle method (Fig. 5-38):

1. From center point P, measure and mark off five consecutive and equal angles of 72°. This can be accomplished with a protractor, an adjustable triangle, or the head of a drafting machine.

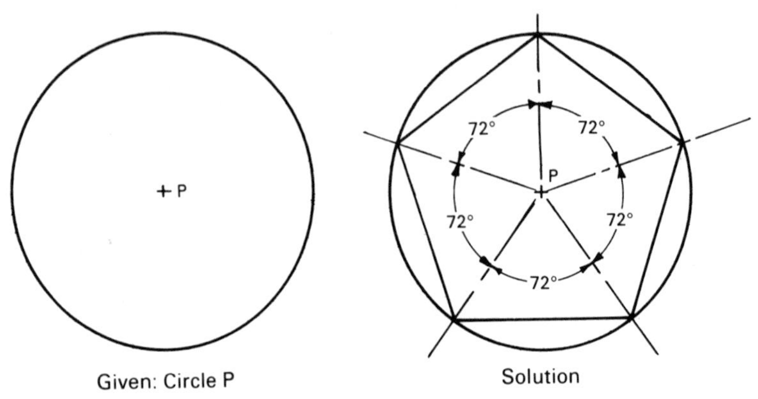

Fig. 5-38. Angle method for constructing a regular pentagon.

2. Draw lines from P through the marks so that the lines will intersect the circle's circumference.
3. The regular pentagon will be constructed when the five intersecting lines are joined by straight lines.

Procedure B—estimate method (Fig. 5-39):

1. Using a divider or compass, estimate the length of one side of the pentagon, which will be slightly larger than the circle's radius. Step off this distance about the circumference of the circle.
2. If the five divisions are not exact (the fifth mark will be either short or past the starting point), adjust the divider or compass accordingly and step off the distance about the circumference again. Repeat this procedure until there are five equal sides.
3. The pentagon will be constructed when the five marks are connected by straight lines.

Procedure C—geometric method (Fig. 5-40):

1. Bisect the radius of circle P to find point A.
2. Using A as the center point, strike an arc whose radius will be equal to distance AB, so that it will intersect the center line at point C.
3. Using B as the center, strike an arc so that it will pass through point C and the circumference of the circle.

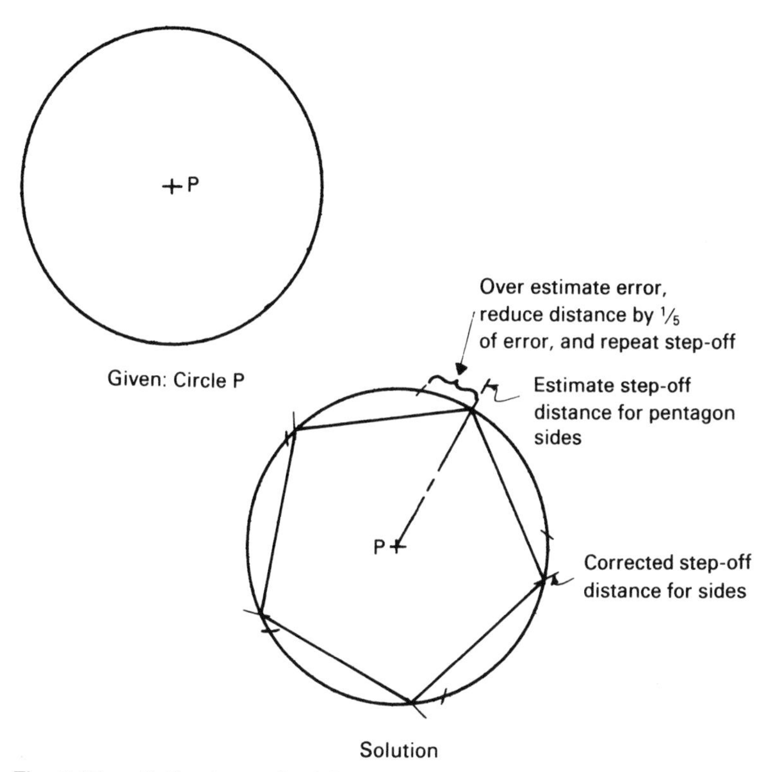

Fig. 5-39. Estimate method for developing a pentagon.

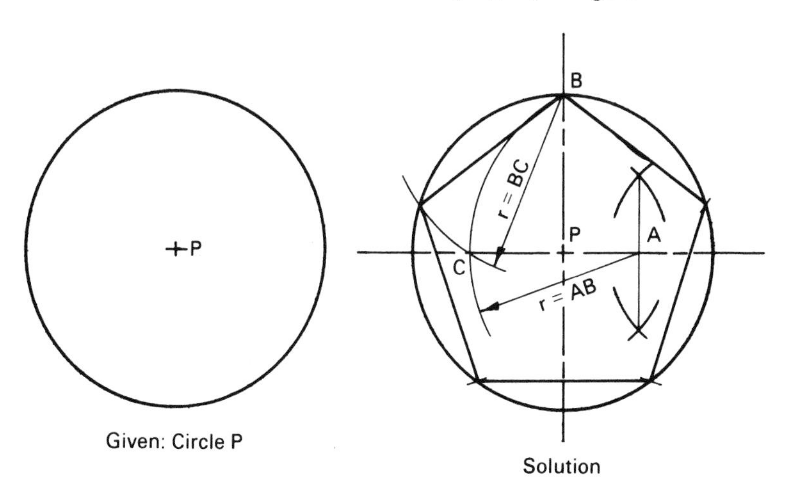

Fig. 5-40. Geometric method for constructing a regular hexagon.

4. The distance from B to the circumference intersection is equal to one side of the pentagon.
5. The pentagon will be constructed once the sides have been stepped off with a divider or compass and the points connected.

DEVELOPING A REGULAR HEXAGON

As in pentagon construction, several procedures are available for constructing a hexagon. One of the most common is the first technique described; the second is a modification of the first. The third procedure is usually limited to drafting practices, while the fourth should be used when the distance across the hexagon's flats are known.

Given: Circle Q

Procedure A—radius method (Fig. 5-41):

1. From point Q, determine the radius of the circle; this distance will be equal to the length of the regular hexagon's side.
2. With a divider or compass, step off the sides of the hexagon about the circumference of the circle.
3. The hexagon will be developed once the six points have been joined by straight lines.

Procedure B—arc method (Fig. 5-42):

1. Draw one center line through point Q so that it will intersect the circumference at points A and B.

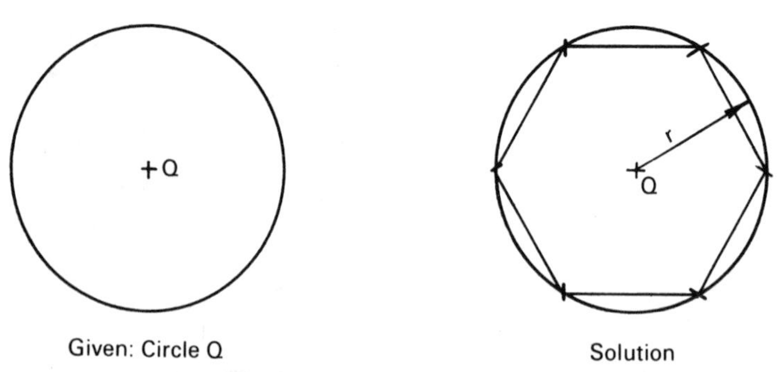

Fig. 5-41. Radius method for constructing a regular hexagon.

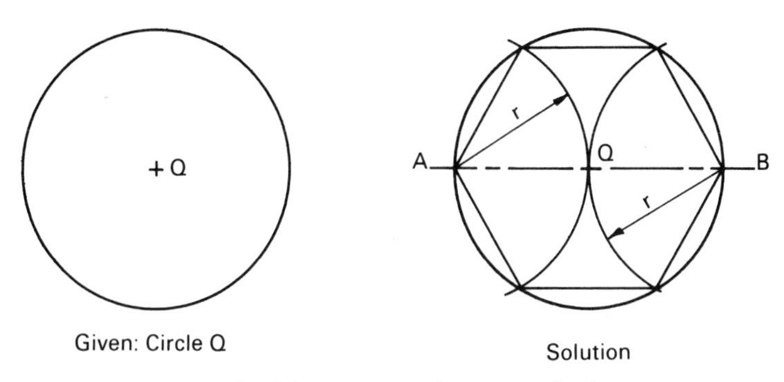

Fig. 5-42. Arc method for constructing a regular hexagon.

2. With A and B as the center points, strike arcs, whose radius will be equal to the circle's radius, so that they intersect the circle at two locations.
3. The developed hexagon will be constructed when the points of intersection are joined by straight lines.

Procedure C—across corners method (Fig. 5-43):

1. Through point Q, draw two perpendicular center lines so that they will intersect the circle's circumference.
2. Draw two 30° diagonals through Q until they intersect the circumference.
3. The hexagon will be constructed when all points of intersection are connected by a straight line.

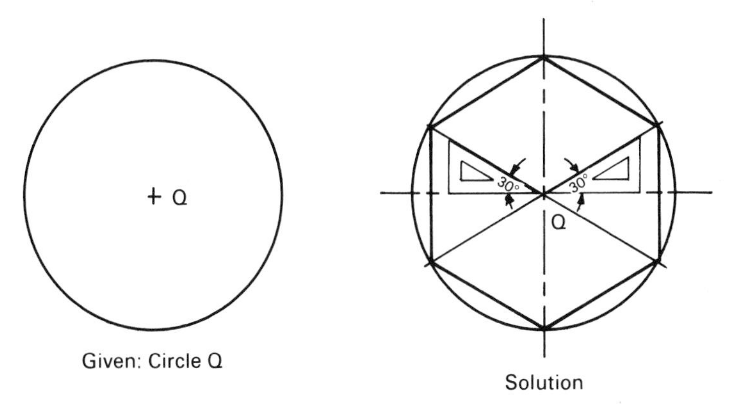

Fig. 5-43. Across corners method for constructing a regular hexagon.

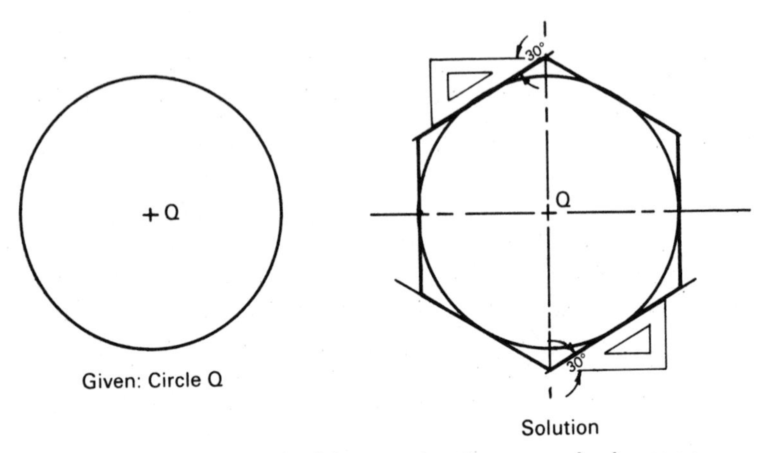

Fig. 5-44. Tangent method for constructing a regular hexagon.

Procedure D—tangent method (Fig. 5-44):

1. Draw two perpendicular center lines through point Q.
2. Using a 30°-60° triangle and a horizontal straightedge, draw six tangents about the circle, so that they will be 60° to one another. Two tangent lines will be perpendicular to one center line, while each of the other four will form a corner at the other center line.
3. The constructed hexagon will be completed when all tangents are drawn to each other.

CONSTRUCTING A REGULAR HEXAGON GIVEN ONE SIDE

Sometimes a regular hexagon must be constructed when the only information available is the length of the sides. When it is not advantageous to use a circle for construction purposes, the technique presented here may be used.

Given: Side AB of a regular hexagon

Procedure (Fig. 5-45):

1. Draw a 60° line from points A and B until they intersect one another and extend to a convenient length.

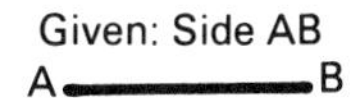

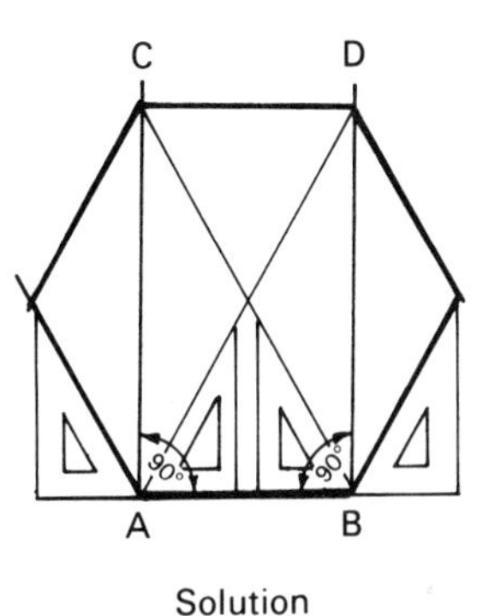

Solution

Fig. 5-45. One side method for constructing a regular hexagon.

2. Draw two perpendicular lines from A and B until they intersect the 60° lines. Label these points of intersection C and D.
3. Draw 60° lines from the four labeled points until they intersect each other. The constructed hexagon will be completed when all the lines are drawn.

DEVELOPING A REGULAR OCTAGON

There are two procedures for developing a regular octagon. The first should be used when the distance across the flats of the octagon is known, and the second when it is not advantageous to draw a circle. Note that the diameter of the circle will be the same as the across flats distance.

Given: Circle R or across flats distance AB

Procedure A—tangent method (Fig. 5-46):

1. Draw vertical and horizontal tangents (perpendicular to each other) about the circle.
2. With a 45° triangle, draw four additional tangents so that they are 45° to the vertical and horizontal tangents.

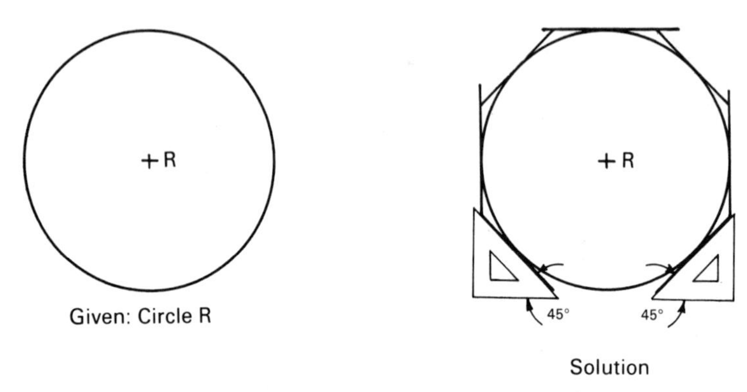

Fig. 5-46. Tangent method for constructing a regular octagon.

3. The constructed octagon will be completed when the tangents are drawn to one another.

Procedure B—across flats method (Fig. 5-47):

1. Draw a square whose sides are equal to distance AB.
2. Draw two diagonals from each corner of the square.
3. Using the corners as center points and a radius equal to the distance from the corner to the intersection of the diagonals, strike an arc so that it intersects the sides of the square.
4. The octagon will be constructed when the intersected arcs are connected by a straight line.

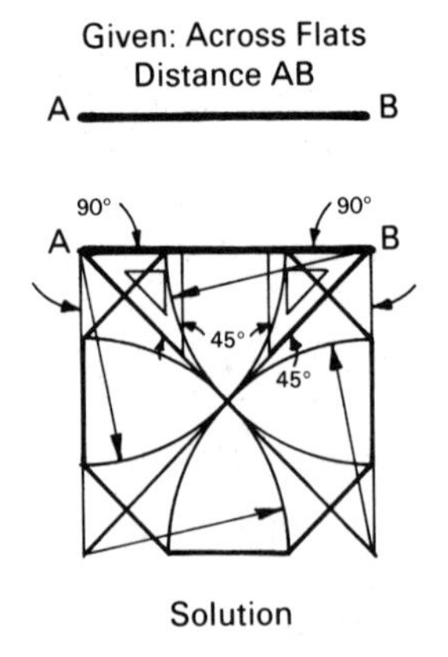

Fig. 5-47. Across flats method for constructing a regular octagon.

TRANSFERRING A POLYGON

When drawings must be copied, or plans drawn for existing parts, it may be necessary to reconstruct a polygon. There are two basic procedures used equally by industrial drafters.

Given: A polygon

Procedure A—triangulation method (Fig. 5-48):

1. Divide the polygon into triangles.
2. Using the procedures illustrated in Fig. 5-33, reconstruct each triangle in its appropriate location.
3. The polygon will be constructed when all triangles are drawn in their appropriate locations.

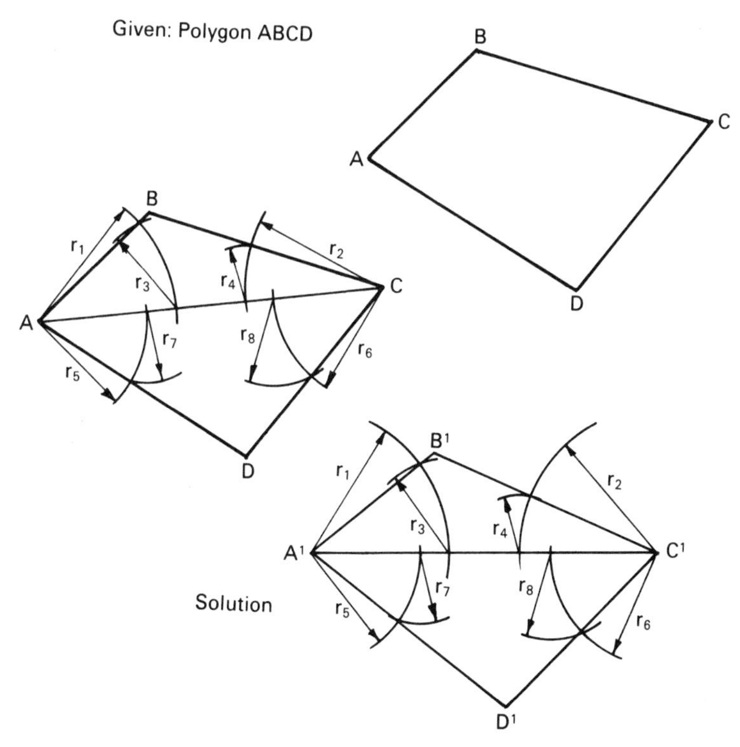

Fig. 5-48. Triangulation method for transferring a polygon.

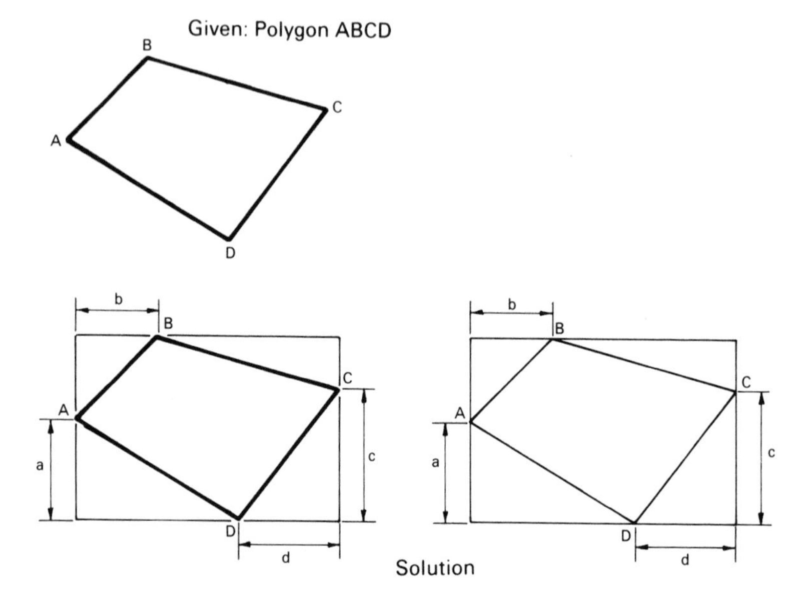

Fig. 5-49. Rectangular method for transferring a polygon.

Procedure B—rectangulation method (Fig. 5-49):

1. Circumscribe a rectangle about the polygon.
2. Measure and record the distances from each corner of the rectangle to the corner points of the polygon.
3. Draw another rectangle of equal size in the desired location, and locate the corner points of the polygon from the measurements taken.
4. The polygon will be constructed when the located points are joined by straight lines.

Circles and Curves

Almost all drawings that have circular and curved surfaces show them in geometric relationship to straight lines and other curved surfaces. To draw these relationships accurately, drafters must often incorporate geometric construction techniques. These constructions involve drawing circles and curves through points, establishing tangents, and approximating the length of curved surfaces.

DRAWING A CIRCLE THROUGH THREE POINTS

There are times when a circle or a circular segment must be drawn through three established locations, as in the placement of instruments and machinery. To solve these problems, there is one basic construction technique that is used by drafters.

Given: Points A, B, and C

Procedure (Fig. 5-50):

1. Using a straightedge, connect A to B and B to C with a straight line.
2. Bisect lines AB and BC so that their bisectors will intersect at point D.
3. Using point D as a center, draw a circle whose radius is equal to DA (which is equal to DB and DC).

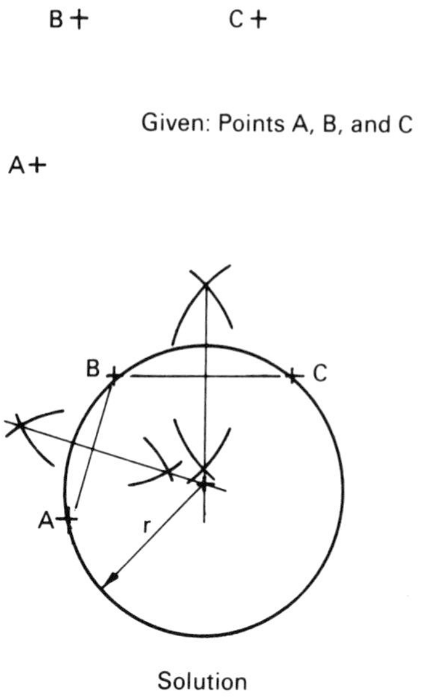

B+ C+

Given: Points A, B, and C

A+

B C A r

Solution

Fig. 5-50. Drawing a circle through three points.

LOCATING THE CENTER OF A CIRCLE

A problem sometimes encountered by drafters is one in which a circle has been identified with no center point. Of the two construction procedures used for this problem, the first method requires less time than the second, and is therefore usually preferred.

Given: A circle

Procedure A—perpendicular chord method (Fig. 5-51):

1. Draw chord EF.
2. From points E and F, draw chords EG and FH perpendicular to EF.
3. Draw a line from G to F and from H to E.
4. The center of the circle will be at the intersection of GF and HE.

Procedure B—chord bisector method (Fig. 5-52):

1. Draw two nonparallel chords.
2. Bisect the two chords so that the bisectors will intersect one another.
3. The center of the circle will be the point of bisector intersection.

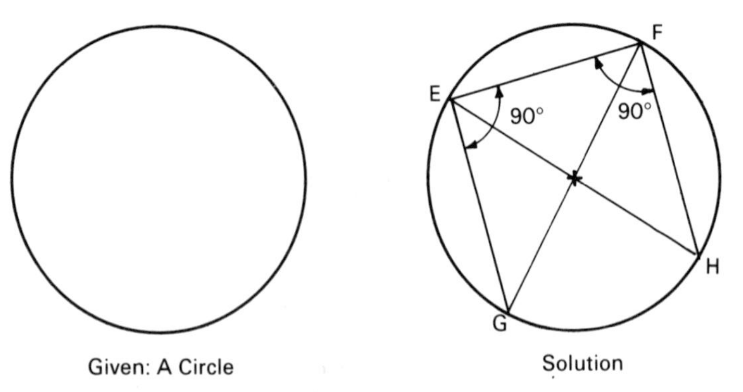

Fig. 5-51. Perpendicular chord method for locating center of a circle.

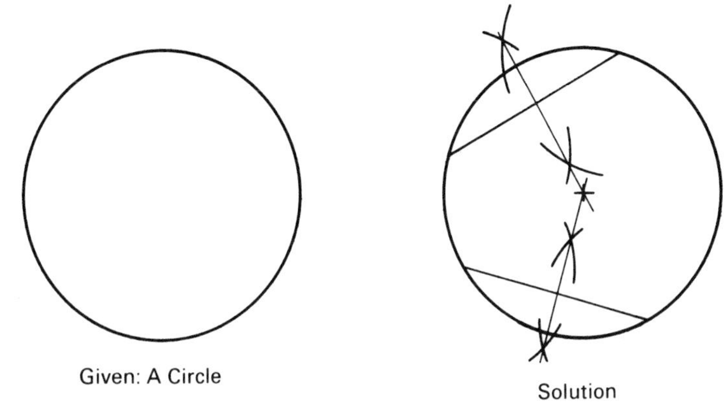

Fig. 5-52. Chord bisector method for locating center of a circle.

DRAWING A CIRCLE TANGENT TO A LINE

A common problem in many drawings is drawing a circle or arc tangent to a line. Tangency is defined as the point where the tangent line is perpendicular to the radius of the circle or arc. Based upon this characteristic, one easy construction method is used by industrial drafters.

Given: Line JK

Procedure (Fig. 5-53):

1. Locate a point on line JK where the point of tangency (T) is desired.
2. From T, draw a perpendicular line equal in length to the radius of the circle.
3. The circle will be drawn tangent to line JK when a circle is drawn with its center at the end of the perpendicular and its radius equal to that perpendicular.

DRAWING A LINE TANGENT TO A CIRCLE

If the center of a circle or arc is known, the procedure used for drawing a line tangent to it is quite simple. To solve this problem, one primary procedure is used by drafters.

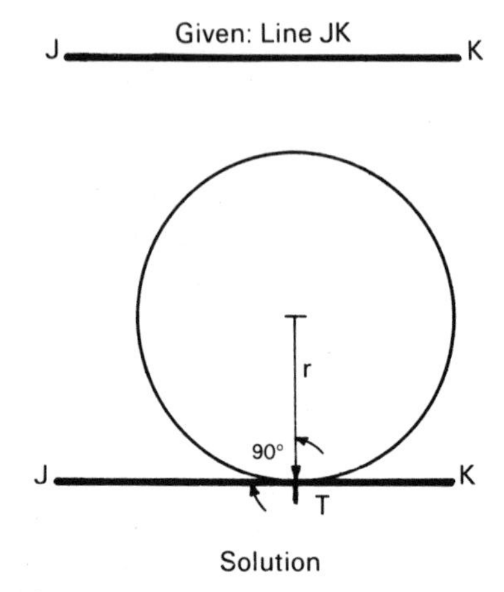

Fig. 5-53. **Drawing a circle tangent to a line.**

Given: Circle O

Procedure (Fig. 5-54):

1. Locate a point (T) on the circumference of circle O for the desired point of tangency.
2. Draw a radius line from O to T.
3. The line of tangency will be constructed when a perpendicular is drawn to OT through point T.

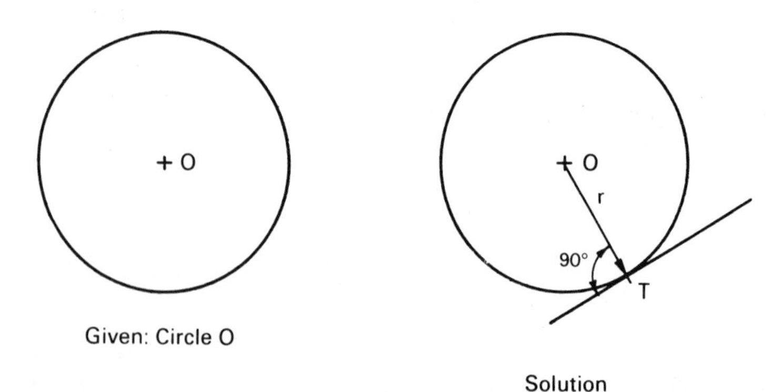

Fig. 5-54. **Drawing a line tangent to a circle.**

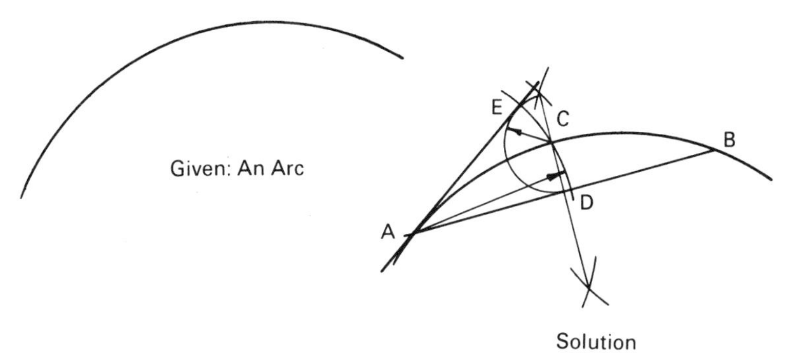

Fig. 5-55. Drawing a line tangent to an arc with an unknown center.

DRAWING A LINE TANGENT TO AN ARC WITH AN UNKNOWN CENTER

Another tangency problem, a variation of the previous two, is one in which a line must be drawn tangent to an arc or circle whose center point is unknown. Again, one basic procedure is used for this problem.

Given: An arc

Procedure (Fig. 5-55):

1. Locate a point of tangency (A) on the circumference of the arc.
2. Draw chord AB, and construct a perpendicular bisector.
3. From A, swing an arc through the point where the bisector intersects the arc (C) and continue until it intersects the chord at point D.
4. Using C as a center and CD the radius, swing another arc to establish point E (the intersection of the two arcs).
5. The line of tangency will be constructed when a line is drawn through points A and E.

DRAWING TANGENTS TO TWO CIRCLES

To draw tangents to two circles there are two possible configurations. In the first, the tangents are on the outside of the circles, and in the second, the tangents cross. The first procedure presented here can be used for both outside and crossing tangents, the second only for outside tangents, and the third only for crossing tangents. The first procedure requires less time, but is not as accurate as the second and third.

Given: Circles C_1 and C_2

Procedure A—inspection method (Fig. 5-56):

1. Position a straightedge until its ruling edge is visibly tangent to the two circles.
2. Keeping the straightedge fixed, position the triangle until it passes through the center of C_2 with the right-angle side. This side should be exactly in line with the point of tangency. If it is not, make the appropriate adjustments, and mark the point of tangency. Repeat for C_1.
3. The lines of tangency will be constructed when a line is drawn from the tangency marks on C_1 to C_2.

Procedure B—outside tangent method (Fig. 5-57):

1. Using point C_2 as the center, draw an arc whose radius is equal to the difference between the radii of C_1 and C_2, or $r = r_1 - r_2$.
2. From C_2, draw a tangent to the arc using the procedure illustrated.

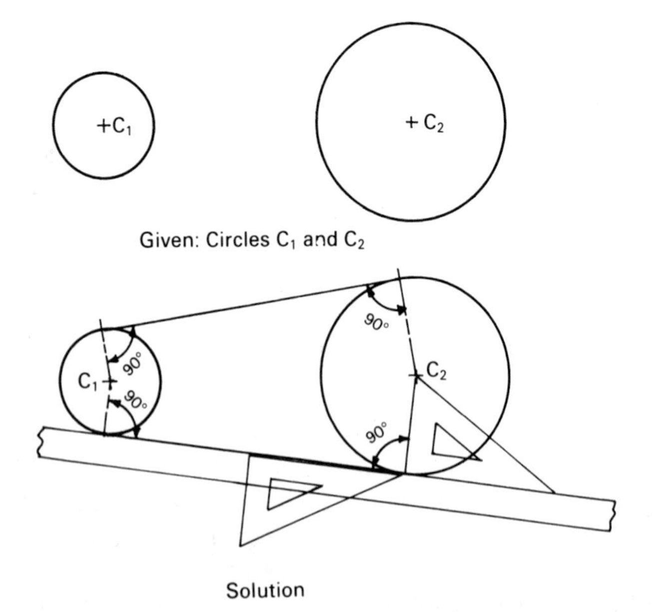

Fig. 5-56. Inspection method for drawing tangents to two circles.

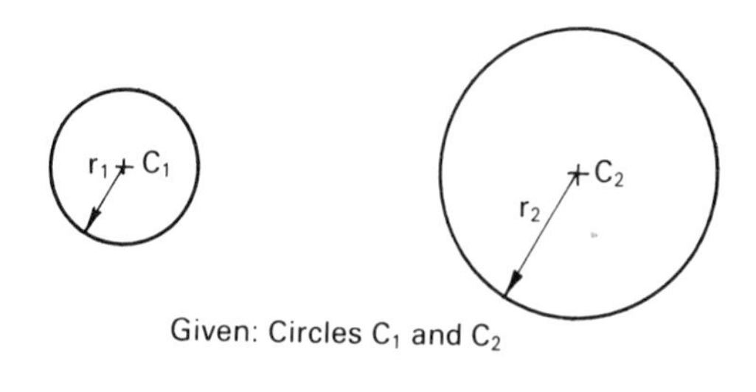

Given: Circles C_1 and C_2

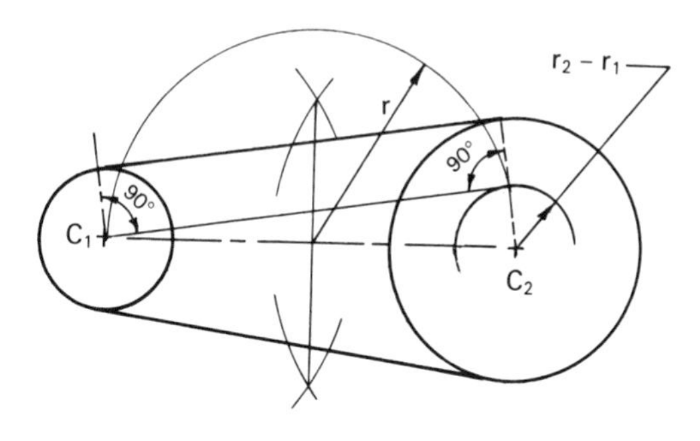

Solution

Fig. 5-57. Outside tangent method for drawing tangents to two circles.

3. From C_2 and the point of tangency on the arc, draw a perpendicular until it intersects the circumference of C_1 and C_2.
4. The line of tangency will be found when a line is drawn through the intersection points of C_1 and C_2.

Procedure C—cross-tangent method (Fig. 5-58):

1. Using point C_2 as the center, draw an arc whose radius is equal to the sum of radii C_1 and C_2, or $r = r_1 + r_2$.
2. From C_1, draw a tangent to the arc using the procedure illustrated.
3. From C_1 and C_2, draw a line perpendicular to the tangency line until it intersects the circumference of circles C_1 and C_2.
4. The line of tangency will be found when a line is drawn through the intersection points of the two circles.

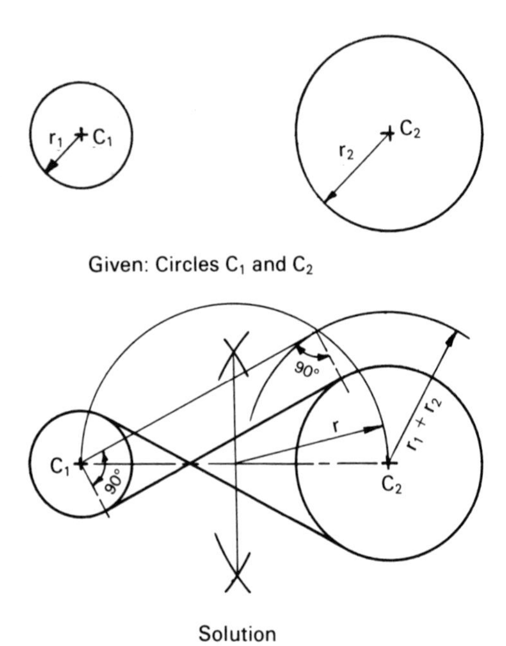

Fig. 5-58. Cross-tangent method for drawing tangent to two circles.

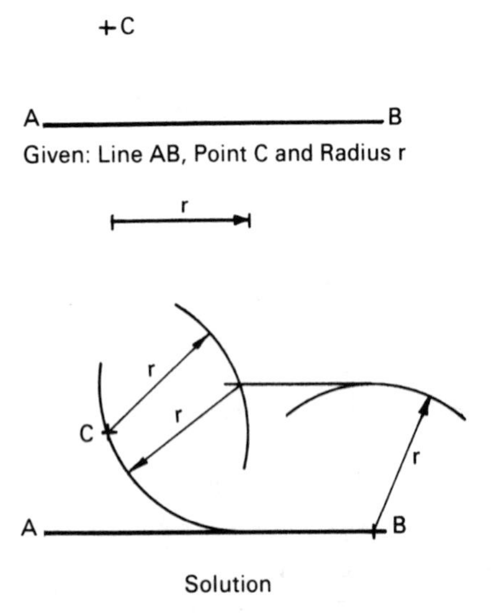

Fig. 5-59. Drawing an arc tangent to a line and through a point.

DRAWING AN ARC TANGENT TO A LINE AND THROUGH A POINT

Two procedures are available to drafters for solving this problem; either can be used with similar speed and accuracy.

Given: Line AB, point C, and arc radius

Procedure A—radius method (Fig. 5-59):

1. Draw a line parallel to AB at a distance equal to the arc radius.
2. From point C, draw an arc equal to the arc radius, so that it will intersect the parallel line at point D.
3. The tangency will be constructed when an arc with the given radius is drawn from point C.

Given: Point C and line AB

Procedure B—bisector method (Fig. 5-60):

1. Identify a desired point of tangency on line AB, and label it D.
2. Draw line CD.
3. From point D, draw a perpendicular.
4. Bisect line CD so that its bisector will intersect the perpendicular at point E.
5. The arc of tangency will be constructed when it is drawn from point E.

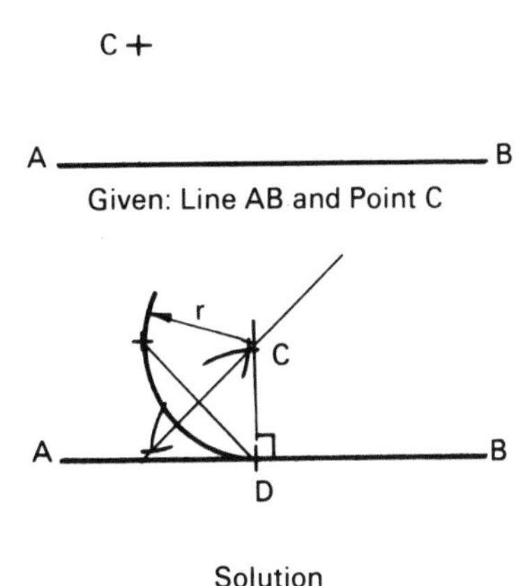

Fig. 5-60. Bisector method of drawing an arc tangent to a line and through a point.

DRAWING AN ARC TANGENT TO ANOTHER ARC AND THROUGH A POINT

There is one primary geometric construction procedure used for this problem.

Given: Arc F, point G, and arc radius

Procedure (Fig. 5-61):

1. From point G, strike an arc equal to the arc radius.
2. From F, strike an arc with a radius equal to the sum of the arc radius plus the radius of arc F, so that it will intersect the arc drawn from points G and H.
3. The arc tangency will be constructed when it is drawn from point H.

DRAWING AN ARC TANGENT IN A RIGHT ANGLE

A simple geometric construction that has many applications in industrial drafting is the drawing of an arc tangent to two perpendicular lines. One procedure is recommended for this problem.

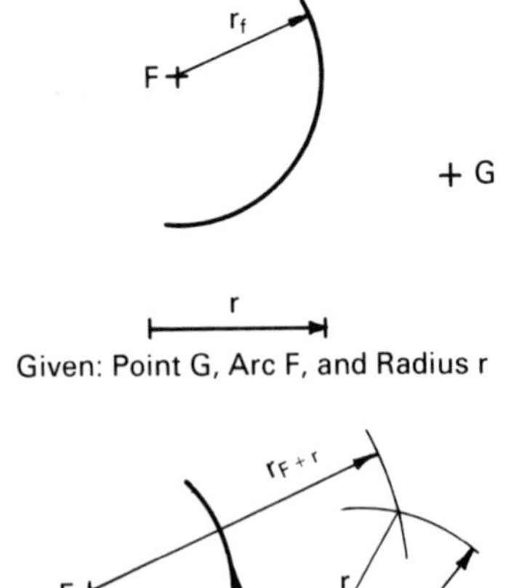

Given: Point G, Arc F, and Radius r

Solution

Fig. 5-61. Drawing an arc tangent to an arc and through a point.

Given: Right angle BAC and arc radius

Procedure (Fig. 5-62):

1. Strike an arc from point A, with a radius equal to the given arc radius, so that it intersects lines AB and AC at points D and E, respectively.
2. From D and E, strike two arcs with the same given radius, so that they intersect at point F.
3. To locate the exact points of tangency, draw a perpendicular from F to AB and AC.
4. The arc of tangency will be constructed when the given arc radius is drawn from point D to the points of tangency.

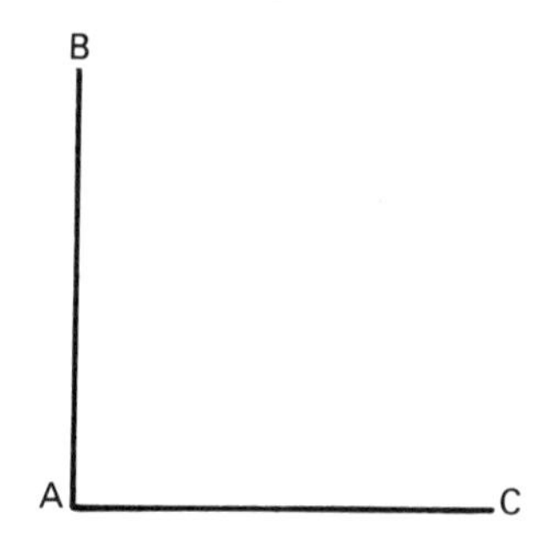

Given: Right Angle BAC and Radius r

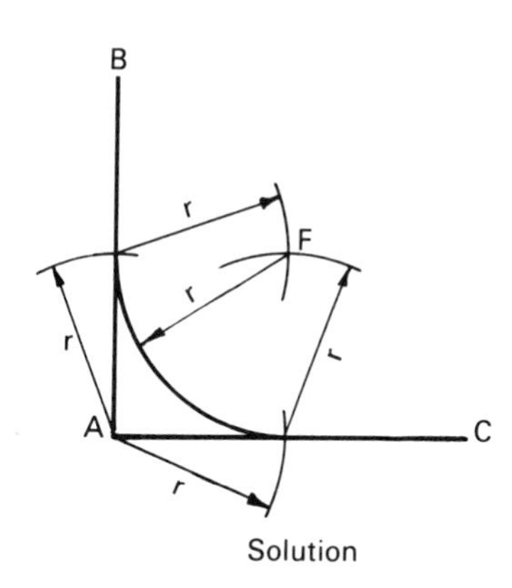

Solution

Fig. 5-62. Drawing an arc tangent to a right angle.

DRAWING AN ARC TANGENT IN A NON-RIGHT ANGLE

This problem is similar to that in Fig. 5-62. The one procedure outlined here can be used with either acute or obtuse angles, as illustrated in the examples (Figs. 5-63a and b).

Given: Angle EDF and arc radius

Procedure (Fig. 5-63):

1. Draw lines parallel to the two legs of the angle so that they intersect at point T. The parallel lines must be a distance equal to the given arc radius away from the angle legs.
2. To locate the exact points of tangency, draw perpendiculars to DE and DF from point T.
3. The arc of tangency will be constructed when the given arc radius is drawn from T to the points of tangency.

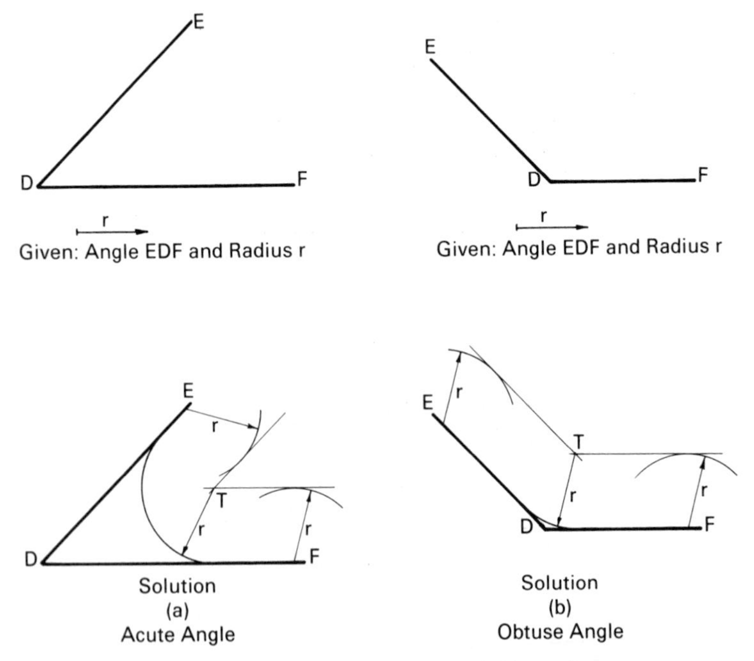

Fig. 5-63. Drawing an arc tangent to nonright angles.

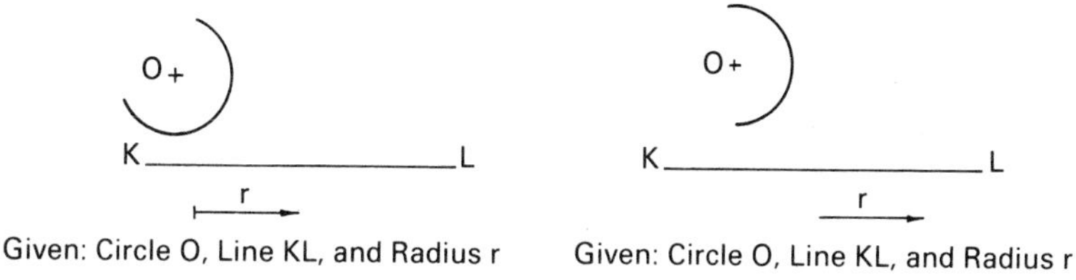

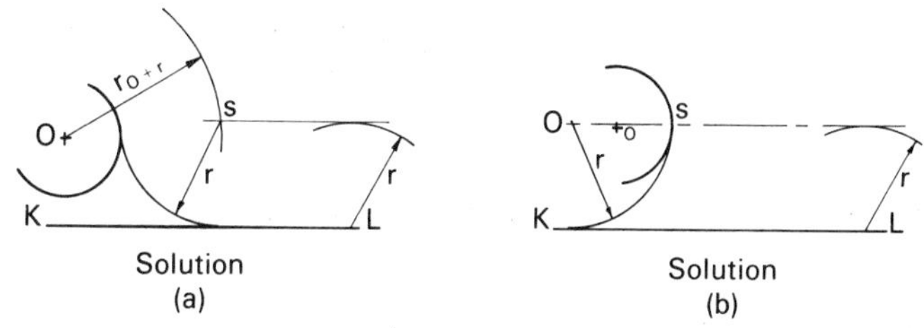

Fig. 5-64. Drawing an arc tangent to a line and an arc.

DRAWING AN ARC TANGENT TO AN ARC AND LINE

Another problem that involves the use of parallel lines is the drawing of an arc tangent to another arc and line. There are two possible situations, each of which is illustrated in the following procedure.

Given: Line KL, arc with center O, and arc radius

Procedure (Fig. 5-64):

1. Draw a line parallel to KL and an arc parallel to arc O, so that they are equidistant from the given arc radius.
2. Where the parallel line and arc intersect at point S, determine the points of tangency by drawing a perpendicular to KL and a line from O to S.
3. The arc of tangency will be constructed when the given arc radius is drawn from point S to the points of tangency.

DRAWING AN ARC TANGENT TO TWO ARCS

Again, there are two possible types of drawing problem that can arise when drawing an arc tangent to another arc, both of which are addressed in the following procedure.

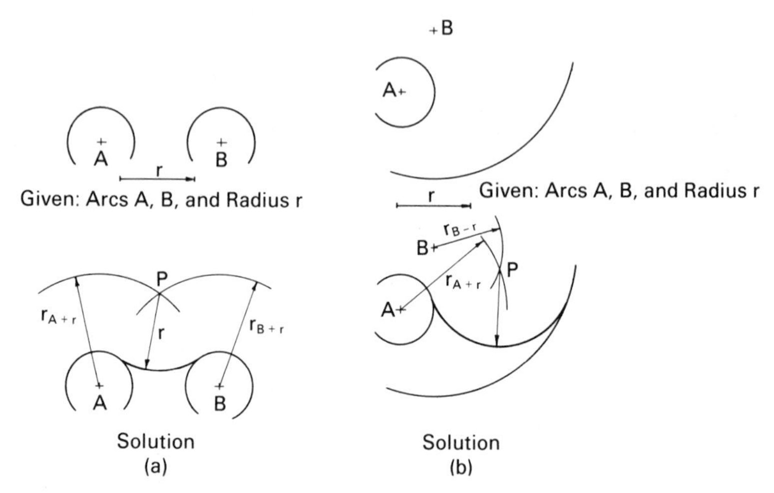

Fig. 5-65. Drawing an arc tangent to two arcs.

Given: Two arcs with centers A and B, and arc radius

Procedure (Fig. 5-65):

1. From points A and B, draw two parallel arcs at a distance equal to the given arc radius, so that they intersect at point P.
2. To locate the exact points of tangency, draw a line from P to points A and B. Where this line intersects the arcs will be the points of tangency.
3. The arc of tangency will be constructed when the given arc radius is drawn from point P to the points of tangency.

DRAWING A REVERSE OR OGEE CURVE

There are two primary configurations that employ the drawing of a reverse curve. A reverse curve, also known as an ogee, is one that changes direction so that it takes on the shape of an S or backward S. Presented here are two possible construction procedures.

Given: Parallel lines AB and CD

Procedure (Fig. 5-66):

1. Draw line BC, and determine the location where the curve is to change direction. Label that point F.
2. Bisect BF and CF.

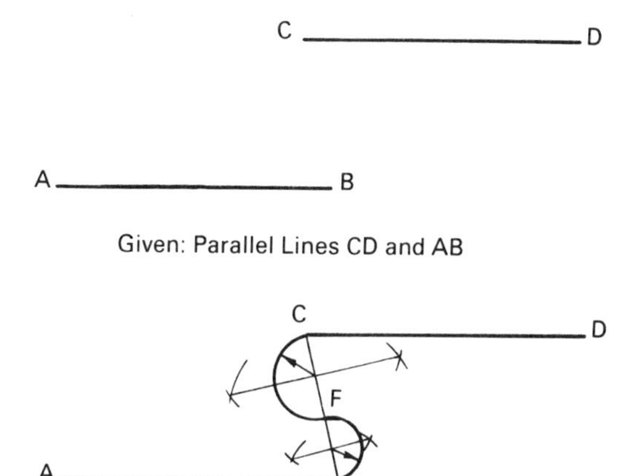

Fig. 5-66. Drawing a reverse curve given two parallel lines and arc radius.

3. Draw perpendiculars from B and C so that they intersect the bisectors at points R and S.
4. The reverse curve will be constructed when arcs are drawn from R at radius RB, and S at radius SC.

Given: Parallel lines EF and GH, and arc radius

Procedure (Fig. 5-67):

1. Draw perpendicular FJ at a length equal to the given arc radius.
2. Draw a line parallel to GH at a distance equal to the given arc radius.
3. Strike an arc from point J with a radius equal to twice the given arc radius, so that it will intersect the parallel line at point K.
4. The reverse curve will be constructed when arcs are drawn at the given arc radius from points J and K.

DRAWING ARCS TANGENT TO THREE LINES

A more complex tangency problem requires the drafter to draw arcs tangent to three lines. This may appear difficult, but in reality it is quite simple. This problem is solved by the use of one basic procedure.

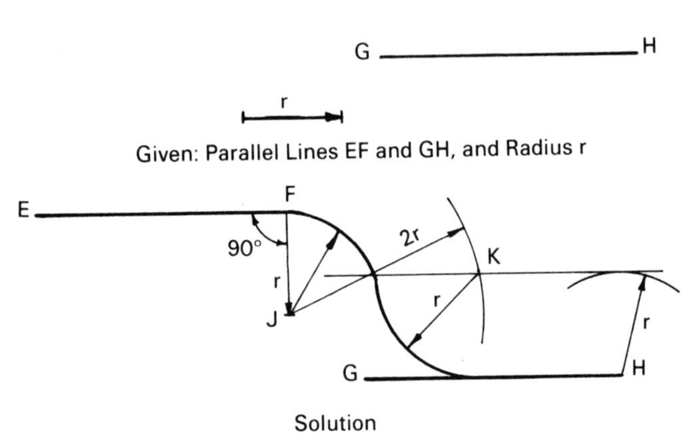

Fig. 5-67. Drawing a reverse curve given two parallel lines and arc radius.

Given: Lines PQ, RS, and TU

Procedure (Fig. 5-68):

1. Identify a convenient point of tangency on RS and label it T_1.
2. Locate T_2 and T_3 by striking an arc from R with a radius equal to distance RT_1, and another arc from S with a radius equal to ST_1.
3. Draw perpendiculars at T_1, T_2, and T_3 so that they will intersect at points O_1 and O_2.
4. The curve tangents will be constructed when arcs are drawn from O_1 and O_2 to the three lines.

APPROXIMATING A SERIES OF ARCS

There are times when a drafter may not wish to use an irregular curve to draw a given curve; the curve will then be drawn with a compass. This technique involves approximating a series of arcs, which make up the curve. This estimation of a curve is accomplished by the use of one geometric construction technique.

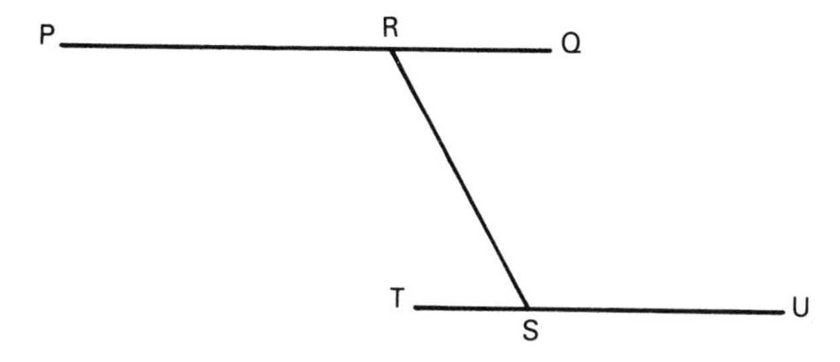

Given: Lines PQ, RS, and TU.

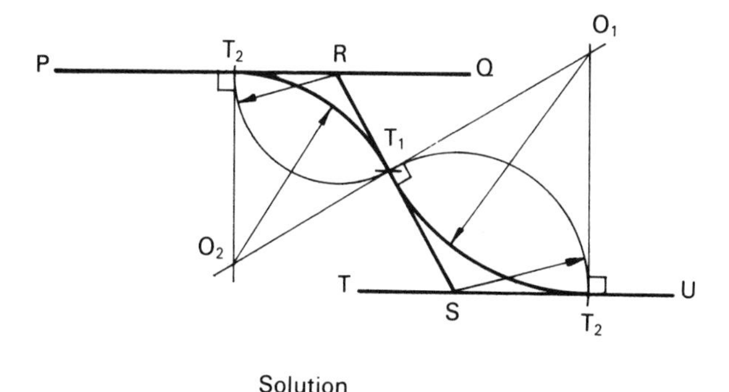

Solution

Fig. 5-68. Drawing tangents to three lines.

Given: A sketched smooth curve

Procedure (Fig. 5-69):

1. Divide the curve into convenient arc sections.
2. Estimate the radius and the center of each arc section.
3. The series of arcs will be constructed when an arc is drawn from each center, with a compass, so that the successive arcs meet in a smooth curve.

RECTIFYING CIRCULAR ARCS

To rectify a circular arc is to lay out the true length of that arc along a straight line. Three basic procedures are available to the drafter. Each is considered an approximation of the arc's true length, but is accurate for most instrument drawings.

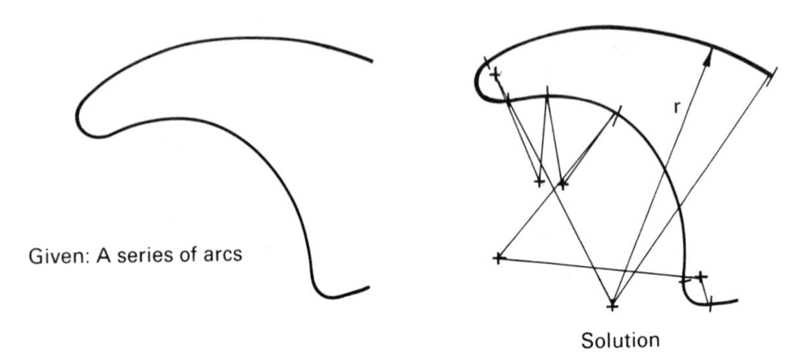

Fig. 5-69. Approximating a series of arcs.

Given: Quadrant AOB of a circle

Procedure (Fig. 5-70):

1. Draw a tangent to the circle at point A.
2. From B, draw a 60° line so that it intersects the tangent at point C.
3. The rectified quadrant will be distance AC, with an error of plus and minus 0.004.

Given: Arc CD

Procedure (Fig. 5-71):

1. Draw a tangent to the arc at point D.
2. Draw chord CD, and bisect it.
3. Extend chord CD to E, with DE being equal to half of CD.
4. The rectified arc will be constructed when an arc is drawn from C to point F, with E as the center point.

Given: A circle

Procedure (Fig. 5-72):

1. Draw a tangent to the circle at point A and extend it to B so that AB is equal to three times the circle's diameter (3D).
2. From C, draw an arc equal to the circle's radius so that it will intersect the circumference at point D.
3. With a radius equal to distance DB, strike an arc that passes through D to establish point E on the circle's center line.
4. The rectified circumference will be distance EB.

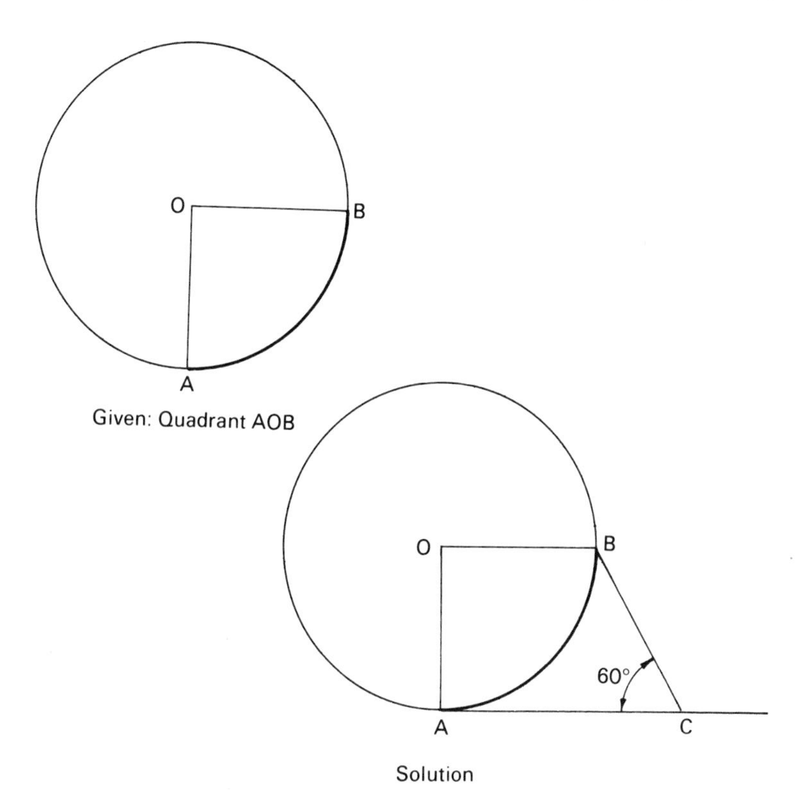

Fig. 5-70. Rectifying a quadrant of a circle.

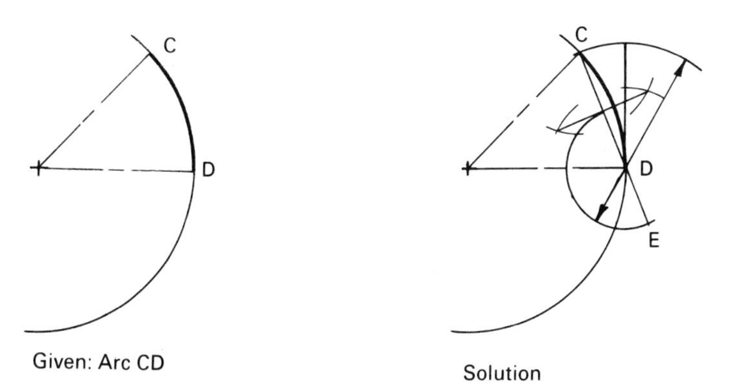

Fig. 5-71. Rectifying an arc.

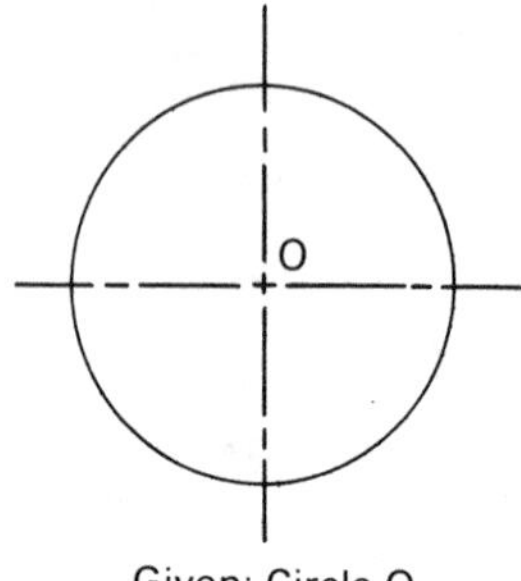

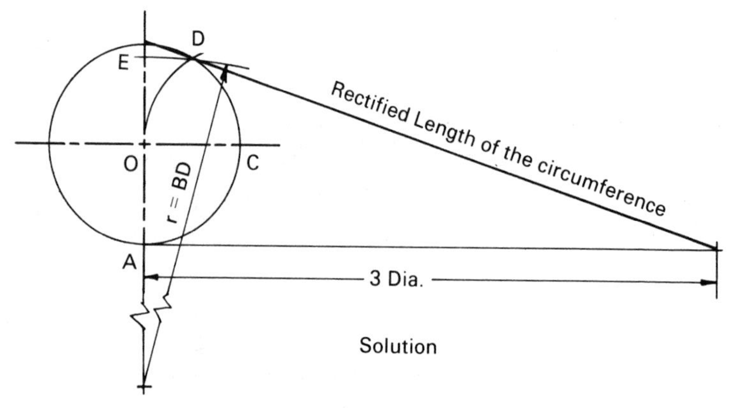

Fig. 5-72. Rectifying a circle's circumference.

Ellipses

An ellipse is made up of a major and a minor axis. The major axis is the longer of the two. As shown in Fig. 5-73, two foci lie on the major axis, and are located by striking an arc from the end of the minor axis, with a radius of one-half the major axis, to the major axis at two locations. As a result, the location of every point along the circumference of an ellipse can be described in relationship to the major axis. That is, the sum total distance of a point along an ellipse from the two foci will equal the length of the major axis.

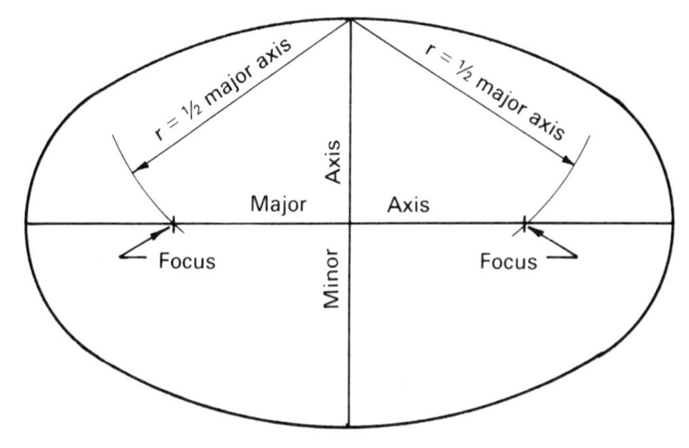

Fig. 5-73. Ellipse axes and foci.

DRAWING A FOCI ELLIPSE

This geometric construction solution is based upon the mathematical relationship just described, and it is sometimes used to develop larger ellipses.

Given: Major and minor axes AB and CD

Procedure (Fig. 5-74):

1. Locate foci F_1 and F_2 by swinging an arc from C or D with a radius equal to half the length of AB.
2. Between F_1 and the intersection of the two axes, mark off and number a series of points that will correspond to the number of points that will be used to plot each quadrant of the ellipse. Note that it is recommended that the spacing between each point lessen as one approaches the focus.
3. Beginning with point 1, strike arcs from F_1 and F_2 through 1, so that they intersect in the respective quadrant. Repeat this operation for each succeeding point until all points are plotted in all four quadrants.
4. The foci ellipse will be constructed when the plotted points are connected by a smooth curved line.

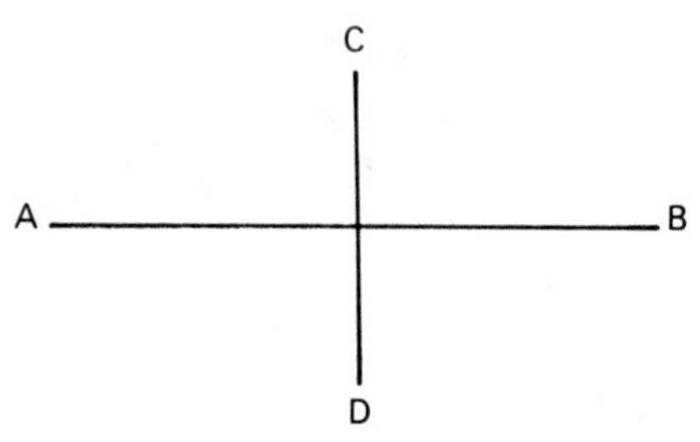

Given: Major Axis AB, and Minor Axis CD

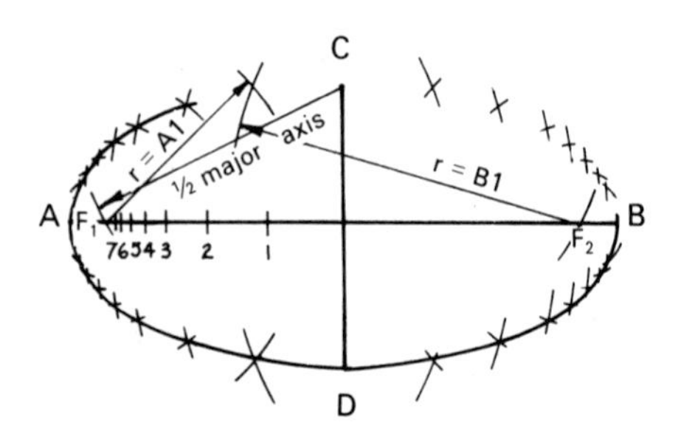

Solution

Fig. 5-74. Drawing a foci ellipse.

DRAWING A TRAMMEL ELLIPSE

One of the quicker methods used to construct an ellipse is by use of a trammel. Trammels used in this construction method can be made from a stiff piece of paper or cardboard.

Given: Major and minor axes EF and GH

Procedure (Fig. 5-75):

1. Make a trammel by marking off distances PG and GS, representing one-half of the major and minor axes' distances.
2. To plot the points of the ellipse's circumference, place the trammel across the axes so that E falls over the minor axis and C falls over the major axis. In this position, the location of P will mark one point on the ellipse's circumference.
3. Keep moving the trammel along the axes until a sufficient number of points have been plotted.

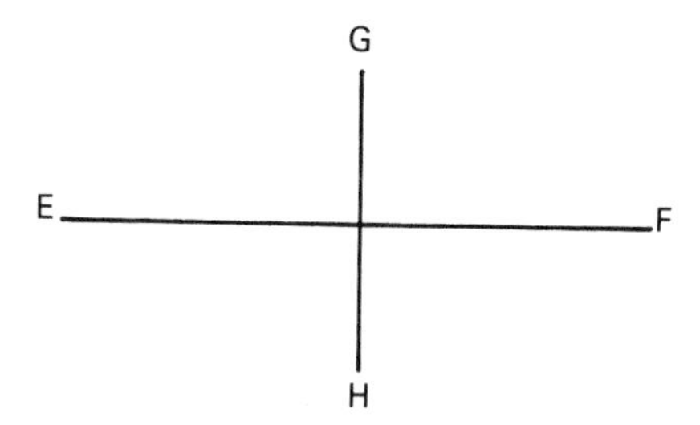

Given: Major Axis EF and Minor Axis GH

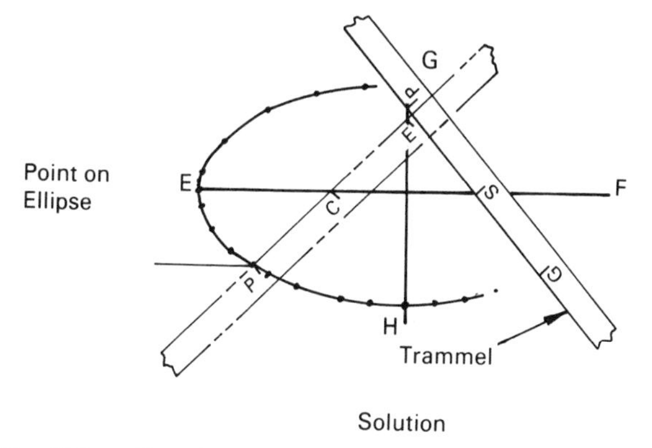

Solution

Fig. 5-75. Drawing a trammel ellipse.

4. The trammel ellipse will be constructed when all points are connected by a smooth curved line.

DRAWING A CONCENTRIC-CIRCLE ELLIPSE

Concentric-circle ellipses are developed from a pair of circles whose diameters are equal to its major and minor axes. One basic construction procedure is used for this problem.

Given: Major and minor axes JK and LM

Procedure (Fig. 5-76):

1. Draw two concentric circles whose diameters are equal to the major and minor axes.

2. Draw a diagonal through the common center point so that it intersects the circumferences of both circles.
3. From the points of intersection on the smaller circle, draw a line parallel to the major axis. From the points of intersection on the larger circle, draw a line parallel to the minor axis until it intersects the other parallel lines.
4. The points of intersection of the parallel lines locate specific points on the ellipse's circumference. Draw as many diagonals as necessary to plot the ellipse accurately.
5. The concentric-circle ellipse will be constructed when all points are connected by a smooth curved line.

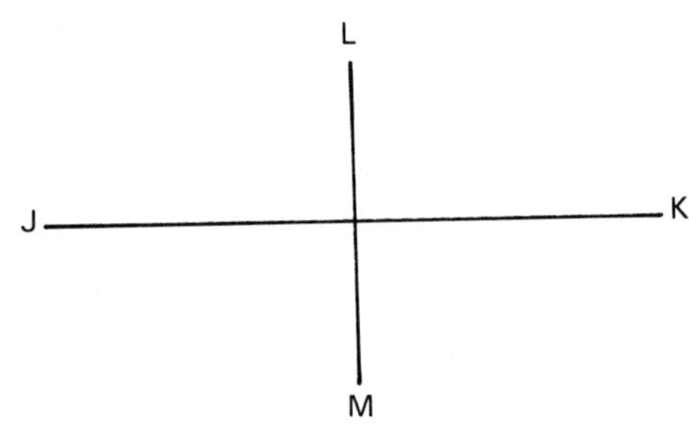

Given: Major Axis JK, and Minor Axis LM

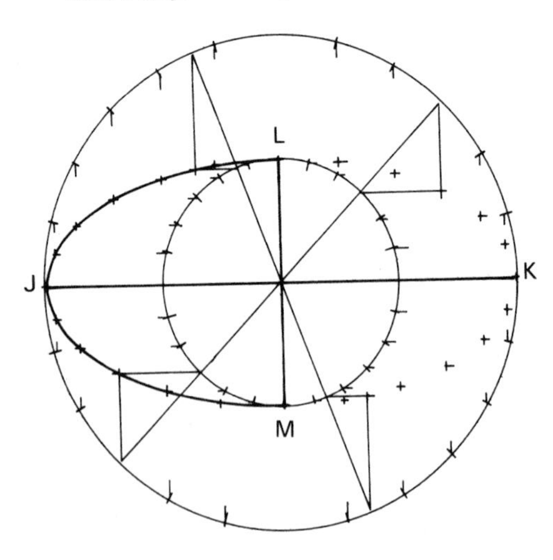

Solution

Fig. 5-76. Drawing a concentric circle ellipse.

DRAWING A PARALLELOGRAM ELLIPSE

Parallelogram ellipses require a procedure that can be implemented when the major and minor axes are either perpendicular or nonperpendicular to one another. When perpendicular, the parallelogram will take the form of a rectangle. When not perpendicular, the ellipse is said to have *conjugate diameters.* Regardless of the configuration of the axes, there is but one geometric construction technique that is applicable to both situations.

Given: Major and minor axes AB and CD

Procedure (Fig. 5-77):

1. Draw a parallelogram so that its sides are parallel to the axes of the ellipse.
2. Divide AP and BP, and AG and BH into the same number of equal parts, and number each.
3. From D, draw a line through each division on the major axis.
4. From C, draw a line to each division on the parallelogram's side so that it will intersect the similarly numbered line. The points of intersection are the points on the ellipse's circumference.
5. The parallelogram ellipse will be constructed when all points are connected by a smooth curved line.

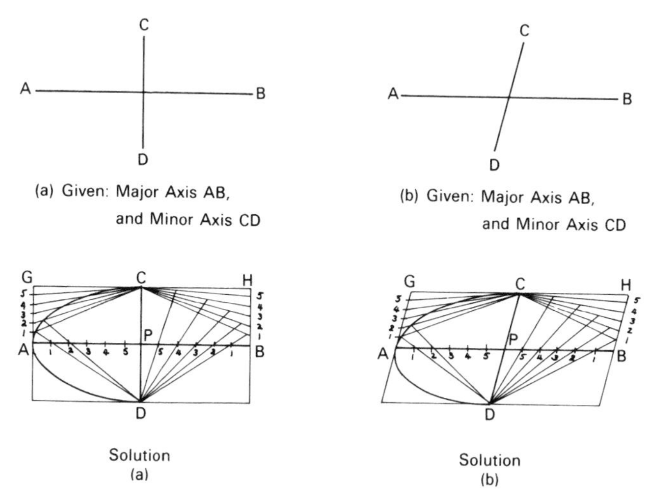

Fig. 5-77. Drawing parallelogram ellipses.

DRAWING AN APPROXIMATE ELLIPSE

One of the fastest and most popular methods for constructing an ellipse is the approximation method. Also known as the four-center technique, this geometric construction not only reduces drawing time, but also provides for a symmetrical ellipse.

Given: Major and minor axes EF and GH

Procedure (Fig. 5-78):

1. Draw line EG.
2. Using Q as the center, and distance QE as the radius, strike arc EJ.
3. Draw line EG.
4. With Q as center, and distance QG as the radius, strike arc GK.
5. From G, strike an arc with a radius equal to distance KE, to establish point L.
6. Bisect line EL, and extend the bisector so that it will intersect the major axis at point 1 and the extended minor axis at point 2.
7. Locate points 3 and 4 relative to 1 and 2.
8. The approximate ellipse will be constructed when arcs are drawn from points 1, 2, 3, and 4 (note that the points of tangency are located at the point where the arcs meet the lines connecting the four centers).

DRAWING A TANGENT AT ANY POINT ON AN ELLIPSE

As in drawing tangents to circles, there are times when tangents must be drawn to ellipses. Two construction techniques are available, each designed for a particular type of ellipse.

Given: Tangency point T on a concentric-circle ellipse

Procedure (Fig. 5-79):

1. Draw a line through T that is parallel to the minor axis, so that it will intersect the outside circle at point L.
2. Draw a tangent to the outside circle at point L, and extend it until it intersects the major axis at point M.
3. The line of tangency will be constructed when a line is drawn from M through T.

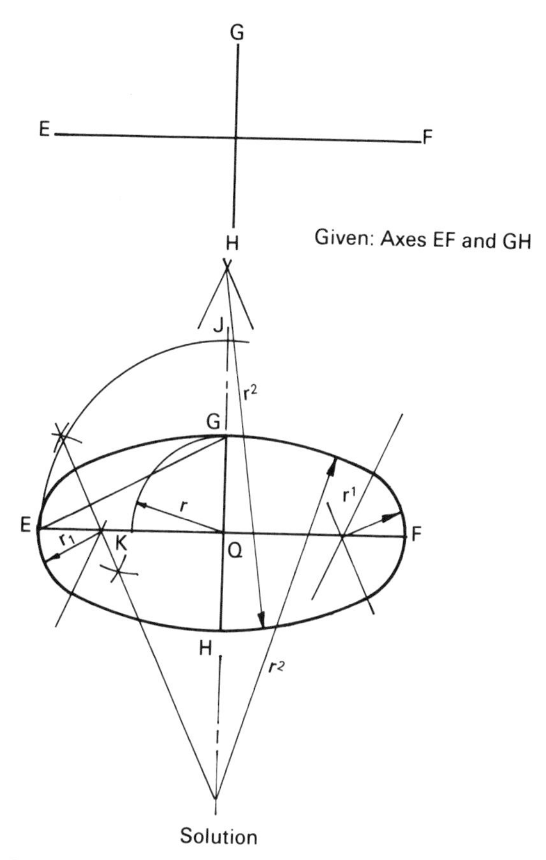

Fig. 5-78. Drawing an approximate ellipse.

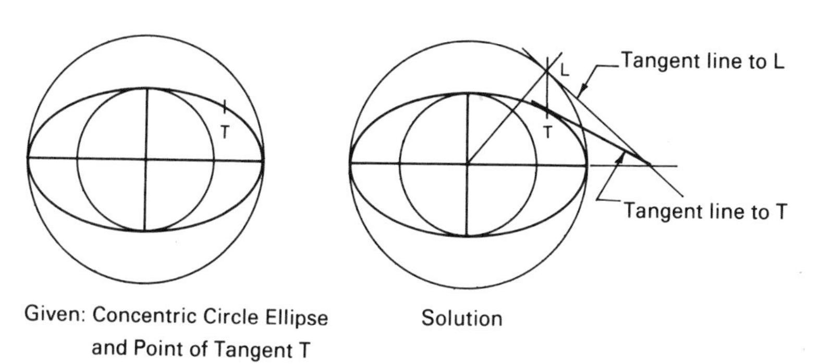

Fig. 5-79. Drawing a tangent at any point on an ellipse by concentric circle method.

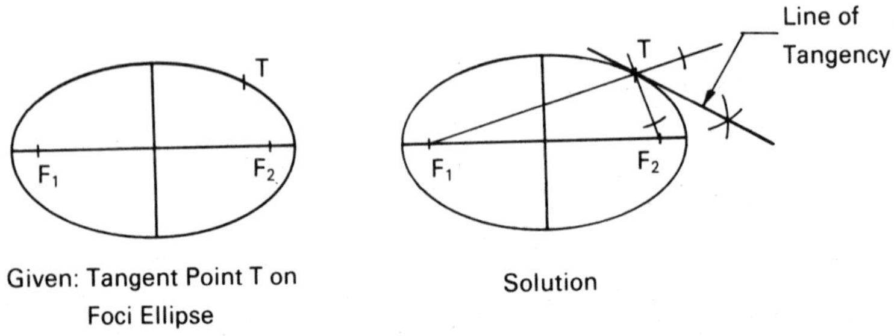

Fig. 5-80. Drawing a tangent to a point on a foci ellipse.

Given: Tangency point T on a foci ellipse

Procedure (Fig. 5-80):

1. Draw a line from the two focal points F_1 and F_2 to point T.
2. Extend F_1T beyond point T so that an angle is established at the point of tangency.
3. The line of tangency will be constructed when the established angle is bisected.

DRAWING A TANGENT FROM A POINT OUTSIDE AN ELLIPSE

Again, drawing a tangent to an ellipse from a point located in space can be done in two ways, depending upon the type of ellipse. Presented here are the two construction procedures.

Given: Point A and a concentric-circle ellipse

Procedure (Fig. 5-81):

1. Draw a line through A parallel to the minor axis so that it will intersect the path of the major axis at point P.
2. Draw a line from A to G, establishing point L.
3. Draw a line from H through L, establishing point M. From M, draw a line tangent to the outside circle, identifying the point of tangency as Y.
4. From Y, draw a line parallel to the minor axis until it intersects the ellipse at T.
5. The tangency will be constructed when a line is drawn from A through T.

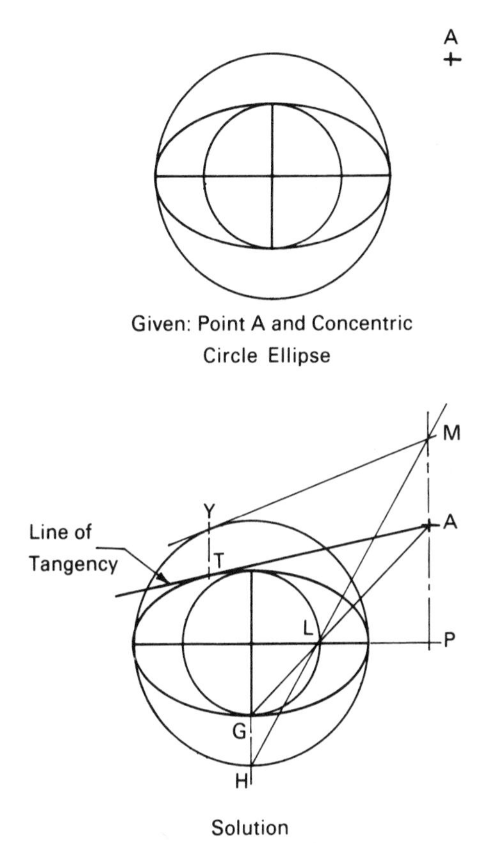

Fig. 5-81. Drawing a tangent to a concentric circle ellipse from an outside point.

Given: Point A and a foci ellipse

Procedure (Fig. 5-82):

1. Strike an arc from A with a radius equal to distance AF_2.
2. Strike a second arc from F_1 with a radius equal to the length of the major axis, so that it will intersect the first arc at points B and C.
3. From F_1, draw a line to B and C that will intersect the ellipse at points T_1 and T_2.
4. The tangencies will be constructed when a line is drawn from A through T_1 and T_2.

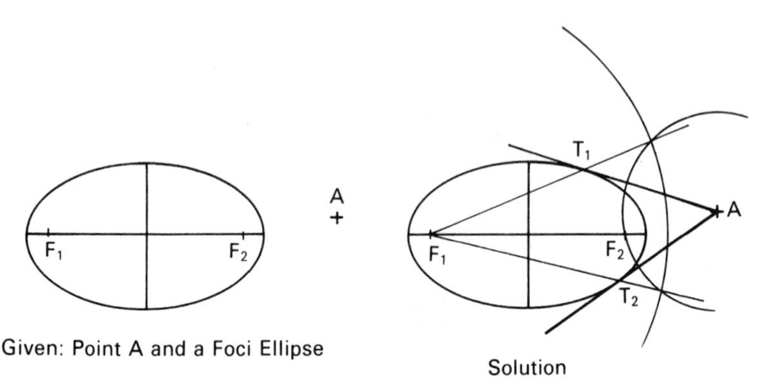

Fig. 5-82. Foci method for drawing tangents from a point outside an ellipse.

Other Curves

The geometric constructions presented in this last section include curved forms other than arcs, circles, and ellipses. Construction procedures will be described for the parabola, hyperbola, spiral, helix, involute, and cycloid. These curves are used in drawings for highway layouts, architectural structures, structural products, sheet material products, industrial equipment, and the graphing of data. Owing to the complex nature and variety of these curves, few templates are available to drafters that fit all situations, and it is often necessary to construct them.

DRAWING A PARABOLA

The drawing of a parabola requires the construction of a plane curve that is formed by a point moving in such a manner that its distance from a fixed point (focus) will be equal to its distance from a fixed line (directrix). Three basic procedures are available, each applicable to particular situations.

Given: Directrix AB and focus F

Procedure (Fig. 5-83):

1. Draw line CD parallel to the directrix.

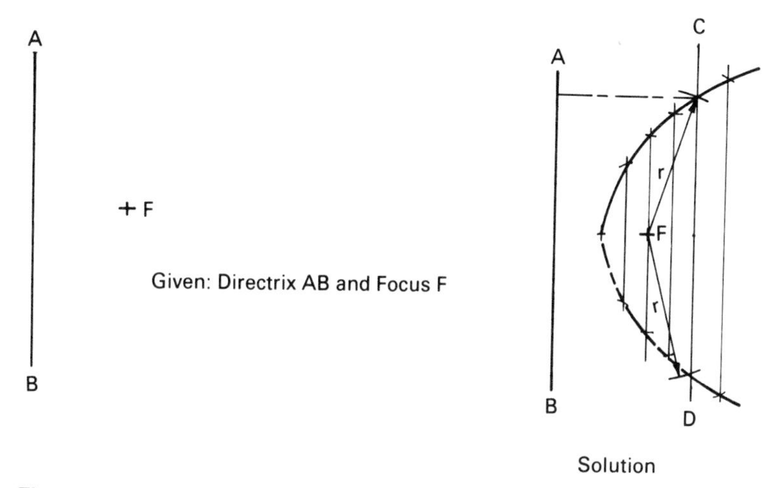

Fig. 5-83. Directrix method for drawing a parabola.

2. From F, strike an arc with a radius equal to the distance between CD and the directrix, so that the arc intersects CD at two points.
3. Repeat the procedure until a sufficient number of points have been plotted.
4. The parabola will be constructed when all points have been connected by a smooth curved line.

Given: The rise and fall (Span) of a parabola

Procedure (Fig. 5-84):

1. Draw vertical line DE with center point O, and horizontal line EF.
2. Divide OE into a number of equal divisions.
3. Divide EF into a number of equal units so that the divisions will equal the square of that number from starting point E (e.g., division 1 = 1 unit, division 2 = 4 units, division 3 = 9 units, and so on).
4. Plot each point along the parabola by determining the point of intersection between like numbered divisions along DE and EF.
5. The parabola will be constructed when all plotted points have been connected by a smooth curved line.

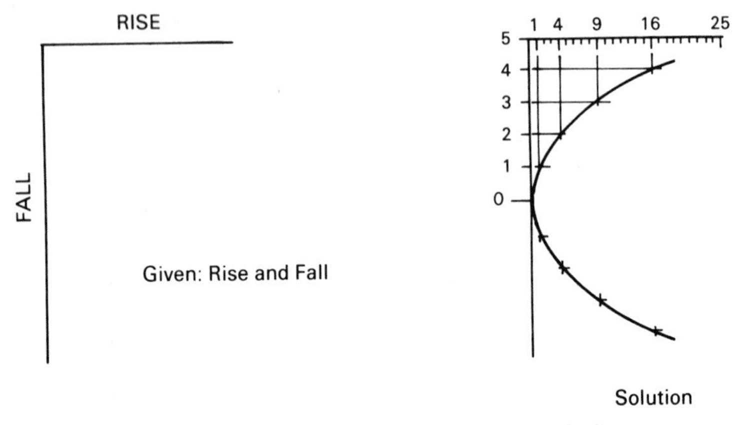

Fig. 5-84. Rise and fall method for drawing a parabola.

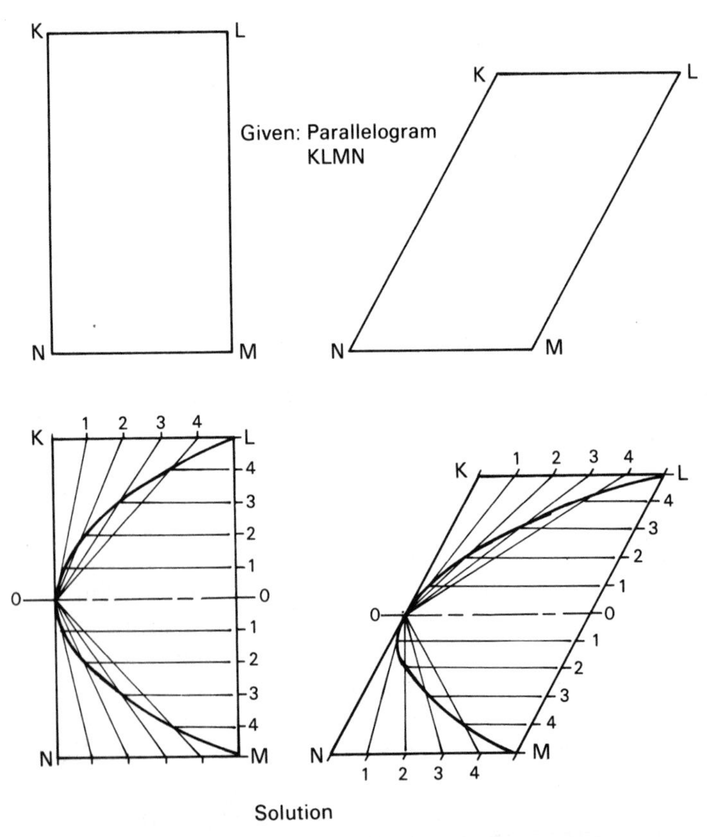

Fig. 5-85. Parallelogram method for drawing a parabola.

Given: Parallelogram KLMN

Procedure (Fig. 5-85):

1. Divide LM into an even number of equal parts, with a zero point in the center. Number each point in succession from the zero point.
2. Divide KL and MN into half as many equal parts, and number each.
3. From each point along LM, draw horizontal and parallel lines across the rectangle.
4. From each point along KL and MN, draw a line to the center point on KN. The intersections of the corresponding numbered lines will be the points about the parabola's circumference.
5. The parabola will be constructed when all plotted points have been connected by a smooth curved line.

DRAWING TANGENTS TO A PARABOLA

To draw a line through a tangent point of a parabola is a simple and quick procedure. Regardless of how the parabola was constructed, one procedure is applicable to all drawings.

Given: A parabola, focus, and directrix

Procedure (Fig. 5-86):

1. Select point T as an appropriate point of tangency.

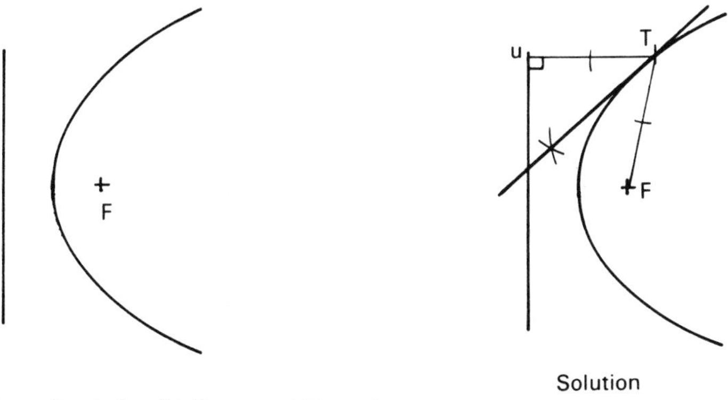

Fig. 5-86. Drawing a tangent to a parabola.

2. Draw a line from focus F to T, and a perpendicular line from the directrix to T, forming angle FTU.
3. The line of tangency will be constructed when angle FTU is bisected.

LOCATING THE FOCUS OF A PARABOLA

At times a parabola is given without a focus point. For various drawing procedures, this focus must be located. The geometric construction procedure used to locate the focus of parabolas is described here.

Given: A parabola

Procedure (Fig. 5-87):

1. Select point A at a convenient location on the parabola's circumference.
2. from A, draw a vertical line to locate its counter point B.
3. Construct AB's bisector and extend it to a convenient distance beyond the end of the parabola.
4. Draw a line tangent to A so that it will intersect AB's bisector at point C.
5. Bisect line AC.
6. The focus of the parabola will be located where AC's bisector intersects AB's bisector at point F.

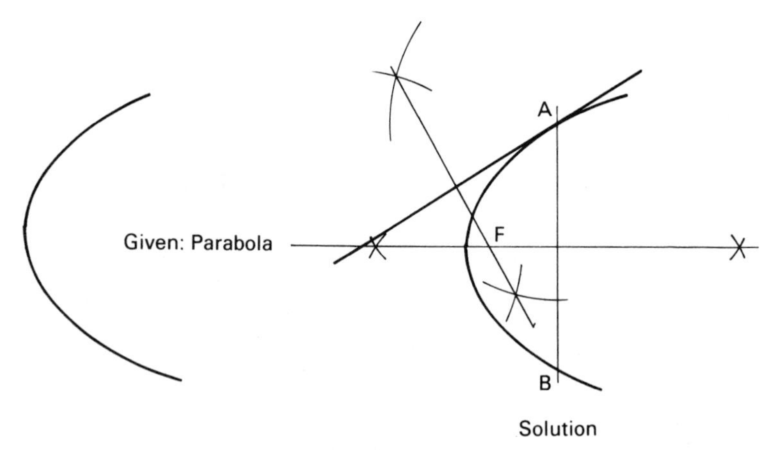

Fig. 5-87. Locating the focus of a parabola.

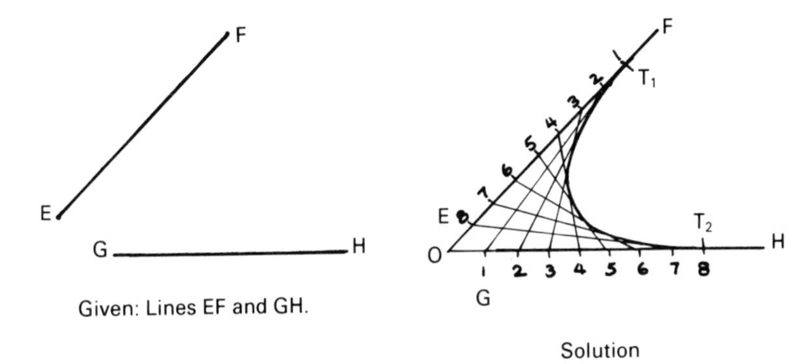

Fig. 5-88. Drawing a parabola tangent to two lines.

DRAWING A PARABOLA TANGENT TO TWO LINES

One geometric construction is used for drawing parabolas tangent to a pair of lines that are either intersecting or nonintersecting.

Given: Lines EF and GH

Procedure (Fig. 5-88):

1. Extend lines EF and GH so that they intersect at point O.
2. Locate convenient points of tangencies, T_1 and T_2, on OF and OH.
3. Divide lines OT_1 and OT_2 into an equal number of equal parts, and number in reverse directions.
4. Connect corresponding point numbers by a straight line.
5. The parabola will be tangent to lines EF and GH when a smooth curved line is drawn tangent to the intersecting lines.

DRAWING A HYPERBOLA

A hyperbola is a plane curve produced by a moving point in such a way that the difference of the distances from two foci will be constant. In addition to the construction of a hyperbola, the concepts of tangency and asymptotic lines will also be presented.

Given: Traverse axis BC and foci F_1 and F_2

Procedure (Fig. 5-89):

1. Select point A as a convenient point on the traverse axis path.

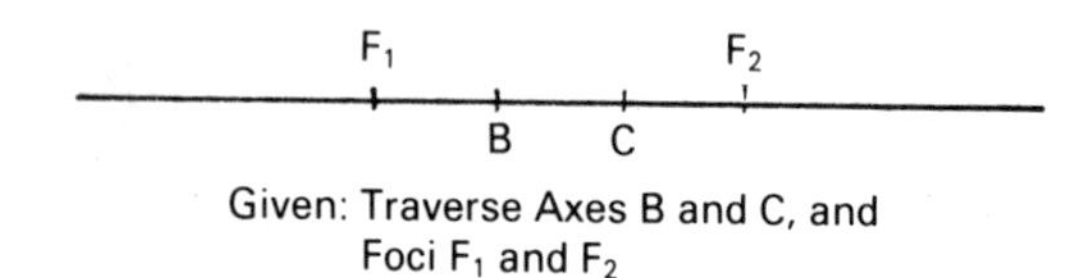

Given: Traverse Axes B and C, and
Foci F_1 and F_2

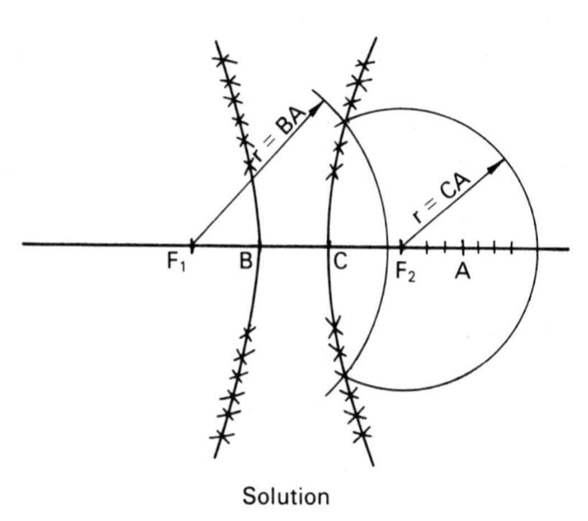

Solution

Fig. 5-89. Drawing a hyperbola.

2. From F_1 and F_2, draw two arcs, the first having a radius equal to distance AB and the second with a radius equal to AC, so that they intersect each other at points 1, 2, 3, and 4.
3. Plot as many points as required to construct the hyperbola.
4. The hyperbola will be constructed when all plotted points are connected by a smooth curved line.

To construct a tangent to a hyperbola, the angle between the focal radii must be bisected. The bisector will then become the tangency to the hyperbola.

Asymptotes are lines that limit the position of hyperbolas, so that they never become tangent with one another—though the hyperbola always moves closer to the asymptotes into infinity. To construct the asymptotes (Fig. 5-90), draw a circle that passes through the two foci and has a diameter equal to the distance between them. Construct perpendiculars from points B and C on the traverse axis until they intersect the circle at points R, S, T, and U. The hyperbola's asymptotes are the drawn lines SU and TR.

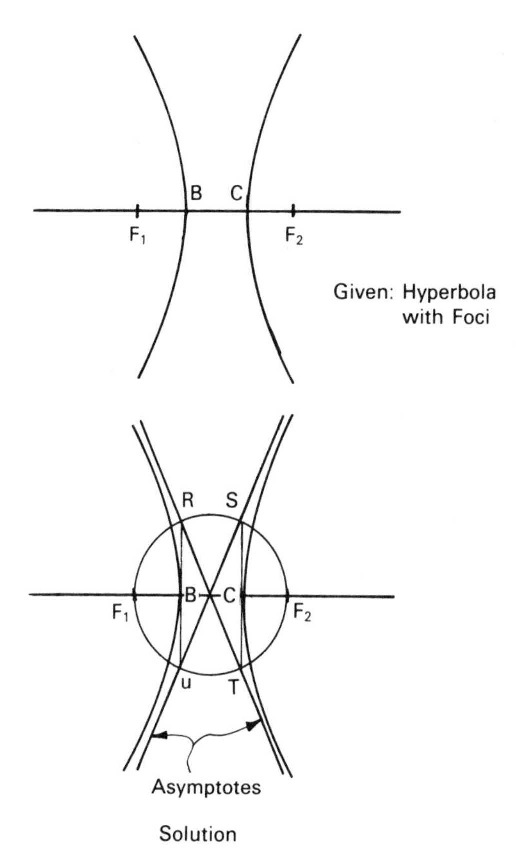

Fig. 5-90. Drawing asymptotes of hyperbolas.

DRAWING AN EQUILATERAL HYPERBOLA

An equilateral hyperbola is characterized by its uniform symmetry. One of the basic assumptions used in its geometric construction is that the hyperbola's asymptotes are perpendicular to one another. The drawing of an equilateral hyperbola can be accomplished by the use of one of two construction procedures.

Given: Asymptotes QA and AB, and a point on the hyperbola

Procedure A—chord method (Fig. 5-91):

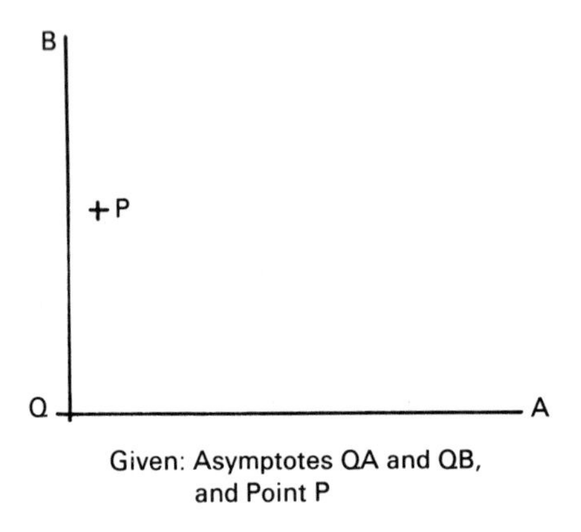

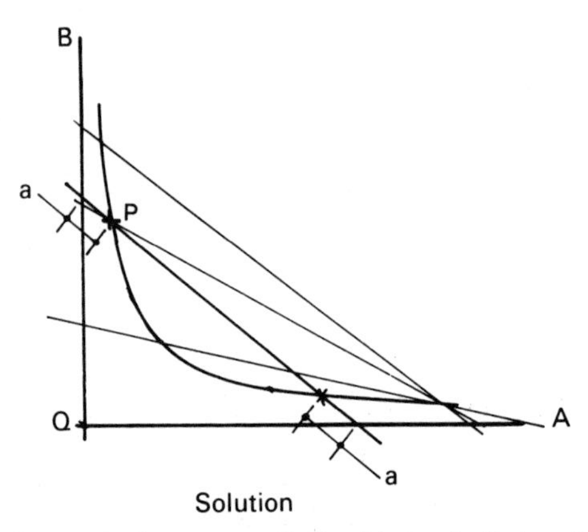

Fig. 5-91. Constructing an equilateral hyperbola by chord method.

1. Pass a chord through point P until it intersects QA and QB at points 1 and 2, respectively.
2. The intercepts between the curves and axes will be equal. Therefore, distance P_1 will be equal to the distance from 2 to a point on the hyperbola.
3. Pass additional chords through P until a sufficient number of points have been plotted (note that all chords do not have to pass through P; other points may be used as they are plotted).
4. The equilateral hyperbola will be constructed when all points are connected by a smooth curved line.

Procedure B—coordinate method (Fig. 5-92):

1. Through P, draw lines 1C and 2D so that they are parallel to QA and QB, respectively.
2. From Q, draw any diagonal until it intersects lines 1C and 2D at X and 3, respectively.
3. Draw a line from X, parallel to QA, and a line from 3, parallel to QB, until they intersect at point C, which lies on the circumference of the hyperbola.
4. Draw additional diagonals until a sufficient number of points have been plotted.
5. The equilateral hyperbola will be constructed when all points are connected by a smooth curved line.

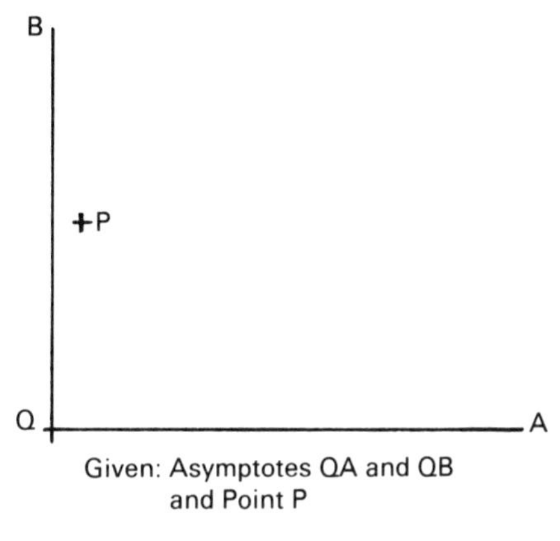

Given: Asymptotes QA and QB and Point P

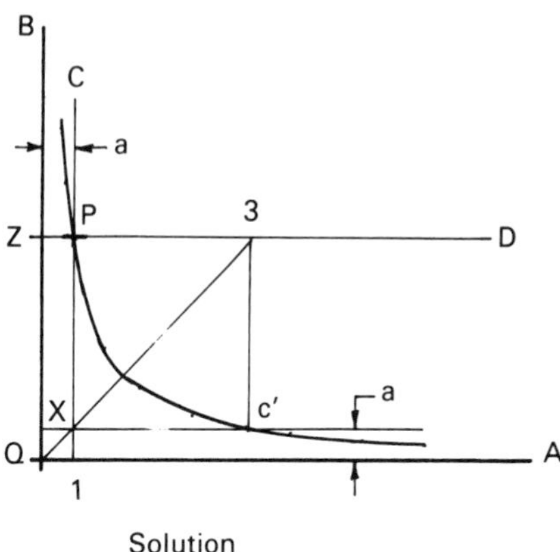

Solution

Fig. 5-92. Chord method for constructing an equilateral hyperbola.

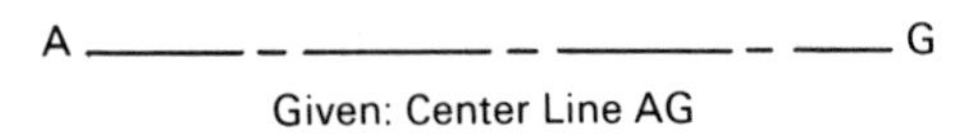

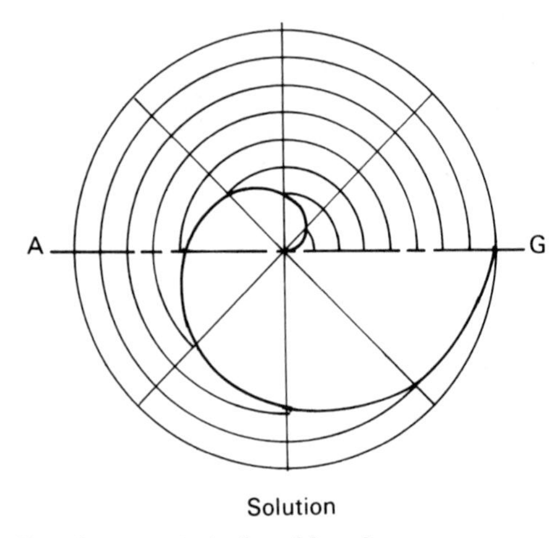

Fig. 5-93. Drawing a spiral of archimedes.

DRAWING A SPIRAL OF ARCHIMEDES

The spiral of Archimedes is a curved form that traces the path of a point as it moves around an axis while constantly receding from it or approaching it. The name of this spiral, "Archimedes," is derived from a first-century B.C. mathematician who used this concept to invent a tubular screw, similar to an auger, to move water along a path.

Given: Central axis AG

Procedure (Fig. 5-93):

1. Draw radial lines through A at equal intervals.
2. On one of the radials, mark off equal unit distances starting from point A and moving outward.
3. Begin plotting the spiral by moving one unit out the first radial, two units out the second, and so on, until all unit distances have been plotted.
4. The spiral of Archimedes will be constructed when all plotted points are connected by a smooth curved line.

DRAWING A HELIX

Helixes are constructed by tracing the path of a point as it rotates around a cylinder or cone in an ascending or descending path. Helixes that are constructed about a cylinder are known as cylindrical helixes, while those that are constructed about a cone are conical helixes.

The distance traversed by the moving point is called the lead, and is drawn parallel to the axis. If the path of the point is in a clockwise direction, the helix will be right-handed. A counterclockwise path generates a left-handed helix. The procedure described here can be used for any type of helix.

Given: Two views of a cylinder and cone

Procedure (Fig. 5-94):

1. Divide the top view (circle) into an equal number of parts and number each.
2. In the front (rectangular and triangular) views, divide the lead into the same number of equal parts and number each.
3. Plot each point along the helix by projecting each numbered division in the top view to the corresponding number in the front view.
4. The helix will be constructed when all plotted points are connected by a smooth curved line.

DRAWING AN INVOLUTE

An involute is a figure that is generated by the path of a line as it unwinds from a given line or plane figure. Thus, an involute is constructed by successive increasing radii that are derived from a line or the sides of a plane figure. Presented here are the procedures used to construct an involute from a line, triangle, square, and circle. The basic technique used in each case can also be applied to any other geometric plane figure.

Given: Line AB

Procedure (Fig. 5-95):

1. With B as center and AB radius, draw semicircle AC.

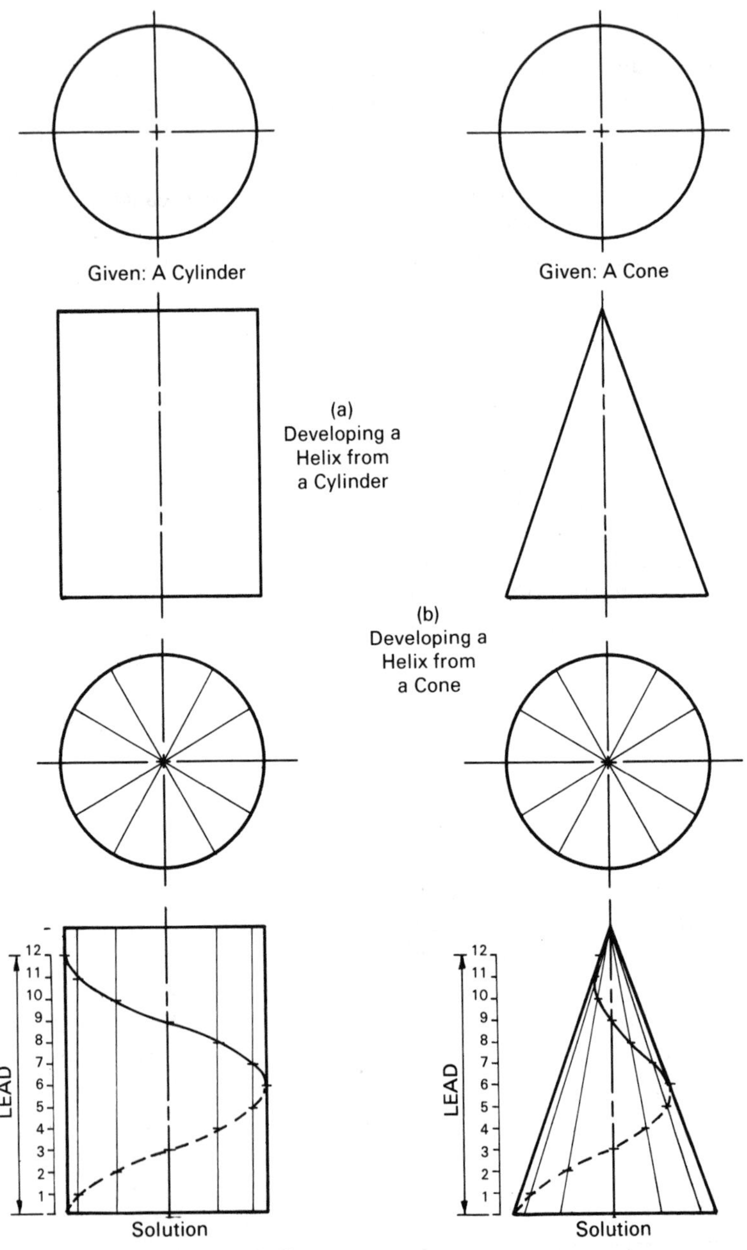

Fig. 5-94. Drawing a helix.

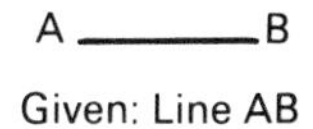

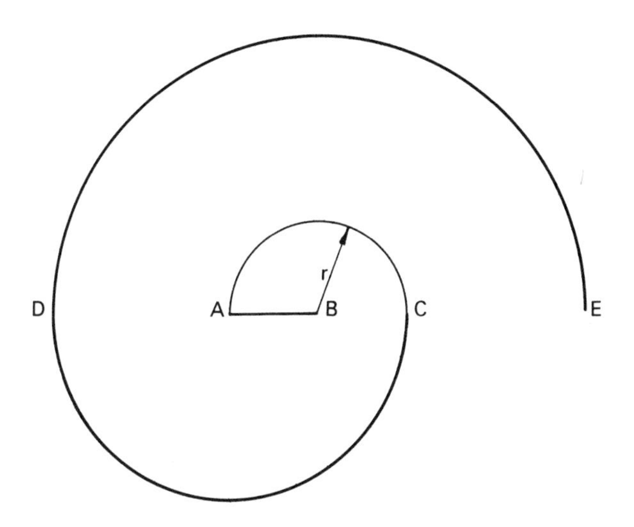

Fig. 5-95. Involute of a line.

2. With A as center and AC radius, draw semicircle CD.
3. Continue to draw semicircles from alternating centers and increasing radii until the involute has been constructed to the desired size.

Given: Triangle CDE

Procedure (Fig. 5-96):

1. With D as center and CD radius, draw arc AF.
2. With E as center and EF radius, draw arc FG.
3. With C as center and CG radius, draw arc GH.
4. Continue drawing arcs from succeeding corners and increasing radii until the involute has been constructed to the desired size.

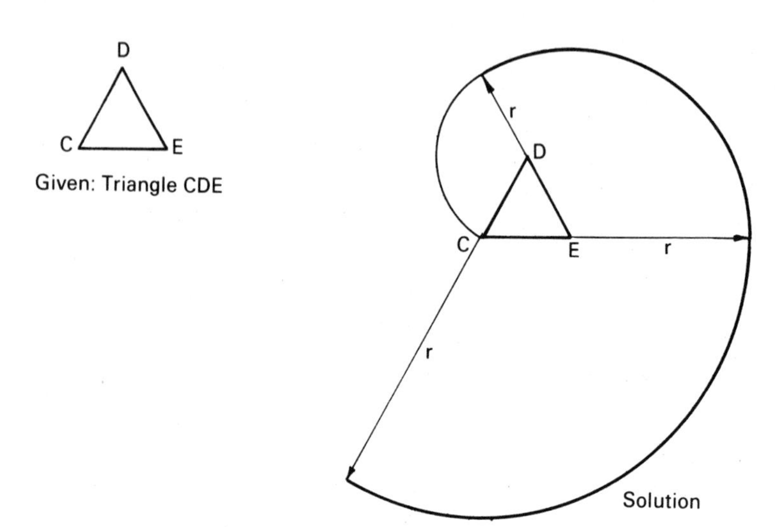

Fig. 5-96. **Involute of a triangle.**

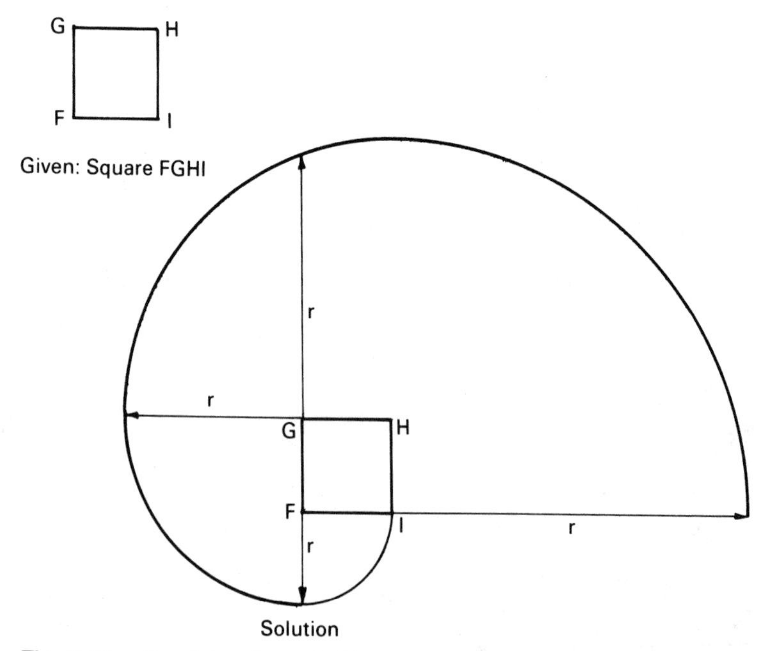

Fig. 5-97. **Involute of a square.**

Given: Square FGHI

Procedure (Fig. 5-97):

1. With G as center and FG radius, draw arc GJ.
2. With H as center and HJ radius, draw arc HK.
3. Continue drawing arcs from succeeding corners and increasing radii until the involute has been constructed to the desired size.

Given: Circle O

Procedure (Fig. 5-98):

1. Divide the circle into a number of equal parts.
2. Draw a tangent to each division point, and mark off the length of the corresponding circular arc (i.e., tangent 1 = one unit length, tangent 2 = two unit lengths, etc., where one unit is equal to the chord length across one circle division).
3. Continue measuring each succeeding tangent at increasing lengths until the involute has been constructed to the desired size.

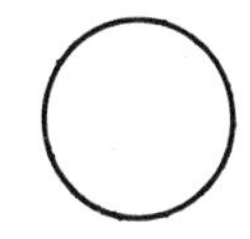

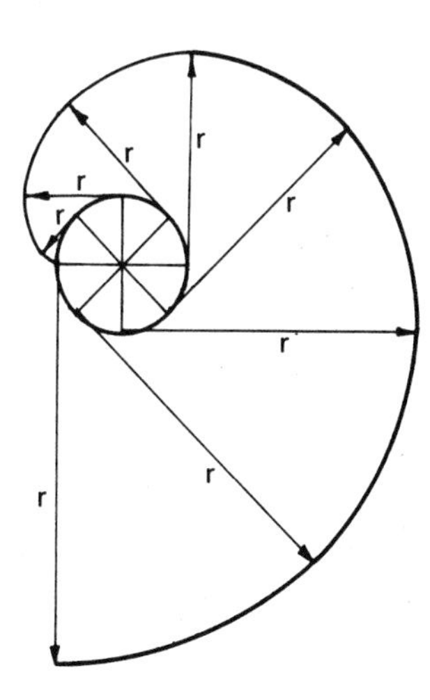

Fig. 5-98. Involute of a circle.

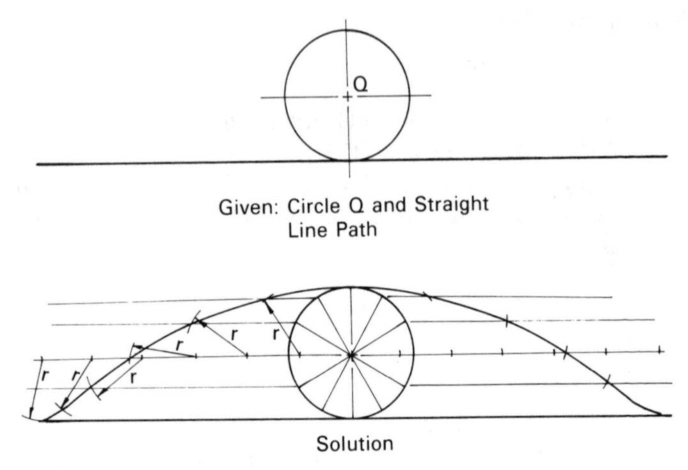

Fig. 5-99. Drawing a cycloid.

DRAWING A CYCLOID

A cycloid is developed by a point that is located on the circumference of a circle. As the circle rolls along a straight path, the point will generate a plane curve called a cycloid.

Given: Circle Q and straight-line path

Procedure (Fig. 5-99):

1. Divide circle Q into an equal number of parts, and number each.
2. Lay off the circumference of the circle on the straight-line path and divide it into the same number of equal parts, and number each.
3. Through each division on the circle, draw a line parallel to the straight-line path.
4. From each division on the straight-line path, draw a perpendicular until it intersects its corresponding numbered line.
5. The cycloid will be constructed when all plotted points have been connected by a smooth curved line.

DRAWING AN EPICYCLOID AND HYPOCYCLOID

The construction of epicycloids and hypocycloids employs procedures similar to that used for the construction of a cycloid. Epicycloids are curves that are generated by a point on the circumference of a circle as it rolls on the *outside* of a fixed circle. Hypocycloids, on the other hand, are curves formed by a point on the circumference of a circle as it rolls on the *inside* of a fixed circle. The procedure presented here can be used for either situation.

Given: Circle O and circular path

Procedure (Fig. 5-100):

1. Divide circle O into a number of equal parts, and number each.
2. Measure the circumference of circle O along the circular path, and divide it into the same number of equal parts and number each.
3. From the center point of the circular path, draw parallel arcs through each division on circle O.
4. From the center point of the circular path, draw a straight line through each numbered division on the path until it intersects its corresponding numbered arc.
5. The epicycloid, or hypocycloid, will be constructed when all points have been connected by a smooth curved line.

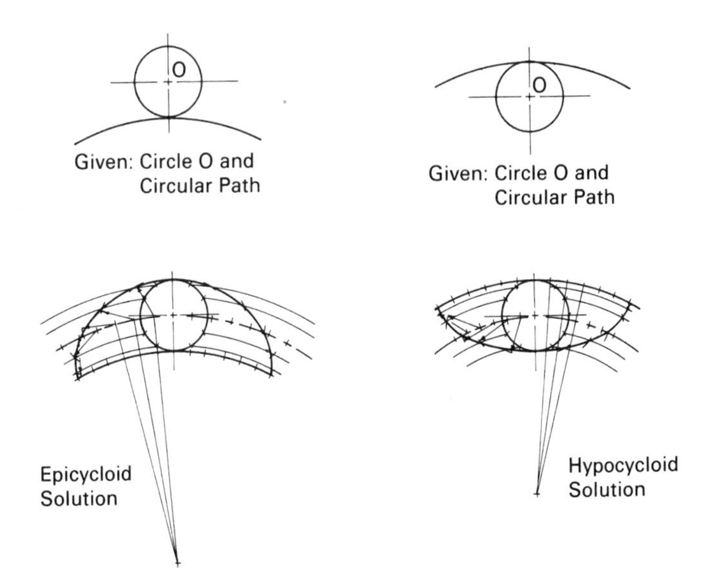

Fig. 5-100. Drawing an epicycloid and hypocycloid.

Exercises

5.1 Draw a line 5.5 inches long and divide it proportionally in the ratio 1:2:3.

5.2 Construct a regular hexagon and octagon having an across-flats distance of 4 inches.

5.3 Inscribe a regular pentagon in a circle with a diameter of 70 mm.

5.4 Draw two vertical lines 3 inches apart. Locate two points, one on each line, that are 4 inches apart vertically. Draw an ogee curve tangent to the parallel lines.

5.5 Construct two ellipses, both having a major diameter of 90 mm and a minor diameter of 50 mm. For the first ellipse, use the trammel method, and the concentric circle method for the second ellipse.

5.6 Construct a parabola with a vertical axis where the focus will be 25 mm from the directrix. Select any point on the parabola and draw a tangent to it.

5.7 Construct a hyperbola that has a transverse axis of 1.5 inches and foci 2 inches apart.

5.8 Construct the asymptotes for the hyperbola in problem 5.7.

5.9 Construct the involute of a regular pentagon having sides equal to 20 mm.

5.10 Construct the involute of a circle whose diameter is 3 inches.

5.11 Construct a cycloid generated by an 80-mm diameter circle.

5.12 Construct the hypocycloid generated by a 2-inch circle rolling on a 5-inch circle.

5.13 Construct an epicycloid generated by the same circle in problem 5.12.

CHAPTER 6

Orthographic Projection

- Principles of Orthographic Projection
- Orthographic Projection Theory
- ISO Standards
- Relationship of Points, Lines, and Surfaces
- Arrangement of Views
- Interpreting Drawings
- Exercises

Industrial drafting departments are frequently involved in the design, development, and construction of machines, structures, and products. To communicate these designs to engineers, technicians, and machinists, descriptions must be prepared that give every physical detail for all parts of the machine, structure, or product. Because of the complexity of this requirement, drawings are used as the primary method of communication.

Product shapes and specifications are presented in drawings by using a graphic procedure known as *orthographic projection*. This projection technique is the basis upon which industrial drawings are prepared world-wide. This chapter will discuss orthographic projection theory and techniques as applied to the drafting process.

Principles of Orthographic Projection

Graphic projection is a process whereby a drawing is made by rays of sight that are "projected" onto a viewing or picture plane. The only exception is in the graphic science of *projective geometry*, in which nonplane surfaces are used. The method of projection used by drafters will be determined by what type of drawing will best represent the design (e.g., orthographic or pictorial).

Three basic methods of projection are used by drafters. The first and most common is *orthographic*, in which the rays of sight are perpendicular to the picture plane; "orthographic" is derived from the Greek word *orthogonios*, which means to be perpendicular. Thus, orthographic views are placed perpendicular to one another.

If the rays of sight are nonperpendicular to the picture plane, the projection is classified as *oblique*, and shows a pictorial presentation of the object. Rays of sight that are projected to a given station point are classified as *perspective projection*, and present the object as it would appear to the human eye. This chapter, however, will focus upon orthographic methods of drawing.

Orthographic Projection Theory

To prepare industrial working drawings correctly, drafters must have a clear understanding of orthographic projection theory. In a typical working drawing (Fig. 6-1), the object is presented in a set of two or more separate views. Each view presents the object as it is observed from a different direction (usually perpendicular to each other). All the views taken together should describe the object completely.

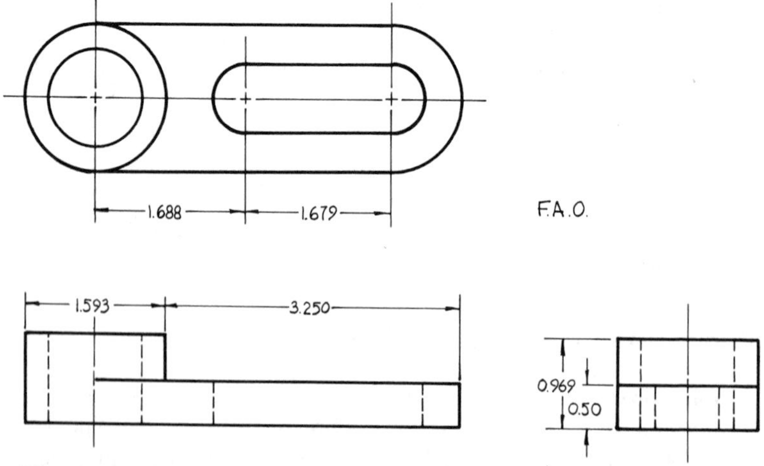

Fig. 6-1. Typical working drawing using orthographic projection techniques.

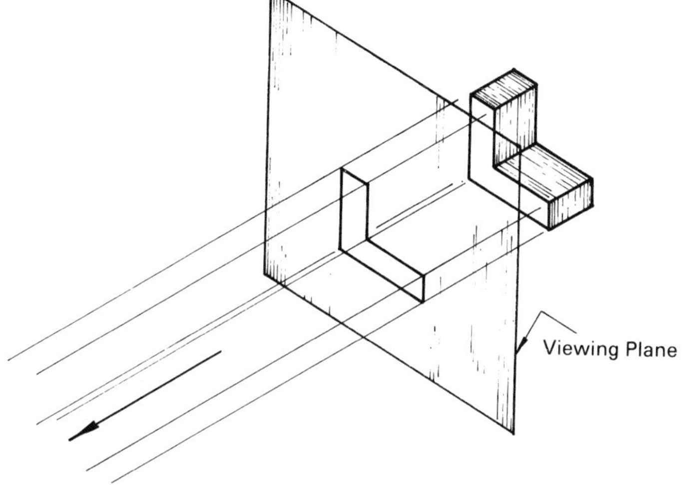

Fig. 6-2. Orthographic projection onto a viewing plane.

The concept of orthographic projection is illustrated in Fig. 6-2. Here an object is placed behind a transparent viewing plane. If one were to project lines of sight perpendicular to the viewing plane, these rays would be parallel to one another. Where they intersect the picture plane would be where the orthographic view would be observed.

During actual drawing procedures, projection rays are discarded and the view is *thought* of as being developed by the use of perpendiculars to the picture plane. The "hinge" or cutting plane lines that represent the intersection of the viewing planes are also eliminated from most drawings. It should be noted, however, that sometimes these lines are necessary for solving graphic problems.

An orthographic view that shows the object as viewed from the front is projected onto a frontal plane. The frontal view will give information relative only to length and height measurements. Alone, the frontal view would not communicate sufficient information to describe a three-dimensional form, since it does not give information as to the object's width or details along surfaces perpendicular to the frontal plane. To acquire this information, additional picture planes are used (Fig. 6-3).

The *horizontal plane* is located perpendicular to the frontal plane, and above the object. The horizontal view, sometimes referred to as the top view, is obtained by projecting rays of sight 90° from the ob-

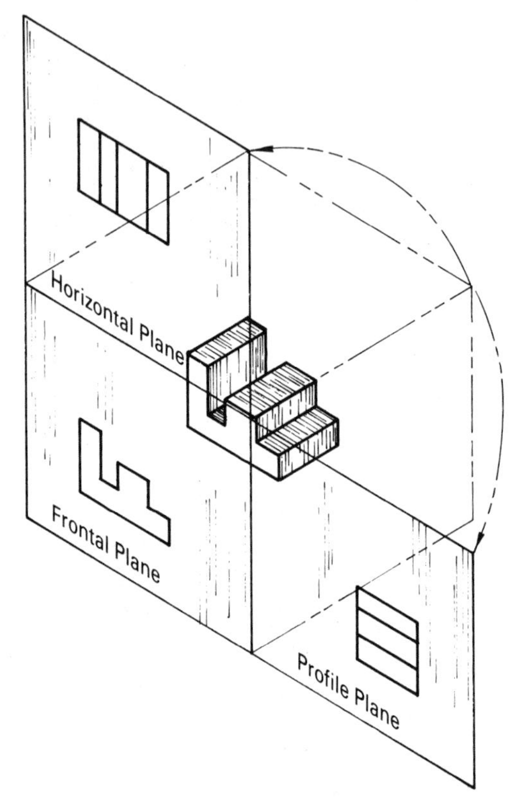

Fig. 6-3. Use of additional planes for interpreting object presentation.

ject to that plane, and presents both length and width (depth) dimensions. Another plane incorporated in drawings is the *profile plane.* This plane is perpendicular to both the horizontal and frontal planes and presents both height and width dimensions. When all three views are used, all the information relative to the three-dimensional shape of the object will be presented.

Principal Views of Projection

If we carry the concept of projection a bit further, we see that it is possible to develop a total of six planes of projection and views that represent each side of the viewing box: frontal, horizontal, bottom, rear, right profile, and left profile planes (Fig. 6-4). To obtain all six

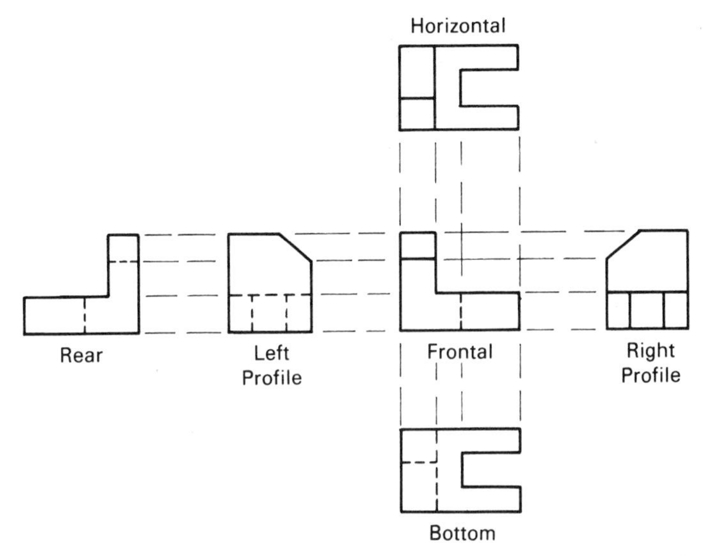

Fig. 6-4. Relative positions of the six principal views of projection.

views, the same projection procedure would be used as described for the frontal, horizontal, and profile views (i.e., perpendicular projection rays to the viewing planes). These six views are known as the principal views of projection.

The most common combination of views used in industrial drawings is front, top, and right side views (Fig. 6-5). Sometimes, however, other views can better describe the shape of the object. When other views are employed, their correct placement is critical for understanding the drawing. View placement therefore *must* follow the principles established by the theories of orthographic projection. These can be summarized as follows:

1. The top and bottom views should always be placed above and below, respectively, the frontal view.
2. If the frontal view is not used, the top and bottom views should be placed above and below, respectively, the rear view.
3. The right and left profile views should always be placed to the right and left, respectively, of the frontal view.
4. If the frontal view is not used, the right and left profile views should be placed to the left and right, respectively, of the rear view.

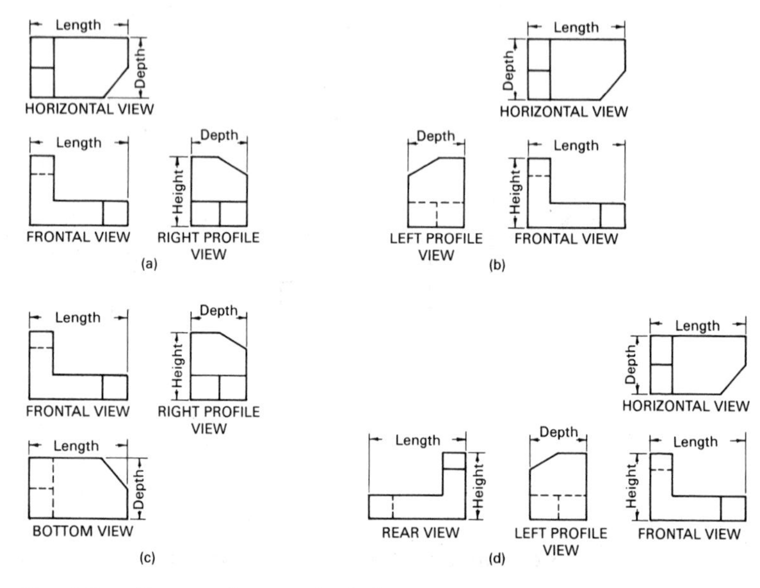

Fig. 6-5. Proper placement of views.

ISO Standards

The method of orthographic projection presented thus far is the principal system used in the United States, Canada, and most industrialized countries. This projection system is frequently referred to as *third angle orthographic projection.* It is also known as *American projection,* since it was first widely used in the United States. Until the recent updating of ISO standards (ISO 128-1982 [E]), this system was also referred to as *method A projection.*

To review briefly, third angle projection places the object in a viewing box whose sides are made up of projection planes. Each view is then developed by projecting rays perpendicular to the planes. Since there are two projection techniques equally acceptable by ISO standards, it is necessary to use the appropriate ISO projection symbol on the drawing, so that the person reading it will be able to interpret the drawing according to that particular technique. The two ISO symbols used to indicate projection are shown in Fig. 6-6.

The second method of projection that is recognized by the ISO is known as *first angle projection.* This technique of orthographic

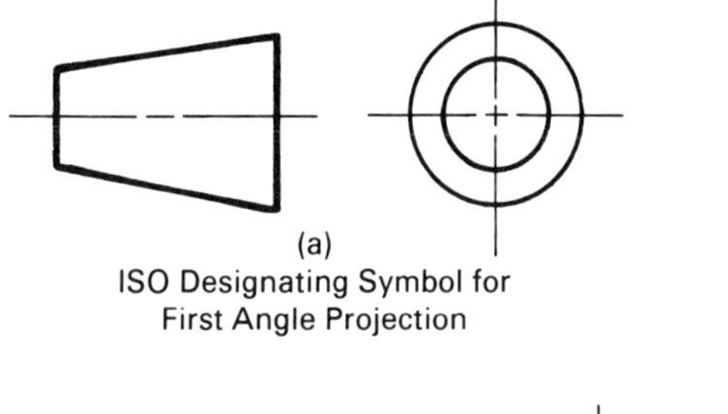
(a)
ISO Designating Symbol for
First Angle Projection

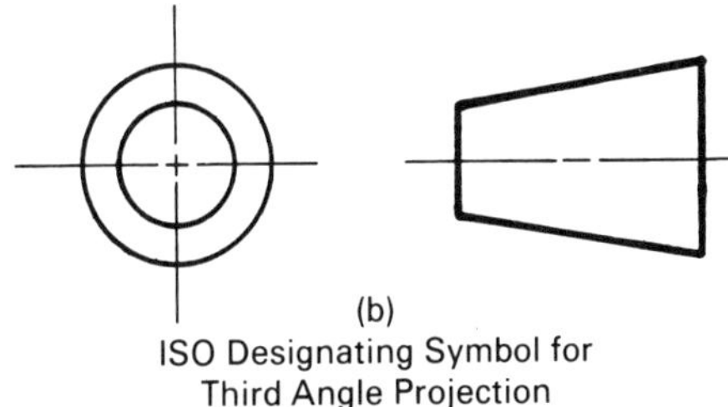
(b)
ISO Designating Symbol for
Third Angle Projection

Fig. 6-6. Designating symbols specified by ISO standards.

drawing is most commonly used in Europe and some of the developing countries. Here, the object is "rolled over" from one side to the next, so as to have the right profile view on the left side, and the top view on the bottom. Shown in Fig. 6-7 and described in Table 6-1 are comparisons of third and first angle projection systems for the same object.

Table 6-1 Comparison of First and Third Angle Projection Systems as Specified by ISO 128-1982 (E)

	Proper Placement of View with Reference to Front View	
	Projection System	
View	*First Angle*	*Third Angle*
view from above	placed underneath	placed above
view from below	placed above	placed underneath
view from left	placed on the right	placed on the left
view from right	placed on the left	placed on the right
view from rear	placed on the left or right	placed on the left or right

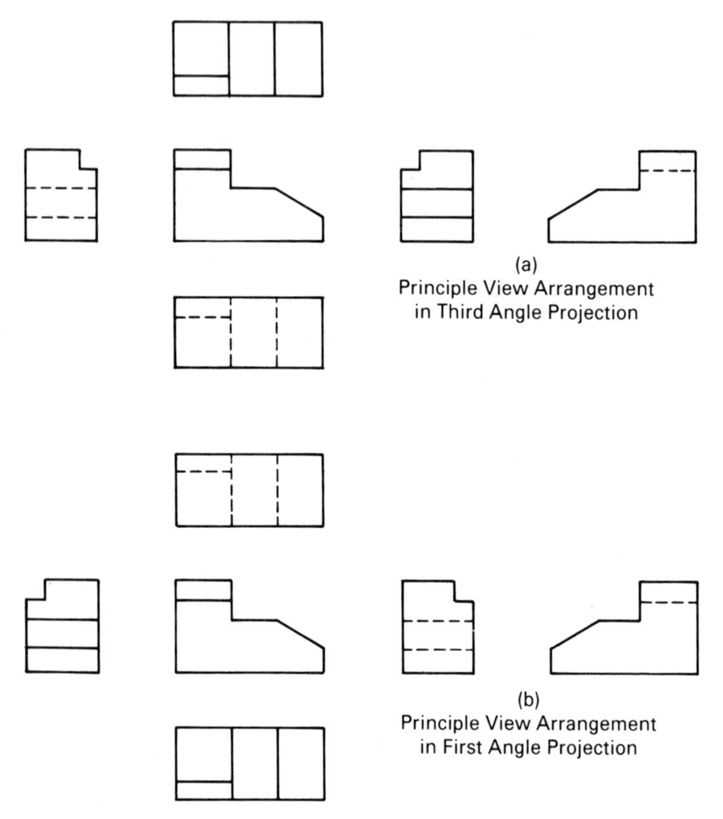

Fig. 6-7. First and third angle projection methods compared.

The terms "first angle" and "third angle" projection refer to the *quadrant number* in which the object is placed in for viewing. Fig. 6-8 shows how the views are developed for each projection method. In third angle projection, the object is placed in the third quadrant so that the views are projected to the appropriate planes. In first angle projection, however, the viewing planes are reversed, so that to view the top and right side of the object, the left profile and bottom planes, respectively, must be used.

Of the two methods of projection, third angle is by far the most widely used. In fact, this technique is recognized and used in many European industries, even though first angle projection was the standard in Europe for many years. Some international companies are

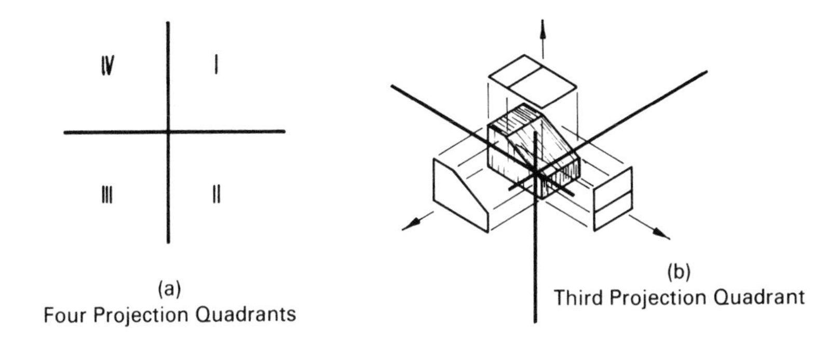

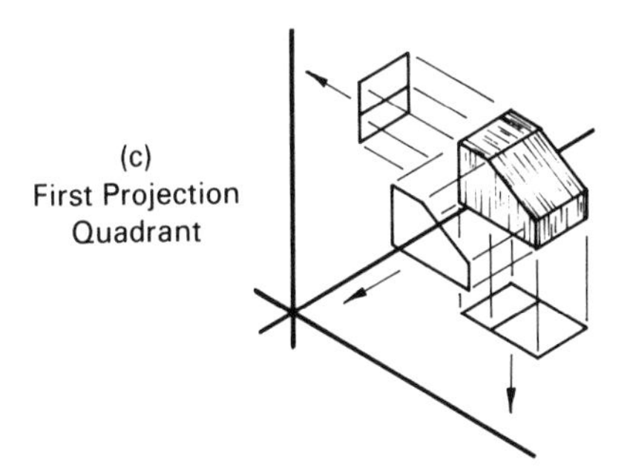

Fig. 6-8. Theory of first and third angle projection.

in the process of converting all their first angle projection drawings to third angle. Within the ISO community, however, the two orthographic projection methods are of equal standing. Drafters working for firms involved in international work should thus be familiar with both techniques.

Relationship of Points, Lines, and Surfaces

To solve drawing problems, drafters must be able to visualize the problem and its solution—that is, how the object appears, and its relationship to the principal planes of projection. The solution of drawing problems, then, is a combination of drawing procedures and visualization.

Points and Lines

A point is used to indicate or show a location in space. When shown in a drawing, this nondimensional element should be labeled appropriately. For example, assume that we have point H, which is shown in the frontal, horizontal, and profile views. The proper notation for this point in each view is H_F, H_H, and H_P, respectively.

Like points, lines must also be properly identified and noted. Since lines have dimension (length), it is critical to note their relationship to the planes of projection. Thus, a view of a line may or may not present it in its true length; in many cases, lines will appear foreshortened. When a line does appear in its true length, it should be labeled "TL" for True Length. This label should be used when solving graphic problems, but is not used in the preparation of most detail and assembly drawings.

It is easy to determine if a line is true length in a given view; simply draw the hinge or cutting plane line. If a given line appears parallel to the drawn cutting plane line, then it will be true length in the adjacent view. Sometimes a line appears as a point; then the line will appear true length in the adjacent view (Fig. 6-9).

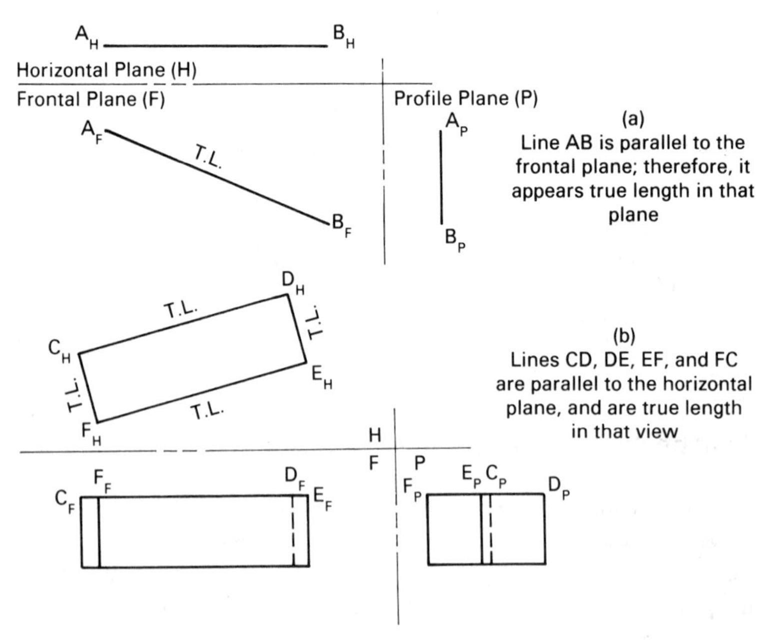

Fig. 6-9. True length of lines.

The notation used for lines will follow the same logic as used for point notations. Hence, line K_HL_H represents the view of line KL in the horizontal plane. Lines are often classified according to their true length specification, that is, the view in which they will be viewed as true length. For example, if a line is true length in the front view, then it must be parallel to the frontal plane of projection and will be called a frontal line. There are six classifications of lines (Fig. 6-10):

1. *Principal lines* are those lines that will appear true length in at least one of the principal views of projection.
2. *Horizontal lines* are those that are true length in the horizontal view.
3. *Frontal lines* are lines that are true length in the frontal view.
4. *Profile lines* are those lines that will appear true length in the profile view.
5. *Vertical lines* are lines that will appear true length in all vertical views of projection (frontal, left and right profile, and rear views), and will be shown as points in the horizontal and bottom views.
6. *Oblique lines* are lines that are not shown in their true length in any of the principal views of projection.

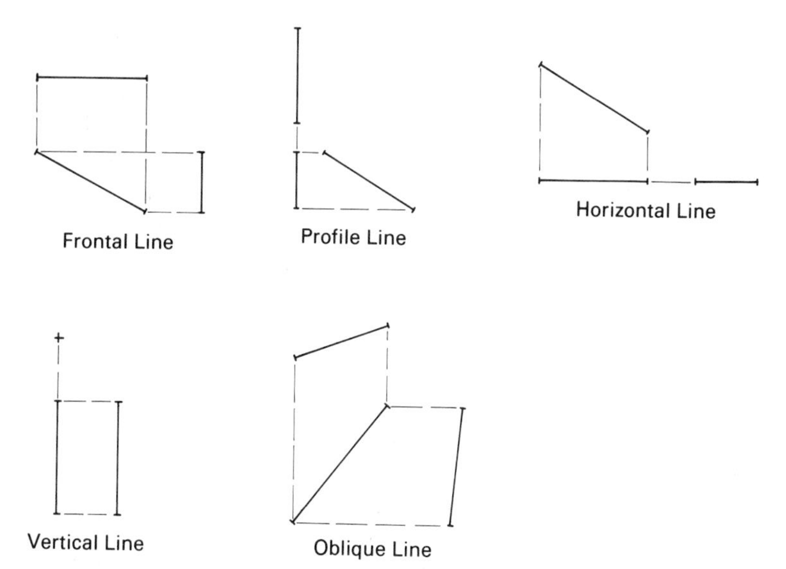

Fig. 6-10. Classification of lines.

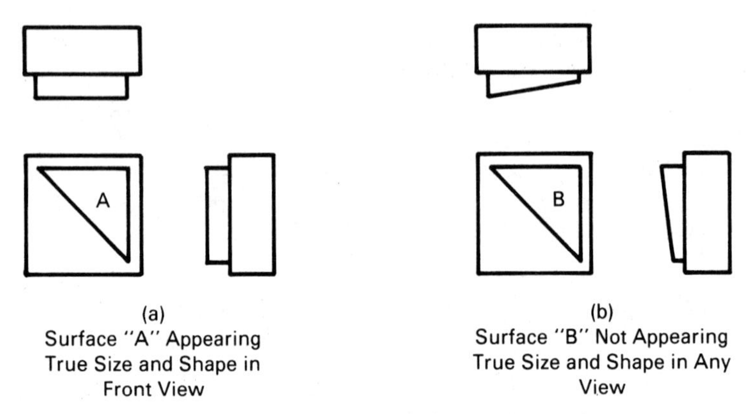

(a)
Surface "A" Appearing True Size and Shape in Front View

(b)
Surface "B" Not Appearing True Size and Shape in Any View

Fig. 6-11. Determining true size and shape of a surface.

Surfaces

Like lines, surfaces have dimensional qualities. Their relationship to the views of projection are therefore critical to their interpretation. When a surface (plane) is parallel to a view of projection, it will appear true size and shape, and is sometimes noted as "TS&S" or "TS."

One quick method for determining if a surface appears in its true size and shape in a particular view is to look at an adjacent view. If that surface appears as a straight line, then it will be true size and shape in the projection plane to which it is parallel. For example, in Fig. 6-11, surface A appears as a line in the horizontal and profile views, meaning that it is parallel to the frontal plane of projection, and it will appear in its true size and shape in the front view. By comparison, surface B does not appear as a line in any of the views, since it is not parallel to the planes of projection. Surface B, then, is not seen in its true size and shape in any of the views presented.

Arrangement of Views

The selection and arrangement of views for a product are important for easy and accurate drawing interpretation. An improperly selected view can confuse any engineer, technician, or machinist, since it has no logical relationship to the drawing.

The first thing that drafters must decide is which view of the object best shows the product. Whichever view this is should be made the front view. On the drawing sheet the front view is usually placed in the lower left portion of the page. Once that is placed, all other views may be positioned.

Guidelines that are usually followed in the selection of the front view include:

1. The view that best shows the physical characteristics of the product should be designated as the front view.
2. The front view should also be the view that gives the most detail and information about the object's features.
3. In many cases, the longest surface is also the most important; the front view will often represent this dimension.
4. Objects that have a specific functional position should be shown in that position in the front view whenever possible.
5. Exact placement of the front view will be influenced by the other views selected for illustration.

Once the front view is identified, the next to be determined is the top or bottom view. Only when an object is very complicated and intricate should both top and bottom views be used; usually only one will be drawn. If only the top view is used, then the front view will be located in the lower section of the drawing page. On the other hand, if the bottom view is chosen, then it should be placed in that location.

Remember that all profile views must be located to the side of the front view. A rule of thumb for selecting either the right or left profile view is: the profile view should be that view presenting the best features of the object and having the fewest number of hidden lines. The majority of drawings will use the right profile view, though it may sometimes be necessary to use the left profile view. Shown in Fig. 6-12 is an example of an object whose left profile view is preferable to the right profile view.

There are times when it is necessary to show both left and right profile views (Fig. 6-13a). When the information presented in two views is exactly the same, as in Fig. 6-13b, the duplicate view should be eliminated. This is typically the case with symmetrical objects.

Interpreting Drawings

There is no "one right way" of reading drawings, owing to the wide range of types of drawings employed in industry. There is, however,

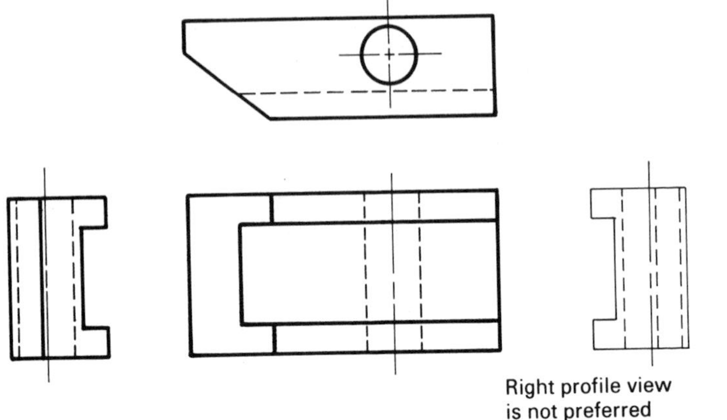

is not preferred
because of the additional

which adds more confusion

Fig. 6-12. Preference of left profile view for clearer interpretation of drawing.

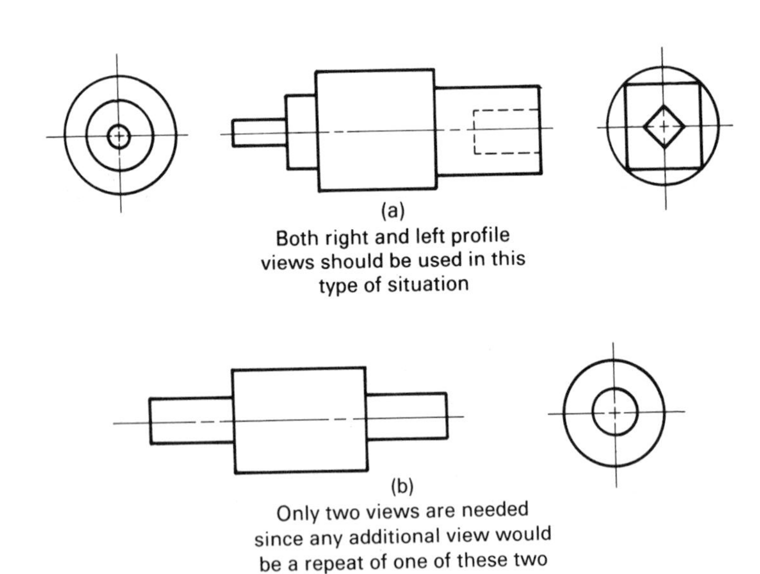

(a)
Both right and left profile
views should be used in this
type of situation
(b)
Only two views are needed
since any additional view would
be a repeat of one of these two

Fig. 6-13. Examples of view selection.

a guideline that should be followed in reviewing a set of drawings. It comprises six basic steps, as follows:

1. Look at the overall appearance of the drawing, and identify the views presented.
2. Determine the general shape of the object. Look at each view and try to visualize the object by studying its major features and shapes, and their relationships.
3. Read all features and shapes, beginning with the most important and proceeding to lesser ones. It is also easier to start visualizing the simpler features first, and then progressing to the more complex. Pay particular attention to holes and slots, and feature and location dimensions.
4. Characteristics that are unfamiliar, complex, or difficult to interpret should be read carefully. You should be able to find the location of every point, line, and surface in all views.
5. Identify the functional and structural relationships between all the elements of the object.
6. Any portion of the drawing that is not understood or is unclear should be reread and studied until all the features of the object can be clearly visualized.

Interpretation of Lines

A line that appears in a view will have one of three meanings; it can be:

1. The edge of a surface or plane
2. The intersection of two or more surfaces
3. The limit of a surface

Since a line can represent any of these three conditions, it is important that the drafter and reader make use of other views for interpretation.

Fig. 6-14 illustrates the need for consulting more than one view. In the first drawing (Fig. 6-14a), line AB in the front view cannot be properly defined until the profile view is examined. AB can now be identified as an edge view of a surface. In the second drawing (Fig. 6-14b), by comparison, line CD in the front view is a surface. Even though the front and top views in both drawings appear the same, they represent two different objects, as is made clear in their profile views.

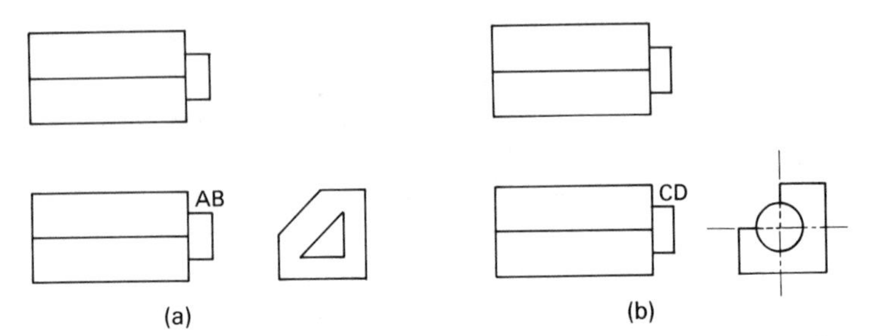

Fig. 6-14. Consulting more than two views for proper interpretation.

The front views of lines AB and CD in the two drawings also appear to be exactly alike. When the profile views of each drawing are examined, however, it is apparent that line AB in the first drawing is a corner formed by the intersection of two surfaces, while CD in the second drawing is the limit of a surface.

Interpretation of Surfaces

In this discussion, the term "surface" pertains to any area, contour, plane, or combination of surfaces that can be shown in orthographic views. It should be remembered that a surface that appears in its true size and shape in a view will appear as a line in its adjacent orthographic view. Furthermore, every surface, regardless of its characteristics, will appear as either a line or an area in all principal views of projection. Shown in Fig. 6-15 are examples of four different surface presentations.

It is impossible for two or more adjacent surfaces to be located in the same plane, because there would be no visible separation or boundary; one would be a continuation of the next. When adjacent surfaces are in different planes, their relationship will be constant in all views.

Reading Drawings by Sketching and Modeling

Experienced drafters, engineers, technicians, and machinists have little difficulty in interpreting working drawings. From years of drawing, reading, and interpretation, they are able to read a drawing and quickly interpret its meaning and the physical characteristics of the object. For the beginner, however, this often proves difficult and frustrating.

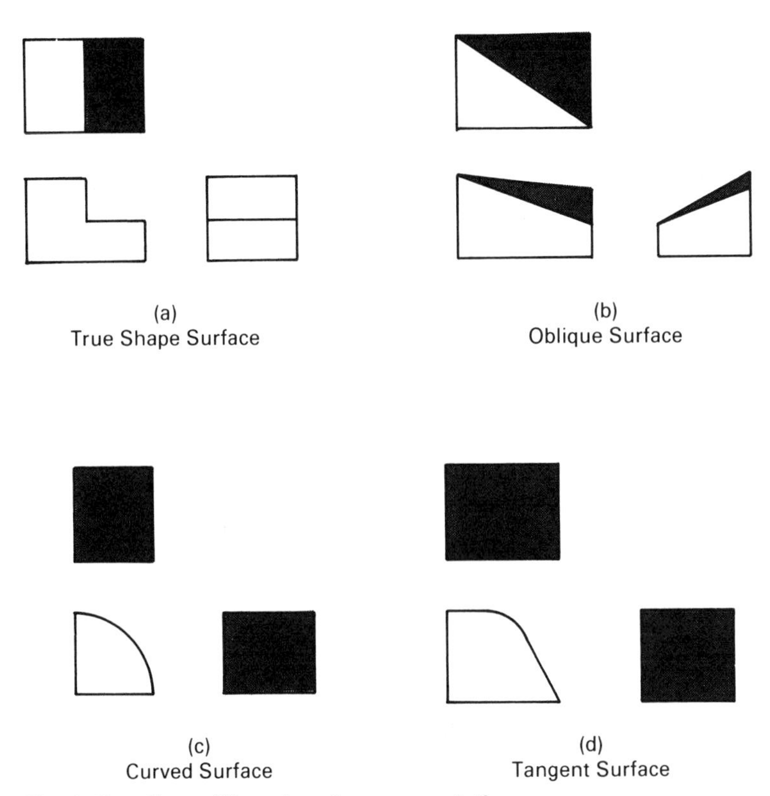

Fig. 6-15. Four different surface presentations.

There are two techniques that are sometimes used by both students and beginners alike. The first method is making a pictorial sketch of how the object appears when the information provided in all views is used. Before outlining the sketching procedure, it is first necessary to discuss several preliminaries. The first is the use of a pictorial box drawn from three axes, one vertical and the other two at 30° to it (Fig. 6-16). Each axis is measured off at the appropriate measurements that correspond to the overall length, width, and height of the object.

When the box is drawn, it is now possible to make the sketch. The steps to follow are shown in Fig. 6-16, and include:

1. On the top surface of the box, lightly draw in the same lines that appear in the top view of the drawing.

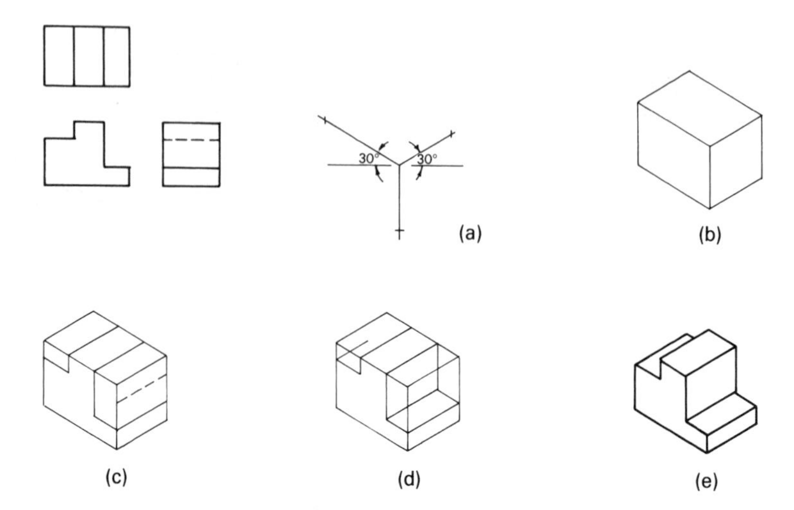

(a) Draw axis at 30° to vertical
(b) Sketch block to proportion
(c) Draw views on appropriate faces
(d) Begin to "Cut-Away" areas
(e) Complete by erasing any lines or edges that do not appear in finished object

Fig. 6-16. Using pictorial sketching for interpreting orthographic views.

2. Repeat the same procedure for the front and profile surfaces by drawing the lines shown in the front and profile views of the drawing.
3. Note which lines will not be in the same plane as shown in each view. Start to "cut" into the box, making the visible corners and edges dark.
4. Erase any lines or edges that do not appear as visible object lines in the sketch. If hidden surfaces are needed to interpret the object, they may be drawn in lightly as hidden lines.
5. Review the sketch to see that it corresponds to the views given in the drawing, and make any necessary corrections.

The second method used for interpreting drawings is modeling, a technique often employed in apprenticeship training programs, in which the object is modeled out of clay or modeling wax. The con-

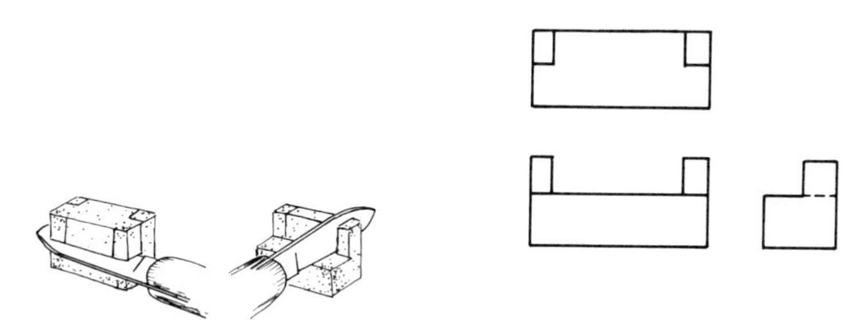

Fig. 6-17. Using clay to create three-dimensional Model from Orthographic Views.

cept here is the same as that used in sketching, except that a three-dimensional form is used (Fig. 6-17). The procedures are as follows:

1. Begin with a rectangular block of clay or modeling wax that has the same proportions as the outside dimensions of the object.
2. Using a knife or carving razor, "draw" in the lines on all surfaces as they appear on the drawing.
3. Begin to remove the material from the block of clay in successive cuts until the model is completed.

In addition to using clay or modeling wax, it is possible to build models by using various geometrically shaped blocks. A major advantage of modeling over sketching is that it will give a better representation of how the object will appear, including areas that are hidden from view.

Exercises

6.1 Sketch the proper orthographic views for the six objects illustrated in Fig. 6-18.

6.2 Draw the necessary views for the four objects illustrated in Fig. 6-19.

6.3 Prepare a pictorial sketch for each of the drawings shown in Fig. 6-20.

6.4 Make a clay model for each of the drawings shown in Fig. 6-20, and compare them with the sketches drawn in problem 6.3.

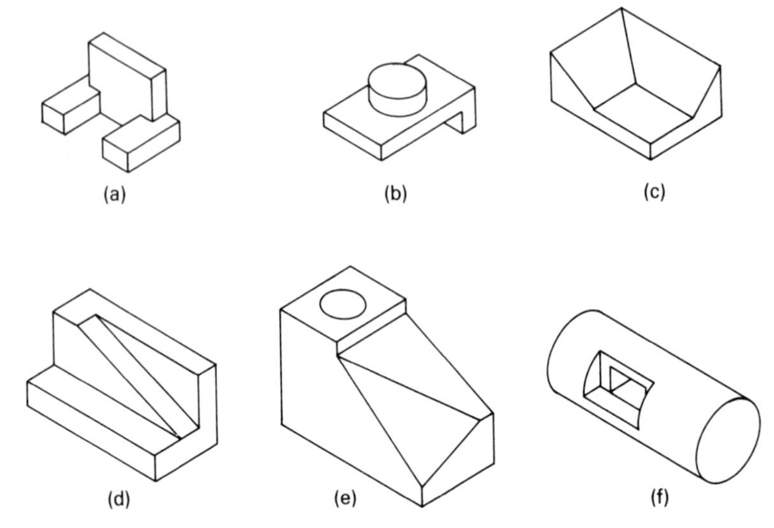

Fig. 6-18. Problem 6.1.

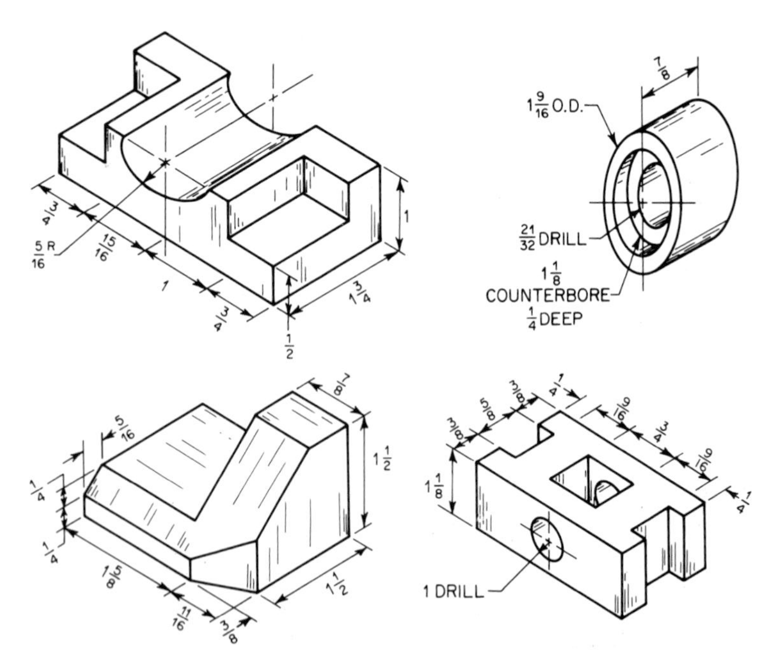

Fig. 6-19. Problem 6.2.

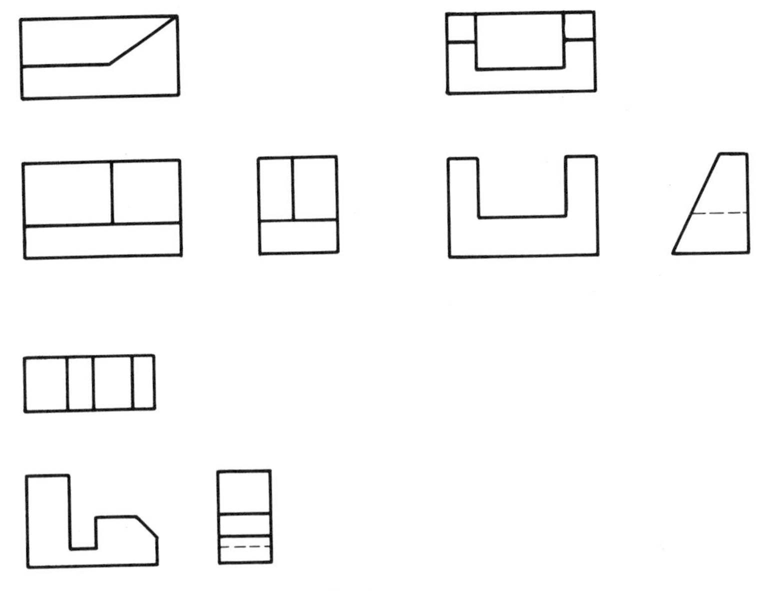

Fig. 6-20. Problems 6.3 and 6.4.

CHAPTER 7

Auxiliary Views

Many products and machine parts have surfaces that are not perpendicular to any of the principal planes of projection. As mentioned in the previous chapter, these surfaces are known as either inclined (sloped) or oblique planes. In each of the principal viewing planes, these surfaces would appear as foreshortened or not of true size and shape.

When the characteristics of oblique and inclined surfaces are important and their true size and shape must be determined, a special viewing technique must be employed. This form of presentation is known as an auxiliary view. This chapter will discuss the principles and practices used by industrial drafters when drawing auxiliary views.

Principles of Auxiliary Views

The basic theory employed in projecting principal plane views also applies to auxiliary views. By definition, an auxiliary view is an imaginary projection plane that is drawn at a given angle to a point, line, or surface. Thus, an auxiliary view would show how an object would appear if observed from that angle.

Basic Concepts

A line will be observed in its true length when the direction of view is 90° to it. For example, in Fig. 7-1, line AB is shown in its true length

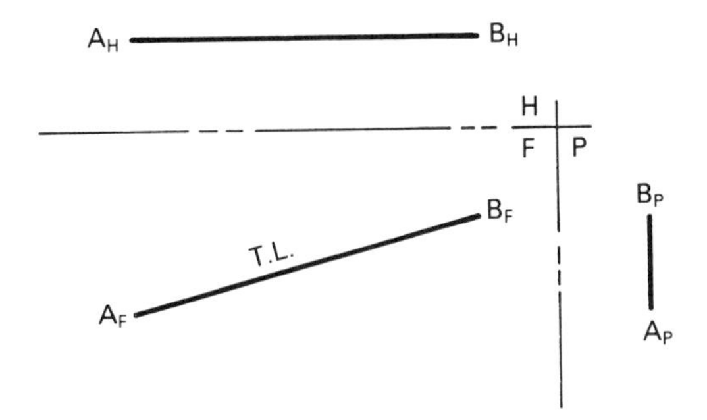

Fig. 7-1. True length line presentation.

in the frontal view because it is viewed perpendicularly to that plane. This can easily be determined by noting that AB is parallel to reference plane lines HF and PF.

A common situation encountered in industrial drawings is presented in Fig. 7-2. In this example a line is oblique, yet we must determine its true length. Since the true length of line CD will not be observed in any of the principal planes of projection, an auxiliary plane is employed to project its true length. The procedures used here are as follows:

1. Select a view from which an auxiliary view is to be projected (frontal, in this case).
2. Since a true-length view of a line is shown when the cutting plane is parallel to it, draw auxiliary reference plane line, F-1, parallel to the frontal view of line CD.
3. Project C and D perpendicular to line F-1.
4. To find the location of C and D in the auxiliary view, measure the distance of these points behind the frontal plane in either the horizontal or profile view, and transfer to the auxiliary view. That is the distance from F-H or F-P to points C and D.
5. The drawn line C_1D_1 will be the true-length view of CD.

In addition to finding the true length of lines, auxiliary views are also used to obtain the true shape of a surface. An example of the auxiliary view procedure used to obtain the true size and shape of a surface is shown in Fig. 7-3. Here we have three views of an object that has an inclined surface ABCD. Owing to the angle of inclination, surface ABCD will not appear in its true shape in any of

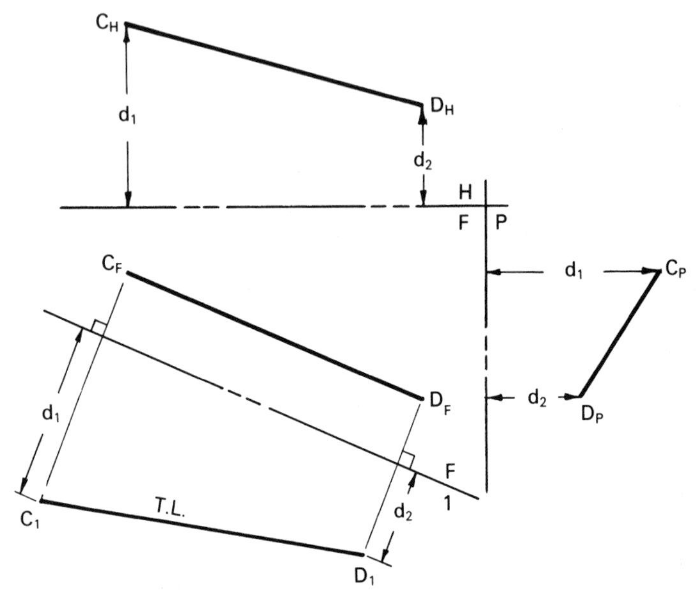

Fig. 7-2. Finding the true length of an oblique line.

the principal views of projection. To find the true shape of this surface, it is necessary to use an auxiliary plane where the lines of projection are perpendicular to it.

The procedure employed to find the true size and shape of surface ABCD is based upon the same principles as those used in determining the true length of a line. What must be remembered here is that a surface will appear in its true shape in a viewing plane that is parallel to an edge view of the surface. Since surface ABCD already appears as an edge view in the frontal plane, all that is needed to find its true shape are the following procedures:

1. Draw the auxiliary reference plane line parallel to the edge view of ABCD, and label it F-1.
2. Project A, B, C, and D into the auxiliary view so that the projectors will be 90° to F-1.
3. Determine the distance of A, B, C, and D behind the frontal plane by measuring its distance into the horizontal or profile planes from reference plane lines F-H or F-P.
4. When A_1, B_1, C_1, and D_1 are connected, the true size and shape of surface ABCD will be shown.

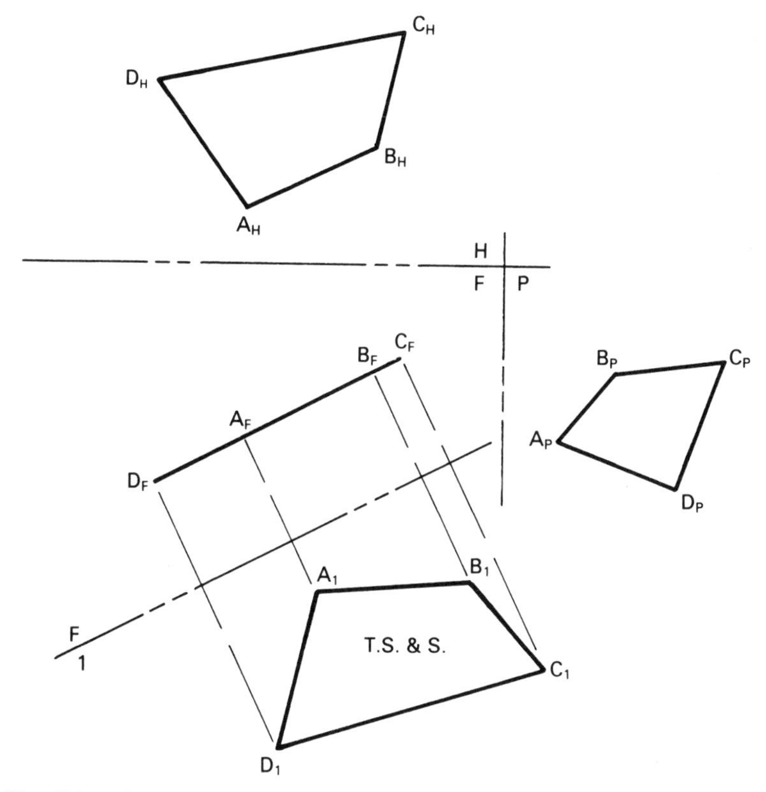

Fig. 7-3. Finding the true size and shape of an inclined plane.

Types of Auxiliary Views

There are three basic types of auxiliary view: frontal, horizontal, and profile. Each type is dependent upon the principal plane of projection that is used to develop the auxiliary view. Fig. 7-4a shows an auxiliary view projected from the frontal view. Here the distances used to determine points behind the frontal plane can be taken from either the horizontal or profile view. In this case, a projection plane reference line is not used.

Fig. 7-4b is an example of the second type of auxiliary view. Here the auxiliary plane is perpendicular to the horizontal plane and inclined in the frontal plane. The auxiliary view is projected from the horizontal view, and the measurements used to determine distance behind the auxiliary plane line are taken from the frontal view.

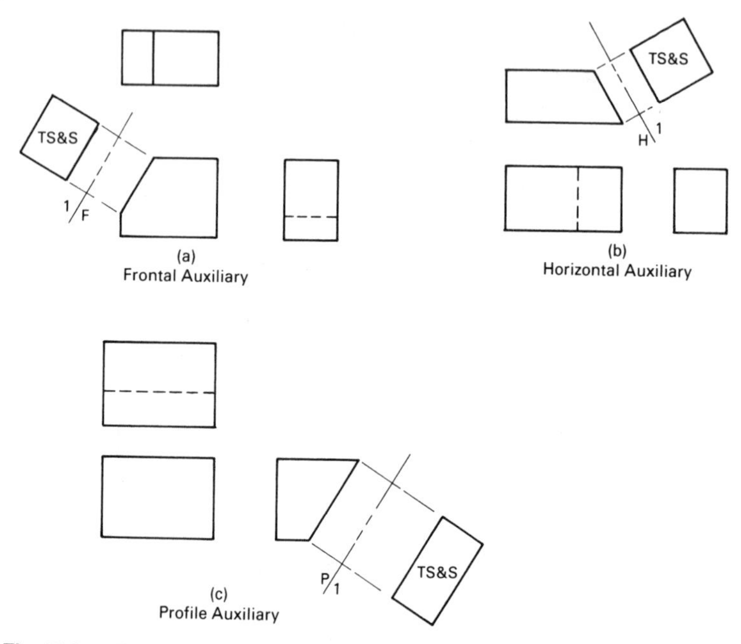

Fig. 7-4. Three types of auxiliary view used to find the true size and shape of an inclined surface.

The third type of auxiliary view is shown in Fig. 7-4c. The auxiliary view is projected from the profile plane of projection, and its distances are measured from the frontal plane. The construction procedure used is similar in all three cases, and is based upon the principal viewing plane to which it is perpendicular.

Primary Auxiliary Views

All surfaces appearing as inclined in principal viewing planes will appear as true size and shape in a primary auxiliary view. Industrial drawings incorporating auxiliary view will normally show only a *partial auxiliary view* that shows only the inclined surface. The reason for this is that the projection of the entire object into the auxiliary view would not only be time-consuming, but would rarely add anything useful to the description of the object. Fig. 7-5 illustrates the advantage of a partial auxiliary view.

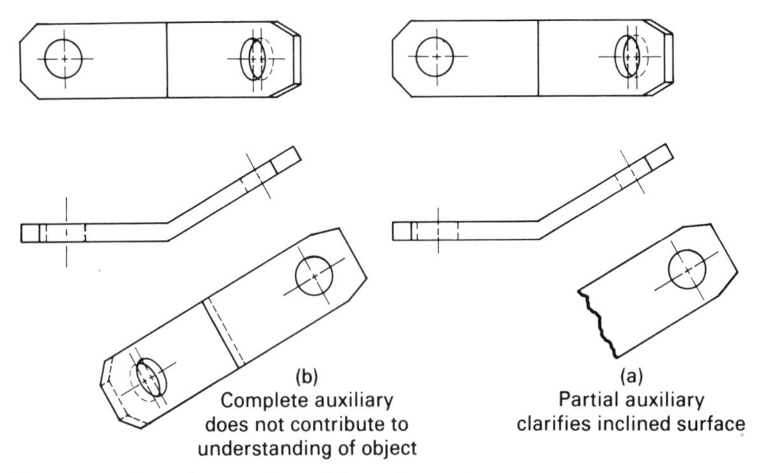
(b)
Complete auxiliary
does not contribute to
understanding of object
(a)
Partial auxiliary
clarifies inclined surface

Fig. 7-5. Complete and partial auxiliary views.

Symmetrical Auxiliary Views

A view of an object will present it as either symmetrical or unsymmetrical. Symmetrical auxiliary views differ from unsymmetrical views in that they are drawn about a center or reference line, such as a circle or ellipse. There will thus be "matching halves" on each side of the center line.

When an inclined surface is known to be symmetrical, it is drawn in reference to the center line. In other words, the center line becomes the reference plane line. Since the center line functions as a reference line for the auxiliary view, it may be located at any distance from the principal view. It is advisable, however, to locate it so that the finished view will be approximately the same distance from the adjoining view as the distance between all principal views.

Before drawing a symmetrical auxiliary view, the drafter must clearly visualize how the surface appears in true shape. For novice drafters, it is good practice to try and "see" how the object would appear if it is placed in an enclosed glass box, with the sides serving as planes of projection. After the initial visualization, the procedures recommended for drawing a symmetrical auxiliary view are as follows (Fig. 7-6a):

1. Select a location, and draw the reference center line parallel to the inclined surface.

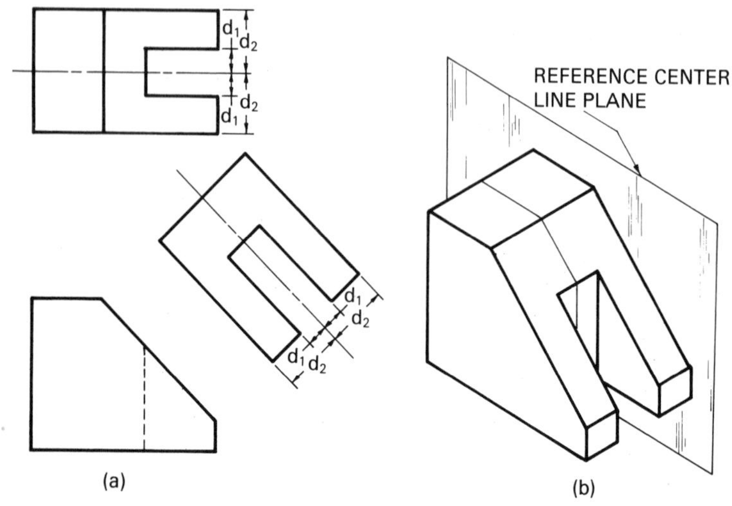

Fig. 7-6. Symmetrical auxiliary view.

2. Draw light projection lines from the primary view to the auxiliary view and perpendicular to the reference center line.
3. Locate each point on the surface by transferring distances from the adjoining primary view to the auxiliary view. These distances should be taken in reference to the center line.
4. Connect the points and darken in the view to complete the auxiliary presentation.

After reviewing this procedure, one should make careful note of the relationship of the reference center line to the primary views (Fig. 7-6b). Here the center line of the horizontal view also serves as the reference plane line of a surface that is parallel to the frontal plane.

Unilateral and Bilateral Auxiliary Views

Two other auxiliary view techniques make use of unilateral and bilateral procedures. Both types of auxiliary view are determined by the placement of the reference auxiliary plane line. Unilateral auxiliaries place the view entirely on one side of the reference line. As shown in Fig. 7-7, the reference line is in contact with the outer sur-

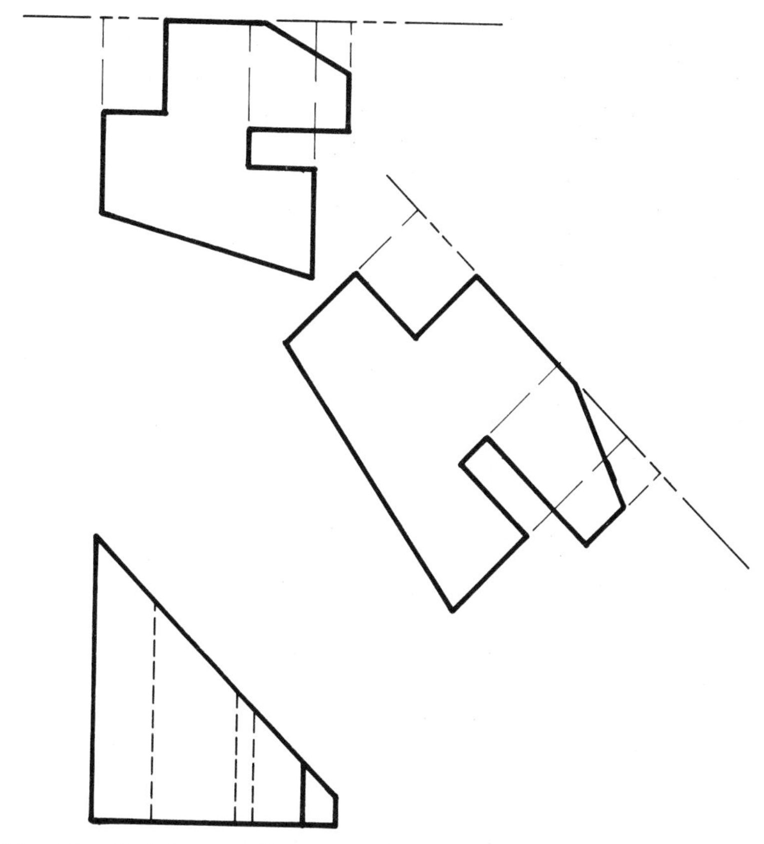

Fig. 7-7. **Unilateral auxiliary.**

face of the object. All the points of the auxiliary are projected in the same manner as in other auxiliary techniques, the primary difference in this example being that the distance measurements are made relative to their location *in front* of the reference line.

Bilateral auxiliary views are similar to unilateral views except that measurements are made on both sides of the reference line (Fig. 7-8). In fact, all symmetrical views are bilateral, but not all bilateral views are symmetrical. The placement of the auxiliary reference line is, again, based upon the proper location and spacing between views.

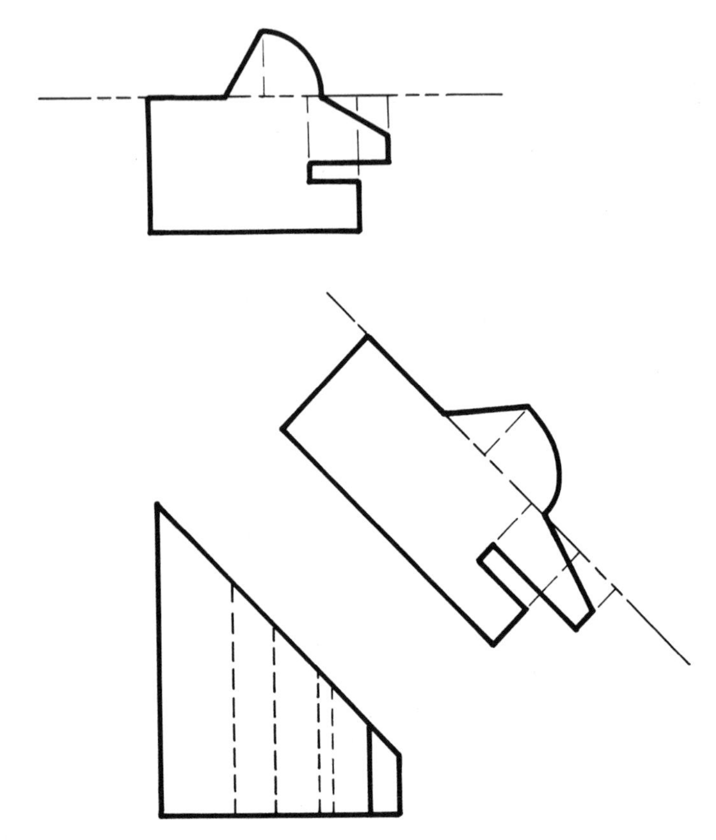

Fig. 7-8. Bilateral auxiliary.

View Placement

Sometimes an auxiliary view cannot be placed in its proper (projected) location. This can arise when there is insufficient room on the drawing paper owing to oversized objects or complex details. If the direction of viewing is different from standard third or first angle projections, reference arrows should be used.

An example of how reference arrows should be used is shown in Fig. 7-9. Regardless of the direction of viewing, capital letter referencing should be presented in a position that is normal to the direction of reading (e.g., not upside down or sideways).

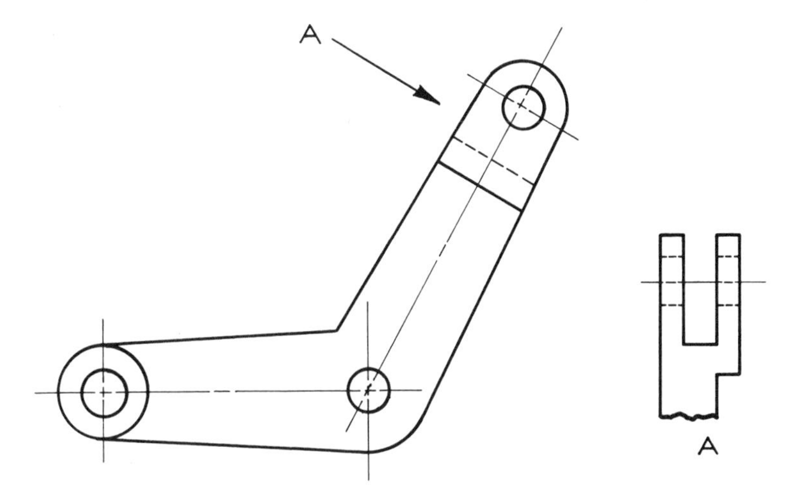

Fig. 7-9. Use of reference arrow for view placement.

Auxiliary View Problems

A wide variety of individual problems can be solved by using auxiliary view projection techniques. Some of these have already been discussed, such as partial, symmetrical and unsymmetrical, and unilateral and bilateral auxiliary views. Presented here are two other procedures that are applicable to many drawing problems.

CURVED LINES

The drawing of an auxiliary view for a curve requires that the drafter identify points along the perimeter. Once accomplished, the points are projected to the auxiliary plane. An example of such a problem is shown in Fig. 7-10, and described as follows:

1. Divide the circle in the profile view into a given number of equal parts, numbering each, and project them to the frontal view.
2. Using a symmetrical auxiliary view technique, draw the reference center line parallel to the inclined curved surface.
3. Project the division points to the auxiliary view.

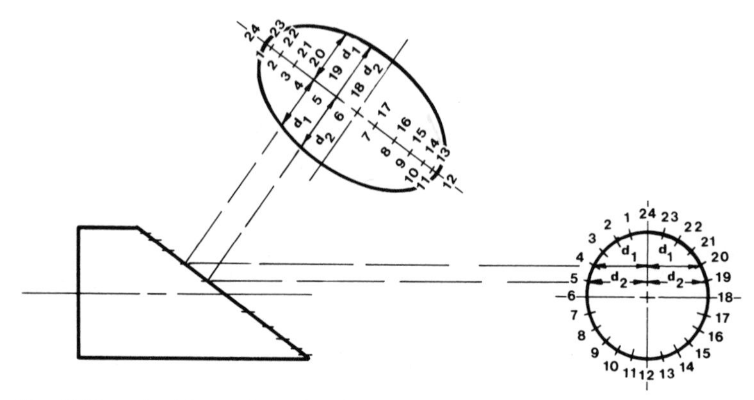

Fig. 7-10. Auxiliary view development for a circular surface.

4. Obtain distance measurements of the points about the center line from the profile view, and transfer them to the auxiliary view.
5. Once all points are plotted, connect them with a smooth, dark curved line.

DETERMINING ANGLES BETWEEN PLANES

There are times when an auxiliary view is used to determine the true size of an angle (dihedral angle) between two surfaces. Shown in Fig. 7-11 is a drawing for a fixture with a V-slot. As drawn, none of the primary views will show the true size and shape of the dihedral angle. To solve this problem, an auxiliary view is positioned so that the two intersecting surfaces will appear as edge views. The procedures used here are:

1. Draw an auxiliary plane reference line perpendicular to the sides of the V-slot.
2. Project the angle's edges into the auxiliary plane so that they are 90° to the reference line.
3. Obtain distance measurements from the frontal view to establish the position of the angle.
4. Connect the points and measure the dihedral angle.

LINES OF INTERSECTION

A frequently encountered drawing problem is finding the lines of intersection between two surfaces. In cases where the surfaces are in-

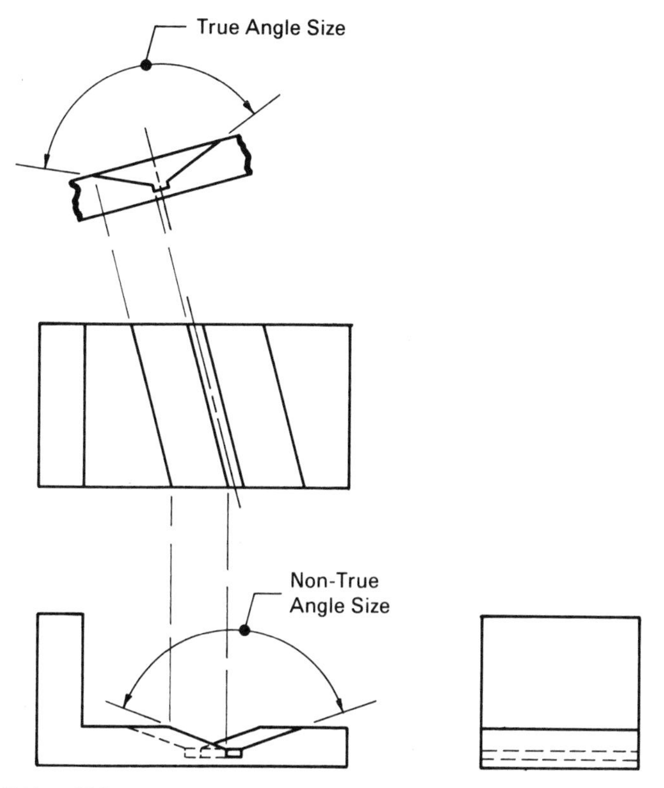

Fig. 7-11. Using an auxiliary view to find a dihedral angle.

clined, the true points of intersection cannot be accurately determined in the primary views. Therefore, an auxiliary view projection must be used.

In this procedure, the characteristic of the auxiliary view is known, but not the primary view. As seen in Fig. 7-12, the intersecting points of two planes cannot be determined in the two primary views provided; therefore one must work in a reverse procedure to complete the primary view. The procedures used here are:

1. Establish reference points about the auxiliary view of the planes, and number each.
2. Project these points to each of the primary views.
3. Using the center line as a reference point, establish the location of each point in the auxiliary view and transfer them to the appropriate primary view.

4. Connect the points with a smooth curved line to establish the intersection of the two planes.

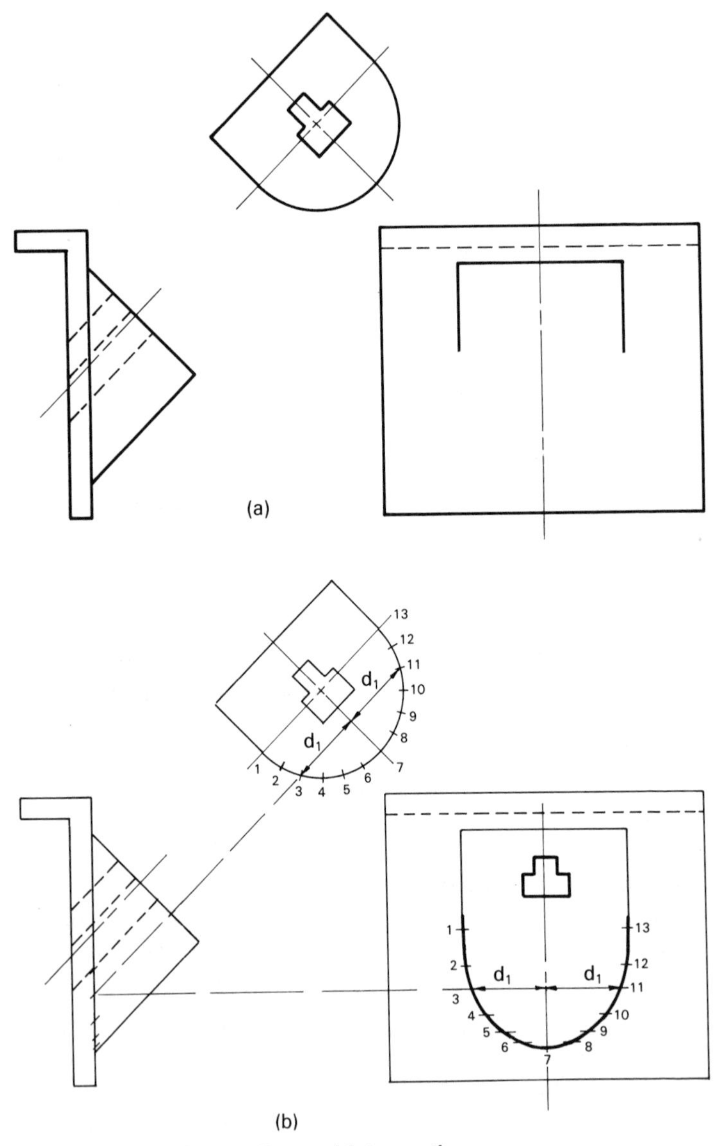

Fig. 7-12. Determining lines of intersection.

Secondary Auxiliary Views

All auxiliary view problems presented thus far have dealt with inclined surfaces in which one view presented the surface as a line or edge view. In many drawings, however, oblique surfaces are developed— that is, surfaces that do not appear as an edge view in any of the primary planes of projection. In such cases, primary auxiliary views alone will not provide the necessary true size and shape presentation.

To develop a view that will show an oblique surface in its true shape, two auxiliary views must be used, the second of which is known as a secondary auxiliary view. The function of the primary auxiliary is to present the oblique surface as an edge view. From this auxiliary view, a second auxiliary plane is placed perpendicular to the edge view so that it can be seen in its true size and shape.

An example of this procedure is shown in Fig. 7-13. Here surface ABC is an oblique plane. To find its true size and shape, a secondary auxiliary view procedure must be used. This procedure is shown in Fig. 7-13, and described as follows:

1. To find the edge view of the oblique plane, it is first necessary to determine the true length of one of the lines lying in the plane. Since none of the perimeter lines appear true length in any of the views, a reference line is used, and labeled RF.
2. RF is drawn parallel to the frontal plane in the horizontal plane, with its position projected to the front view. RF, therefore, appears true length in the front view.
3. The first reference plane is drawn perpendicular to true-length line RF, with all points of plane ABC projected to that auxiliary view.
4. The position of $A_1B_1C_1$ is determined by measures taken to the horizontal plane. The result is line $A_1B_1C_1$.
5. Draw the second reference plane line parallel to $A_1B_1C_1$, and project all points of plane ABC to the secondary auxiliary view.
6. The positions of points A, B, and C in that plane are measured from line F-1 to the frontal view, and transferred from line 1-2 along the plane's projectors.
7. When A_2, B_2, and C_2 are connected, the true size and shape of plane ABC will be shown.

This same procedure should be followed for all oblique problems. It should be remembered that if one of the lines in an oblique plane

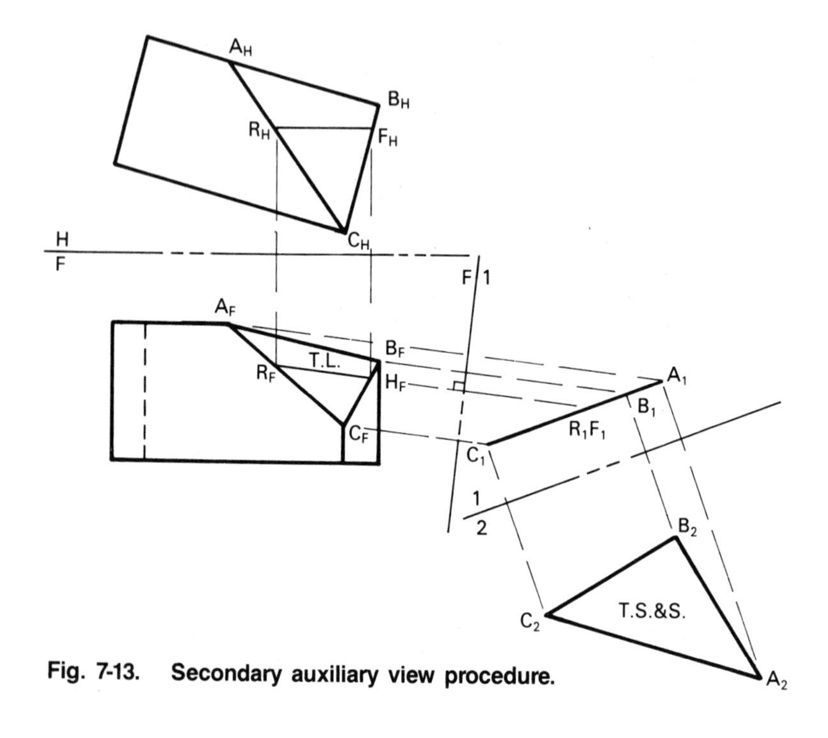

Fig. 7-13. Secondary auxiliary view procedure.

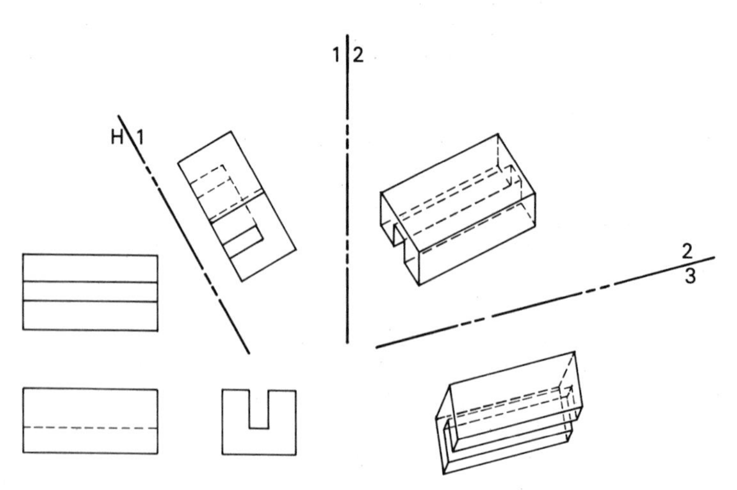

Fig. 7-14. Multiple auxiliary views.

appears true length in a view, then a reference line will not be necessary. As in primary views, since the oblique surface is the main concern of the drawing, only a partial view is normally required.

To carry secondary auxiliary methods a step further, it is possible to have an entire series of auxiliary views. Such views are sometimes used to show how the object would appear if rotated at various angles. An example of this is shown in Fig. 7-14.

Exercises

7.1 Draw the views of the objects presented in Fig. 7-15, and draw an auxiliary view of their inclined surfaces.

7.2 Determine the true size and shape of the objects illustrated in Fig. 7-16.

7.3 Draw the necessary primary views for the bracket presented in Fig. 7-17, and develop auxiliary views that will show the true size and shape of the two tabs.

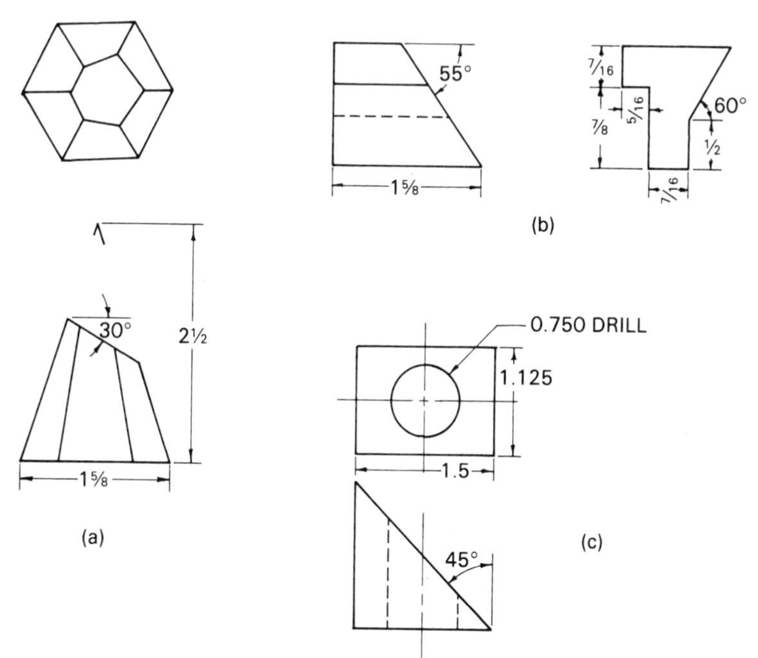

Fig. 7-15. Problem 7.1.

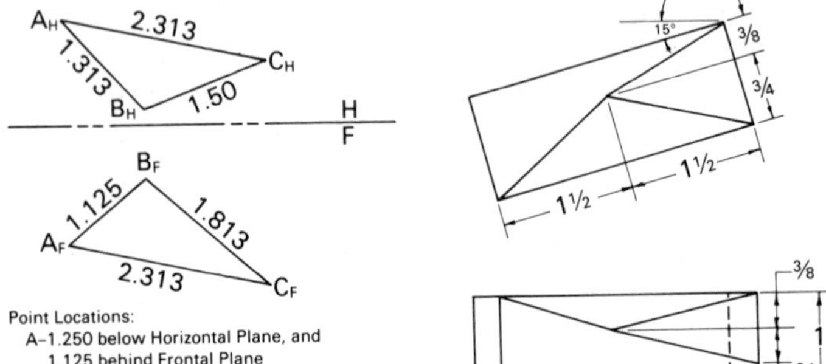

Point Locations:
A–1.250 below Horizontal Plane, and 1.125 behind Frontal Plane
B–0.563 below Horizontal Plane, and 0.125 behind Frontal Plane
C–1.750 below Horizontal Plane, and 0.750 behind Frontal Plane

(a)
Determine True Size and Shape of Plane ABC

(b)
Determine True Size and Shape of the Two Oblique Triangular Surfaces Shown in the Two Views

Fig. 7-16. Problem 7.2.

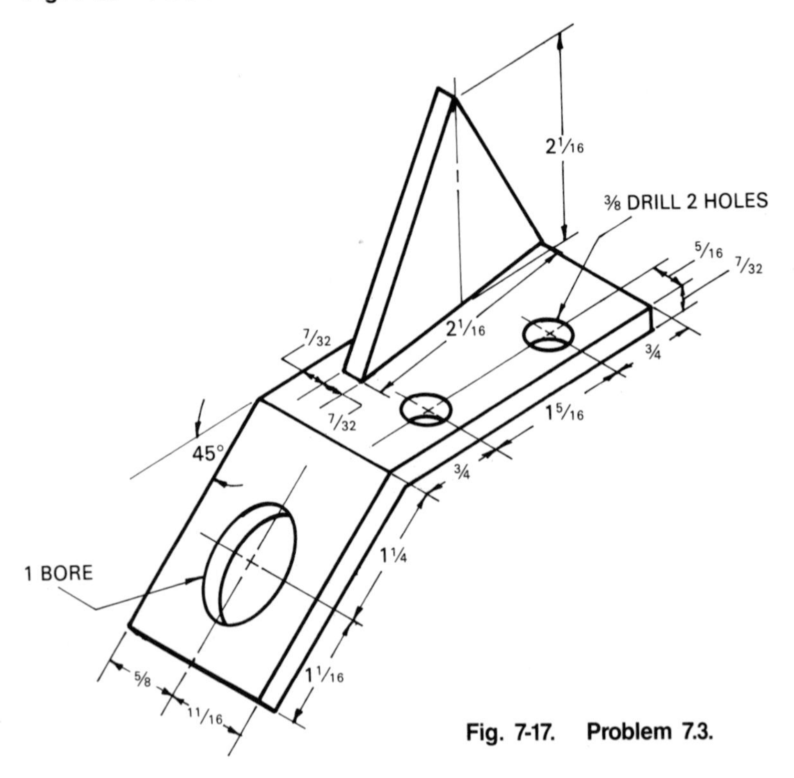

Fig. 7-17. Problem 7.3.

CHAPTER 8

Section Views

- Principles of Sectioning
- Types of Section View
- Section View Drawing Practices
- Exercises

Orthographic projection techniques can be effectively used to illustrate the external, or surface, features of complex objects. When interior features must be clearly presented, standard projection techniques may prove ineffective. This is particularly true for products that have complex cavities or subassemblies located behind an exterior wall or shell. This can be clearly illustrated in the drawing shown in Fig. 8-1a, where the typical use of hidden lines is confusing and difficult to interpret.

Such situations require a viewing technique that shows the object as it would appear if a portion or section were to be cut away to reveal the internal features. This is known as a *section view*. Section view drawings enable drafters to use an imaginary cutaway section to adequately represent the interior of an object. As seen in Fig. 8-1b, the section view clearly shows the interior features of the object.

Principles of Sectioning

Section views can be presented in a number of different ways. To draw them correctly, however, it is necessary to visualize clearly how the exposed interior surfaces will appear. Such drafting competency requires a sound understanding of the principles of sectional views. This chapter will discuss the basic principles involved in sectioning procedures.

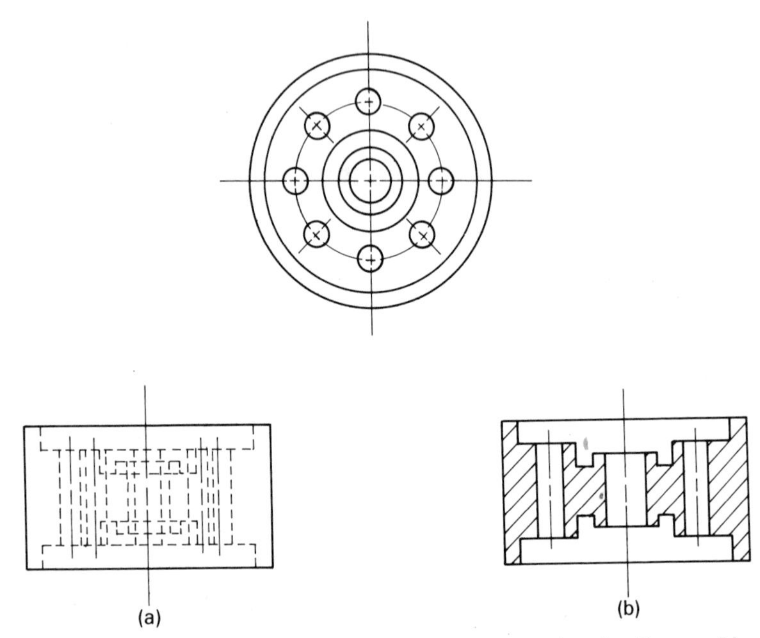

Fig. 8-1. Advantage of section view over conventional orthographic view.

Cutting Plane Lines

A section view appears as if a saw had been used to cut away a section or portion of the object in order to reveal its internal features. This "saw" is referred to as a *cutting plane*, and is presented as a line (straight, curved, or multidirectional) in the drawing. Hence the term "cutting plane line." See Fig. 8-2.

A cutting plane line is used to show the path that the imaginary cutting plane takes as it exposes the object's interior surface. As a line, it represents an edge view of the cutting plane. Arrowheads are used and drawn at the ends of these lines to show the viewing direction. The letter notation (e.g., A-A or Y-Y) is usually placed at each arrowhead to identify the section view that corresponds to the appropriate cutting plane.

Shown in Fig. 8-3 are two common cutting plane line and notation forms used in industrial drawings. The first (Fig. 8-3a) is employed

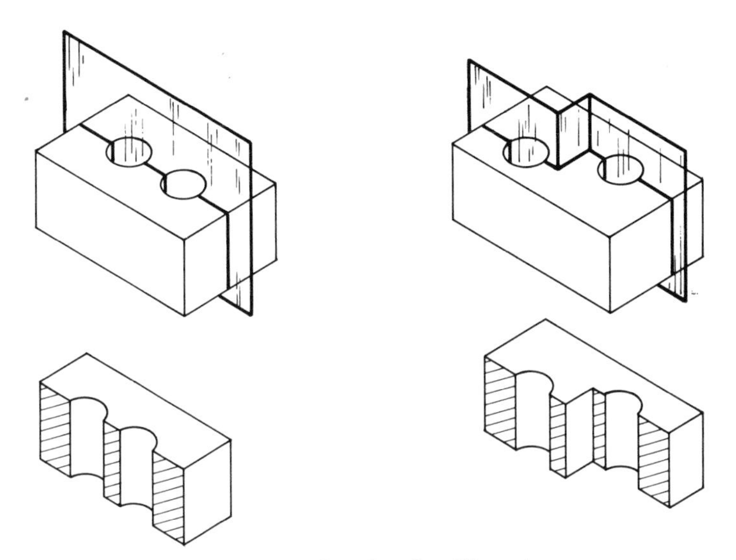

Fig. 8-2. **Straight and multidirectional cutting planes.**

by most drafters preparing drawings meeting American (ANSI) and Canadian (CSA) standards. The second (Fig. 8-3b) is used by drafters preparing drawings according to ISO standards.

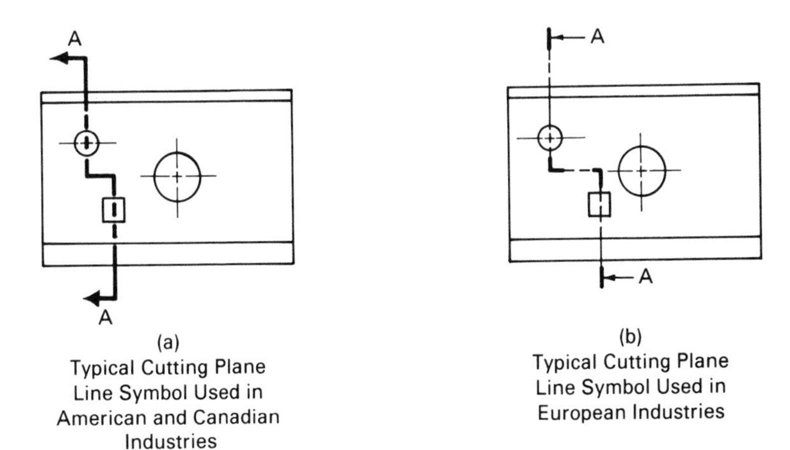

(a)
Typical Cutting Plane Line Symbol Used in American and Canadian Industries

(b)
Typical Cutting Plane Line Symbol Used in European Industries

Fig. 8-3. **Cutting plane lines.**

Occasionally arrowheads and notations are not used in drawings, but this should be the case only when there is absolutely no question as to the location or viewing of the section. If there is any possibility of misinterpretation, arrowheads and letter notations should be used.

Visibility

The visibility of a line is related to the way the section view should be viewed and presented in the drawing. Fig. 8-4 is a drawing of an object that has been sectioned to show the details of a drilled and countersunk hole. Unlike typical orthographic views, the section view should not show hidden lines. There are exceptions to this, however. Hidden lines may be used when they are necessary to understand the object's characteristics, and when they have a direct bearing on the function of the cutaway section. It is assumed that all hidden surfaces can be interpreted by other view presentations.

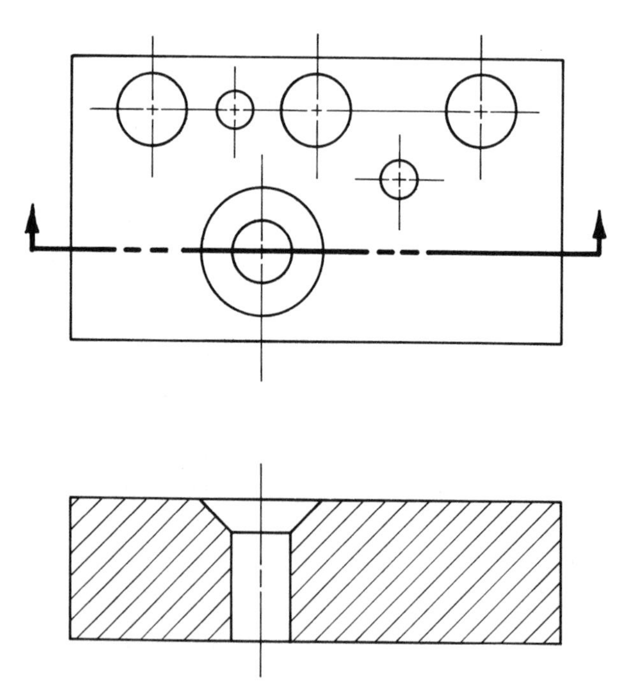

Fig. 8-4. Though there are hidden edges behind this section view, they should not be drawn.

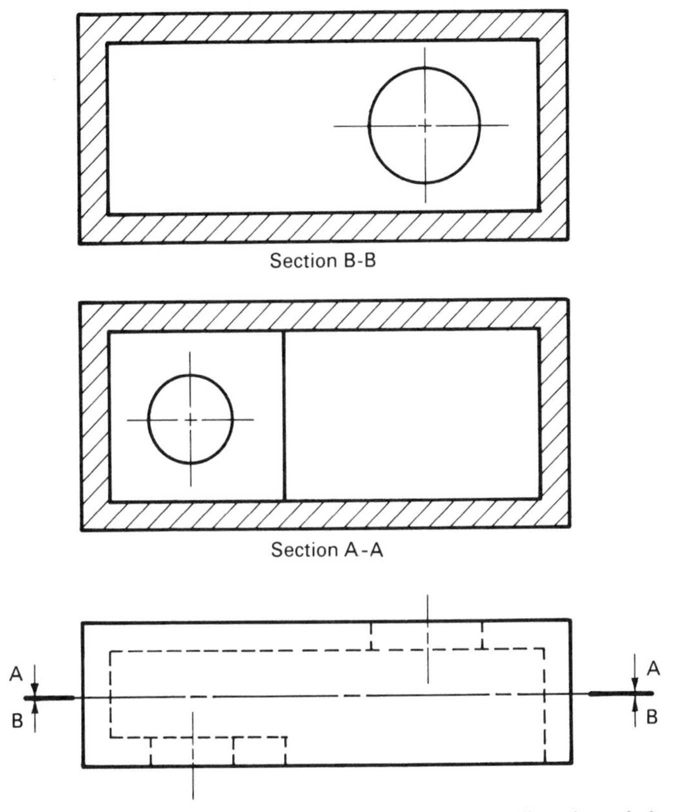

Fig. 8-5. Visibility of section views dependent upon direction of viewing arrows.

The actual visibility of a section view is directly dependent upon the direction in which the viewing arrows are pointed. Shown in Fig. 8-5 are two section view drawings of an object that is cut along the same path; the only difference is that the viewing arrows are pointed in opposite directions. Note the difference in the two section views. Care must be taken to ensure that the best viewing position is selected and appropriately indicated on the drawing.

Section Lines

A cutting plane theoretically "cuts" through solid material as it passes through an object. To illustrate the difference between cut and uncut portions and various other parts, symbols are used. In section views, these symbols are known as *section lines*. Section lines have been standardized among the various industries to show the type of material of which an object or part is made.

Presented in all illustrations thus far has been the hatching symbol most frequently used in section drawings: the general-purpose hatching or section line. This symbol is used throughout the industry, and is the same as for cast iron. In its simplest form, general-purpose hatchings are drawn as thin parallel lines, at a convenient angle—usually 45°. Other commonly used angles are 30° and 60°.

Areas that are not next to each other should show as section lines (hatching in the same direction and angle). When two components are adjacent, their section lines should be drawn in different directions. If three or more components are adjacent, their section lines should be drawn in different directions and angles, as shown in Fig. 8-6.

Spacing between the section lines on most drawings will be 0.09 inch (2 mm). The spacing may be drawn as little as 0.03 inch (0.8 mm) between section lines in smaller drawings, or as much as 0.12 inch (3 mm) for larger areas.

Crosshatching, or section lining, can be modified to meet special drawing problems, for instance, as in exceptionally large areas that have to be sectioned. Here the usual crosshatching would prove to be time-consuming and costly; an alternate technique, known as *outline sectioning* is therefore frequently used, which consists of crosshatching around the boundary of the sectioned area (Fig. 8-7).

Frequently different types of material are used for the various parts of a product, and a general crosshatching line will not suffice. In these cases, different section lining symbols can be used to represent different materials. Fig. 8-8 shows a number of common material symbols available for the drafter to use.

These various section line symbols represent general types of material, such as steel and cast iron. Modern industries, however, make use of a wide variety of engineering materials that cannot be found in any general list of material symbols. In fact, there are a variety of submaterials in each general material category. For example, though there is a general section line symbol for steel, it gives no indication as to the type of steel required. There are literally hundreds of different types of steels (e.g., low carbon, structural carbon, high-strength low alloy, ultraservice low-carbon alloy, maraging, stainless, boron, tool, forging, and spring steels).

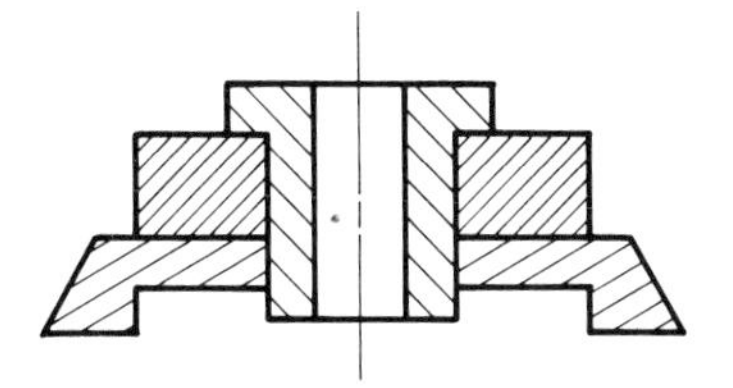

Fig. 8-6. Section lining for adjacent parts.

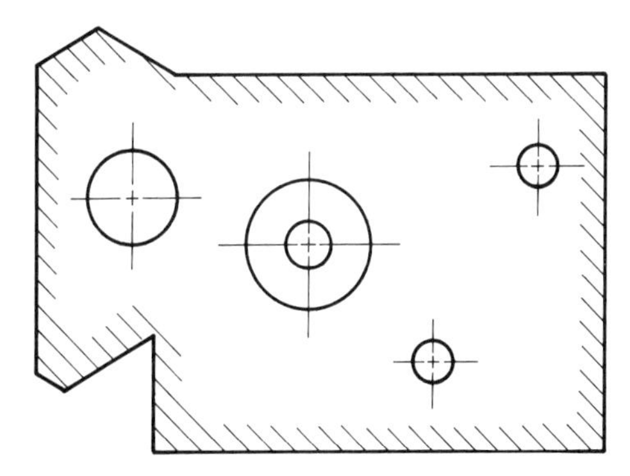

Fig. 8-7. Outline sectioning.

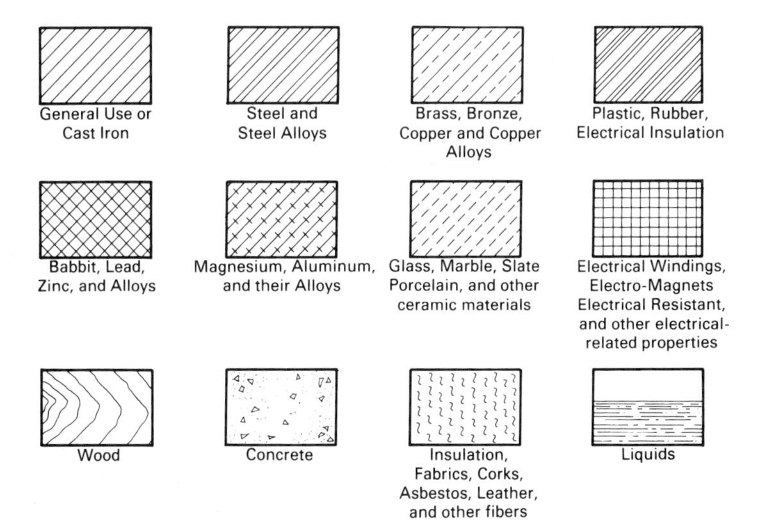

Fig. 8-8. Common material symbols.

The exact material specification is usually given on the drawing as a notation. In some cases, a material symbol is "made up" and referenced in a drawing key. Because special material symbols normally take additional time to draw, most firms recommend that the general-purpose section line be used, regardless of material specification, and that the material be identified by notation.

It should also be mentioned that different crosshatching symbols may be used as alternative symbols for several components in adjacent positions. Instead of drawing section lines at different angles and directions, different crosshatching symbols may be used for each component. When this technique is used, be sure to note the material that each component is made of.

Types of Section View

There are various types of section view drawing that can be used to illustrate the interior characteristics of an object. Each type addresses a specific situation or problem; the name given to each is therefore descriptive of the section view's characteristic. As in conventional orthographic views, the type of section view drawing used is not noted on the drawing itself (e.g., full section, half section, etc.). It is assumed that the person reading the drawing will understand what section view is presented.

Full Section

A full section is one in which the cutting plane line passes through the entire object. Here, the cutting plane line often passes through the central axis of the object in a straight-line path. Sometimes, however, a straight-line path is insufficient, and if used, will fail to illustrate certain important features. In these situations, the cutting plane line should follow a path that will best describe the interior features of the object. Fig. 8-9 shows cutting plane line paths that are used in full sections.

In most cases, one full section will be sufficient to reveal the internal features of an object. There are times, however, when a single offset cutting plane line will not be enough to reveal these features. When such a problem arises, it might be necessary to make use of two or more full sections. Fig. 8-10 shows how three full sections can be effectively used for describing an object.

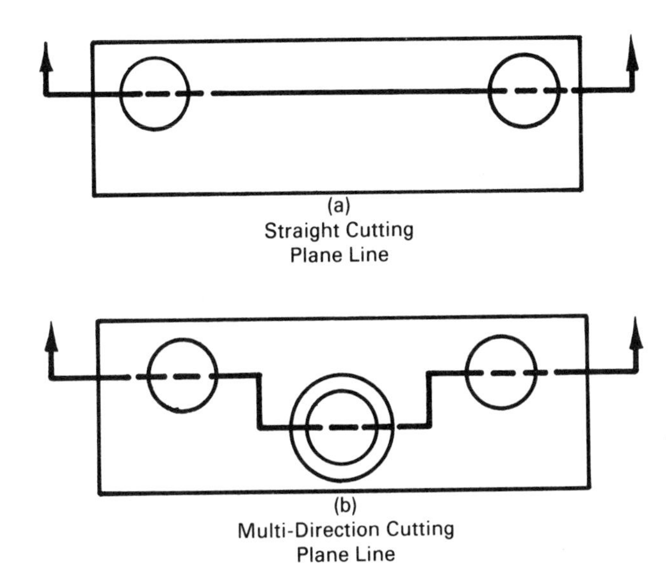

Fig. 8-9. Straight and multi-directional cutting plane lines.

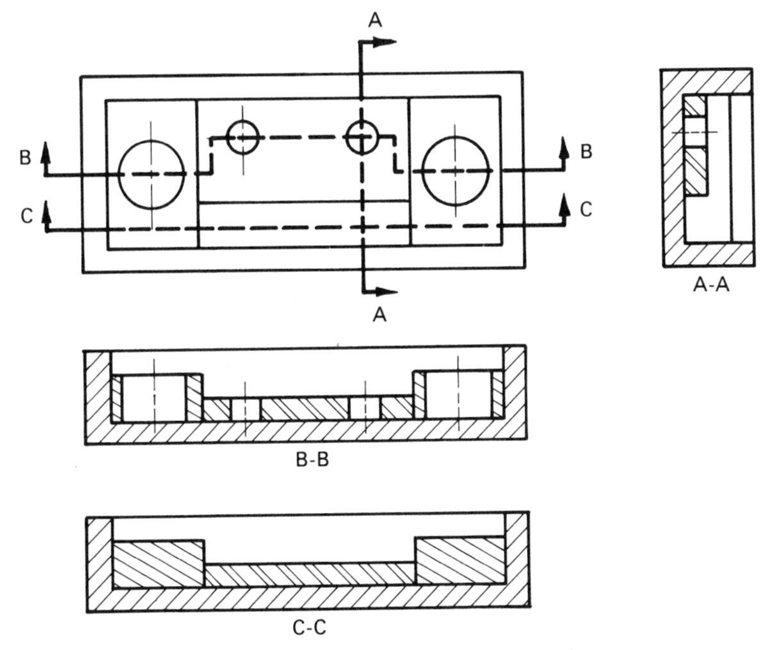

Fig. 8-10. Use of three cutting plane lines.

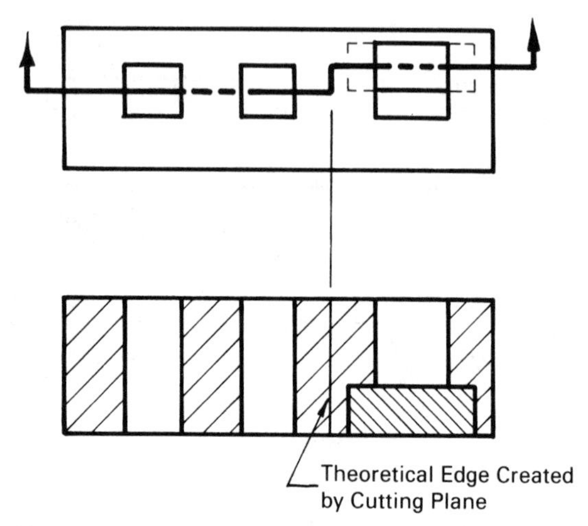

Fig. 8-11. Multidirectional cutting planes create a theoretical edge, but it is not shown in the drawing.

Note that the two sections are treated as unrelated views as they present two different observations. When multiple section views are drawn, care should be taken to label each cutting plane line properly and to note the corresponding section view. Though the section view may be anywhere on the drawing, it is good practice to locate it along the alignment of the cutting plane line.

Using multiple and offset cutting plane lines sometimes creates problems for the beginning drafter or for persons who are not familiar with reading procedures. To minimize confusion, not only should letter notations be used to identify the cutting plane line, but arrowheads should also be clearly drawn to identify the viewing direction.

Because offset cutting plane lines change directions, an edge or corner is theoretically created. However, this edge should never be shown in the section view. Fig. 8-11 shows the correct use and representation of an offset cutting plane line.

Half Section

A half section is made when the cutting plane line passes halfway through the object, then changes direction at a 90° angle, removing one-fourth of that object. This technique is best used with symmetrical objects, since the features of symmetrical products will be the same on both sides of the center line or axis, and the use of a full section would only duplicate the information needed. The objective here is to simplify the drawing and eliminate any redundancy.

Shown in Fig. 8-12 are examples of half-section views. Unlike the full section, half sections are drawn in an adjacent orthographic view, rather than as a separate view. The practice of not showing hidden lines should again be followed for half sections, except when they are needed for drawing interpretation or dimensioning.

The basic difference between the two section views illustrated in Fig. 8-12 is the cutting plane line representation. Section view (a) makes use of a representation that is commonly used in the United States and Canada, while section view (b) is typically found in Euro-

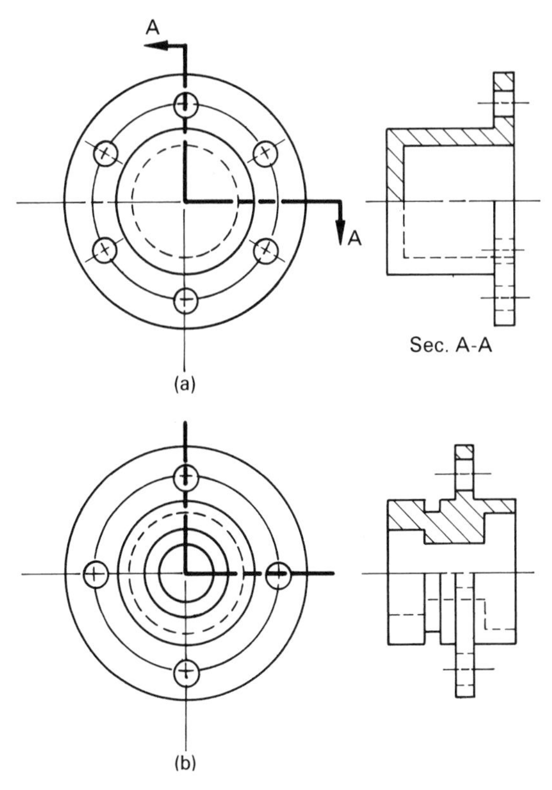

Fig. 8-12. Two examples of half sections.

pean industries. Another representation that could be used is to draw the center line along the cutting plane path in a dark and heavy line weight (thick chain).

If there is any question as to which cutting plane line representation should be used, it is recommended that the thick chain representation be used, since it is equally recognized by both ANSI and ISO standards. The reason for this is that the cutting plane line is an imaginary line, and no edge will be created by that plane. Thus, no edge will exist on the object at the center line position, so the center line will take priority over any other line. If there is still confusion, drafters should follow company policy.

Broken-out Section

A broken-out section, also known as a *partial section*, is used to show a small portion of a product's interior surface. It is used when a full or half section would remove important external elements or features, but a section is still required to visualize internal features. As shown in Fig. 8-13, the sectioned portion is defined by a *break line* that is drawn only as far as needed for adequate illustration. The break line should be drawn at the same weight as other cutting plane lines, but as a solid irregular line.

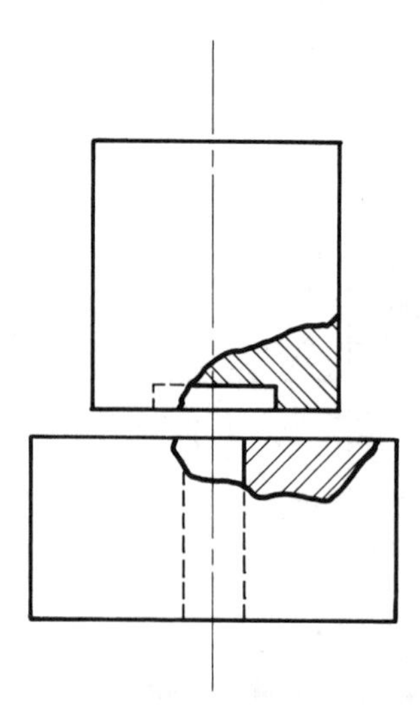

Fig. 8-13. Broken-out section.

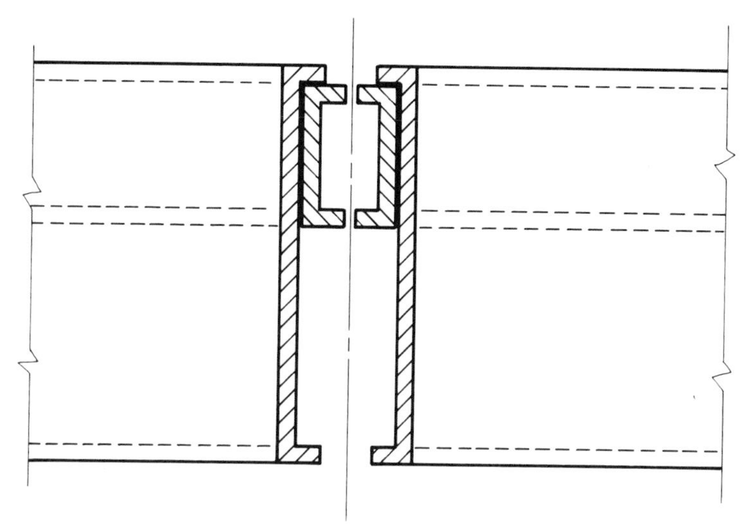

Fig. 8-14. Revolved section.

Revolved Section

A revolved or rotated section employs a cutting plane line that slices a thin section of an object at a location that would be difficult to illustrate with other sectioning techniques. As shown in Fig. 8-14, the section is rotated or revolved where its profile or edge view (not to be confused with a profile plane of projection) is located. In all cases, the revolved section will be drawn about a center line or axis.

To draw a revolved section, the cutting plane line must be perpendicular to the axis or center line of the object. At the cutting plane position the cross-sectional shape of the part is then revolved 90° about the center line and perpendicular to the object's axis. If desired or needed for clarity, the revolved section may be "brokenout."

Removed Section

A removed section is similar to a revolved section, but instead of revolving a section at the location of the cutting plane intersection (i.e., on the object itself), the section is removed to an adjacent location on the drawing. This type of sectional view should be used when there is a limited amount of space on the object for the section or dimensioning, or when an important surface detail would be hidden.

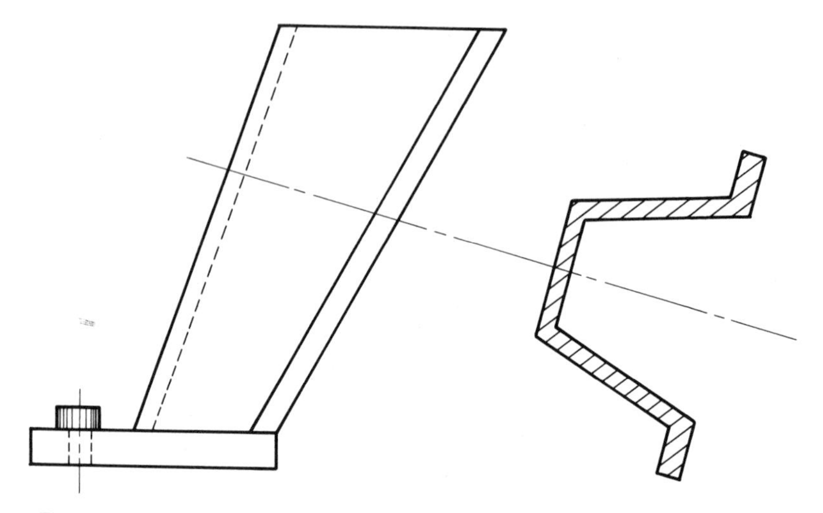

Fig. 8-15. Removed section.

A convenient method for presenting a removed section is shown in Fig. 8-15. Since the section view is connected by the center line, there is no need for viewing arrowheads and reference letters. When the section view is in a different position, cutting plane lines, viewing arrowheads and reference letters should be used.

Industrial drawings occasionally require a large number of section views, which makes it impractical or impossible to include all sections on a single drawing. In such cases, a special notation procedure can be used that references the section view to another page. A common reference notation is shown in Fig. 8-16. Here the numerator A-A identifies the section, and the denominator ME 4 refers to the sheet on which the section view can be found. It is recom-

Fig. 8-16. Section view reference notation.

mended, however, that whenever possible the removed section be placed on the same sheet as the object and in alignment to the cutting plane line.

Another common use for removed sections is in drawing details for small parts. The advantage here is that the section can be drawn to a larger scale than the object itself (e.g., twice or three times the size). The internal features can be more easily interpreted, and dimensioning is also easier.

Auxiliary Section

Auxiliary sections are used to show the true size and shape of internal surfaces that are drawn on an incline. Since the surface is on an incline, the cutting plane line often functions as an auxiliary plane so that the section view is projected perpendicular to it. It is therefore essential that the drafter attempt to align the section view so that the projection lines are 90° to the cutting plane line.

Fig. 8-17 shows an auxiliary section. Note the alignment of the section view relative to the cutting plane line, and the ease of interpretation.

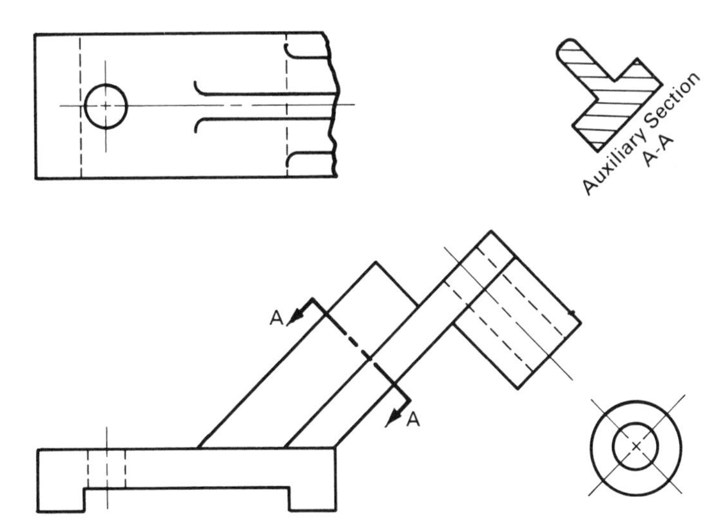

Fig. 8-17. Auxiliary section.

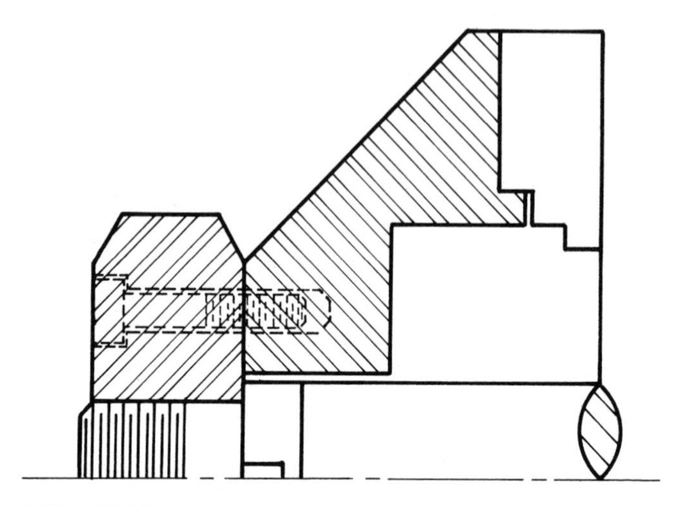

Fig. 8-18. Hidden section view showing fastening of two components.

Hidden Section

The hidden section is not frequently encountered by industrial drafters, but is a valuable technique for special situations. There are two primary drawing problems in which hidden sections can be used: when a section would either remove or hide an important feature on the external surface of the object. In such cases, the hidden section will show the object's general outline and external details as visible object lines, while the section view is drawn with hidden (dash) lines (see Fig. 8-18).

Phantom Section

Another sectional view, which is similar to the hidden section, is known as the phantom section. This section is used when the location of a sectioned part must be shown relative to another. Unlike conventional phantoms, ANSI standards do not make use of phantom lines.

When ISO standards are used, the adjacent part is only outlined in phantom line presentation. This differs from ANSI standards in that solid object lines are used for visible lines and hidden dash lines are used for section lines. A comparison of the two standards is shown in Fig. 8-19.

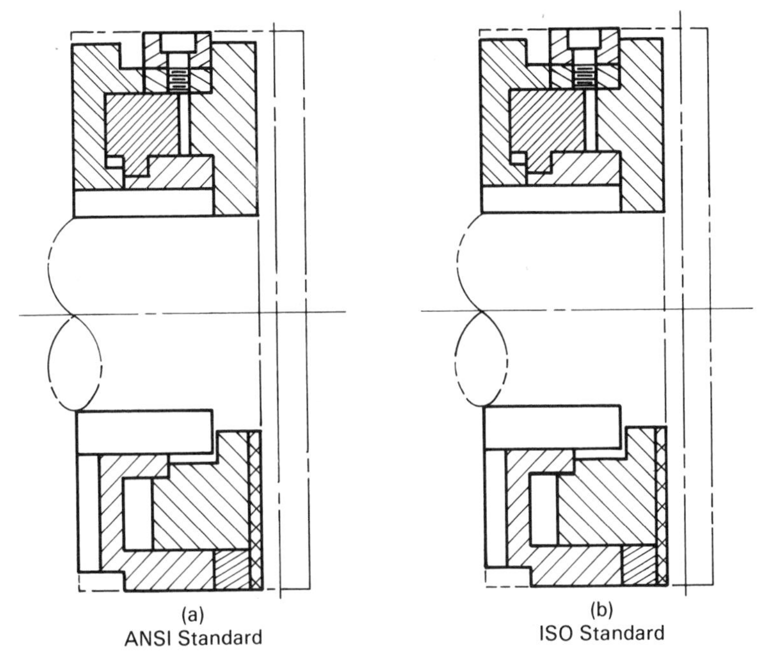

Fig. 8-19. Phantom section.

Assembly Section

The purpose of an assembly section is to show the internal features of a product that is made up of a number of different parts. The cutting plane of an assembly section is frequently offset so that all necessary features can be shown in the section view. All previously presented sections may be used as assembly sections when they facilitate the understanding and interpretation of the assembly.

Fig. 8-20 is an example of an assembly section. Only interior parts and details that need to be identified and dimensioned should be shown. In addition, when clearances or spaces exist between mating parts (e.g., it is not unusual to have as much as 1/16-inch spacing between a bolt and its mating hole), they should not be shown in the drawing. The reason for this is that such clearances may be exaggerated and increase confusion about the assembly.

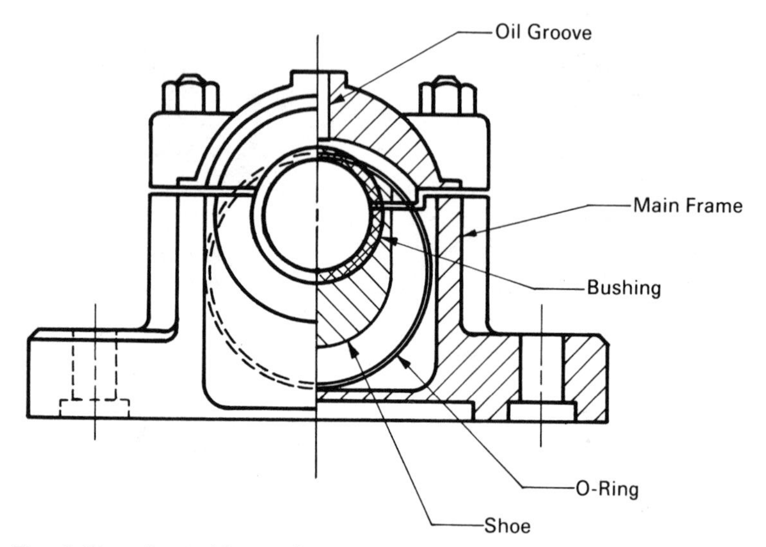

Fig. 8-20. Assembly section of a rigid ring-oiling pillow block.

Section View Drawing Practices

Owing to the variety of products and their designs, some unique problems may be encountered by drafters that a series of drawing practices have been developed to meet. Some of these are discussed in this section.

Section Lining

The angle and direction of standard section lines has already been discussed earlier in this chapter. There are times, however, when conventional angles of 30°, 45°, and 60° will not be appropriate. When unusually shaped objects must be sectioned, it is important to avoid drawing crosshatchings that are parallel to the view's object lines. Thus, it may be necessary to draw the section lines horizontally, vertically, or at 15° increments. Examples of alternate section lining techniques are shown in Fig. 8-21.

Another crosshatching problem that is sometimes encountered occurs when the sectioned material is very thin and conventional crosshatching cannot be used, because in the cross-sectional area there

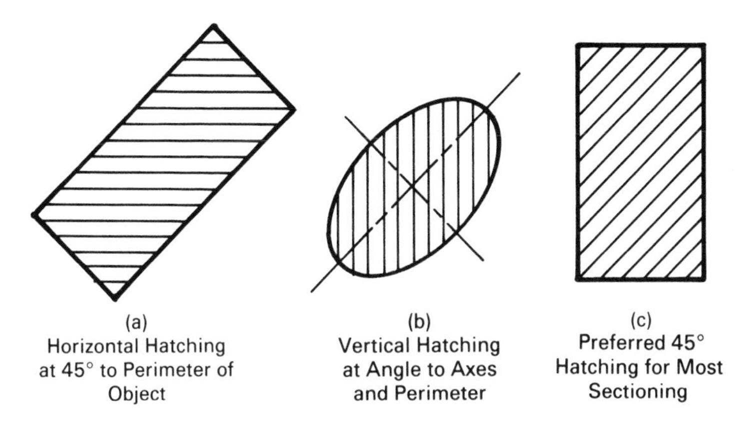

(a)
Horizontal Hatching at 45° to Perimeter of Object

(b)
Vertical Hatching at Angle to Axes and Perimeter

(c)
Preferred 45° Hatching for Most Sectioning

Note: The simplest form of hatching is often the best, and is based upon lines drawn at convenient angles

Fig. 8-21. Alternate section lining techniques.

is insufficient room to draw the hatchings. Thin sections, such as glass, gaskets, sheet metal, and structural parts, may be shown as solid black. When two or more thin sections are placed next to each other, a thin white space may be shown between them. An example of this practice is shown in Fig. 8-22.

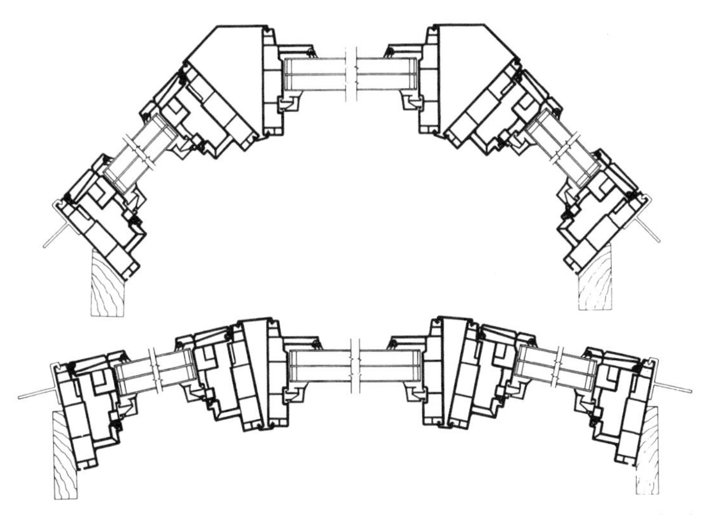

Fig. 8-22. Thin sections of window casements shown entirely black. *(Courtesy of Paxson Building Products)*

Sectioning of Spokes, Arms, and Ribs

The practice of sectioning spokes and arms is somewhat different from that for other sectioned parts. For most parts, all elements that have been cut by a cutting plane line must be shown with section lines in the section view. This is not always true, however, for circular parts. As a rule, any portion of a circular part that is not solid around the axis of that part is not drawn with crosshatchings in the section view.

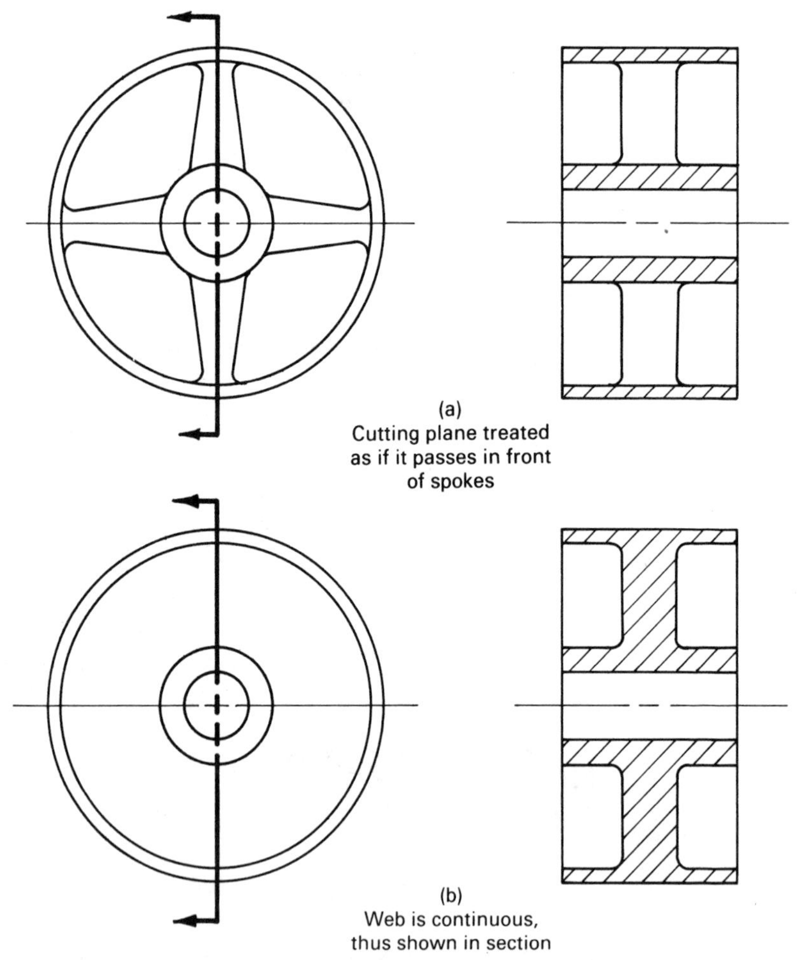
(a)
Cutting plane treated as if it passes in front of spokes
(b)
Web is continuous, thus shown in section

Fig. 8-23. Spokes and webs in section.

Two examples of sectioning for spokes are shown in Fig. 8-23. Note that even though two spokes are cut by the cutting plane line, it is not shown in section because it is not continuous. In the second example, a pulley wheel has a solid web that is continuous. Therefore, it will be drawn with section lines in the sectional view.

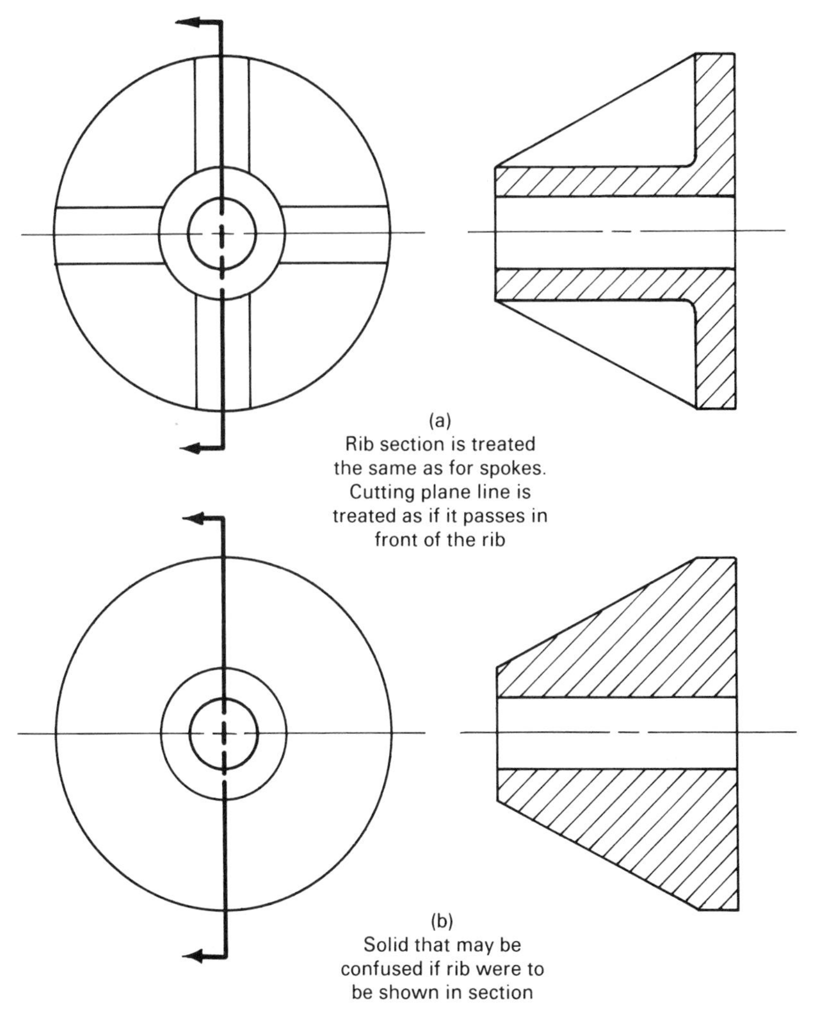

(a)
Rib section is treated the same as for spokes. Cutting plane line is treated as if it passes in front of the rib
(b)
Solid that may be confused if rib were to be shown in section

Fig. 8-24. Rib in section.

Ribs drawn in section incorporate the same principles as used for spokes and arms. If the rib is continuous, then it will be shown sectioned; otherwise, no sectioning should be used. In noncircular parts, it may sometimes be necessary not to show a rib in section, especially if the sectioned rib will add to the confusion of the drawing. An example of such a situation is shown in Fig. 8-24.

Alignment in Sections

There are times when the true projection of a section view would tend to confuse a drawing rather than clarify it. In such cases, it may be necessary to align the section so that the drawing would be better understood. An example of this is shown in Fig. 8-25.

Here, as is the case with any part with an odd number of spokes, arms, or ribs, that will produce an unsymmetrical section. The preferred section is one in which the arms are drawn in alignment, in such a way that the section is rotated to the path of the cutting plane line and projected to the appropriate view. Any additional arms that may be projected should be eliminated if they do not add to the clarity of the drawing.

Products with other unsymmetrical features should also be presented in alignment. Fig. 8-26 gives several examples of how cutting plane lines and acceptable alignment techniques should appear in drawings.

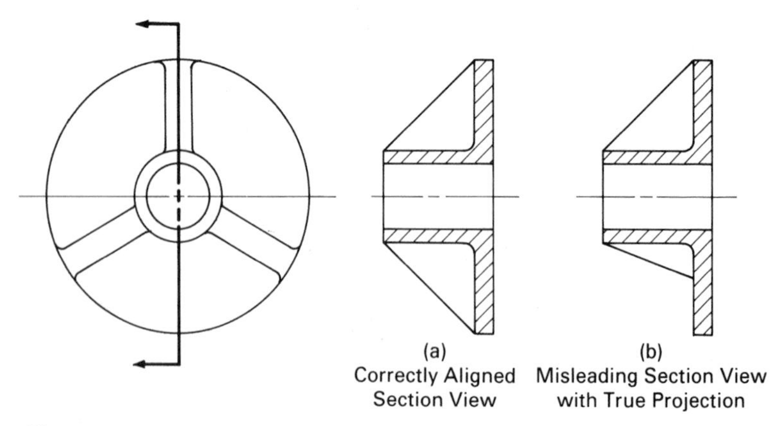

Fig. 8-25. Alignment procedure in section views.

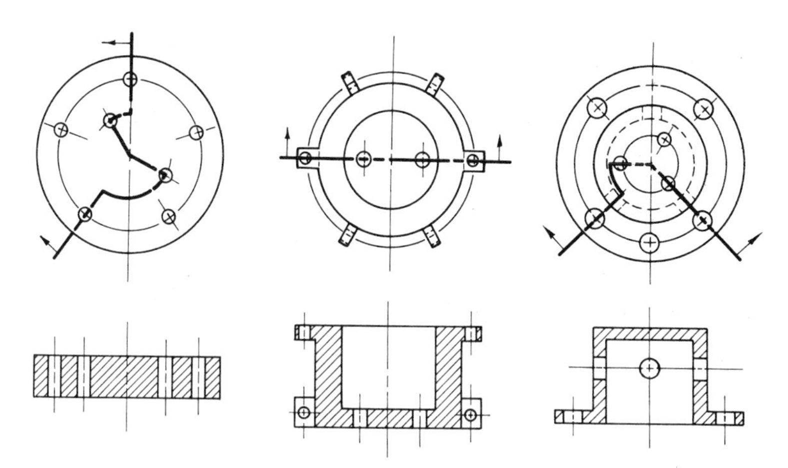

Fig. 8-26. Cutting plane lines and hole alignments.

Exercises

8.1 Draw the two views of the object shown in Fig. 8-27, and draw the full sections indicated by the cutting plane line.

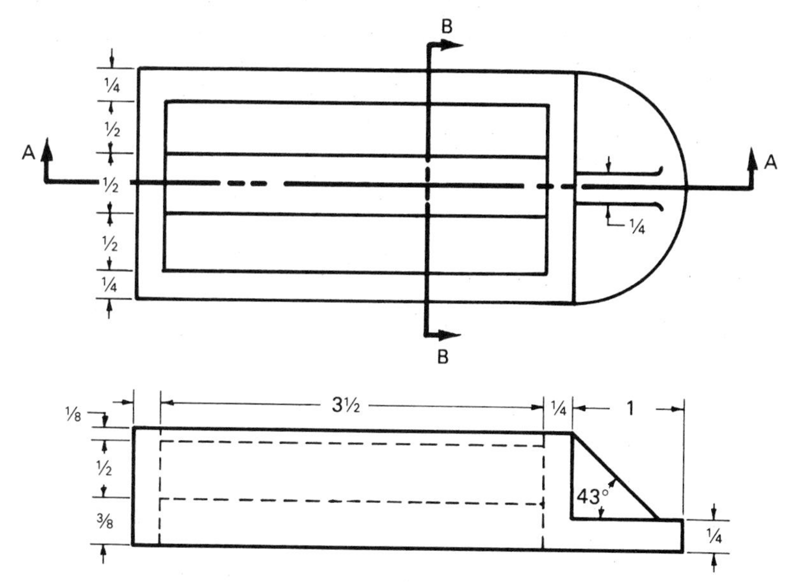

Fig. 8-27. Problem 8.1.

8.2 Draw a half section for the V-pulley wheel shown in Fig. 8-28.

8.3 Draw the horizontal view for the object in Fig. 8-29, and develop a full section of the front view.

8.4 Draw an auxiliary section for the brace shown in Fig. 8-30.

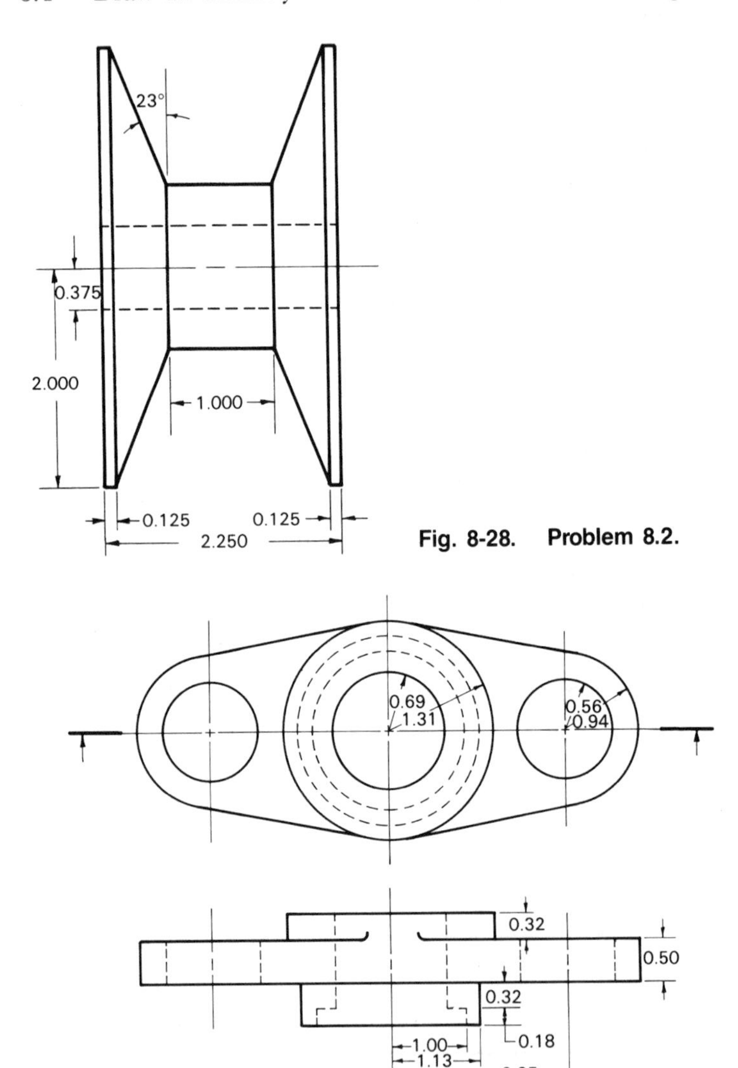

Fig. 8-28. Problem 8.2.

Fig. 8-29. Problem 8.3.

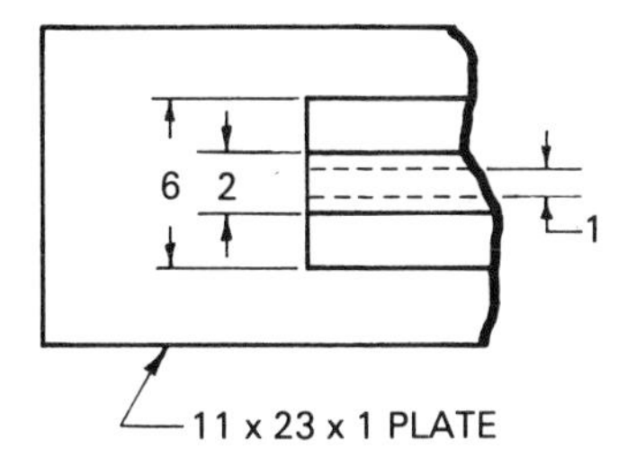

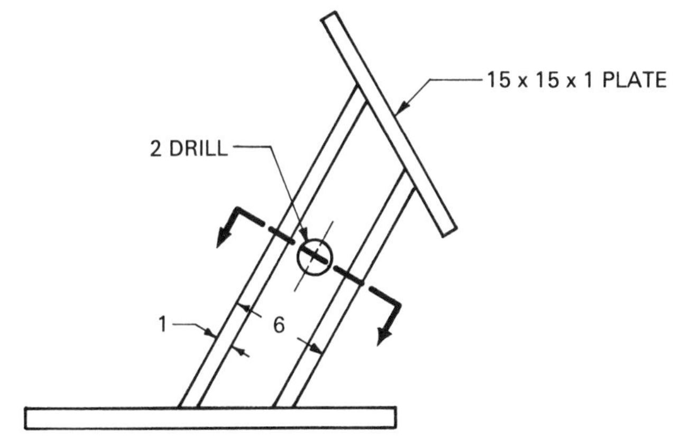

Fig. 8-30. Problem 8.4.

CHAPTER 9

Dimensions and Notations

- Dimensioning
- Notations
- Exercises

Orthographic projections and alternate viewing techniques provide valuable information as to the characteristics of objects, yet additional data must be presented concerning an object's size and specifications. Generally, when the description of an object by a graphic representation is coupled with size descriptions, a complete set of data has been provided for producing that object.

Size information is presented on drawings in the form of dimensions, while other information is provided by using special notations. It must be understood that the dimensions and notations found on drawings are not always those necessary to make the drawing, but rather those that are needed to produce, install, or service the object. This chapter will discuss acceptable dimensioning and notation procedures that are found throughout the industry.

Dimensioning

Before it is possible to understand and use dimensioning practices in drawings, one must have a firm grasp of the basic factors that influence dimensioning practices. The first is a thorough knowledge of and drawing competence in the line weights and spacing of lines for dimensions and notations. Second is the ability to select the proper distances to be dimensioned. These distances, in turn, are determined by what is needed to produce, install, or service the product.

Next is the orderly placement of dimensions so that the reader will be able to understand clearly the intent of the information given. Fourth is the requirement to dimension all standard features such as angles, chamfers, and geometric shapes. Fifth is the dimensional interrelationship of parts, which is given in terms of precision, limits, and tolerances. Finally, information pertaining to production methods, such as casting, forging, machining, and extrusion, must be provided on the drawing.

Dimensioning Lines and Symbols

There are two basic methods by which distance measures can be given: dimension and notation. Of primary concern here is the dimension. Dimensions are used to give the distance between two locations, which may represent points, lines, planes, solids, or any combination thereof. A numerical value is placed between these locations to indicate the size of that length.

Dimension, extension, and leader lines are all drawn with the same line weight: thin, full lines. Their weight should be the same as for center lines, so that they can be differentiated from the outlines of the object and other object or visible lines. In all cases, these lines should be drawn with the aid of a straightedge. Fig. 9-1 shows a simple illustration of these lines.

Arrowheads are drawn in freehand at the ends of the dimension and leader lines. They may be drawn with either one or two strokes of the pencil or technical pen. Two arrowhead styles are acceptable in drafting: open and solid arrowheads (Fig. 9-2a). Of the two, the solid arrowhead is usually preferred; it is drawn with straight sides, and is narrower than the open style.

In general, the width of an arrowhead should not exceed one-third of its length. Thus, if an arrowhead is drawn to a length of 3/16 inch (5 mm), its width should be no more than 1/16 inch (1.5 mm). For most dimensions, an arrowhead 1/8 inch or 3 mm long will suffice. Larger drawings, however, may require arrowheads of 3/16 inch or 5 mm, and longer. These specifications may not seem important to the beginning drafter, but drawing proper arrowheads is essential to maintaining the quality appearance of the drawing. Drafters must

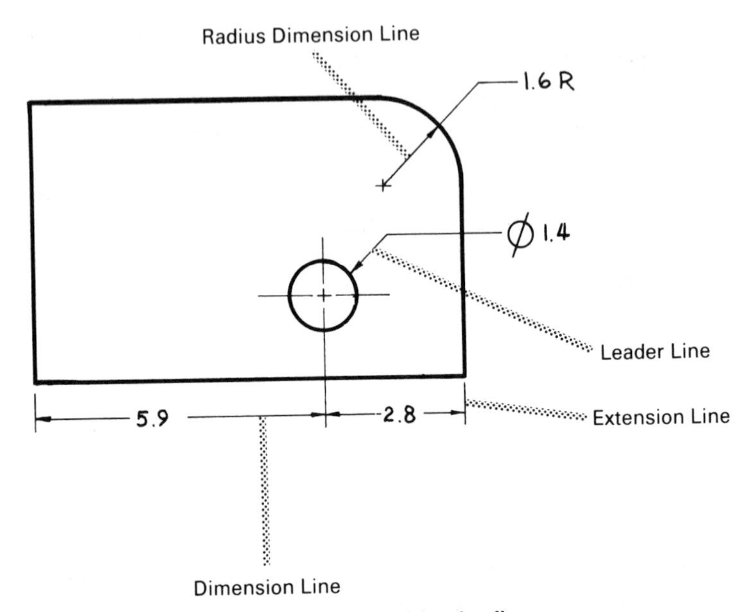

Fig. 9-1. **Extension, dimension, and leader lines.**

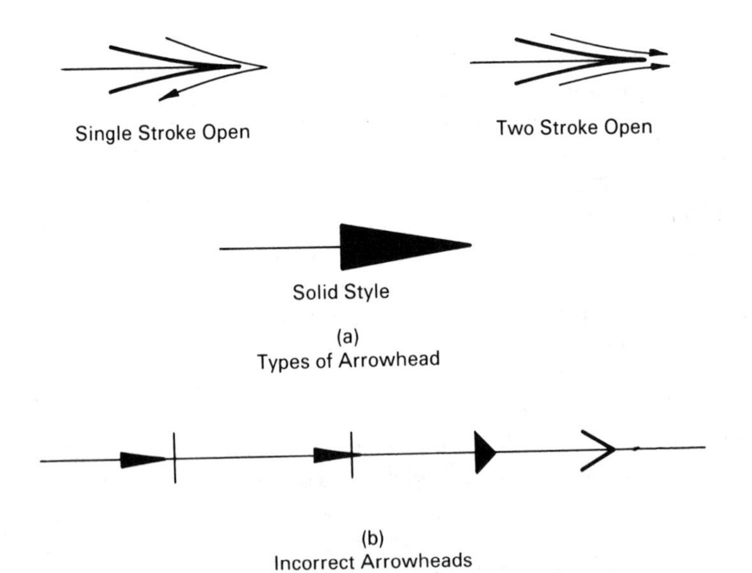

Fig. 9-2. **Arrowhead presentations.**

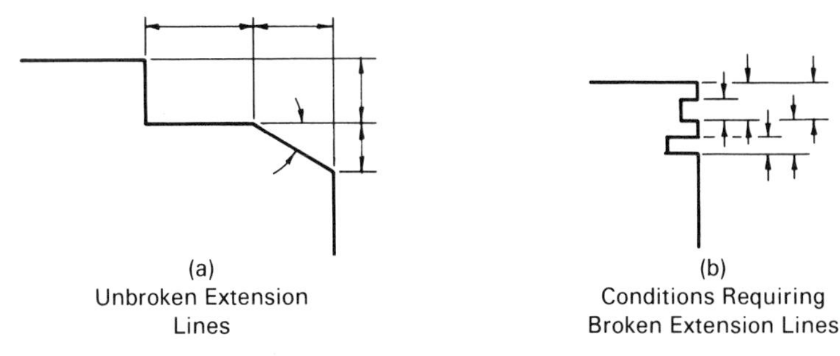

(a)
Unbroken Extension Lines
(b)
Conditions Requiring Broken Extension Lines

Fig. 9-3. Extension lines.

avoid the use of "quick" and incorrect arrowhead presentations, such as those shown in Fig. 9-2b.

Extension lines are used to show the outside limits of a dimension. They should not come in direct contact with the outline of the part that they are dimensioning, and should begin about 1/16 inch or 1 mm from the object and extend approximately 1/8 inch or 3 mm beyond the last dimension line. Once begun, extension lines should not be broken (Fig. 9-3a), except when there is a space limitation (Fig. 9-3b). In addition to extension lines, dimensions may begin or end at object lines or center lines. It is poor practice, however, to dimension to hidden lines.

Leader lines should be drawn as straight rather than curved lines. The function of the leader line is to give a dimensional value or some other descriptive quality to a geometric shape. For example, in Fig. 9-4 the leader lines are used to specify the dimensional sizes of a hole, chamfer, and slot. When dimensioning a circular shape, the leader

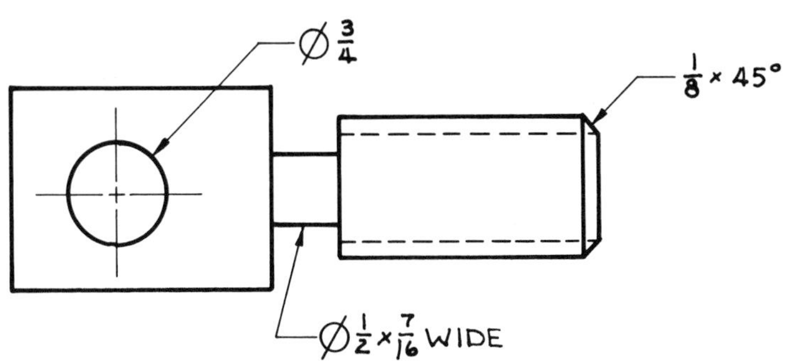

Fig. 9-4. Use of leader lines.

should be drawn preferably at a 45° angle (30° and 60° are also acceptable, as are other 15° increments, when there are space limitations). If it is extended, the leader line should pass through the center of the circular shape.

When dimensioning noncircular shapes, the leader should be drawn at the same angles specified for circles. Arrowheads at the end of the leader should be used when referring to the edge or periphery of the object. When reference is made to the surface, a "dot" is recommended by both ANSI and ISO standards. Whenever possible, avoid the crossing of leaders.

The figure, or dimensional value, is one of the most important dimensional symbols used, for it presents the exact value of the distance. For most drawings, figures should be made between 1/8 inch or 2.5 mm and 3/16 inch or 5 mm in height. Again, when necessary figure heights may vary according to drawing requirements.

Drafters usually leave a space in the middle of the dimension line for placement of the dimensional value. This differs somewhat from international practice, where the value is placed above the dimension line. This second technique, however, is also used throughout structural and in many architectural drawings. Examples of these two techniques are shown in Fig. 9-5.

When lettering fractions, the fraction line should be drawn horizontally, rather than at an angle. Thus, the fraction one-fourth is to be presented as $\frac{1}{4}$, and not 1/4. In most manufacturing industries, all dimensions are given in inches. When metric measures are used, dimensions are frequently given in millimeters. It is therefore preferable to leave all inch (") and millimeter (mm) notations off the dimension.

ANSI standards recommend that all dimensions up to and including 72 inches (6 feet) be expressed in inches. Anything greater

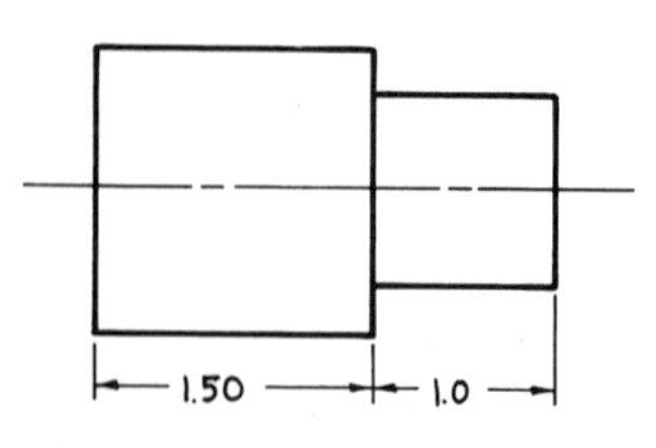

Most Common Dimensioning Placement for ANSI Drawings

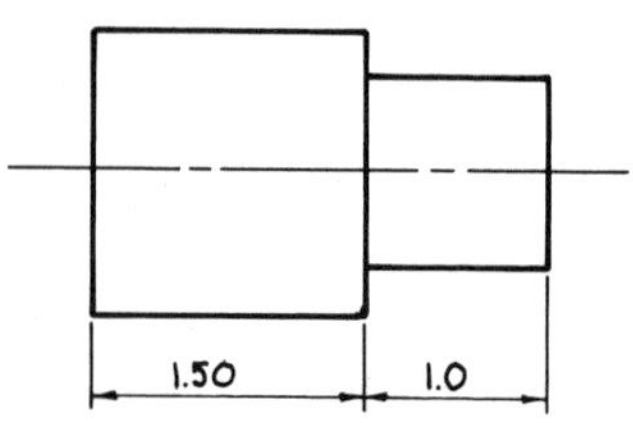

Recommended Dimensioning Placement for ISO Drawings

Fig. 9-5. Placement of dimensions.

should be given in feet and inches. When both units are used, it will be necessary to give the foot (') and inch (") symbol after each value. The only exception to this is in structural drawings, in which plate and beam sizes are given in inches and no inch markings are used, even though the dimensions are in feet and inches.

Dimensioning Principles

It is the objective of all good drafters to produce high-quality drawings in the simplest and most readable form. Therefore, one should strive to minimize the number of dimensions on a drawing. This can only be accomplished by understanding how dimensions should be selected for a given product; a drawing prepared for a service technician will be somewhat different from one drawn for product manufacturing. In either case, several basic dimensioning principles are generally followed by competent drafters. First, however, there are four important definitions:

1. *Finished product drawing* pertains to the complete description of the object in the form in which it is to be used; in other words, the requirements necessary for the product to function within acceptable standards. This includes detailing of parts that are subassemblies or smaller parts of a larger product.
2. *Features* are individual characteristics of a product. They pertain to the geometric shapes and forms that make up the product.
3. *Functional features* are those product elements that are essential in the performance or serviceability of that part. This may be the location of an operating part (e.g., key, spigot, or pin), or the nature of a working surface (e.g., precision ground ways, or the bore of a bearing).
4. *Functional dimensions* are those dimensions that must be known for the proper functioning of the product. In comparison, a *nonfunctional dimension* may be necessary for manufacturing, but has little to do with the way it works, and an *auxiliary dimension* is used for general information only, and is presented without tolerance. Examples of these dimensions are shown in Fig. 9-6.

The first, and most basic, principle of dimensioning is that only those dimensions that are necessary to provide for the accurate use

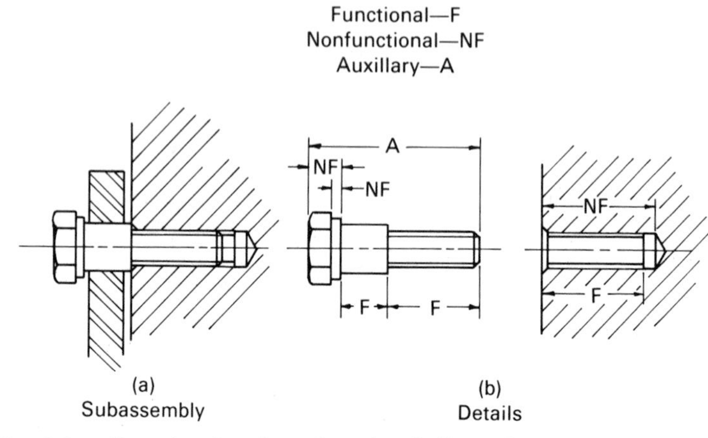

Fig. 9-6. Functional and nonfunctional dimensions.

of the part should be given directly on the drawing. These include any additional information that may be necessary to describe the finished part. What must be kept in mind is that the drawing is to be used for both manufacturing and inspection (quality control) purposes. In relationship to this, the following practices should be observed:

1. It is poor practice to show the same dimension two or more times on the same drawing. Duplicate dimensioning should be avoided, unless absolutely necessary.
2. Though the number of dimensions should be kept to a minimum, all functional dimensions must be given. Functional dimensions must not be extrapolated or deduced from other dimensions or by direct reading on the drawing with a scale.
3. Dimensions should be located in the view that shows the referenced feature most clearly.

The second basic principle is that only those dimensions necessary to describe the finished product should be given. In addition, no more than one toleranced dimension should be used in any one direction for the location of a feature. There are, however, exceptions to these practices: First, there are times when it will be necessary to give dimensions that are used for intermediate processes, such as for the size of a casting before machining, or of a feature before heat treating and finishing to size—both processes resulting in a change or difference in product dimension.

Second, auxiliary dimensions are often desirable information, but are not toleranced. If auxiliary dimensions are used, it is recommended that they be enclosed in brackets, which will indicate that they are not directly tied to the tolerances necessary for product acceptability (see Fig. 9-6).

Third, functional dimensions should be made directly on the drawing itself. If a general tolerance notation is used, to calculate it from other dimensions often results in closer tolerances than are required. An example of this is shown in Fig. 9-7. This, however, should not discourage the drafter from dimensioning distances between center lines for holes, even though the functional dimension will be from one edge to another.

Fourth, nonfunctional dimensions, when given, should be presented in a view and location that is most convenient and readable to the manufacturer and quality control technicians. Though these dimensions are not needed for the functioning of the part, they must be known for production purposes, and should be included in detailed drawings.

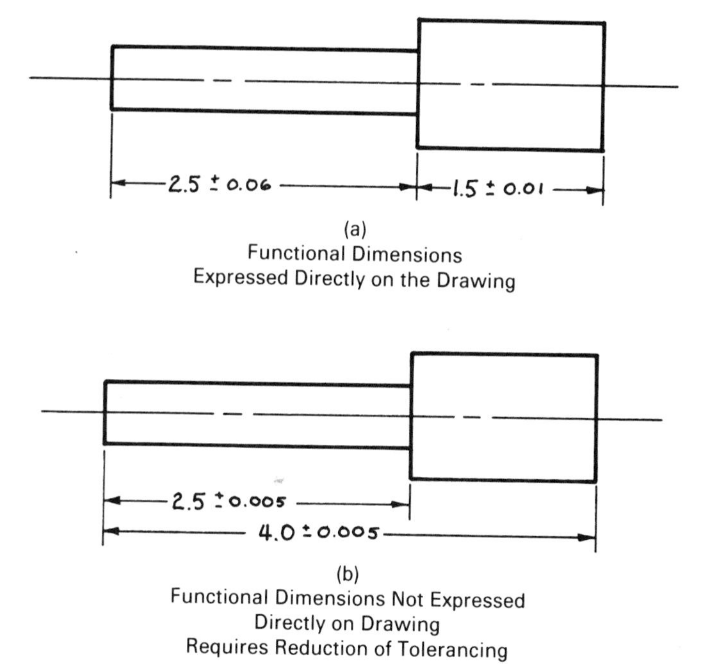

Fig. 9-7. Expressing dimensions.

Fifth, for products that have no established (standardized) tolerances or manufacturing practice guarantees, tolerances should be given. Tolerances should also be presented even when wide variations of dimensional limits are permitted. Specification notes should be used where the tolerance over a series of dimensions is less than the sum of tolerances for individual dimensions—calling attention to the unique situation.

Sixth, wherever practical, standardized sizes and surface finishes should be used. This will not only reduce the number of special notations and tolerances, but can also reduce production costs. For example, it would be much easier and less expensive to drill a hole with a standard 0.250-inch drill than to machine a 0.2578-inch hole.

Last, in most drawings it is not necessary to specify a manufacturing process or inspection method for the product. This is usually left to the discretion of the production engineer or manufacturing staff, so that satisfactory interchangeability, processing, and functioning can be maintained. However, this does not preclude specifying machine tool sizes (e.g., drill and ream sizes), nor does it apply to process drawings.

Dimensioning Methods

Consistency and standardization of dimensioning procedures should be the guiding rule for dimensioning methods. Drafters must adopt a dimension method that is standard within their industry, and use it consistently in all their drawings. This section discusses dimensioning methods that are commonly used in industrial drawings.

SYSTEMS OF DIMENSIONAL VALUES

One of the most basic concerns in dimensioning is the system used to write the dimensional value. Dimensions may be based upon either fractional (e.g., 1/8, 1/4, 1/2, etc.) or decimal (e.g., 0.185, 0.25, 0.50, etc.) values. From these two types of expressions, one may select one of three systems for writing dimensional values:

1. *Common fraction systems* are used in a variety of industries, including manufacturing, architectural, and structural. Here all dimensional values will be expressed as units and common fractions, such as 4 3/4, 1 1/8, 3/16, and 5/64. When this system is

used, fractions are normally expressed in graduations no smaller than sixty-fourths of an inch. The reason for this is that machinists' rules and steel scales use sixty-fourths as their smallest graduation.

2. *Combined common fraction and decimal fraction systems* are primarily used in machine drawings. In this system, both fraction and decimal expressions are found on the drawing. Here measures with fractional values that are smaller than sixty-fourths of an inch (e.g., one hundred twenty-eights, or two hundred fifty-sixths) will be expressed in decimal fraction form. A second situation in which the decimal fraction would be used is with measures requiring tolerances or accuracy greater than sixty-fourths of an inch. In this last case, the decimal dimension will be expressed in as many places as necessary for the required level of accuracy.
3. *Decimal fraction systems* use decimal values exclusively. According to ANSI standards, a two-place decimal should be used for dimensions in which common fraction values would typically be used, such as 0.18, 0.25, and 0.50. Dimensions requiring greater accuracy may be expressed by the required number of decimal places.

When using decimal fraction systems, it is often necessary to convert common fraction values into decimal equivalents. For example, one-fourth inch converts to the decimal equivalent of 0.250, and three-fourths inch to 0.750. When the decimal equivalent exceeds three decimal places, the value should be rounded off to the nearest three places. Thus, 5.557682 would be rounded off to 5.558, and 4.322107 to 4.322.

Some industries (e.g., aerospace and automotive drafting standards) require that if the last digit to be rounded off is 5, followed by zeros, then the preceding numbers would remain the same. Thus, the value 1.234500 would round off to 1.234. Before this procedure is adopted, company policies and standards should be checked first to see whether or not it is acceptable.

EXTENSION, PROJECTION, AND DIMENSION LINES

There is a slight difference in the presentation of dimensions as specified by ANSI and ISO drafting standards. As already mentioned, ANSI

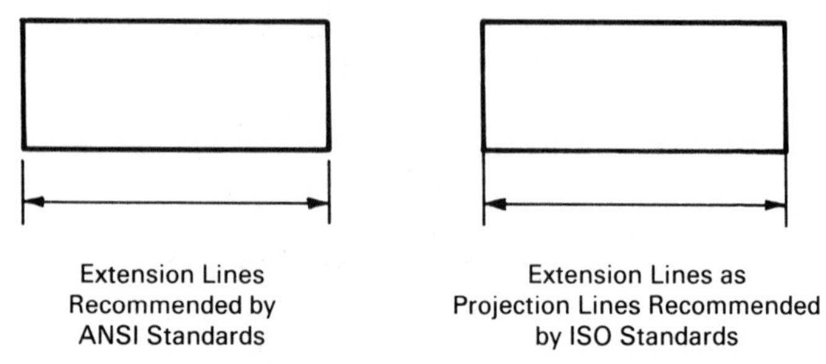

Fig. 9-8. Extension and projection lines.

extension lines do not come in contact with the edge that is being dimensioned. ISO standards, however, treat extension lines as if they were projection lines.

When extension lines are treated as projection lines, they will emanate from the edge that is being dimensioned; in other words, the projection (extension) line will touch the dimensioned edge. Once drawn, the dimension line is placed between the projection lines in the same manner as extension lines. Fig. 9-8 compares these two standards.

Sometimes it is necessary to determine the dimension of a point not located on the object itself. A typical example of this is shown in Fig. 9-9, in which a nonexistent corner must be located and dimensionally described. In such cases, construction lines should be projected and slightly extended beyond their point of intersection. From this point, the extension line may be drawn.

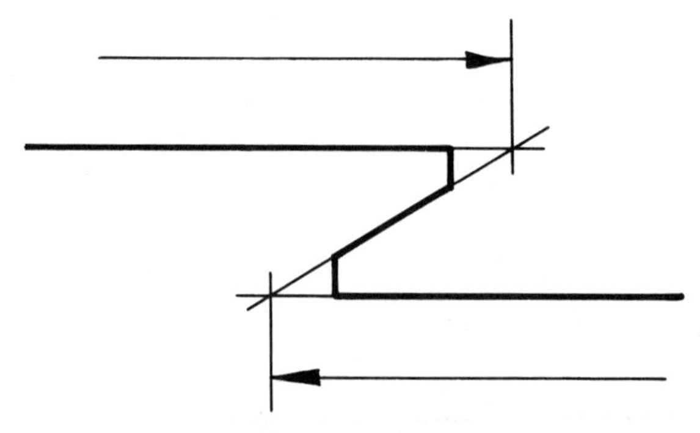

Fig. 9-9. Using construction lines to identify dimension points.

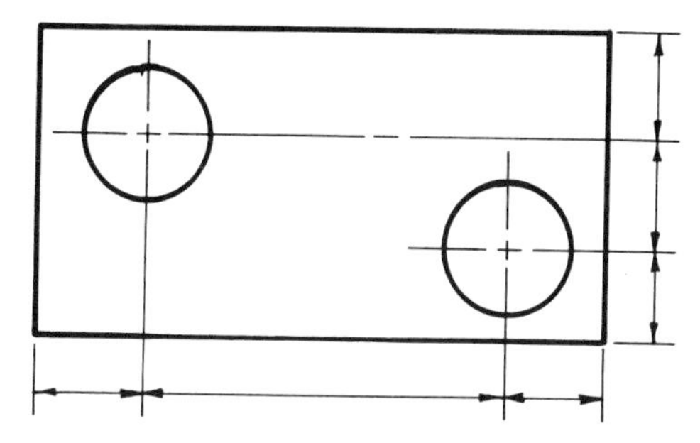

Fig. 9-10. **Using center lines as extension lines.**

It is acceptable to use an axis or object outline as an extension line (Fig. 9-10). However, such lines should never be used as dimension lines. Extension and dimension lines should not cross over one another. Sometimes, however, this is unavoidable, but the drafter is expected to keep it to a minimum. Where there is a series of dimension lines, it is recommended that the first dimension line be drawn approximately 3/8 inch or 9 mm from the object, with each succeeding dimension line spaced every 1/4 inch or 6 mm. Again, this may be modified according to the size of the drawing and the available space.

Extension lines should be drawn 90° to the feature being dimensioned. Usually this is a simple rule to adhere to, but there are times when it would confuse the reader and clutter the drawing. When necessary, the extension lines may be drawn at an angle to the feature (Fig. 9-11), but they should be parallel to each other.

To ensure the proper interpretation of chords, arcs, and angles, care must be taken that extension lines are properly placed. Fig. 9-12

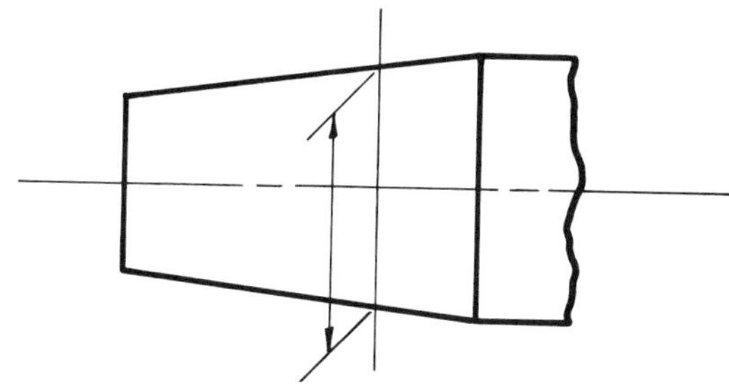

Fig. 9-11. **Proper drawing extension lines at nonperpendicular angles to dimensioned feature.**

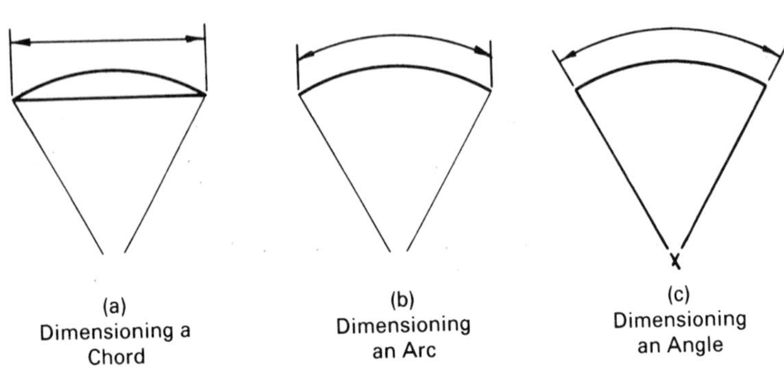

Fig. 9-12. Dimensioning circular elements.

shows the proper method for extension and dimension line placement for each type of feature.

Partial views and sections of symmetrical parts pose a unique dimensioning problem. Since the entire object (view) cannot be used to identify the limits of the dimension, the segment of the dimension line shown should extend a short distance beyond the axis. Once the extension line is drawn, an arrowhead should be drawn only on the end of the dimension line that is in contact with the extension line (Fig. 9-13), eliminating the second arrowhead.

Some product designs require that an object's outline be drawn at such a radius that the center point falls outside the limits of the drawing surface. When this occurs, the radius line should be broken or interrupted. A broken radius line is used when a center has been located, while an interrupted line may be drawn when a center is not required. See Fig. 9-14.

ARROWHEADS

The standard width-length proportion of 1:3 for arrowheads should be maintained on all drawings. The exact size, however, should be in proportion to the size of the drawing and the thickness of the lines of the drawing. Generally, the thicker the line, the larger the arrowhead.

Arrowheads should be drawn at the ends of the dimension line. When dimensioning a radius, one arrowhead at the arc end of the line is sufficient. If there is insufficient space to draw the arrowhead properly, it may be drawn outside the extension line, but along the

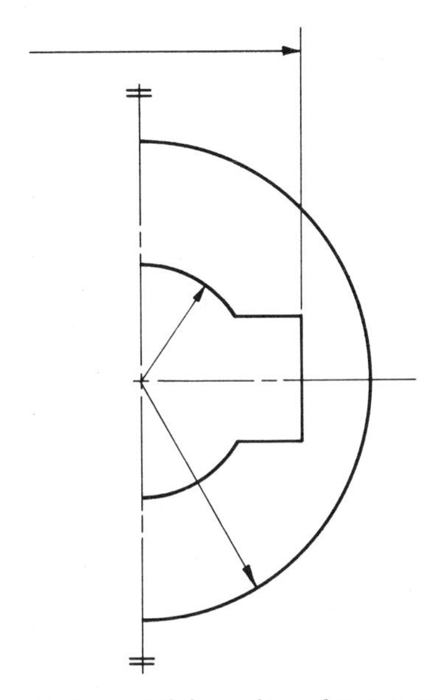

Fig. 9-13. Dimensioning partial section of symmetrical object.

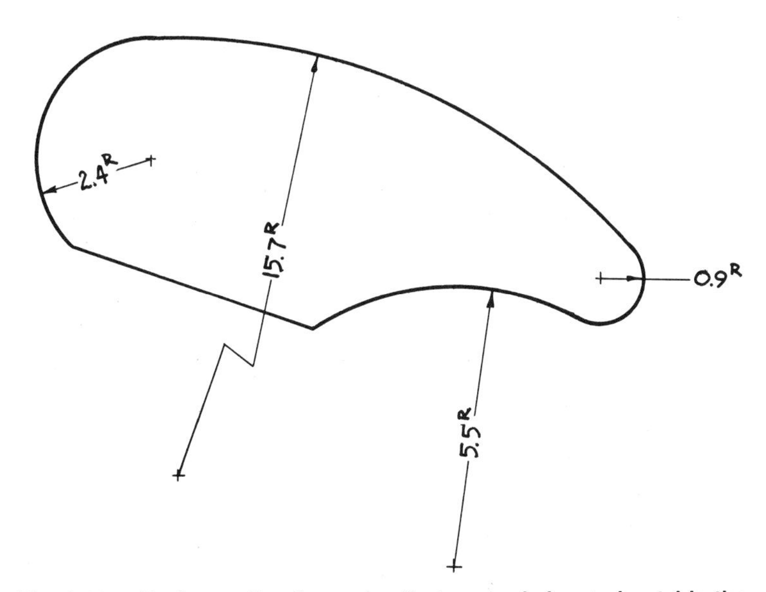

Fig. 9-14. Broken radius line notes that center is located outside the limits of available space.

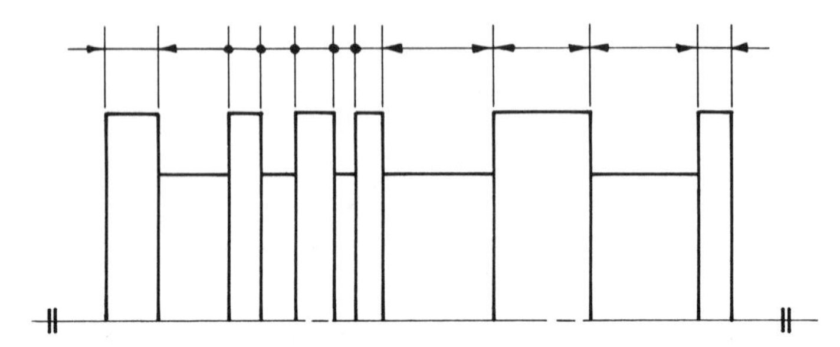

Fig. 9-15. Proper use of arrowheads and dots in dimensioning.

path of the dimension line. In some cases, a dot may be used to replace a single arrowhead. This latter method is commonly used for a succession of dimensions where inadequate spacing is available for arrowheads. See Fig. 9-15.

PLACEMENT OF DIMENSIONS

All figures and letters should be made large enough so that they can be clearly read and interpreted. Preferably, dimensions should be placed in the middle of the dimension line, and away from any other line that may cross it. Occasionally, space constraints make it difficult to avoid superimposing figures. In such cases, the dimension line may be shortened with one arrowhead, as shown in Fig. 9-16.

If there is insufficient space between extension lines for the dimension, it should be located on the right-hand side of the extension line (Fig. 9-17). When a series of dimensions prohibits the placement to the side of the extension line, a leader line may be drawn to the dimension line, as shown in Fig. 9-17.

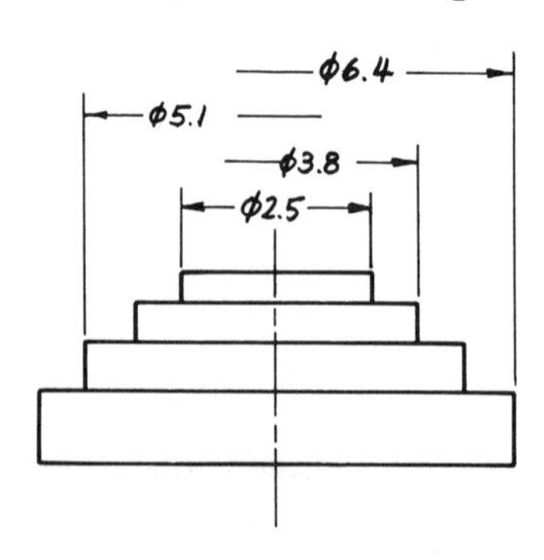

Fig. 9-16. Technique for avoiding superimposing dimensions in limited space.

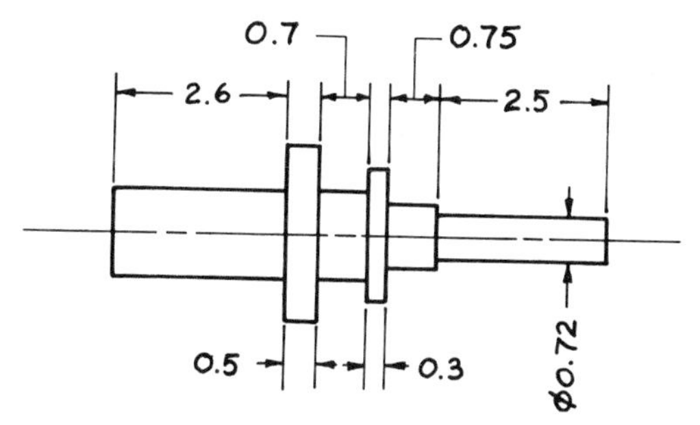

Fig. 9-17. Dimensions and arrowhead placement on object with limited space.

Sometimes a view is not drawn to scale. In this situation, the view should be labeled NOT TO SCALE, or drawn so that each dimension will be underlined.

To ensure easy reading of dimensions, figures should be situated on the drawing so that they can be read from the bottom or the right of the drawing. Unless there is no alternative, dimensions should not be placed in the 30° zone identified in Fig. 9-18a. Angular dimensions can be written either at an angle or horizontally (Fig. 9-18b).

USE OF LETTERS AND SYMBOLS

Special letter and symbol designations are used in the dimensioning of figures. Relative to geometric shapes, the following indications (Fig. 9-19) should be made with the dimension:

1. The dimensioning of a diameter should be prefaced with the symbol ∅, for example, ∅6.25.
2. Before the dimension of a radius, the letter R should precede the numerical value, as in R 2.50.
3. Before the dimension of a square section, the symbol □ should be used; for example, □ 1.75.
4. Before the dimension of a sphere, the word "sphere" or letter S should precede the diameter symbol and measurement. If SR precedes the measurement, this refers to the sphere's radius.

Sometimes it is necessary to dimension a series of repeated shapes with the same dimension. Instead of repeating the same dimension a number of times, reference letters should be used, and referenced

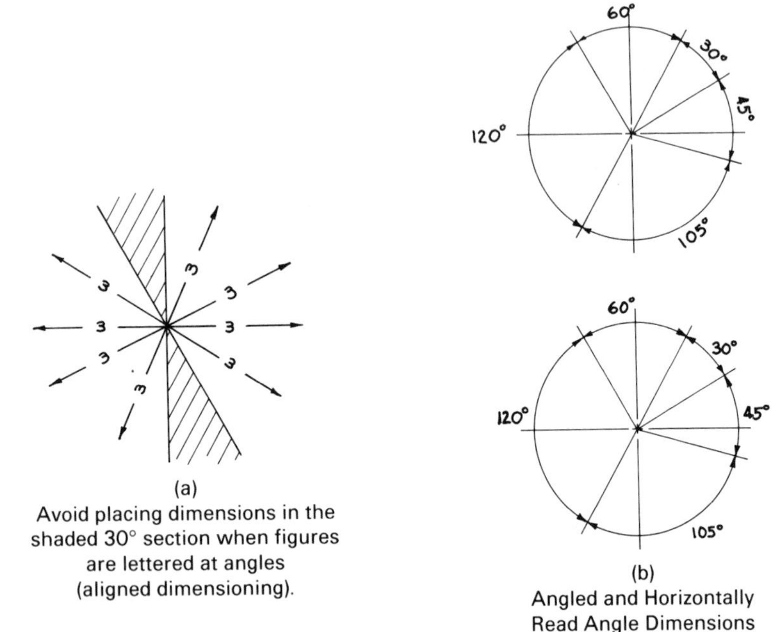

(a)
Avoid placing dimensions in the shaded 30° section when figures are lettered at angles (aligned dimensioning).

(b)
Angled and Horizontally Read Angle Dimensions

Fig. 9-18. Ensuring the readability of dimensions.

in an explanatory table or note. Another technique, which has recently been adopted by ANSI, is the use of X to denote the multiplication symbol. Thus, the notation 6X ∅ 0.250 refers to six holes (circles) with a diameter of 0.250.

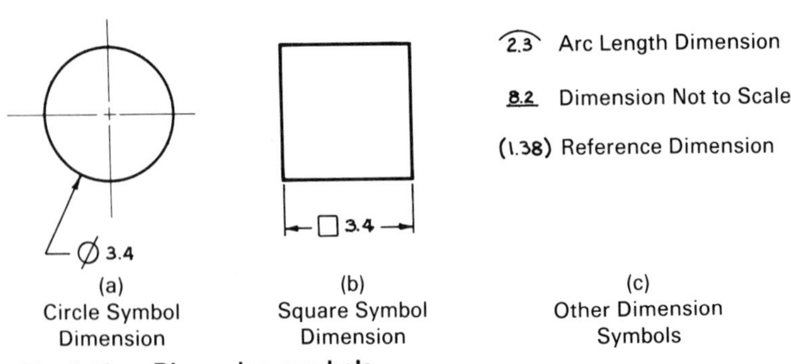

(a)
Circle Symbol Dimension

(b)
Square Symbol Dimension

(c)
Other Dimension Symbols

Fig. 9-19. Dimension symbols.

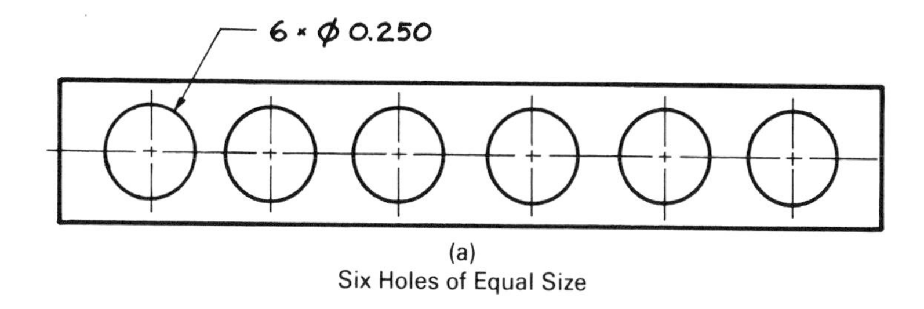

(a)
Six Holes of Equal Size

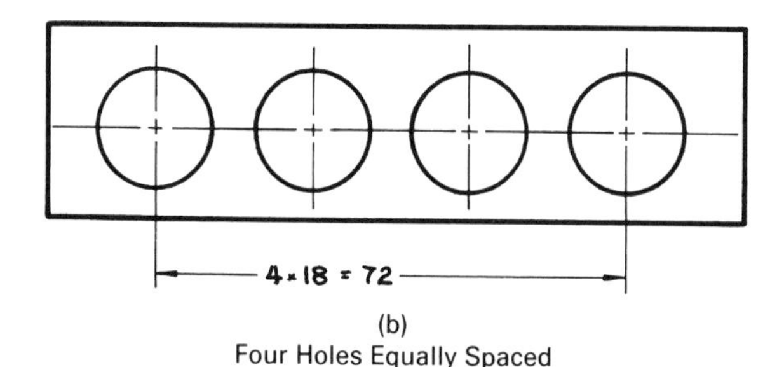

(b)
Four Holes Equally Spaced

Fig. 9-20. Simplified dimensioning for repeated measures.

A series of holes that are equidistant from one another can also be dimensioned by a simplified procedure. For example, if four holes were spaced equally over a distance of 72 inches, extension lines would be drawn at that distance with the following dimension: 4× 18 = 72. This means that the four holes are spaced equally at 18 inches over a distance of 72 inches. The two simplified dimensioning methods are shown in Fig. 9-20.

Arrangement of Dimensions

One of the most difficult problems faced by beginning drafters is the proper arrangement of dimensions. Generally, there are two types of dimension: size and location. Size dimensions are used to describe the various features of the part, such as holes, slots, squares, circles, arcs, rounds, surfaces, and corners. Location dimensions are used to specify the position of features relative to one another. Perhaps the most common form of position reference is the center line. Shown in Fig. 9-21 is a comparison of size and location dimensions.

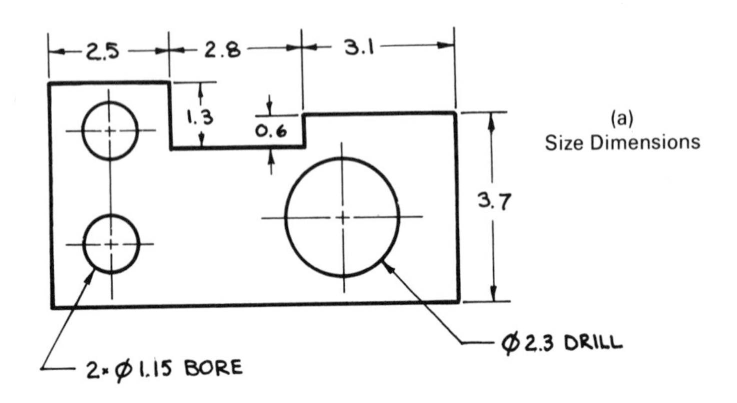

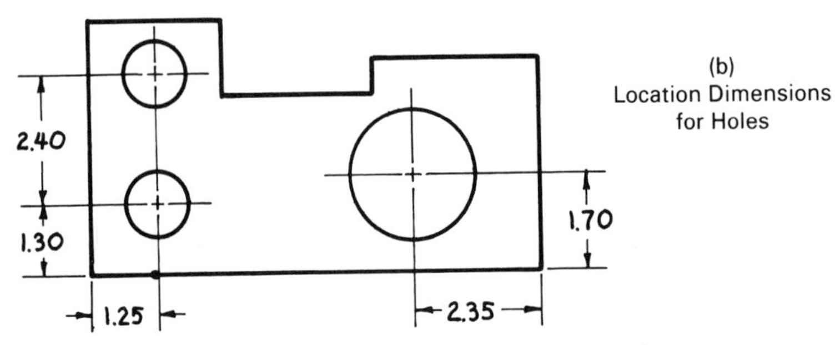

Fig. 9-21. Comparison of size and location dimensions.

DIMENSIONING RULES

Before drafters can successfully determine the specific arrangement of dimensions, they must first be aware of several basic rules for dimensioning a drawing. The rules presented here apply to most drawing problems:

1. Dimensions located closest to the view should be those of the smallest feature.
2. Dimension placement should then follow for each successively larger feature, until the last dimension given represents the largest dimensional measure. Thus, a dimension giving the overall size of the view should be the farthest away from that view.
3. A feature should be dimensioned in the view that shows it most clearly.

4. Normally, at least two dimensions are required to describe a feature adequately.
5. Dimensions should not be placed on the object, unless it is unavoidable.
6. Most of the dimensions should be placed between views.
7. Dimensions should be made from object lines. It is considered poor practice to dimension from hidden lines. If a hidden feature must be dimensioned, it may be advantageous to use a broken-out section to do so.
8. Good dimensioning practice dictates that dimensions be made from finished surfaces (e.g., machined surfaces or surfaces that are not to be processed), center lines, or base lines.
9. The dimensioning of holes and complete circles should be given as diameters.
10. The location of all circular features should be made with reference to their center point.
11. All circular features, excluding full circles, should be dimensioned by radii (e.g., arcs, rounds, fillets, etc.).
12. When two or more parts are drawn in subassemblies or assembly drawings, dimensions should be grouped in relationship to each part.

PLACEMENT OF DIMENSIONS

Two general methods are used for the placement of dimensions. The first is known as *aligned dimensioning*. Shown in Fig. 9-22, this method arranges dimensions so that they can be read from the bottom *and* the right side of the drawing. The aligned system of dimensioning is the older of the two forms of dimensioning.

The second method for dimension placement is *unidirectional dimensioning*. This procedure places the dimensions in such a manner that they are all read from the bottom of the drawing (see Fig. 9-23). Drawings employing this technique are easier to read and interpret; unidirectional dimensioning is therefore the method most commonly used by drafters today.

ARRANGEMENT PRACTICES

Based upon the rules of dimensioning, a standardized system of practices has been established for arranging dimensions on drawings. The practices presented here will cover most of the dimensioning problems

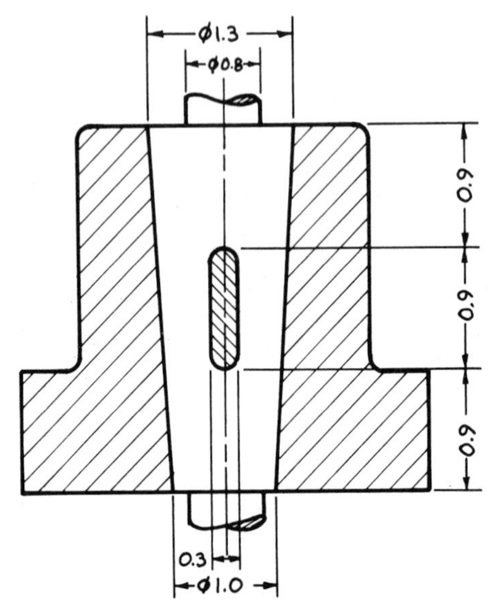

Fig. 9-22. Aligned dimensioning.

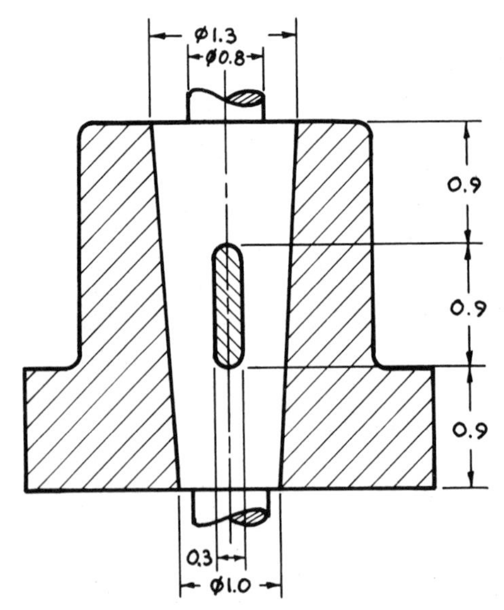

Fig. 9-23. Unidirectional dimensioning.

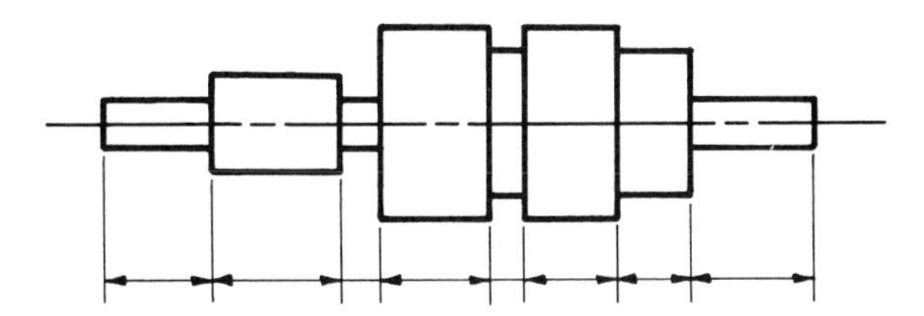

Fig. 9-24. Chain dimensioning.

encountered by the drafter. Should an unusual problem arise, it is recommended that the drafter consult company policy pertaining to such situations. If none exists, the drafter should attempt to solve the problem based upon the basic rules and practices described herein.

A dimensioning practice that is used when the accumulation of tolerance is not threatened is known as *chain dimensioning* (Fig. 9-24). If there is any chance that processing procedures may violate tolerances so as to endanger the functional qualities of the part, this technique should be avoided.

An alternative practice to chain dimensioning is *parallel dimensioning*. Sometimes referred to as *rectangular coordinate dimension-*

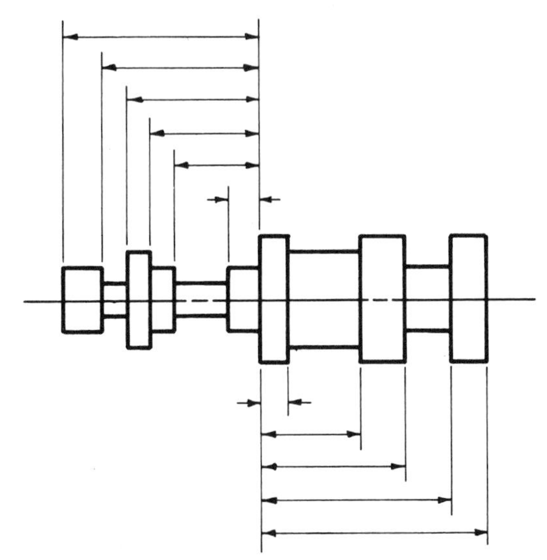

Fig. 9-25. Parallel dimensioning.

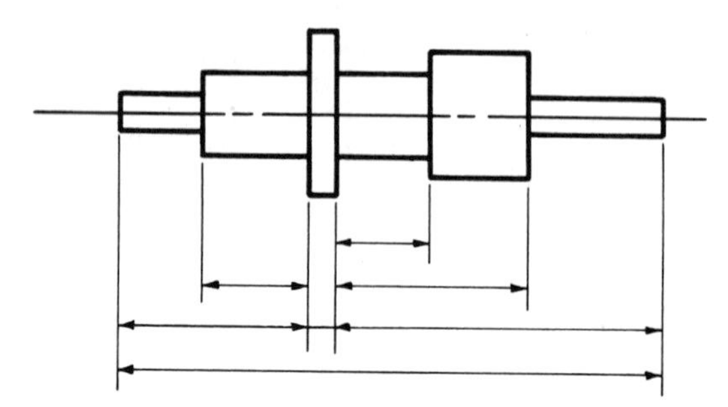

Fig. 9-26. Combined chain and parallel dimensioning.

ing, this technique provides a series of dimensions that have a common reference or datum feature (Fig. 9-25). This technique is used to hold processing tolerances within tighter limits by avoiding the possibility of accumulative error.

Another form of parallel dimensioning is used when there is no risk of confusion. Here the datum line is noted by a dot and a zero sign. The dimension itself is placed in line with each extension line along the path of the dimension line. Note that only one arrowhead is used per dimension span (although two arrowheads per dimension length may be used by some companies).

Because chain and parallel dimensioning are designed with specific parameters in mind, it is seldom necessary to prepare a drawing that uses only one or the other technique; a more common practice is to use a combination of the two. Fig. 9-26 shows the use of both parallel and chain techniques, which offers greater flexibility in dimensioning problems.

A geometric feature that is commonly encountered in many drawings is the chamfer. Two techniques can be used to dimension such features. The first (Fig. 9-27a) is the most traditional form, in which the width and angle of the chamfer are individually dimensioned. The second method (Fig. 9-27b) provides a simplified and less cumbersome technique by combining the two measures in one dimension.

PRECISION WORK DIMENSIONING

The need for producing precision parts has placed an added burden upon engineers, technologists, and drafters. To meet tolerance requirements demanded by product designs, care must be taken to en-

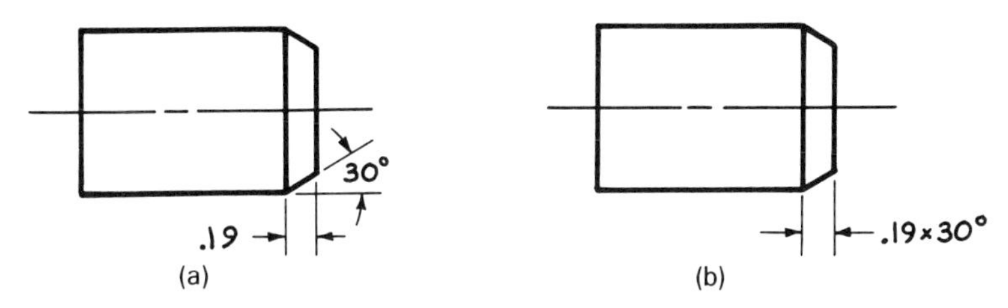

Fig. 9-27. Dimensioning chamfers.

sure that correct dimensioning procedures be used so that processing personnel can interpret and meet design criteria. Three dimensioning methods are used for this function.

The first method is known as *base line dimensioning*, and requires the use of finished surfaces as a reference. These surfaces, normally at right angles to each other, function as base lines and are sometimes called datum planes. Illustrated in Fig. 9-28, base line dimensioning is similar to parallel dimensioning in that datums are used for computing the location and relationship of a product's features.

Datums are identified by boxed-in letters, and may be surfaces or any other recognized feature that is assumed to be exact. Since datums are used for measurements, they must be accessible and can-

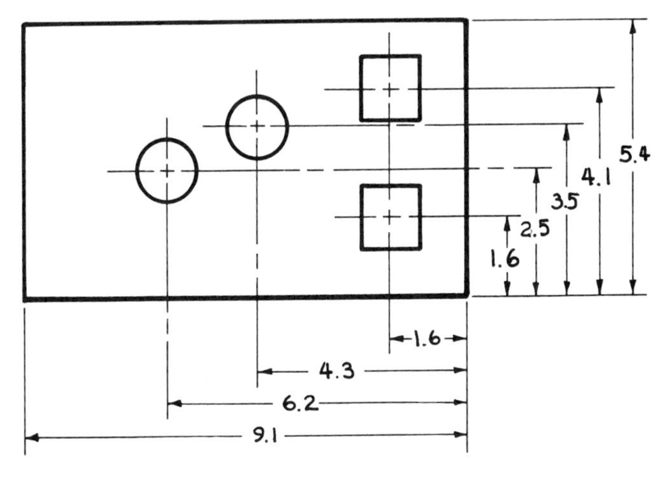

Fig. 9-28. Base line dimensioning.

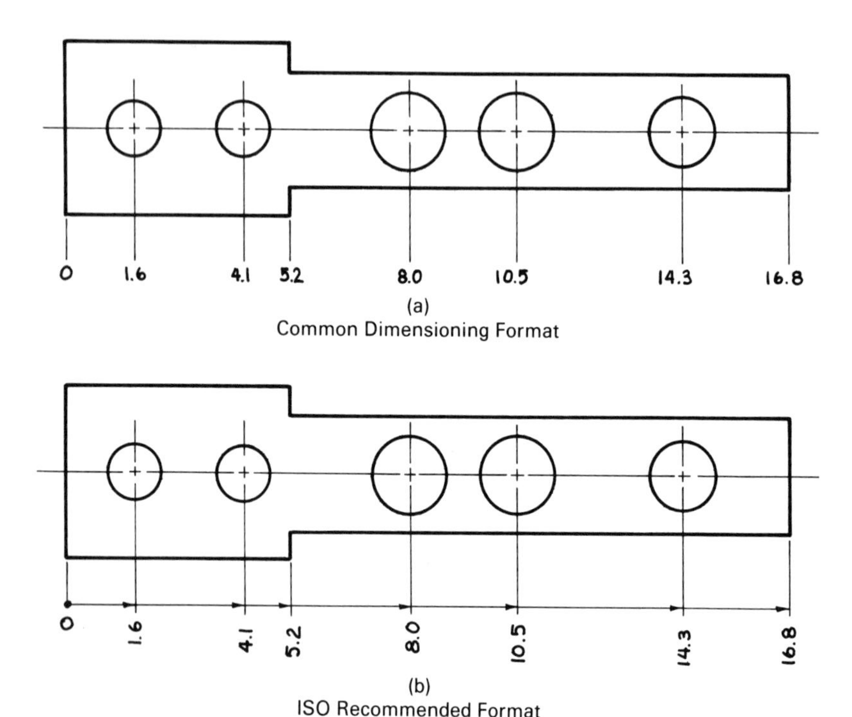

Fig. 9-29. Ordinate dimensioning.

not be an imaginary point. In subassemblies and assembly drawings, datums may be joining or connecting features of mating parts.

A second method used for dimensioning precision parts is *ordinate dimensioning*. A newer dimensioning method, ordinate dimensioning uses two or three datum planes, which are labeled with zeroes. Two techniques can be used for specifying dimensions from the datum planes. The first is recommended by ANSI standards, and makes no use of dimension lines or arrowheads. Here the dimension is placed at each extension line. The second technique is presented in ISO standards, and is similar to the simplified parallel dimensioning procedure. The two are compared in Fig. 9-29.

Features that are located on the surface of an object may be dimensioned in one of two ways. The first is by direct dimensioning on the object itself, and the second is by reference to a table.

The last precision dimensioning technique is another form of base line dimensioning, and is known as *tabular dimensioning*. This tech-

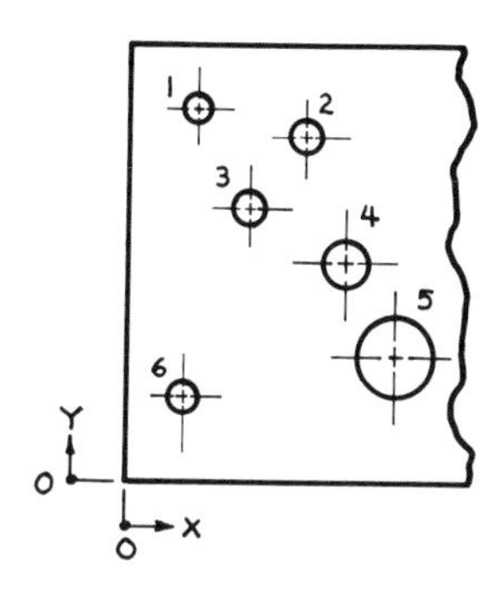

	1	2	3	4	5	6
X	1.1	2.8	2.0	3.5	4.3	0.9
Y	5.7	5.3	4.2	3.4	1.9	1.3
Z	THRU	THRU	THRU	1.5	2.0	THRU
Ø	0.5	0.6	0.6	0.8	1.3	0.5

Fig. 9-30. Tabular dimensioning.

nique is based upon an X, Y, and Z coordinate system that is extensively used in numerical control machining and CAD/CAM systems. Here the X and Y coordinates pertain to length and width measures, while Z coordinates correspond to depth measures. In this procedure, all measures are referenced to a table. See Fig. 9-30.

As an example of the use of tabular dimensioning, take a hole with the following specifications:

X = 2.204
Y = 1.355
Z = 0.125
Ø = 0.500

This is interpreted as meaning that the center of the hole is located 2.204 inches from the datum point along the X axis and 1.355 inches along the Y axis. It has a diameter of 0.500 inch and a depth of 0.125 inch.

Notations

Most companies have established a standard title block that provides information relative to the drawing. Data such as part name, drawing scale and tolerances, dates, drafter, and checker are used to provide readers with the basic information needed for drawing interpretation. In addition to the title block, other notations, in the form of symbols and words, are used to clarify the drawing further. These notations are particularly important for processing, installment, and servicing operations. This section will discuss those symbols and word notations that are used in industrial drawings.

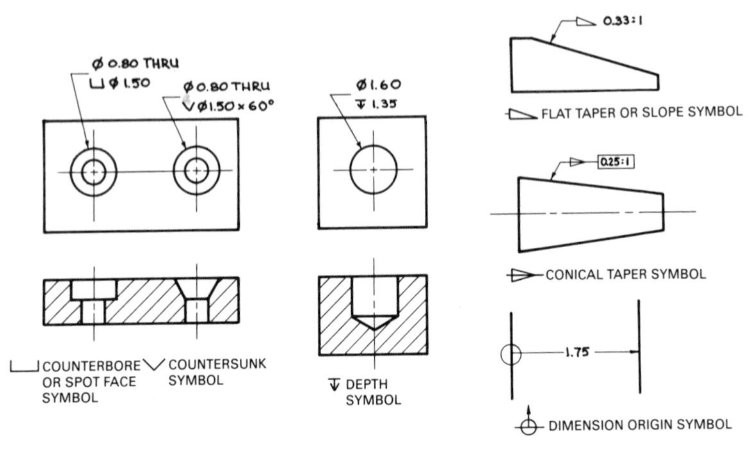

Fig. 9-31. ANSI symbols adopted in 1982.

Notation Symbols

Prior to 1982, machining and dimension notations were given in words. For example, if a 3/4-inch hole was to be drilled, the following notation would be given on the drawing: 3/4 DRILL. This same procedure would be used for other processes, such as countersinking and counter boring.

Today, traditional machining and dimensional notations have given way to a series of standardized symbols. Already presented in this chapter were the symbols used for circle and square shapes. A summation of machining and dimension symbols is given in Fig. 9-31.

Word Notations

Even with the adoption of symbol notations, there still persists the need for general word notations. Words should be used to simplify and/or complement the dimensioning process. There are two basic types of word notations used on drawings:

1. *General notes* are used to refer the entire drawing or object. When lettered, they should be located in an easily identifiable position, preferably a central location on the drawing. In many

cases, preprinted drawing paper will provide a notation column. Examples of general notes are:

- FINISH ALL OVER
- ALL RADII 1/4 UNLESS OTHERWISE SPECIFIED
- TOLERANCES ON LINEAR DIMENSIONS (INCL. HOLES) 0.010, ANGULAR 1/2°, EXCEPT 5° ON ALL CHAMFERS
- ALL BEND RADII 0.095

2. *Local notes* are used for local features or requirements only. They are usually connected by a leader to the feature to which the information applies. Typical examples of local notes include:

   - 3/8 DRILL, 4 HOLES
   - WELD C.R.S. STRIP TO BOLT
   - 10-32 TAP THRU
   - 1/4 × 45° CHAMFER

Many different abbreviations are used in general and local notations, primarily to save time and space. Common abbreviations such as DIA, R, THD, MAX, MIN, ID, OD, and DP are usually understood by a wide cross section of industries. Others, such as BRZ (bronze), SCR (silicon-controlled rectifier), and IPS (iron pipe size), should be used with the utmost care so as to avoid any misinterpretation. Any abbreviation used should conform to the standard of the industry. As a rule of thumb, when in doubt, spell it out.

Exercises

9.1 The drawing shown in Fig. 9-32 is printed half size, and is made of cast iron. Draw and dimension this part. All measurements are to be taken to the nearest 1/64 inch, and presented in decimal form.

9.2 Prepare a tabular dimension drawing for the part illustrated in Fig. 9-33.

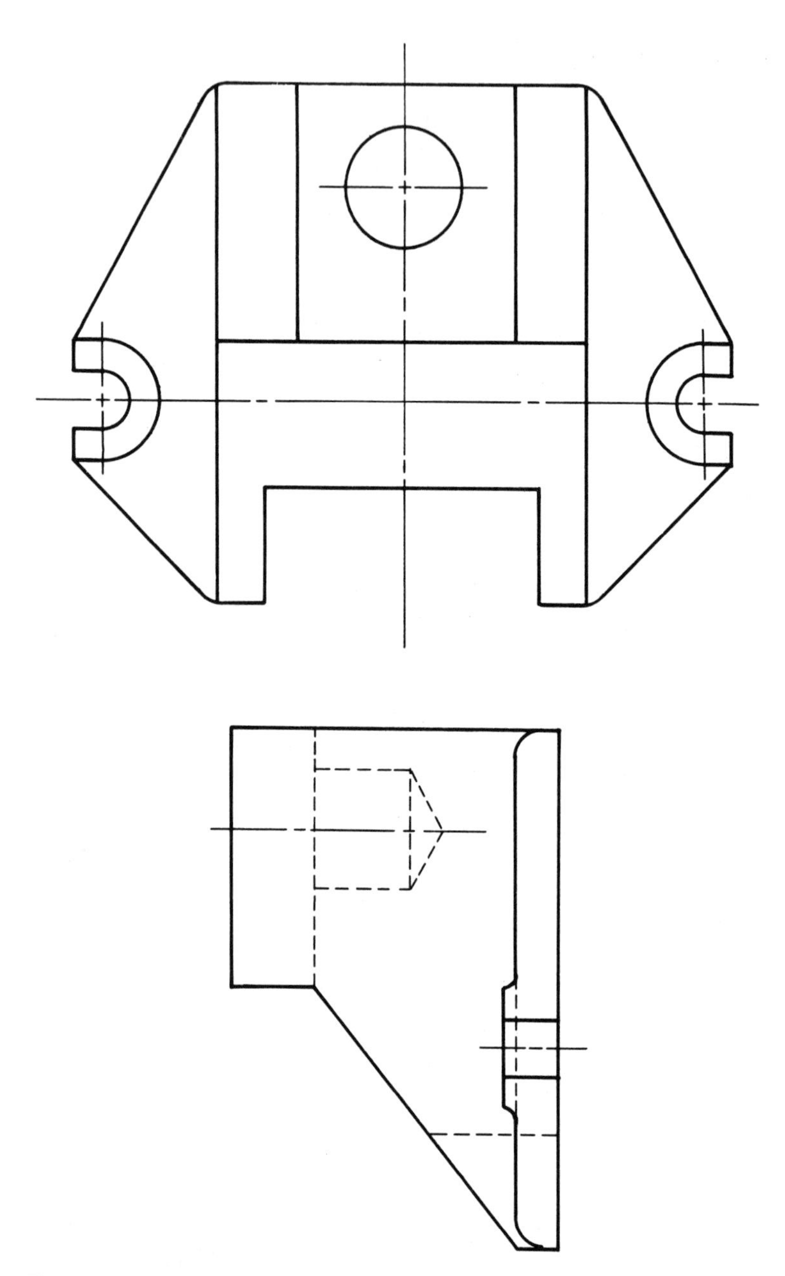

Fig. 9-32. Problem 9.1.

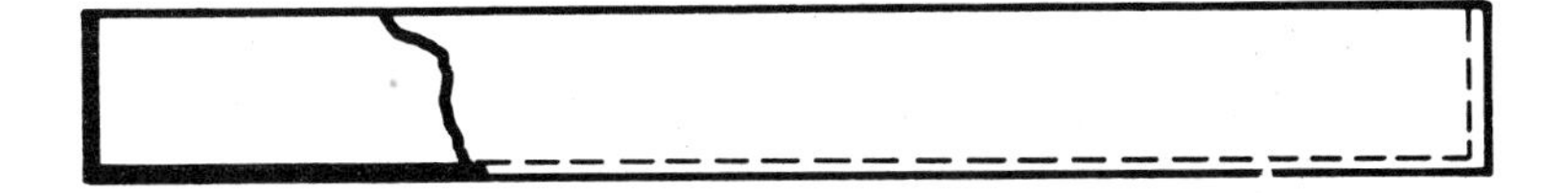

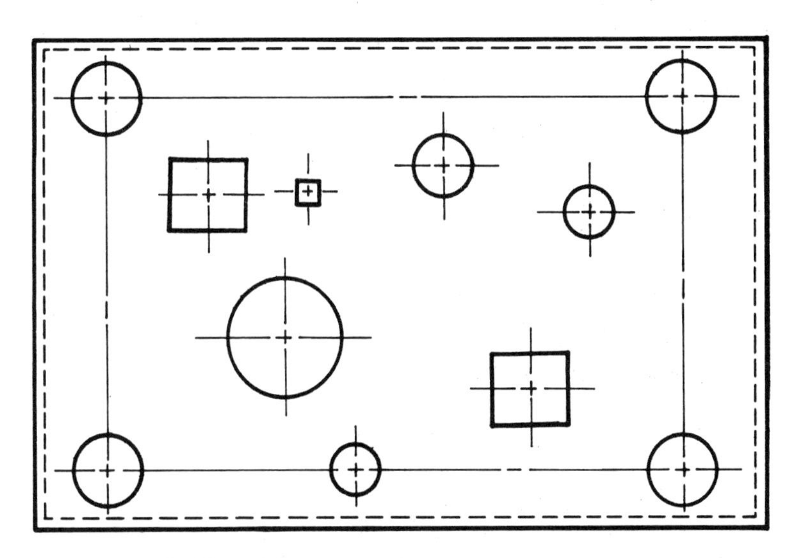

Fig. 9-33. Problem 9.2.

CHAPTER 10

Allowances and Tolerances

The use of an allowance and tolerance system for defining limits and fits is of primary importance in the manufacturing of structures, machines, and machine elements. The degree of precision and care that is used in the manufacturing of a product will determine its quality relative to competing products, and will contribute to both production and selling costs. Since the manufacture of most products requires that some degree of precision be maintained, it is essential that drafters have a clear understanding and working knowledge of how precision is expressed in drawings.

The specification of precision on drawings is made in terms of allowances and tolerances, and is presented as a dimensional expression. Hence, allowances and tolerances for limits and fits is another aspect of dimensioning. This chapter will discuss those drawing practices that are used by drafters for the expression of precision.

Basic Definitions and Expressions

There are three general concepts that must be understood when dealing with manufacturing accuracy. The first is *precision*. Precision is defined as a degree of accuracy or exactness that is used to ensure the required assembling or functioning of a part or machine element. An example of this is a comparison of mating and nonmating surfaces. Nonmating surfaces do not come in contact with any other surface and will not have any direct effect on the function or assembly of another part or element. These surfaces are usually exposed to the air and environment and require no precision processing. Mating surfaces require significantly greater machining accuracy because they are directly related to the operation or function of other surfaces or parts; they are therefore machined to a required smoothness and distance from each other.

The second concept to be dealt with is *tolerance*. Tolerance is the amount of dimensional deviation that is allowed for given measurements—it is the method used for achieving precision. The level of tolerance on individual dimensions will vary, depending upon the degree of precision required for that surface. As an example, mating surfaces will require closer tolerances, as small as millionths of an inch (0.000 001 in.), while nonmating surfaces will vary from one inch to hundredths of an inch (1 to 0.01 in.).

The third concept related to allowances and tolerances is that of *limits and fits*. Fits between parts will dictate the proper assembly and performance of a product. *Clearance fits* are used to express the amount of freedom of motion between parts (e.g., shaft and hole), while *Interference fits* are used to fasten or secure parts. The expression of these fits is dimensional, and involves the adoption of manufacturing limits for these parts. Thus, the limits and fits of parts are expressions of manufacturing tolerance.

Technical Terms

A number of technical terms are used in relation to limit and fit dimensioning. The basic terms used in precision dimensioning are listed here, and are illustrated in Fig. 10-1a and b.

Actual size of a part is that size that is determined by direct measurement.

Limits of size pertain to the minimum and maximum sizes that are permitted for a particular part.

Maximum limit of size is the greater of the two limits of size.

Minimum limit of size is the smaller of the two limits of size.

Basic size is a reference dimension that is given to both parts of a fit. From the basic size, limits of deviation (maximum and minimum limits) are derived.

Deviation is the algebraic difference between a size and the corresponding basic size.

Actual deviation is the algebraic difference between the actual size and the corresponding basic size.

Upper deviation is the algebraic difference between the maximum limit of size and the corresponding basic size.

Lower deviation is the algebraic difference between the minimum limit of size and the corresponding basic size.

Fundamental deviation is the one of the two deviations that comes closest to the basic size, and is designated by the letter "H" (e.g., 38H8).

Zero line is a straight-line graphic representation of limits and fits to which the deviations are referenced.

Tolerance is an absolute value with no plus or minus sign that is calculated by finding the algebraic difference between the upper and lower deviations.

Tolerance zone is the graphic representation of tolerance. It is that area bounded by the two limits of tolerance lines, and is defined by the size (magnitude tolerance) and its position relative to the zero line.

Grade of tolerance is a standardized system of limits and fits. It is defined as a group of tolerances that correspond to the same level of precision (accuracy) for all basic sizes.

Clearance is a positive value found by the mathematical difference (prior to assembly) between the sizes of a hole and shaft.

Interference is a negative value found by the mathematical difference (prior to assembly) between the sizes of a hole and shaft.

Clearance fit is the relationship between assembled parts when clearance exists under all tolerance conditions (i.e., when the difference between the hole and shaft is positive).

Interference fit is the relationship between assembled parts when interference exists for all tolerance conditions (i.e., when the difference between the hole and shaft is negative).

Transition fit is a fit that provides either clearance or interference so that the tolerance zones of the hole and shaft will overlap.

Shaft-basis system of fits is that system in which clearance and interference are calculated by associating holes with a single

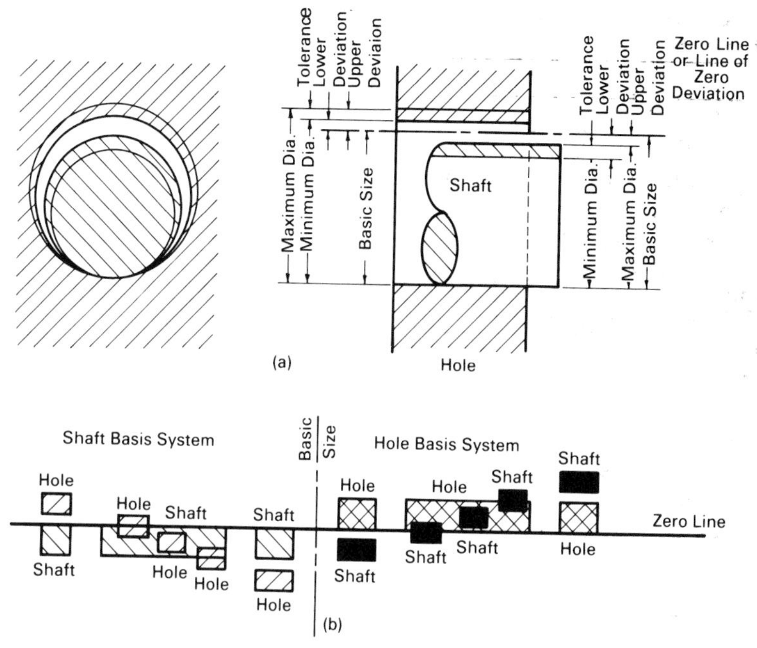

Fig. 10-1. Limits and fits nomenclature.

shaft or shafts of different grades with the same fundamental deviation. Within the ISO system, the basic shaft is defined as a shaft having an upper deviation of zero.

Hole-basis system of fits is that system in which clearance and interference are calculated by associating shafts with a single hole or holes of different grades with the same fundamental deviation. Within the ISO system, the basic hole is defined as a hole having a lower deviation of zero.

Description of Fits

ANSI B4.1 recommends preferred sizes, allowances, and tolerances for fits between cylindrical components. These include bearings, shrink fits, and drive fits. Fits are broken down into three major categories: 1) running and sliding fits, 2) locational fits, and 3) force or shrink fits. A graphic representation of these standard limits and fits is given in Fig. 10-2.

RUNNING AND SLIDING FITS

This category, designated as RC, is designed to provide similar running performance, with appropriate lubrication allowance, throughout the specified range of sizes. RC fits are further subdivided into the following classes:

Class RC1: Close-Sliding Fits

This type of fit is used for the accurate location of parts that must be assembled without perceptible play.

Class RC2: Sliding Fits

Intended for parts that must move and turn easily, but cannot run freely, allowing for greater maximum clearance. With this type of fit, larger parts will tend to lock when subjected to small temperature changes.

Class RC3: Precision-Running Fit

Parts subjected to this fit will meet the closest fit that will still allow them to run freely. Precision-running fits are designed for precision work at slow speeds and light journal pressures. This class of fit is not recommended where there may be any appreciable temperature change.

Class RC4: Close-Running Fits

Intended for running fits on accurate machinery that has moderate surface speeds and journal pressures. Used where accurate location and minimum play is desired.

Class RC5 and RC6: Medium-Running Fits

Used for higher running speeds and/or heavy journal pressures.

Class RC7: Free-Running Fits

Intended for use where accuracy is not essential and/or large temperature variations are likely to be encountered.

Class RC8 and RC9: Loose-Running Fits

For use with materials, such as cold-rolled shafting or tubing, in which wide commercial tolerances may be necessary, together with an allowance on the external member.

LOCATIONAL FITS

All locational fits are designed to determine the location of the parts. The three general classes of locational fits are:

LC. Locational Clearance Fits

These fits are intended for parts that are normally stationary, but are freely assembled and disassembled. They range from snug fits for parts requiring accuracy of location, through medium clearance fits (e.g., spigots), to looser fastener fits, where freedom of assembly is of prime importance.

LT. Locational Transition Fits

This class of locational fit is used where accuracy of location is important, but a smaller amount of clearance or interference is allowable. This fit is a compromise between a location clearance and an interference fit.

LN. Locational Interference Fits

Used for parts in which accuracy of location is critical and which require rigidity and alignment with no special requirement for bore pressure. These fits are designed to transmit friction loads from one component to another by virtue of the tightness of fit.

FORCE FITS

Force or shrink fits (FN) make up a special category of interference fit that is commonly characterized by the maintaining of constant bore pressures through the entire range of sizes. As a result, the interference will vary almost directly with the diameter (the difference between its minimum and maximum is small) to maintain acceptable resulting pressures. In all, there are five classes of force fit:

Class FN1. Light-Drive Fits

Used in thin sections, long fits, cast-iron external members for applications that require light assembly pressure.

Class FN2. Medium-Drive Fits

Suitable for ordinary steel parts or for shrink fits on light sections. Of the five classes, Class FN2 is the tightest fit that can be used on high-grade cast-iron external members.

Class FN3. Heavy-Drive Fits

Applicable for heavier steel components or shrink fits within medium sections.

Classes FN4 and FN5. Force Fits

Intended for parts that may be highly stressed. In situations in which heavy forces are impractical for mounting, shrink fits are used instead of press fits.

ISO Series of Tolerances

The American National Standard Preferred Metric Limits and Fits (ANSI B4.2-1978) describes the ISO system of metric limits and fits (ISO 1962, 1969) for mating parts. The ISO system uses eighteen tolerance grades designated IT 01, IT 0, IT 1, IT 2 . . . IT 16. Within each grade, the specified tolerance will increase with the diameter in accordance with an algebraic formula. The size or magnitude of the tolerance per grade for a given size is approximately 1.6 times greater than the tolerance for the next smaller grade. For grades above IT 6, the tolerance is multiplied by 10 at each fifth step.

The fundamental deviation, in the ISO system, provides twenty eight deviations for sizes up to and including 500 mm. There are another fourteen deviations for larger sizes. Each deviation is designated by the following letters (see Fig. 10-3):

For holes: A, B, C, CD, D, E, EF, F, FG, G, H, J_s, J, K, M, N, P, R, S, T, U, V, X, Y, Z, ZA, ZB, ZC

For shafts: a, b, c, cd, d, e, ef, f, fg, g, h, j_s, j, k, m, n, p, r, s, t, u, v, x, y, z, za, zb, zc.

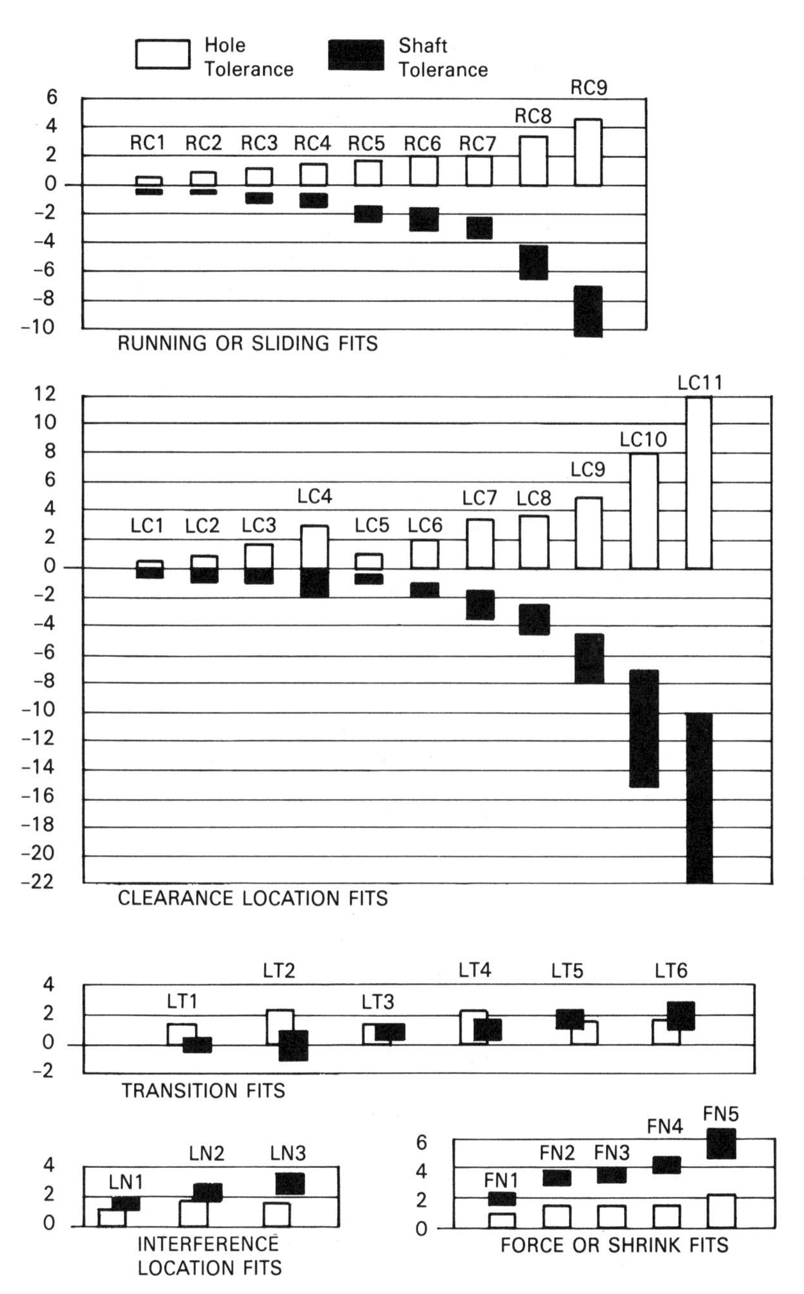

Fig. 10-2. Graphic representation of ANSI standard limits and fits.

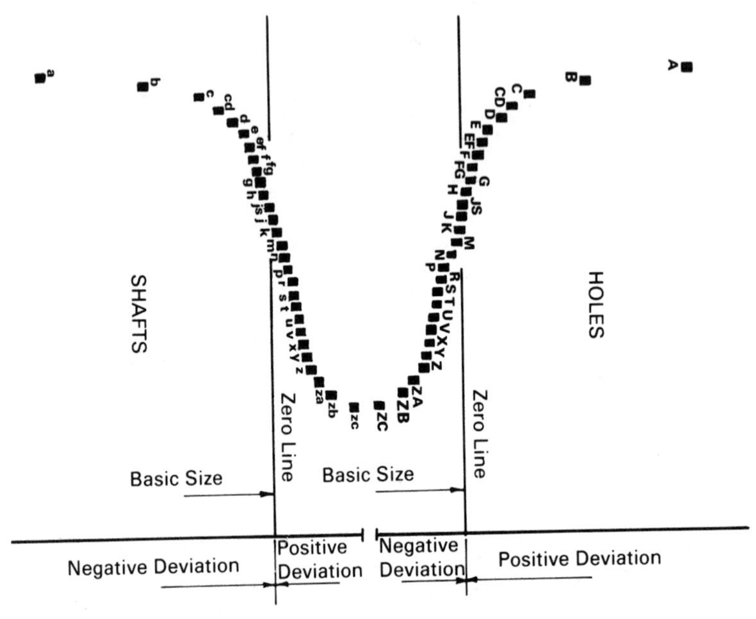

Fig. 10-3. ISO fundamental deviations for holes and shafts.

As shown in Fig. 10-3, deviations A to H are entirely positive and N to ZC are entirely negative. Deviations a to h are entirely negative and deviations n to zc are entirely positive. Deviations J, j, K, k, M, m are partly positive and partly negative. The subscript s used in deviations J_s and j_s notes that they are symmetrical about the zero line.

The ISO system is designed to provide a wide range of fits based upon a large number of variable combination of deviations and tolerances. Preferred fits are described in Table 10-1 and shown in Fig. 10-4.

Determining Tolerances

The selection of proper tolerances is based upon the principle that the clearance or interference between mating parts should not impede their proper functioning or assembly. The mathematical difference between their tightest and loosest fit will produce the sum of the tolerances of both parts. For example, if the parts cannot fit closer than 0.003 inch, owing to lubrication requirements, and must not

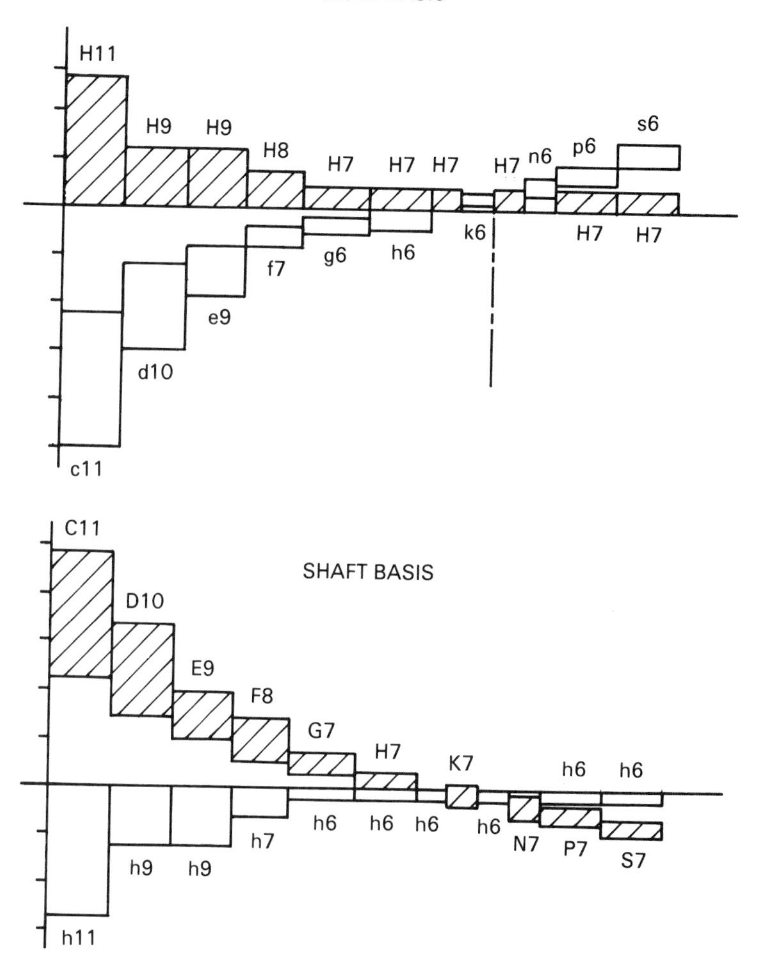

Fig. 10-4. Preferred ISO fits.

have a fit that exceeds 0.006 inch, then the sum of tolerances will be 0.006 − 0.003 = 0.003 inch.

To determine what the tightest and loosest fit should be not only requires engineering information, but also involves the use of engineering tables and mathematical procedures, as well as an appropriate method for expressing and specifying these tolerances. This section will discuss how tolerances can be expressed and calculated.

Table 10-1 Preferred ISO Fits

ISO Symbol		
Hole Basis	*Shaft Basis*	**Applications**
Clearance Fits		
H11/c11	C11/h11	Loose-running fit used for wide commercial tolerances or allowances on external members.
H9/d9	D9/h9	Free-running fit that is not used where accuracy is essential, but is useful where there are large temperature variations, high running speed, or heavy journal pressures
H8/f7	F8/h7	Close-running fit intended for running on accurate machines and where accurate location is necessary at moderate speeds and journal pressures
H7/g6	G7/h6	Sliding fit not designed for parts to run freely, but to move and turn freely and locate accurately.
H7/h6	H7/h6	Locational clearance fit provides for snug fit in the locating of stationary parts that can be freely assembled and disassembled
Transition Fits		
H7/k6	K7/h6	Locational transition fit used for accurate location, and considered to be a compromise between clearance and interference
H7/n6	N7/h6	Locational transition fit intended for more accurate location where greater interference is acceptable.
Interference Fits		
H7/p6	P7/h6	Locational interference fit used for parts requiring rigidity and alignment with prime accuracy of location but without special bore pressure equipment (Note: H7/p6 should be used for basic sizes ranging from 0 through 3 mm.)
H7/s6	S7/h6	Medium drive fit is used for ordinary steel parts or shrink fits on light sections; the tightest fit that can be used with cast iron.
H7/u6	U7/h6	Force fit suitable for components that may be highly stressed or for shrink fits where heavy pressing forces required are not practical

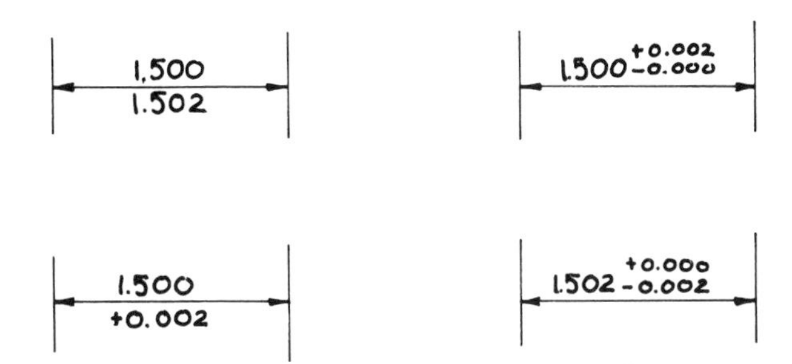

Fig. 10-5. Unilateral tolerancing.

Tolerance Systems

In many instances, necessary tolerances may be expressed by lettering a general note. Such notes make reference to specific dimensions. An example of this is:

ALLOWABLE VARIATION ON ALL DECIMAL DIMENSIONS
IS ± 0.002 UNLESS OTHERWISE SPECIFIED.

This type of notation is applicable to all dimensions when the limits are not given.

In addition to the general notation system, two other systems are used in precision dimensioning. The first is known as the *unilateral tolerance system*. This is a plan whereby a tolerance in which variation is permitted is given in one direction from the design size. Thus, a unilateral tolerance will be expressed in terms of either a positive or a negative value. Unilateral tolerances can be expressed by giving the two dimensional limits or by giving one limiting size accompanied by the tolerance. These two techniques are shown in Fig. 10-5.

The second, and more frequently used, precision dimensioning system is known as the *bilateral tolerance system*. Here the tolerance is expressed in two directions from the design size. A bilateral tolerance will thus be expressed by a positive and a negative value, which are usually equal in magnitude. Examples of bilateral tolerancing are shown in Fig. 10-6.

Fig. 10-6. Bilateral tolerancing.

Tolerance Selection

To make decisions on the precision necessary for a particular part requires a complete understanding of manufacturing processes and the requirements of the product; precision problems and solutions are beyond the scope of this book. It is, however, important for drafters to be aware of how to use particular procedures when precision parameters are defined.

PREFERRED BASIC SIZES AND STANDARD TOLERANCES

Owing to the standardization of manufacturing procedures and parts, companies make frequent use of a series of preferred basic sizes. When this series is used, the specification of fits is selected from a decimal or fractional series. The preferred basic sizes used are given in Table 10.2.

In addition to basic sizes, standard tolerances have been adopted by both ANSI and ISO standards. The ANSI series of standard tolerances are presented in Table 10-3. These tables are so arranged that for any one tolerance grade they represent approximately equal production difficulties for each nominal size. Examples of machining processes that are equated to ANSI tolerance grades are:

Table 10.2 Preferred Basic Sizes
(All dimensions in inches.)

Decimal		
0.010	2.00	8.50
0.012	2.20	9.00
0.016	2.40	9.50
0.020	2.60	10.00
0.025	2.80	10.50
0.032	3.00	11.00
0.040	3.20	11.50
0.05	3.40	12.00
0.06	3.60	12.50
0.08	3.80	13.00
0.10	4.00	13.50
0.12	4.20	14.00
0.16	4.40	14.50
0.20	4.60	15.00
0.24	4.80	15.50
0.30	5.00	16.00
0.40	5.20	16.50
0.50	5.40	17.00
0.60	5.60	17.50
0.80	5.80	18.00
1.00	6.00	18.50
1.20	6.50	19.00
1.40	7.00	19.50
1.60	7.50	20.00
1.80	8.00	.....

Fractional					
1/64	0.015625	2¼	2.2500	9½	9.5000
1/32	0.03125	2½	2.5000	10	10.0000
1/16	0.0625	2¾	2.7500	10½	10.5000
3/32	0.09375	3	3.0000	11	11.0000
1/8	0.1250	3¼	3.2500	11½	11.5000
5/32	0.15625	3½	3.5000	12	12.0000
3/16	0.1875	3¾	3.7500	12½	12.5000
1/4	0.2500	4	4.0000	13	13.0000
5/16	0.3125	4¼	4.2500	13½	13.5000
3/8	0.3750	4½	4.5000	14	14.0000
7/16	0.4375	4¾	4.7500	14½	14.5000
1/2	0.5000	5	5.0000	15	15.0000
9/16	0.5625	5¼	5.2500	15½	15.5000
5/8	0.6250	5½	5.5000	16	16.0000
11/16	0.6875	5¾	5.7500	16½	16.5000
3/4	0.7500	6	6.0000	17	17.0000
7/8	0.8750	6½	6.5000	17½	17.5000
1	1.0000	7	7.0000	18	18.0000
1¼	1.2500	7½	7.5000	18½	18.5000
1½	1.5000	8	8.0000	19	19.0000
1¾	1.7500	8½	8.5000	19½	19.5000
2	2.0000	9	9.0000	20	20.0000
.....	....	...	......	....	....

All dimensions are given in inches.

Lapping and honing: 4 and 5
Cylindrical grinding: 5 through 7
Surface grinding: 5 through 8
Diamond turning and boring: 5 through 7
Broaching: 5 through 8
Reaming: 6 through 10
Turning: 7 through 13
Boring: 8 through 13
Milling and drilling: 10 through 13
Planing and shaping: 10 through 13

BASIC HOLE SYSTEM

The majority of limit dimensions are computed on the basic hole system. Shown in Fig. 10-7 is a typical example of how this procedure is applied. Basically, if the nominal size of the hole is known, the limits of the shaft and hole can be found by adding to or subtracting from the basic sizes the standard tolerance limits.

To solve this problem, see Appendix A, which provides five tables with the standard tolerance limits for the major categories of fits. In

Table 10.3 ANSI Standard Tolerances and Grades

Nominal Size, Inches		Grade									
		4	5	6	7	8	9	10	11	12	13
Over	To	Tolerances in thousandths of an inch*									
0–	0.12	0.12	0.15	0.25	0.4	0.6	1.0	1.6	2.5	4	6
0.12–	0.24	0.15	0.20	0.3	0.5	0.7	1.2	1.8	3.0	5	7
0.24–	0.40	0.15	0.25	0.4	0.6	0.9	1.4	2.2	3.5	6	9
0.40–	0.71	0.2	0.3	0.4	0.7	1.0	1.6	2.8	4.0	7	10
0.71–	1.19	0.25	0.4	0.5	0.8	1.2	2.0	3.5	5.0	8	12
1.19–	1.97	0.3	0.4	0.6	1.0	1.6	2.5	4.0	6	10	16
1.97–	3.15	0.3	0.5	0.7	1.2	1.8	3.0	4.5	7	12	18
3.15–	4.73	0.4	0.6	0.9	1.4	2.2	3.5	5	9	14	22
4.73–	7.09	0.5	0.7	1.0	1.6	2.5	4.0	6	10	16	25
7.09–	9.85	0.6	0.8	1.2	1.8	2.8	4.5	7	12	18	28
9.85–	12.41	0.6	0.9	1.2	2.0	3.0	5.0	8	12	20	30
12.41–	15.75	0.7	1.0	1.4	2.2	3.5	6	9	14	22	35
15.75–	19.69	0.8	1.0	1.6	2.5	4	6	10	16	25	40
19.69–	30.09	0.9	1.2	2.0	3	5	8	12	20	30	50
30.09–	41.49	1.0	1.6	2.5	4	6	10	16	25	40	60
41.49–	56.19	1.2	2.0	3	5	8	12	20	30	50	80
56.19–	76.39	1.6	2.5	4	6	10	16	25	40	60	100
76.39–	100.9	2.0	3	5	8	12	20	30	50	80	125
100.9–	131.9	2.5	4	6	10	16	25	40	60	100	160
131.9–	171.9	3	5	8	12	20	30	50	80	125	200
171.9–	200	4	6	10	16	25	40	60	100	160	250

* All tolerances above heavy line are in accordance with American-British-Canadian (ABC) agreements.

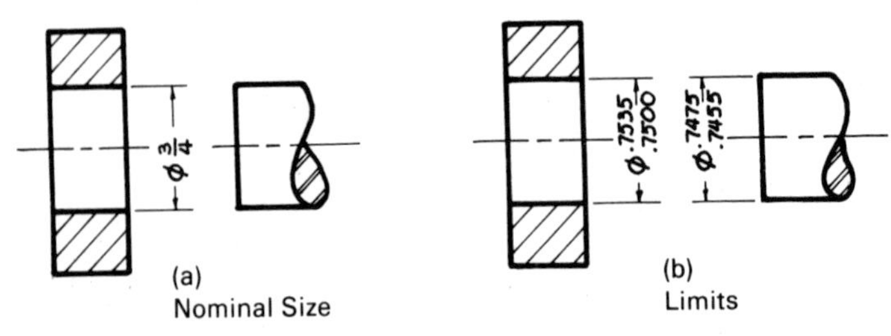

Fig. 10-7. Basic hole system.

our example, a 0.75-inch shaft is to have a class LC8 fit in a 0.75-inch hole. This means that the nominal size of the hole is 0.750 inch and the exact theoretical size is 0.7500 inch.

From Table 2 in Appendix A, it is found that the whole may vary between +0.0035 and −0.0000, while the shaft can vary between −0.0025 and −0.0045. From this we can see that the sum of tolerances for the whole is 0.0035 and 0.0020 for the shaft. The allowance or minimum clearance given in the table is 0.0025 inch.

The limits of the hole are:

$$0.7500 + 0.0035 = 0.7535$$
$$0.7500 - 0.0000 = 0.7500$$

The limits of the shaft are:

$$0.7500 - 0.0025 = 0.7475$$
$$0.7500 - 0.0045 = 0.7455$$

BASIC SHAFT SYSTEM

The basic shaft system is used when a significant number of nominal-sized parts requiring different fits is to be mounted on a shaft. In these situations, it would be easier to adjust the limits for the holes than to machine a number of different holes.

The limits of clearance or interference will be the same as those given in Appendix A for the relevant fits. Symbols used for basic shaft fits will be the same as those used for standard fits, except that the letter S is added. Thus, RC3S specifies a running and sliding fit, class 3, as determined on a basic shaft basis.

Finding the basic shaft limits requires subtracting the value shown for the upper shaft limit from the basic hole limits. This procedure

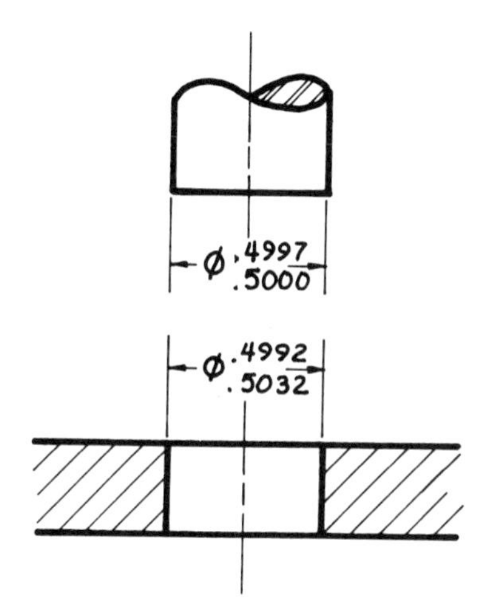

Fig. 10-8. **Basic shaft system.**

is followed in solving for the arrangement in Fig. 10-8, for an FN1S fit with a nominal diameter of 0.50 inch.

The limits of the hole are:

0.5000 − 0.0008 = 0.4992
0.5040 − 0.0008 = 0.5032

The limits of the shaft are:
0.5008 − 0.0008 = 0.5000 (basic size)
0.5005 − 0.0008 = 0.4997

ISO LIMITS OF TOLERANCE

In calculating the limits of tolerance by the ISO system, the deviations and fundamental tolerances provided can be used in any combination necessary to meet the requirements of a fit. For example, the basic hole deviation H and clearance shaft f are often associated, together with one of the ISO tolerance grades IT 01 to IT 16. Appendix A shows ISO Standard Limits and Fits in Table 6, and in Tables 7 and 8 the limits of tolerance for shafts and holes for sizes up through 500 mm.

The ISO system can be used for either hole or shaft basic fit for converting one type of fit to the other, reference is to BS 4500: 1969—ISO limits and fits, BSI (1969); ISO R.286—ISO system of limits and fits, ISO (1962). The specifications of limits of tolerance are coded for a shaft or hole by the use of appropriate letters, where the suffix number is used for the tolerance grade designation. For example, a hole tolerance with deviation H and tolerance grade IT 4 should be noted as H4. Similarly, a shaft with deviation j_s and tolerance grade IT 7 is designated j_s7.

To calculate the limits of size for a shaft, see Table 7 in Appendix A to find the upper deviation es or lower deviation ei. The deviation selected will be determined by the particular letter designation and nominal dimension. If an upper deviation is known, then: $ei = es - IT$, where IT is a value taken from Table 7 for the specified tolerance grade. When the upper hole deviation ES is determined from Table 8, the lower deviation is calculated by: $EI = ES - IT$.

The upper deviations for K, M, and N holes that have tolerance grades through IT 8, plus holes P through ZC with tolerance grades through IT 7, are calculated by the addition of delta values (Δ) found in Table 8.

Geometric Tolerancing

Geometric tolerancing is a method used to specify allowable variations in form or position from true geometry. In reality, this type of tolerance is a dimensional zone within a surface or axis that must lie so as to ensure proper functioning, assemblage, and/or interchangeability. If geometric tolerances are not given on a drawing, it is assumed that the production methods used will be acceptable. Whenever precision parts and products are manufactured, however, geometric tolerancing is usually incorporated into their drawings.

Form and Position Symbols

Standardized symbols and other indications on drawings have been developed for tolerancing form and positions and establishing geometric definitions. These include form, orientation, position, and run-out. The purpose of form and position symbols is to ensure the required functioning and interchangeability of products.

Tolerances of form or position of a geometric form (point, line, or plane) are used to define the "zone" in which the feature is to be contained. Tolerance zones are characterized as one of the following:

1. The *area* within a circle, between two concentric circles, or between two parallel or two straight lines.

2. The *space* between two parallel surfaces or planes, or within a parallelepiped, or sphere, cylinder, or between two coaxial cylinders.

The symbols used to represent the toleranced characteristics fall into three major categories: form of a single feature, orientation of related features, and position of related features. The graphic symbols used here are shown in Fig. 10-9.

The recommended tolerances of form and position symbols are written in a rectangular box, known as a *tolerance frame*, which is

Symbols	Characteristic Toleranced	
⏤	Straightness	FORM OF SINGLE FEATURE
⏥	Flatness	
○	Circularity or Roundness	
⌭	Cylindricity	
⌒	Profile of Any Line	
⌓	Profile of Any Surface	
//	Parallelism	ORIENTATION OF RELATED FEATURES
⊥	Perpendicularity or Squareness	
∠	Angularity	
⌖	Position	POSITION OF RELATED FEATURES
◎	Concentricity or Coaxiality	
⌯	Symmetry	
↗	Run-out	

Fig. 10-9. **Tolerance symbols.**

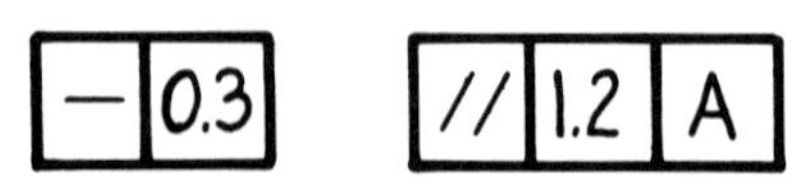

Fig. 10-10. Tolerance frames.

usually divided into two or three compartments (Fig. 10-10). These compartments contain tolerancing information placed left to right, as follows:

1. The symbol of the characteristic to be toleranced (taken from Fig. 10-9).
2. Total tolerance value in the measuring unit for linear dimensions. If the tolerance zone is for a circular, cylindrical, or spherical object, the appropriate symbol should precede the value.
3. When necessary, the letter(s) designating the datum feature(s).

When shown in a drawing, a leader line should originate from the left side (though space requirements allow other locations) of the tolerance frame, and connect to the toleranced feature by an arrowhead. Several examples of this drawing indication are shown in Fig. 10-11.

When tolerancing a datum feature or features, the leader line should be terminated by a solid triangle whose base lies on the outline of the feature or on an extension or projection line originating from the datum feature. Center line axes can also be termination points if they function as the datum line. Examples of these are shown in Fig. 10-12.

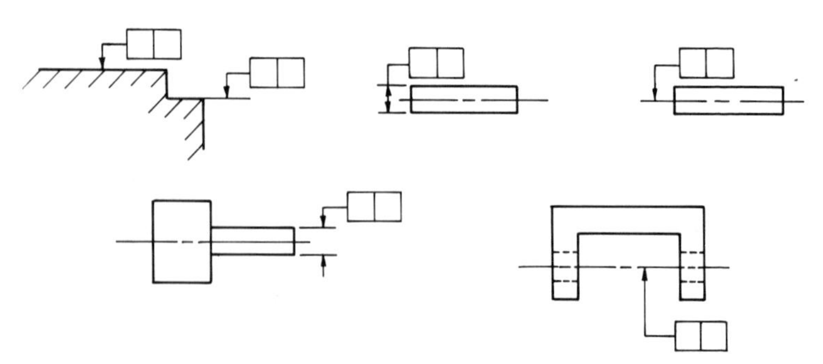

Fig. 10-11. Application of tolerance frames.

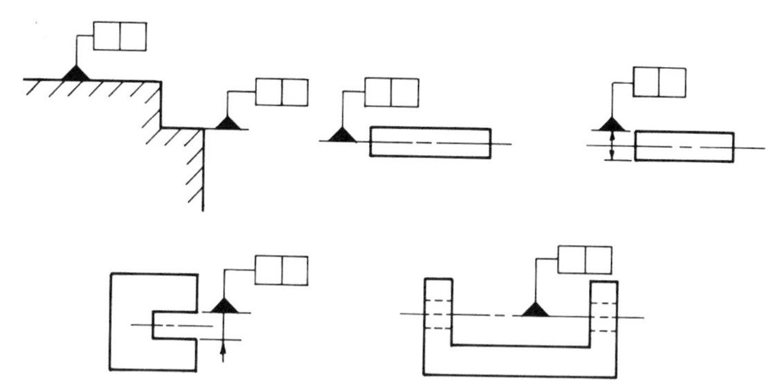

Fig. 10-12. Datum feature indicators.

Solutions of several tolerancing indication problems are shown in Fig. 10-13. In example a, the two features shown are identical, and there is no reason to select one as a datum feature. It is therefore acceptable to tolerance the features with one tolerance frame. In cases in which a tolerance applies to a given length lying anywhere, the value of the length is shown after the tolerance value and is separated from it by a slash, as shown in b. Example c shows how to tolerance two like features when there is insufficient space.

If tolerances of position or profile are specified for a feature, the true position or profile dimensions *must not be toleranced.* The same holds true for tolerances of angularity. To identify these dimensions, they should be enclosed in a box, for example, [45]. Examples of how this is used are shown in Fig. 10-14, which also illustrates an alternate method of showing tolerances by use of a table.

Maximum Material Principle

The assembly and functioning of parts depends upon the combined influence of their finished size and any error of form or position of

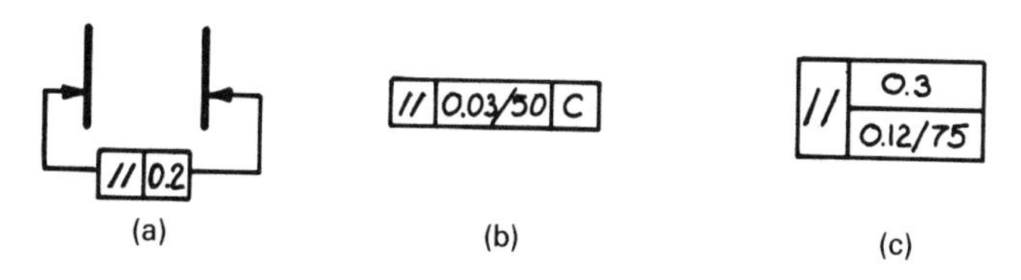

Fig. 10-13. Tolerance indication problems.

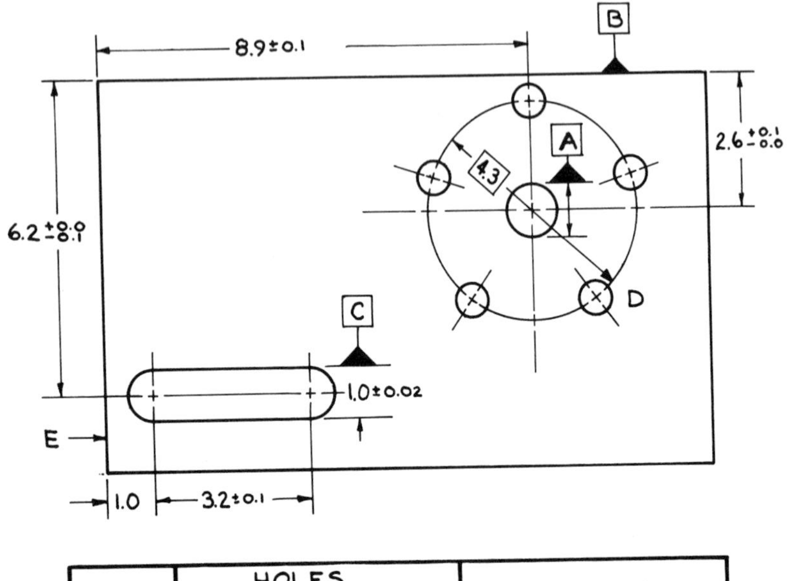

LETTER	HOLES		TOLERANCES	
	DIMEN.	NO.		
A	Ø1.1 +0.1 −0.0	1	DATUM	⌖
B	–	–	DATUM Ⓜ	⊥
C	–	–	DATUM	⌯
D	Ø0.7 +0.02 −0.01	5	TOL. Ø0.02	⌖
E	–	–	TOL. 0.02	⊥

Fig. 10-14. Tolerancing techniques.

mating parts. Minimum clearance always occurs when the object's features are at their maximum material conditions of size and at the highest permissible number of errors of form or position. Thus, the farther away the part is from its maximum material limits of size, the more clearance will exist. Engineering practices therefore take account of the fact that if the actual sizes of the mating features are less than maximum material limits of size, it is possible for form or position tolerances to be exceeded without hindering assembly or function.

The increase of tolerance permitted in the maximum material principle usually helps keep manufacturing process costs to a minimum. However, it may not always be desirable from the point of view of product function. As an example, in positional tolerancing

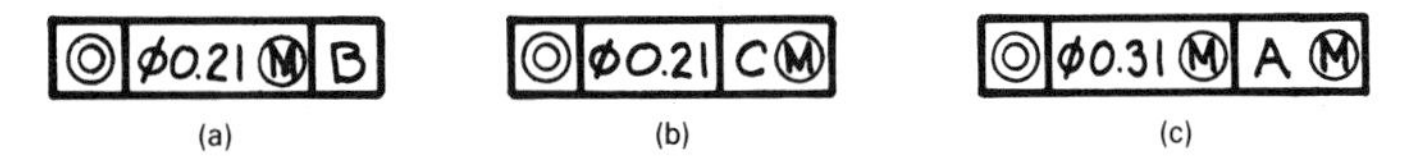

Fig. 10-15. Maximum material condition.

any increase of tolerance is often workable on the center distances of components such as bolt holes, studs, pins, and shafts, but is not always functional in moving or kinematic linkages such as are found at gear centers.

The decision as to whether or not the maximum material principle should be applied in product assembly and manufacturing is usually left to the project engineer or designer. If it is used, a special circle—Ⓜ symbol is added to the corresponding tolerance. In some industries, the maximum material condition is expressed as the *maximum material condition*, which is abbreviated MMC on drawings.

The maximum material condition is shown in symbol form within tolerance frames. Shown in Fig. 10-15 are three situations in which this condition can be noted. In the first, the maximum material condition symbol is placed after the tolerance value; in the second, after the datum letter; the last shows how it is used after both indicators.

There are a wide number of product designs that would warrant the use of a maximum material condition. Fig. 10-16 shows several examples and dimensioning techniques used for this purpose.

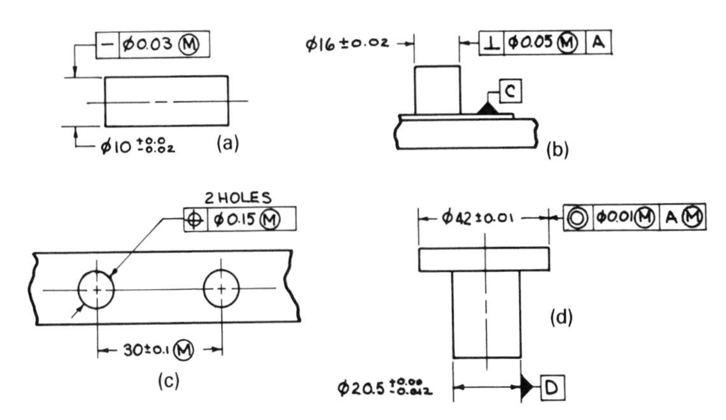

Fig. 10-16. Maximum material condition.

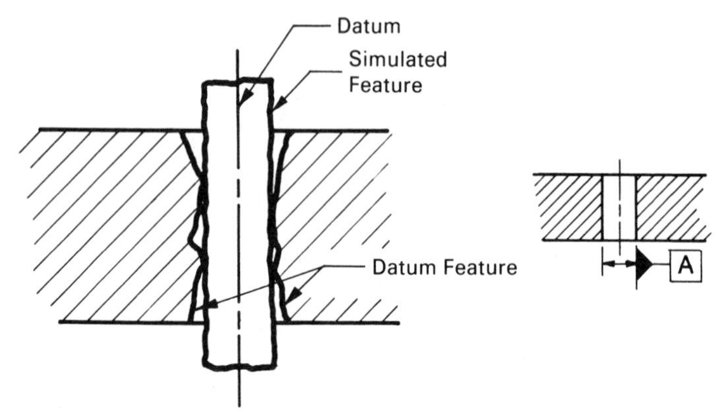

Fig. 10-17. Simulated datum where datum is the axis of a cylinder.

Datums and Datum Systems for Geometric Tolerances

A datum is defined as a theoretically exact geometric reference to which toleranced features are associated. Exactness means that the datum, which may be an axis, a plane, a straight line, or any other feature, is assumed to have no distortions and to be dimensionally exact. When two or more separate datums are used in combination to reference a toleranced feature, they are often referred to as a datum system.

In actuality, datum features will always have inaccuracies resulting from manufacturing and assembly processes. Therefore, the datum feature is usually defined as the contacting surfaces along the feature, and it is often referred to as a *simulated datum* feature (contacting surface). This is shown in Fig. 10-17.

To reference a datum, a capital letter should be placed in a square or rectangular frame. This may be omitted if the frame can be connected with the datum by a leader line. When a datum is drawn for a single feature, it is given as a single letter in the third compartment of the tolerance frame. However, where the common datum is established by two features, two letters separated by a hyphen are placed in the third compartment of the tolerance frame. See Fig. 10-18.

There are times when a datum system must be established by two or more features, resulting in multiple datums. When this occurs, the datum letters will be placed in separate compartments in the tolerance frame. Thus, the primary datum will be in the third

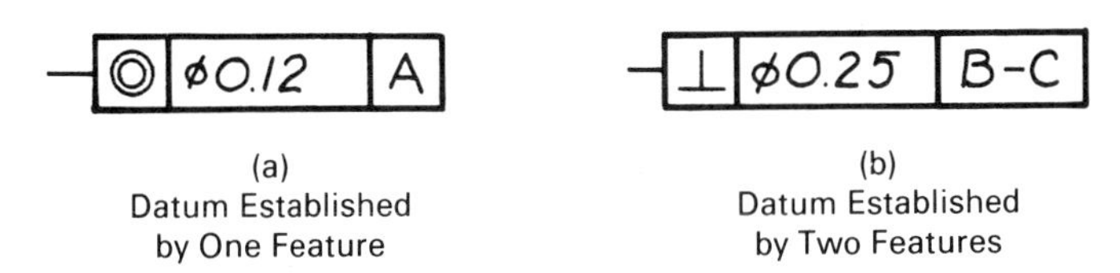

Fig. 10-18. Specification of datum and datum system in tolerance frame.

frame, the secondary datum in the fourth, and the tertiary datum in the fifth frame. Fig. 10-19 shows an example of this.

The last concept to be considered here is the *datum target*. It can be a point, a line, or any limited area on the object (workpiece) that is used for direct contact with manufacturing, processing, or inspection equipment. Datum targets are indicated on drawings by a circular frame divided into two compartments by a horizontal line.

The numerator, or upper compartment, of the datum target symbol is used for such information as the dimensions of the target area. If insufficient room is available in the compartment, the data can be placed outside and connected by a leader line. The lower compartment, or denominator, indicates the datum feature letter and target number.

Datum targets can be represented by three types of features; these features, and the method used to indicate them, are:

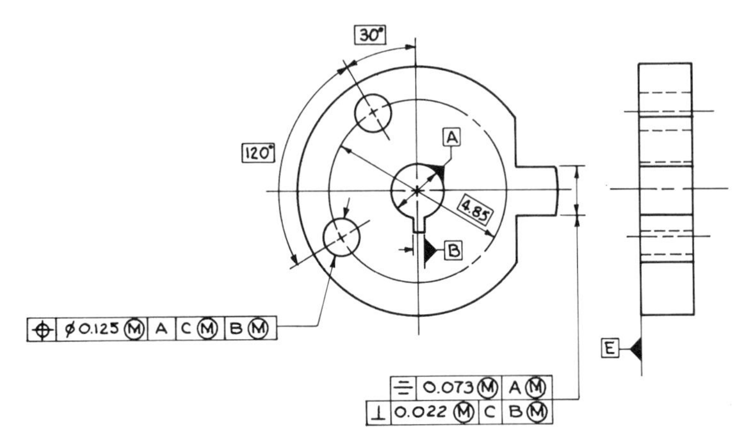

Fig. 10-19. Multiple datum features.

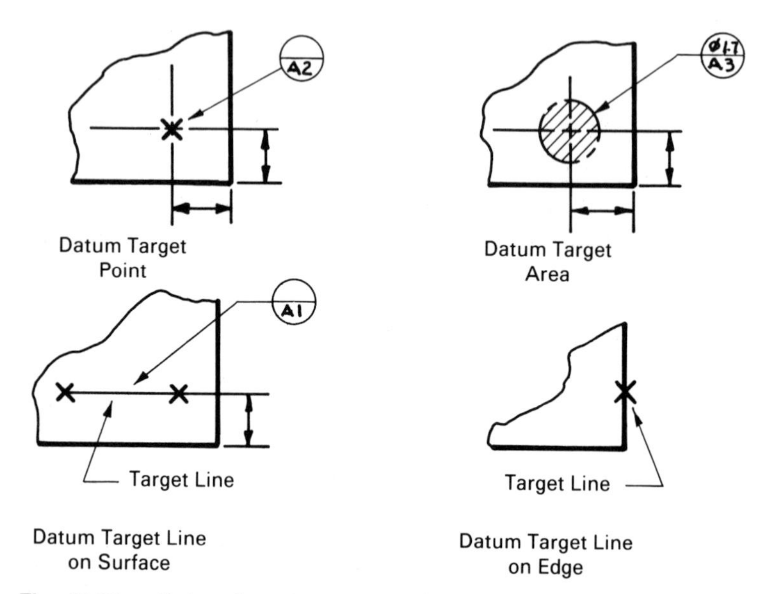

Fig. 10-20. Datum target representations.

1. Datum targets as a point are indicated by a cross.
2. Datum targets as a line are indicated by two crosses connected by a thin line.
3. Datum targets as an area are defined by a phantom line about the hatched area.

Applications of datum targets are shown in Fig. 10-20.

Surface Finishes and Textures

Most products will require some type of machine finish. When properly executed, these finished surfaces will contribute not only to the appearance of the object, but also to its functioning. Drafters should be aware of the drawing procedures used to specify and indicate surface finishes and textures.

Surface Finish

Examples of surface finish requirements can be found throughout most manufacturing industries. Surfaces on rough castings are often ma-

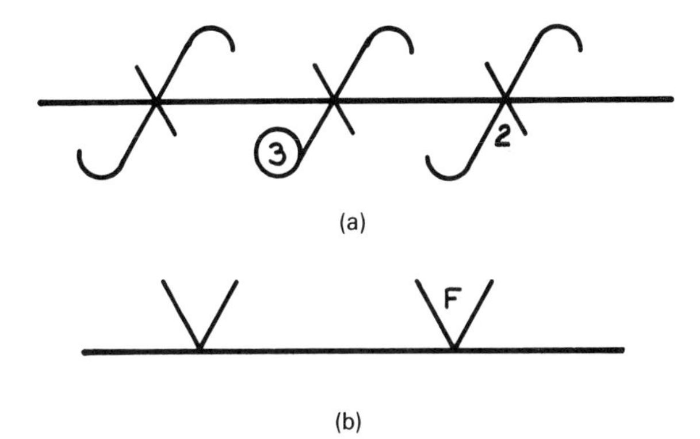

Fig. 10-21. Old finish symbols used to indicate machined surfaces.

chined to produce smooth contoured or flat surfaces, while forgings are frequently ground to produce a desired finish.

Drawings that exhibit such products make use of surface finish symbols that indicate which surfaces are to be machined. Two common symbols are used for this purpose. The first is similar to an italicized *f*, and the second is a V, drawn with legs approximately 60° to each other. When the *f* symbol is used, it is placed crossing the edge view of the finished surface, while the V symbol is located so that its point is just touching the line.

Along with the symbols are alphanumeric codes that are used to define the quality of finish to be given. If these codes are not indicated, it is assumed that the manufacturing process used will be sufficient. The coding used for each is:

f symbol:
1 —rough
2 —ordinary
3 —fine
4 —ground
5 —lapped

V symbol:
R —rough finish
RG —rough grind
G —grind
P —polish
F —file

Because these symbols and codes will often have different qualitative meanings from one industry or shop to another, their use has decreased significantly in the last twenty years. But because they are still incorporated in some industrial drawings, drafters should be aware of their use and meaning. Fig. 10-21 gives examples of how both symbols are used.

Surface Texture

Because of the difficulty of interpreting surface finish symbols, the American National Standard Surface Texture system has been widely adopted in the United States and Canada, in most European (ISO) countries. The basic surface texture symbol consists of two legs of unequal length that resemble a check mark. Shown in Fig. 10-22 are the three symbol forms that are used in drawings, and the position of specifications for surface textures. The first is the basic symbol that is used alone only when it is referenced to a note. The second is a machined surface with no other information, and the last is to note a surface where the removal of material is prohibited.

To the surface texture symbol is added other indicators (Fig. 10-23). The first are indicators of surface roughness. Roughness (R)

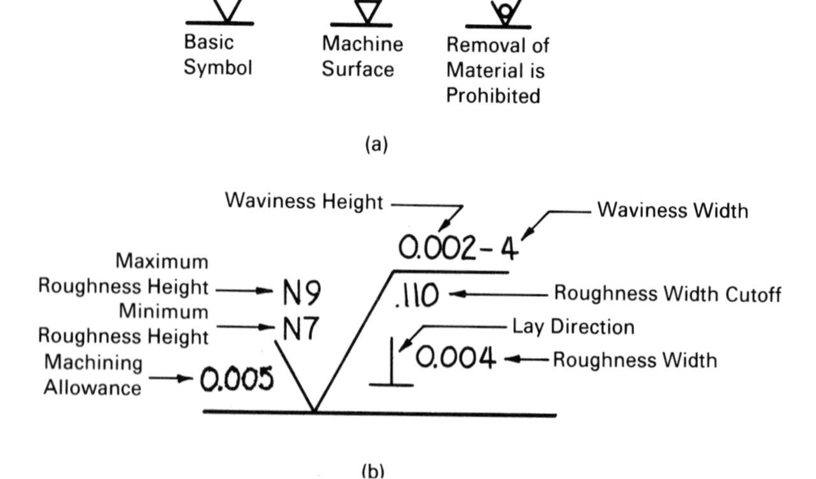

Fig. 10-22. Indicating surface texture.

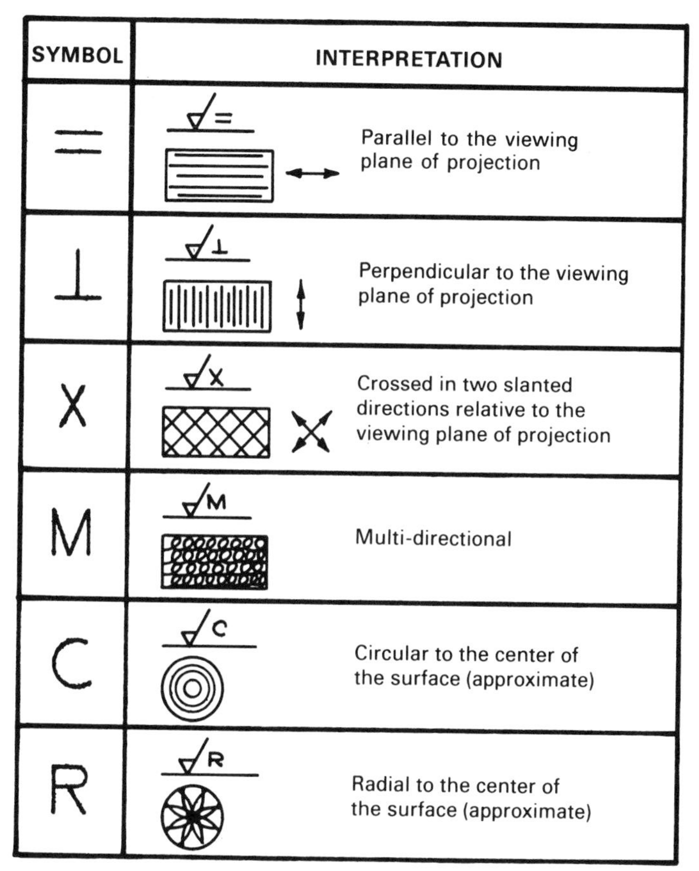

SYMBOL
INTERPRETATION
=
Parallel to the viewing plane of projection
⊥
Perpendicular to the viewing plane of projection
X
Crossed in two slanted directions relative to the viewing plane of projection
M
Multi-directional
C
Circular to the center of the surface (approximate)
R
Radial to the center of the surface (approximate)

Fig. 10-23. Direction of lay symbols.

consists of a series of irregularities that are finely spaced, and are produced by machining operations. Roughness height is given in either microinch (μin) or micrometer (μm) units. Roughness width is the distance between the patterns of roughness. The value of the roughness height and width can be specified either in actual values or in grade numbers. This value is placed over the V section of the surface texture symbol. Table 10-4 gives the equivalencies for grade numbers.

The second indicator is known as waviness. This refers to those surface irregularities that are too far apart to be considered as roughness. This texture is usually caused by either machine or work deflec-

Table 10-4 Roughness Value Equivalencies

Roughness Values R		
μm	μin	Roughness Grade Numbers
50	2,000	N 12
25	1,000	N 11
12.5	500	N 10
6.3	250	N 9
3.2	125	N 8
1.6	63	N 7
0.8	32	N 6
0.4	16	N 5
0.2	8	N 4
0.1	4	N 3
0.05	2	N 2
0.025	1	N 1

tion, vibration, chatter, temperature changes, or internal stresses and/or strains. Waviness height is a peak-to-valley distance, while waviness width is the succession of wave peaks or valleys.

The last indicator is known as lay, and is defined as the direction of the surface irregularity. In many cases, lay is referred to as tool markings. The lay is determined by the direction of the cutting tool, and requires the use of a series of standard symbols that are placed in the lower right side of the surface texture symbol. These symbols are shown in Fig. 10-23.

Exercises

10.1 Using the basic hole system, determine the limits of fits for the following shafts and holes:
a. RC6 fit for a hole and shaft with ½″ nominal size.
b. LC7 fit for a hole and shaft with 4.0″ nominal size.
c. Class LN2 fit for a shaft with 3.5″ nominal size.
d. Class FN2 fit for a 2.25″ nominal size hole and shaft.

10.2 Determine the limits of size for a part 125 mm basic size with a tolerance designation of g^9.

10.3 Interpret the symbol specifications of general form and position tolerance shown in Fig. 10-24.

10.4 Convert the following specifications into symbol form:
a. A surface parallel to B within a 0.004 total.
b. All points on a surface must lie between two planes 0.0035 apart and parallel to datum B.
c. Concentric to surface D within 0.005
d. Angular tolerance, 0.0015 total datum C.
e. Perpendicular to surface F within 0.002 total.

10.5 Explain the meaning of the surface texture symbols illustrated in Fig. 10-25.

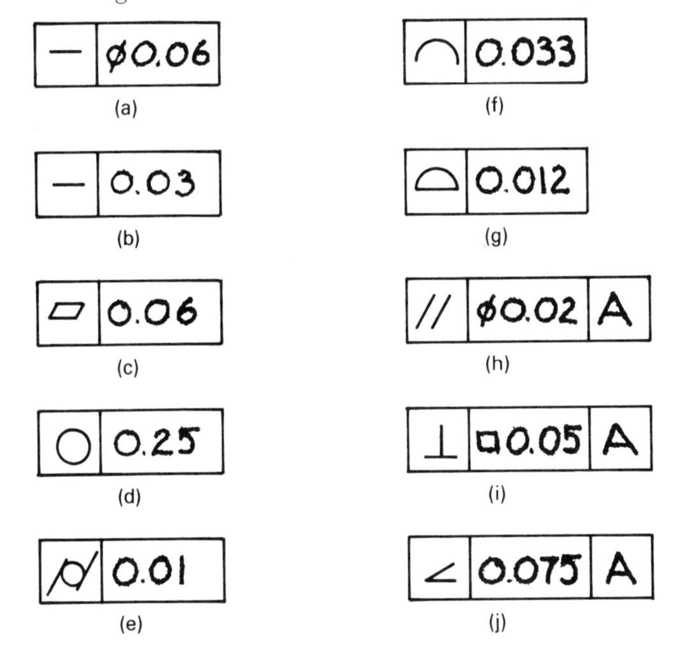

Fig. 10-24. Problem 10.3.

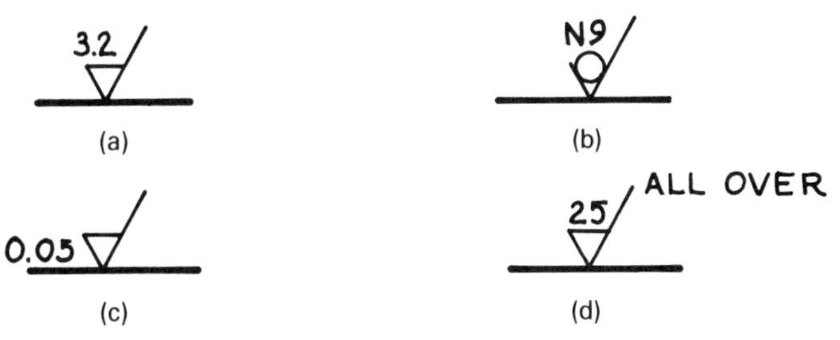

Fig. 10-25. Problem 10.5.

CHAPTER 11

Nonthreaded Mechanical Elements

A nonthreaded mechanical element is a device, component, unit, or material that is part of a machine or mechanical product containing one or more pieces so arranged as to allow for the proper operation of that product. These elements are used for the fastening of parts, as well as for the encouragement and promotion of motion. Nonthreaded mechanical elements are thus widely represented in many industrial working drawings. This chapter briefly discusses the more common elements and their graphic presentation.

Keys, Keyseats, and Splines

Machines and mechanical equipment that require the transmission of power from a shaft to an attached component frequently make use of keyed and spline connections. Keys and keyseats have been standardized in ANSI B17.1, which describes the principal key types and their dimensional characteristics. Splines have been standardized in ANSI B92.1. For metric-sized keys, the reader is referred to the British Standard: BS 4234 *Metric Keys and Keyways*. This section will discuss the way keys, keyseats, and splines are used in machines and other mechanical products and are appropriately represented in drawings.

Keys and Keyseats

A *key* is a demountable piece of material, usually made of steel, that is assembled into a groove known as a *keyseat*. Together they are used for transmitting torque between a shaft and its hub. Keys and keyseats are employed in machines and equipment whose design makes use of gears, pulleys, cranks, and handles. A wide variety of keys are used as mechanical elements, but five major types are most frequently incorporated in machine design.

The major types of key are:

1. *Woodruff keys.* The woodruff key is semicircular in shape and is assembled into a mating keyseat. It is primarily used for the easy removal of pulleys from shafts; it is not used where sliding keys are required. This key is illustrated in Fig. 11-1, with the corresponding dimensional standards given in Table 11.1
2. *Square and flat plain taper keys.* This is perhaps the type of key most commonly found as a machine element today. Its width is approximately one-fourth the shaft diameter. This key is illustrated in Fig. 11-2, with dimensional standards given in Table 11-2.
3. *Gibhead keys.* The gibhead key has the same dimensional standards as the square and flat plane taper key, with the addition of a head (see Fig. 11-3). This head is used for easy removal. Since the key's head protrudes out, it cannot be used in certain machines with safety restrictions. The gibhead dimensions are also given in Table 11-2.
4. *Sunk keys.* Shown in Fig. 11-4, with dimensional specifications in Table 11-3, sunk keys are primarily used for fitting adjacent parts without having either end of the key accessible.
5. *Feather keys.* The major function of feather keys is to prevent adjoining parts from turning or rotating on a shaft, while allowing them to slide lengthwise along the shaft. See Fig. 11-5.

   When large amounts of torque are transmitted along a shaft and hub, it is common practice to use two or more keys, as shown in Fig. 11-6. Of the two configurations shown, (a) allows for easier machining of the keyway. It should be noted that if only one key in configuration (b) is used, then torque can be applied only in one direction.

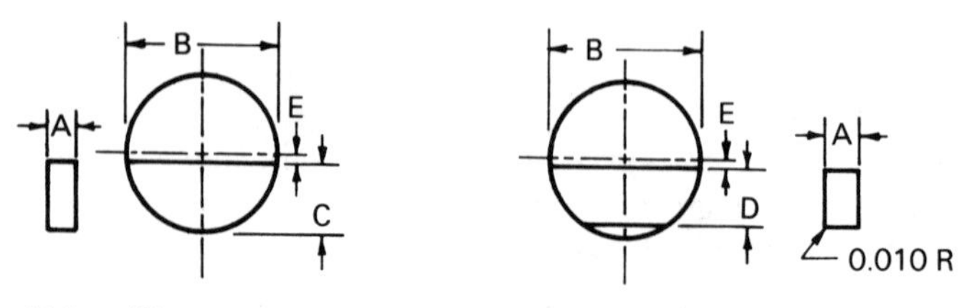

Fig. 11-1. Woodruff key.

The technique used for dimensioning keyways and keyseats will depend on the production method used to make the keyway and keyseat. If a machinist is expected to fit a given key into a keyway and keyseat, dimensions should first specify

Table 11.1 Woodruff Key Dimensional Specifications (All dimensions in inches.)

Key No.	Nominal key size, $A \times B$	Width of key, A		Diam of key, B		Height of key				Distance below E
						C		D		
		Max	Min	Max	Min	Max	Min	Max	Min	
204	1/16 × 1/2	0.0635	0.0625	0.500	0.490	0.203	0.198	0.194	0.188	3/64
304	3/32 × 1/2	0.0948	0.0928	0.500	0.490	0.203	0.198	0.194	0.188	3/64
305	3/32 × 5/8	0.0948	0.0938	0.625	0.615	0.250	0.245	0.240	0.234	1/16
404	1/8 × 1/2	0.1260	0.1250	0.500	0.490	0.203	0.198	0.194	0.188	3/64
405	1/8 × 5/8	0.1260	0.1250	0.625	0.615	0.250	0.245	0.240	0.234	1/16
406	1/8 × 3/4	0.1260	0.1250	0.750	0.740	0.313	0.308	0.303	0.297	1/16
505	5/32 × 5/8	0.1573	0.1563	0.625	0.615	0.250	0.245	0.240	0.234	1/16
506	5/32 × 3/4	0.1573	0.1563	0.750	0.740	0.313	0.308	0.303	0.297	1/16
507	5/32 × 7/8	0.1573	0.1563	0.875	0.865	0.375	0.370	0.365	0.359	1/16
606	3/16 × 3/4	0.1885	0.1875	0.750	0.740	0.313	0.308	0.303	0.297	1/16
607	3/16 × 7/8	0.1885	0.1875	0.875	0.865	0.375	0.370	0.365	0.359	1/16
608	3/16 × 1	0.1885	0.1875	1.000	0.990	0.438	0.433	0.428	0.422	1/16
609	3/16 × 1 1/8	0.1885	0.1875	1.125	1.115	0.484	0.479	0.475	0.469	5/64
807	1/4 × 7/8	0.2510	0.2500	0.875	0.865	0.375	0.370	0.365	0.359	1/16
808	1/4 × 1	0.2510	0.2500	1.000	0.990	0.438	0.433	0.428	0.422	1/16
809	1/4 × 1 1/8	0.2510	0.2500	1.125	1.115	0.484	0.479	0.475	0.469	5/64
810	1/4 × 1 1/4	0.2510	0.2500	1.250	1.240	0.547	0.542	0.537	0.531	5/64
811	1/4 × 1 3/8	0.2510	0.2500	1.375	1.365	0.594	0.589	0.584	0.578	3/32
812	1/4 × 1 1/2	0.2510	0.2500	1.500	1.490	0.641	0.636	0.631	0.625	7/64
1008	5/16 × 1	0.3135	0.3125	1.000	0.990	0.438	0.433	0.428	0.422	1/16
1009	5/16 × 1 1/8	0.3135	0.3125	1.125	1.115	0.484	0.479	0.475	0.469	5/64
1010	5/16 × 1 1/4	0.3135	0.3125	1.250	1.240	0.547	0.542	0.537	0.531	5/64
1011	5/16 × 1 3/8	0.3135	0.3125	1.375	1.365	0.594	0.589	0.584	0.578	3/32
1012	5/16 × 1 1/2	0.3135	0.3125	1.500	1.490	0.641	0.636	0.631	0.625	7/64
1210	3/8 × 1 1/4	0.3760	0.3750	1.250	1.240	0.547	0.542	0.537	0.531	5/64
1211	3/8 × 1 3/8	0.3760	0.3750	1.375	1.365	0.594	0.589	0.584	0.578	3/32
1212	3/8 × 1 1/2	0.3760	0.3750	1.500	1.490	0.641	0.636	0.631	0.625	7/64

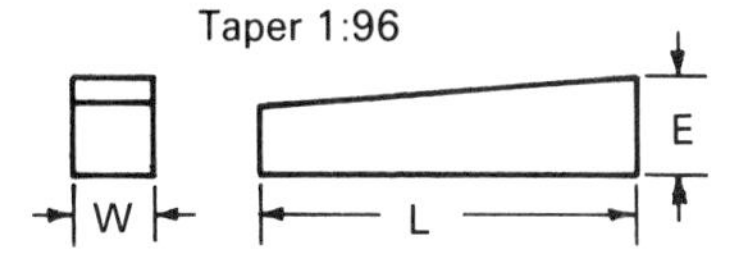

Fig. 11-2. Square and flat plain taper key.

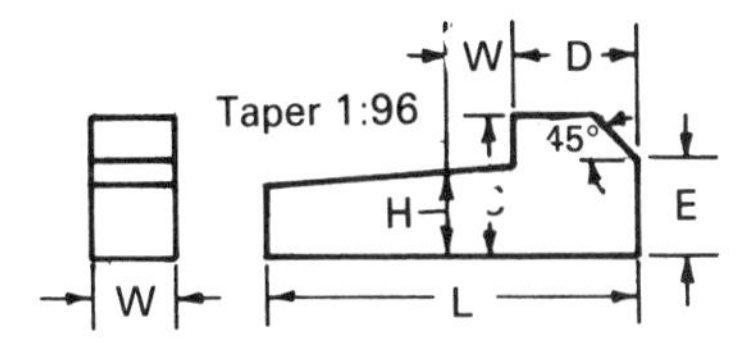

Fig. 11-3. Gibhead key.

the width dimension, followed by the depth dimension. For mass production situations, keyway and keyseat dimensions are specified in limit dimensions. Since most keys used are standard, all that is needed for dimensioning the keyway and keyseat is the key number (e.g., NO. 305 WOODRUFF KEY, or NUMBER 5 SUNK KEY). These three techniques are shown in Fig. 11-7.

Table 11.2 Square and Flat Plain Taper, and Gibhead Keys Dimensional Specifications (All dimensions in inches.)

Shaft diam	Square type					Flat type				
	Key		Gib-head			Key		Gib-head		
	Max width, *W*	Height at large end,† *H*	Height, *C*	Length, *D*	Height edge of chamfer, *E*	Max width, *W*	Height at large end,† *H*	Height, *C*	Length, *D*	Height edge of chamfer, *E*
1/2 – 9/16	1/8	1/8	1/4	7/32	5/32	1/8	3/32	3/16	1/8	1/8
5/8 – 7/8	3/16	3/16	5/16	9/32	7/32	3/16	1/8	1/4	3/16	5/32
15/16–1 1/4	1/4	1/4	7/16	11/32	11/32	1/4	3/16	5/16	1/4	3/16
1 5/16 – 1 3/8	5/16	5/16	9/16	13/32	13/32	5/16	1/4	3/8	5/16	1/4
1 7/16 – 1 3/4	3/8	3/8	11/16	15/32	15/32	3/8	1/4	7/16	3/8	5/16
1 13/16–2 1/4	1/2	1/2	7/8	19/32	5/8	1/2	3/8	5/8	1/2	7/16
2 5/16 – 2 3/4	5/8	5/8	1 1/16	23/32	3/4	5/8	7/16	3/4	5/8	1/2
2 7/8 – 3 1/4	3/4	3/4	1 1/4	7/8	7/8	3/4	1/2	7/8	3/4	5/8
3 3/8 – 3 3/4	7/8	7/8	1 1/2	1	1	7/8	5/8	1 1/16	7/8	3/4
3 7/8 – 4 1/2	1	1	1 3/4	1 3/16	1 3/16	1	3/4	1 1/4	1	13/16
4 3/4 – 5 1/2	1 1/4	1 1/4	2	1 7/16	1 7/16	1 1/4	7/8	1 1/2	1 1/4	1
5 3/4 – 6	1 1/2	1 1/2	2 1/2	1 3/4	1 3/4	1 1/2	1	1 3/4	1 1/2	1 1/4

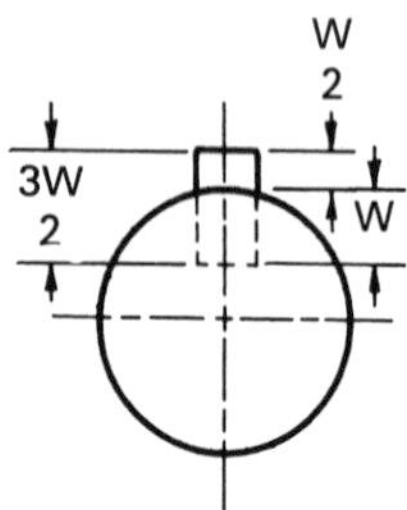

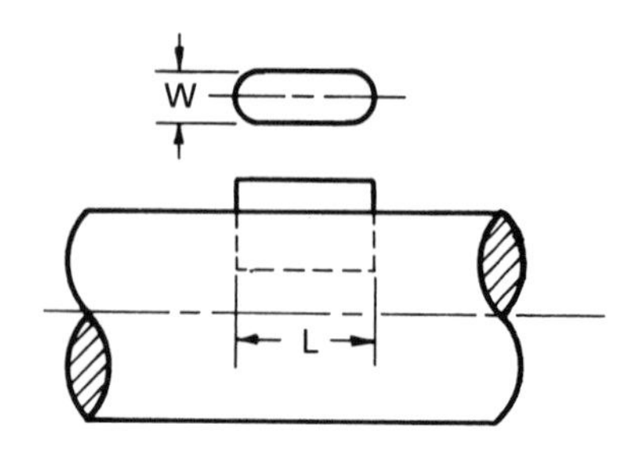

Fig. 11-4. **Sunk key.**

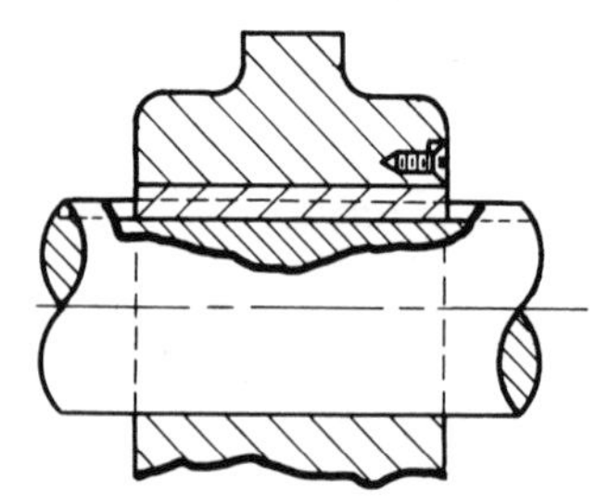

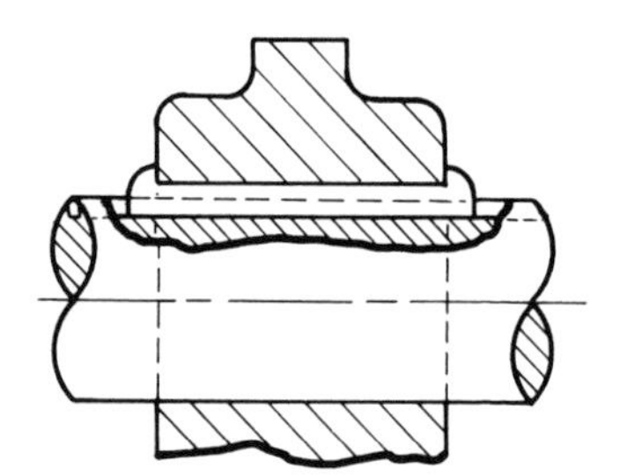

Fig. 11-5. **Feather key.**

Table 11.3 Sunk Keys Dimensional Specifications
(All dimensions in inches.)

Key No.	*L*	*W*	Key No.	*L*	*W*	Key No.	*L*	*W*	Key No.	*L*	*W*
1	1/2	1/16	13	1	3/16	22	1 3/8	1/4	54	2 1/4	1/4
2	1/2	3/32	14	1	7/32	23	1 3/8	5/16	55	2 1/4	5/16
3	1/2	1/8	15	1	1/4	F	1 3/8	3/8	56	2 1/4	3/8
4	5/8	3/32	B	1	5/16	24	1 1/2	1/4	57	2 1/4	7/16
5	5/8	1/8	16	1 1/8	3/16	25	1 1/2	5/16	58	2 1/2	5/16
6	5/8	5/32	17	1 1/8	7/32	G	1 1/2	3/8	59	2 1/2	3/8
7	3/4	1/8	18	1 1/8	1/4	51	1 3/4	1/4	60	2 1/2	7/16
8	3/4	5/32	C	1 1/8	5/16	52	1 3/4	5/16	61	2 1/2	1/2
9	3/4	3/16	19	1 1/4	3/16	53	1 3/4	3/8	30	3	3/8
10	7/8	5/32	20	1 1/4	7/32	26	2	3/16	31	3	7/16
11	7/8	3/16	21	1 1/4	1/4	27	2	1/4	32	3	1/2
12	7/8	7/32	D	1 1/4	5/16	28	2	5/16	33	3	9/16
A	7/8	1/4	E	1 1/4	3/8	29	2	3/8	34	3	5/8

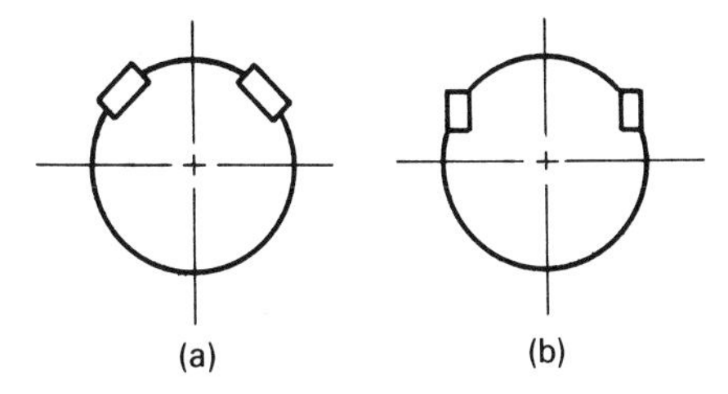

Fig. 11-6. Multiple key arrangements.

Splines

A spline is a series of parallel keys that are an integral part of the shaft. The splined shaft is then mated with a hub or fitting that has corresponding grooves. A key construction that closely resembles a spline would be a shaft with a series of keys or feathers that fit into machined slots in the shaft. The disadvantage of this key configuration is a reduction in torque-transmitting capacity. Thus, splined shafts are capable of carrying heavier loads than keys.

Spline teeth are of two types. The first is known as *parallel-sided splines*. These splines have been standardized by ANSI for 4, 6, 10, and 16 spline fittings, and are shown in Fig. 11-8, with corresponding dimensional specifications in Table 11-4.

The second type of spline is one that has curved-sided teeth, and is known as an *involute spline*. Internal and external involute splines (Fig. 11-9) have a similar appearance to and the characteristics of involute gear teeth. These splines are specified by a fraction where the numerator is the diametral pitch and the denominator is twice the numerator. There are 15 series of involute splines, specified as follows: 1/2, 2.5/5, 3/6, 4/8, 5/10, 6/12, 8/16, 10/20, 12/24, 16/32, 20/40, 24/48, 32/64, 40/80, and 48/96.

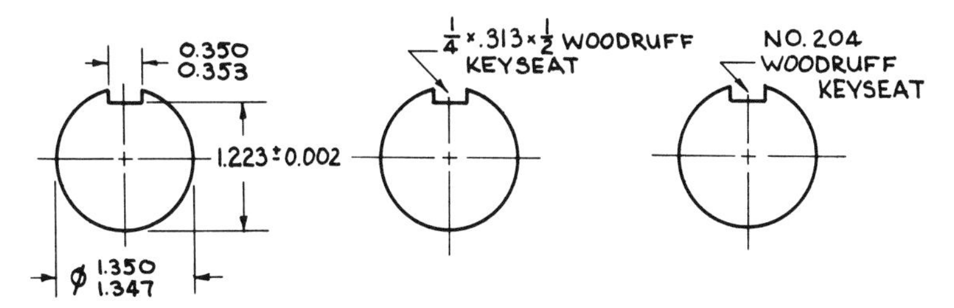

Fig. 11-7. Three methods used to provide dimensional specifications for keys and keyseats.

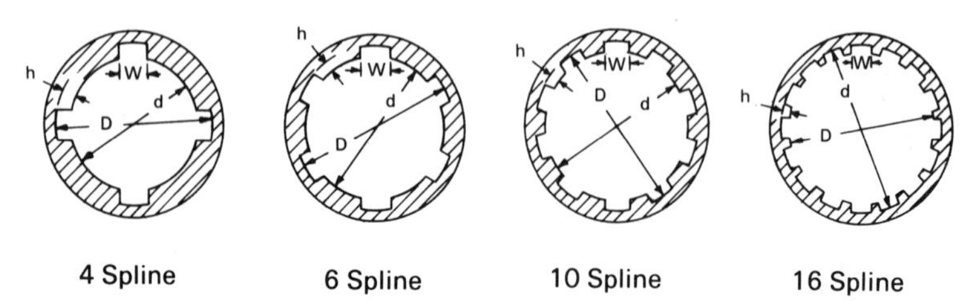

Fig. 11-8. Parallel-sided splines.

Pins and Retaining Rings

In the vast majority of working detailed subassembly and assembly drawings, pins and retaining rings will be selected from standardized sizes and types. As such, it is neither necessary nor recommended that they be fully dimensioned. Instead of conventional dimensions, notations pertaining to the type and size are given on the drawing. There are times, however, when the limits of fit must be given for a hole

Table 11.4 Parallel-sided Spline Proportions (All dimensions in inches.)

Nominal diam	4-spline for all fits		6-spline for all fits		10-spline for all fits		16-spline for all fits	
	D max	*W* max	*D* max	*W* max	*D* max	*W* max	*D* max	*W* max
¾	0.750	0.181	0.750	0.188	0.750	0.117		
⅞	0.875	0.211	0.875	0.219	0.875	0.137		
1	1.000	0.241	1.000	0.250	1.000	0.156		
1⅛	1.125	0.271	1.125	0.281	1.125	0.176		
1¼	1.250	0.301	1.250	0.313	1.250	0.195		
1⅜	1.375	0.331	1.375	0.344	1.375	0.215		
1½	1.500	0.361	1.500	0.375	1.500	0.234		
1⅝	1.625	0.391	1.625	0.406	1.625	0.254		
1¾	1.750	0.422	1.750	0.438	1.750	0.273		
2	2.000	0.482	2.000	0.500	2.000	0.312	2.000	0.196
2¼	2.250	0.542	2.250	0.563	2.250	0.351		
2½	2.500	0.602	2.500	0.625	2.500	0.390	2.500	0.245
3	3.000	0.723	3.000	0.750	3.000	0.468	3.000	0.294
3½	.....	.....	.....	.....	3.500	0.546	3.500	0.343
4	.....	.....	.....	.....	4.000	0.624	4.000	0.392
4½	.....	.....	.....	.....	4.500	0.702	4.500	0.441
5	.....	.....	.....	.....	5.000	0.780	5.000	0.490
5½	.....	.....	.....	.....	5.500	0.858	5.500	0.539
6	.....	.....	.....	.....	6.000	0.936	6.000	0.588

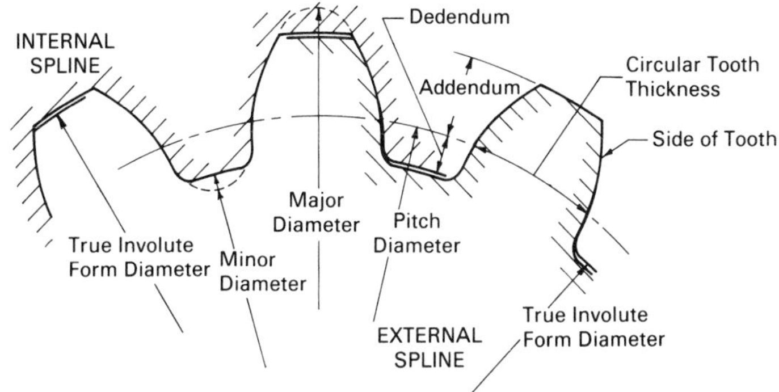

Fig. 11-9. Involute spline.

or shaft that is to receive these items. Where precision is important, the receiving hole or shaft should be appropriately dimensioned.

Pins

Pins are used as inexpensive yet effective fastening devices. They are used for assemblies in which the type of load is primarily shear, and are of two major types. The first is *semipermanent pins,* which require the use of applied pressure for installations and removal. Examples of this type are spring and grooved surface pins. Also included are dowel, clevis, taper, and cotter pins.

The second type of pin is *quick release pins,* which are used for rapid installation and removal. Unlike semipermanent pins, quick release pins require no tools for assembly and disassembly. Examples of quick release pins are push-pull and positive-locking pins.

Within the field of mechanical design, some types of pin are more commonly used than others; these are presented as follows:

1. *Cotter pins* come in 18 standard sizes ranging in nominal diameter size from 1/32 inch to 3/4 inch. They are primarily used with nut and bolt, screw, or stud assemblies, and are also used to hold standard clevis pins in place. Fig. 11-10 shows a cotter pin, with corresponding dimensional specifications given in Table 11-5.
2. *Clevis pins* are available in ten standard sizes ranging from a nominal size diameter of 3/16 inch to 1 inch. This type of pin is used for connecting mating yokes or forks and eye

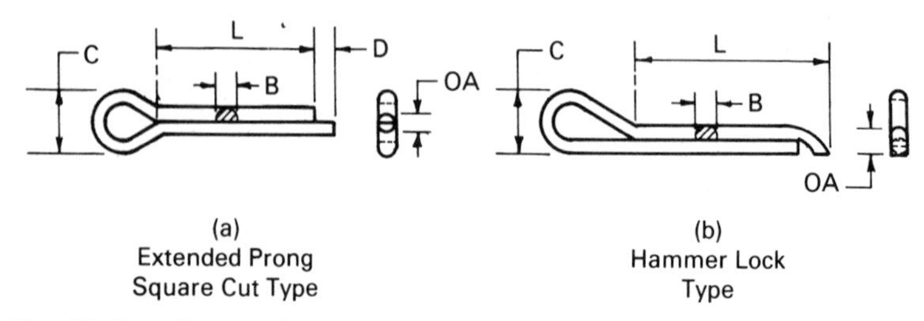

(a)
Extended Prong
Square Cut Type

(b)
Hammer Lock
Type

Fig. 11-10. Cotter pins.

Table 11.5 Cotter Pins Dimensional Specifications (All dimensions in inches.)

Nom. Size	Diam. A* & Width B Max.	Wire Width B Min.	Head Diam. C Min.	Prong L'gth D Min.	Hole Size	Nom. Size	Diam. A* & Width B Max.	Wire Width B Min.	Head Diam. C Min.	Prong L'gth D Min.	Hole Size
1/32	.032	.022	0.06	.01	.047	3/16	.176	.137	0.38	.09	.203
3/64	.048	.035	0.09	.02	.062	7/32	.207	.161	0.44	.10	.234
1/16	.060	.044	0.12	.03	.078	1/4	.225	.176	0.50	.11	.266
5/64	.076	.057	0.16	.04	.094	5/16	.280	.220	0.62	.14	.312
3/32	.090	.069	0.19	.04	.109	3/8	.335	.263	0.75	.16	.375
7/64	.104	.080	0.22	.05	.125	7/16	.406	.320	0.88	.20	.438
1/8	.120	.093	0.25	.06	.141	1/2	.473	.373	1.00	.23	.500
9/64	.134	.104	0.28	.06	.156	5/8	.598	.472	1.25	.30	.625
5/32	.150	.116	0.31	.07	.172	3/4	.723	.572	1.50	.36	.750

members in knuckle joint assemblies. Clevis pins are held in place by a cotter pin or some other fastening device. Fig. 11-11 shows clevis pins, and Table 11-6 gives the corresponding dimensional specifications.

3. *Ground machine dowel pins* come in standard nominal-size diameters ranging from 1/8 to 1 inch. These pins are used to fasten machine parts where alignment tolerances are close,

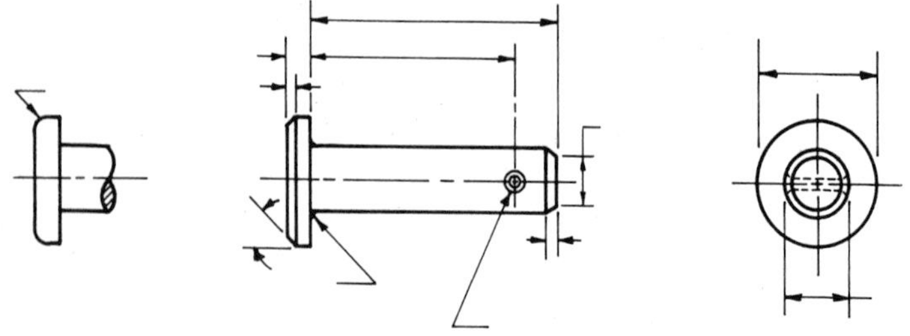

Fig. 11-11. Clevis pins.

Table 11.6 Clevis Pins Dimensional Specifications (All dimensions in inches.)

Nom. Size (Basic Pin Diam.)	Shank Diam. A Max.[1]	Head Diam. B Max.[2]	Head Hgt. C Max.[3]	Head Chamfer D Nom.[4]	Hole Diam. E Max.[5]	Point Diam. F Max.[6]	Pin Lgth. G Basic[7]	Head to Hole Center H Max.[8]	Point Length L Max.	Point Length L Min.	Cotter Pin Size for Hole
3/16	.186	0.32	.07	.02	.088	.15	0.58	0.504	.055	.035	1/16
1/4	.248	0.38	.10	.03	.088	.21	0.77	0.692	.055	.035	1/16
5/16	.311	0.44	.10	.03	.119	.26	0.94	0.832	.071	.049	3/32
3/8	.373	0.51	.13	.03	.119	.33	1.06	0.958	.071	.049	3/32
7/16	.436	0.57	.16	.04	.119	.39	1.19	1.082	.071	.049	3/32
1/2	.496	0.63	.16	.04	.151	.44	1.36	1.223	.089	.063	1/8
5/8	.621	0.82	.21	.06	.151	.56	1.61	1.473	.089	.063	1/8
3/4	.746	0.94	.26	.07	.182	.68	1.91	1.739	.110	.076	5/32
7/8	.871	1.04	.32	.09	.182	.80	2.16	1.989	.110	.076	5/32
1	.996	1.19	.35	.10	.182	.93	2.41	2.239	.110	.076	5/32

to lock components on shafts, and to hold laminated sections together. Because of their close tolerancing, they are installed by a press. These pins are shown in Fig. 11-12, and dimensional specifications are given in Table 11-7.

4. *Taper pins* are standardized to a taper of 1/4 inch per foot. The nominal diameter size is determined by the diameter of the large end of the pin. Taper pins are specified by pin size number and basic pin diameter. There are 21 standard taper pin sizes, ranging from 0.0625 inch (No. 7/0) to 1.5210 inch (No. 1.5210). This pin is illustrated in Fig. 11-13, with dimensional specifications given in Table 11-8.

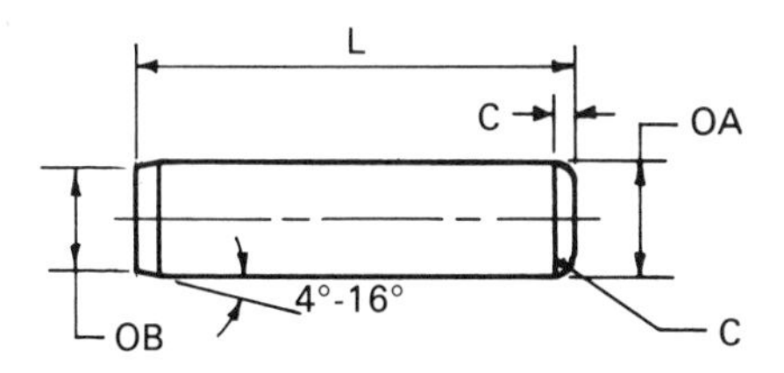

Fig. 11-12. Ground machine dowel pin.

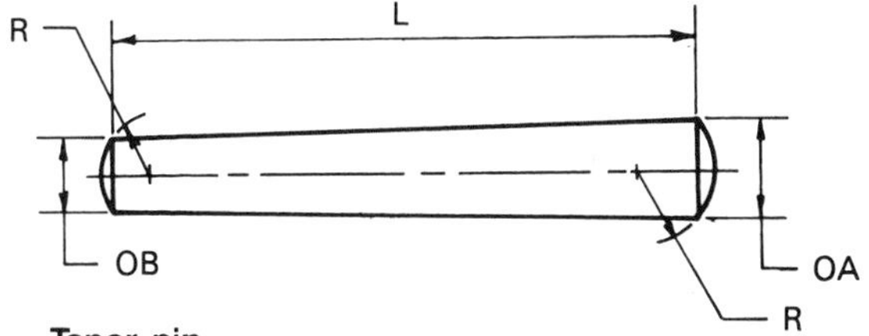

Fig. 11-13. Taper pin.

Table 11.7 Ground Machine Dowel Pins Dimensional Specifications (All dimensions in inches.)

Nominal Size or Nominal Pin Diameter	Pin Diameter, A						Point Diameter, B		Crown Height or Radius, C		Range of Preferred Lengths, L
	Standard Series Pins			Oversize Series Pins							
	Basic	Max	Min	Basic	Max	Min	Max	Min	Max	Min	
1/16 0.0625	0.0627	0.0628	0.0626	0.0635	0.0636	0.0634	0.058	0.048	0.020	0.008	3/16–3/4
5/64 0.0781	0.0783	0.0784	0.0782	0.0791	0.0792	0.0790	0.074	0.064	0.026	0.010	...
3/32 0.0938	0.0940	0.0941	0.0939	0.0948	0.0949	0.0947	0.089	0.079	0.031	0.012	5/16–1
1/8 0.1250	0.1252	0.1253	0.1251	0.1260	0.1261	0.1259	0.120	0.110	0.041	0.016	3/8–2
5/32 0.1562	0.1564	0.1565	0.1563	0.1572	0.1573	0.1571	0.150	0.140	0.052	0.020	..
3/16 0.1875	0.1877	0.1878	0.1876	0.1885	0.1886	0.1884	0.180	0.170	0.062	0.023	1/2–2
1/4 0.2500	0.2502	0.2503	0.2501	0.2510	0.2511	0.2509	0.240	0.230	0.083	0.031	1/2–2 1/2
5/16 0.3125	0.3127	0.3128	0.3126	0.3135	0.3136	0.3134	0.302	0.290	0.104	0.039	1/2–2 1/2
3/8 0.3750	0.3752	0.3753	0.3751	0.3760	0.3761	0.3759	0.365	0.350	0.125	0.047	1/2–3
7/16 0.4375	0.4377	0.4378	0.4376	0.4385	0.4386	0.4384	0.424	0.409	0.146	0.055	7/8–3
1/2 0.5000	0.5002	0.5003	0.5001	0.5010	0.5011	0.5009	0.486	0.471	0.167	0.063	3/4, 1–4
5/8 0.6250	0.6252	0.6253	0.6251	0.6260	0.6261	0.6259	0.611	0.595	0.208	0.078	1 1/4–5
3/4 0.7500	0.7502	0.7503	0.7501	0.7510	0.7511	0.7509	0.735	0.715	0.250	0.094	1 1/2–6
7/8 0.8750	0.8752	0.8753	0.8751	0.8760	0.8761	0.8759	0.860	0.840	0.293	0.109	2, 2 1/2–6
1 1.0000	1.0002	1.0003	1.0001	1.0010	1.0011	1.0009	0.980	0.960	0.333	0.125	2, 2 1/2–5, 6

5. *Grooved pins* come in seven standard forms, ranging from Type A to Type G. This pin is designed with three equally spaced grooves, with an expanded diameter over the crest of

Table 11.8 Taper Pins Dimensional Specifications (All dimensions in inches.)

Pin Size Number and Basic Pin Diameter	Major Diameter (Large End), A				End Crown Radius, R		Range of Lengths, L	
	Commercial Class		Precision Class					
	Max	Min	Max	Min	Max	Min	Stand. Reamer Avail.	Other
7/0 0.0625	0.0638	0.0618	0.0635	0.0625	0.072	0.052	...	1/4–1
6/0 0.0780	0.0793	0.0773	0.0790	0.0780	0.088	0.068	...	1/4–1 1/2
5/0 0.0940	0.0953	0.0933	0.0950	0.0940	0.104	0.084	1/4–1	1 1/4, 1 1/2
4/0 0.1090	0.1103	0.1083	0.1100	0.1090	0.119	0.099	1/4–1	1 1/4–2
3/0 0.1250	0.1263	0.1243	0.1260	0.1250	0.135	0.115	1/4–1	1 1/4–2
2/0 0.1410	0.1423	0.1403	0.1420	0.1410	0.151	0.131	1/2–1 1/4	1 1/2–2 1/2
0 0.1560	0.1573	0.1553	0.1570	0.1560	0.166	0.146	1/2–1 1/4	1 1/2–3
1 0.1720	0.1733	0.1713	0.1730	0.1720	0.182	0.162	3/4–1 1/4	1 1/2–3
2 0.1930	0.1943	0.1923	0.1940	0.1930	0.203	0.183	3/4–1 1/2	1 3/4–3
3 0.2190	0.2203	0.2183	0.2200	0.2190	0.229	0.209	3/4–1 3/4	2–4
4 0.2500	0.2513	0.2493	0.2510	0.2500	0.260	0.240	3/4–2	2 1/4–4
5 0.2890	0.2903	0.2883	0.2900	0.2890	0.299	0.279	1–2 1/2	2 3/4–6
6 0.3410	0.3423	0.3403	0.3420	0.3410	0.351	0.331	1 1/4–3	3 1/4–6
7 0.4090	0.4103	0.4083	0.4100	0.4090	0.419	0.399	1 1/4–3 3/4	4–8
8 0.4920	0.4933	0.4913	0.4930	0.4920	0.502	0.482	1 1/4–4 1/2	4 3/4–8
9 0.5910	0.5923	0.5903	0.5920	0.5910	0.601	0.581	1 1/4–5 1/4	5 1/2–8
10 0.7060	0.7073	0.7053	0.7070	0.7060	0.716	0.696	1 1/2–6	6 1/4–8
11 0.8600	0.8613	0.8593	...	...	0.870	0.850	...	2–8
12 1.0320	1.0333	1.0313	...	...	1.042	1.022	...	2–9

its ridges. The seven types of grooved pin are illustrated in Fig. 11-14, with dimensional specifications given in Table 11-9.

6. *Spring pins* are the last major type of pin to be described here. They are made with a hollow resilient wall that compresses under pressure. They come in two forms, *spiral wrapped* and *slotted tubular pins*, of which the latter is the more common. Both types of spring pin are used to fasten components by the spring pressure exerted against the hole wall. Shown in Fig. 11-15 are the two types of spring pin, with Tables 11-10a and b giving the corresponding dimensional specifications for spiral wrapped and slotted tubular spring pins, respectively.

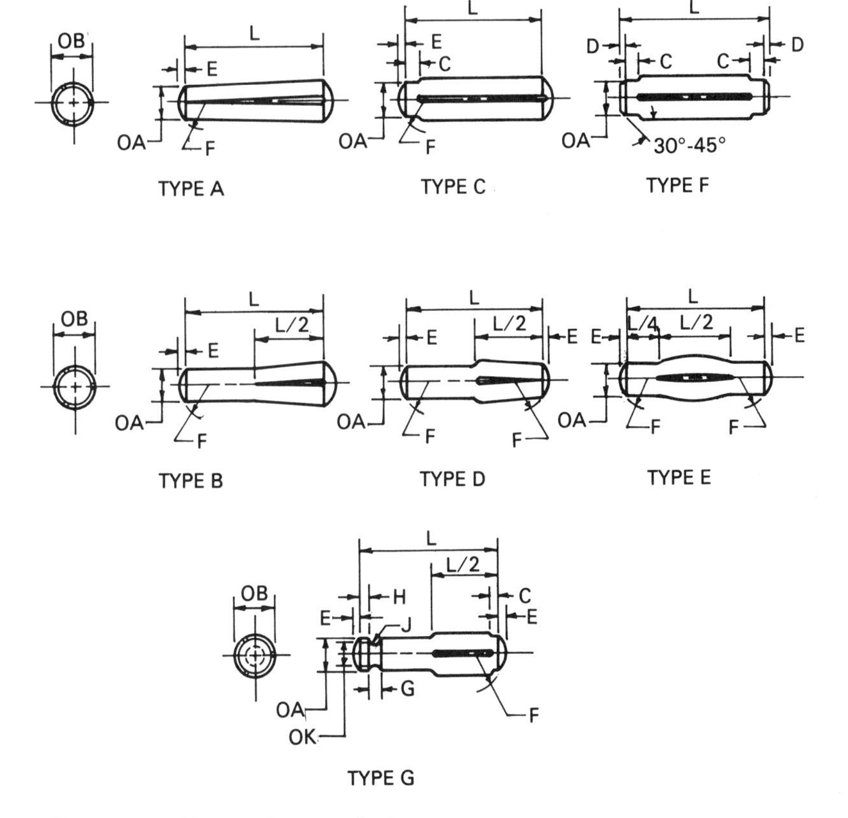

Fig. 11-14. Types of grooved pin.

Table 11.9 Grooved Pins Dimensional Specifications (All dimensions in inches.)

Nominal Size or Basic Pin Diameter		Pin Diameter, *A*		Pilot Length, *C*	Chamfer Length, *D*	Crown Height, *E*		Crown Radius, *F*		Neck Width, *G*		Shoulder Length, *H*		Neck Radius, *J*	Neck Diameter, *K*		Range of Standard Lengths
		Max	Min	Ref	Min	Max	Min	Max	Min	Max	Min	Max	Min	Ref	Max	Min	
1/32	0.0312	0.0312	0.0302	0.015	...	...	...	...	...	...	...	...	...	...	...	...	1/8–1/2
3/64	0.0469	0.0469	0.0459	0.031	...	...	...	...	...	...	...	...	...	...	...	...	1/8–5/8
1/16	0.0625	0.0625	0.0615	0.031	0.016	0.0115	0.0015	0.088	0.068	...	...	...	...	...	...	...	1/8–1
5/64	0.0781	0.0781	0.0771	0.031	0.016	0.0137	0.0037	0.104	0.084	...	...	...	...	...	...	...	1/4–1
3/32	0.0938	0.0938	0.0928	0.031	0.016	0.0141	0.0041	0.135	0.115	0.038	0.028	0.041	0.031	0.016	0.067	0.057	1/4–1 1/4
7/64	0.1094	0.1094	0.1074	0.031	0.016	0.0160	0.0060	0.150	0.130	0.038	0.028	0.041	0.031	0.016	0.082	0.072	1/4–1 1/4
1/8	0.1250	0.1250	0.1230	0.031	0.016	0.0180	0.0080	0.166	0.146	0.069	0.059	0.041	0.031	0.031	0.088	0.078	1/4–1 1/2
5/32	0.1563	0.1563	0.1543	0.062	0.031	0.0220	0.0120	0.198	0.178	0.069	0.059	0.057	0.047	0.031	0.109	0.099	3/8–2
3/16	0.1875	0.1875	0.1855	0.062	0.031	0.0230	0.0130	0.260	0.240	0.069	0.059	0.057	0.047	0.031	0.130	0.120	3/8–2 1/4
7/32	0.2188	0.2188	0.2168	0.062	0.031	0.0270	0.0170	0.291	0.271	0.101	0.091	0.072	0.062	0.047	0.151	0.141	1/2–3
1/4	0.2500	0.2500	0.2480	0.062	0.031	0.0310	0.0210	0.322	0.302	0.101	0.091	0.072	0.062	0.047	0.172	0.162	1/2–3 1/4
5/16	0.3125	0.3125	0.3105	0.094	0.047	0.0390	0.0290	0.385	0.365	0.132	0.122	0.104	0.094	0.062	0.214	0.204	5/8–3 1/2
3/8	0.3750	0.3750	0.3730	0.094	0.047	0.0440	0.0340	0.479	0.459	0.132	0.122	0.135	0.125	0.062	0.255	0.245	3/4–4 1/4
7/16	0.4375	0.4375	0.4355	0.094	0.047	0.0520	0.0420	0.541	0.521	0.195	0.185	0.135	0.125	0.094	0.298	0.288	7/8–4 1/2
1/2	0.5000	0.5000	0.4980	0.094	0.047	0.0570	0.0470	0.635	0.615	0.195	0.185	0.135	0.125	0.094	0.317	0.307	1–4 1/2

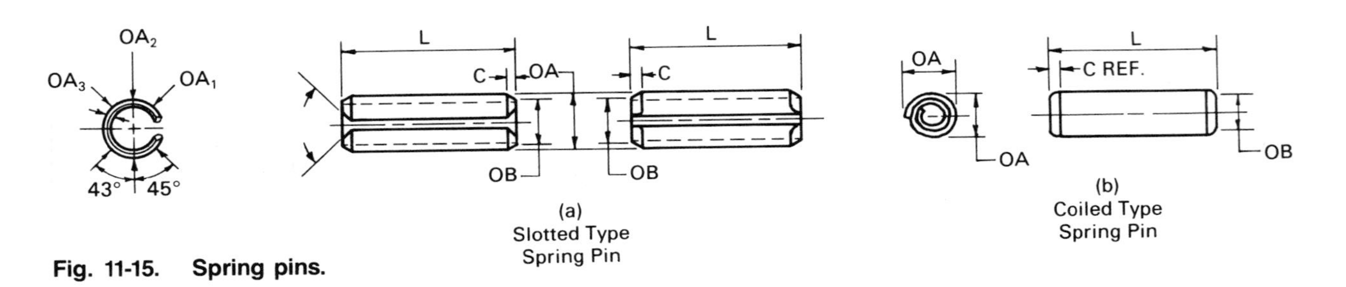

Fig. 11-15. Spring pins.

Table 11.10a Spiral Wrapped Spring Pins Dimensional Specifications (All dimensions in inches.)

Nominal Size or Basic Pin Diameter	Pin Diameter, A — Standard Duty		Heavy Duty		Light Duty		Chamfer — Dia., B	Length, C	Recom-mended Hole Size		Double Shear Load, Min, lb — Standard Duty — AISI 1070-1095 and AISI 420	AISI 302	Heavy Duty — AISI 1070-1095 and AISI 420	AISI 302	Light Duty — AISI 1070-1095 and AISI 420	AISI 302
Diameter	Max	Min	Max	Min	Max	Min	Max	Ref	Max	Min						
1/32 0.031	0.035	0.033	...	...	...	...	0.029	0.024	0.032	0.031	75[2]	60	...	...	...	...
0.039	0.044	0.041	...	...	...	...	0.037	0.024	0.040	0.039	120[2]	100	...	...	...	...
3/64 0.047	0.052	0.049	...	...	...	...	0.045	0.024	0.048	0.046	170[2]	140	...	...	...	...
0.052	0.057	0.054	...	...	...	...	0.050	0.024	0.053	0.051	230[2]	190	...	...	...	...
1/16 0.062	0.072	0.067	0.070	0.066	0.073	0.067	0.059	0.028	0.065	0.061	300	250	450	350	...	135
5/64 0.078	0.088	0.083	0.086	0.082	0.089	0.083	0.075	0.032	0.081	0.077	475	400	700	550	...	225
3/32 0.094	0.105	0.099	0.103	0.098	0.106	0.099	0.091	0.038	0.097	0.093	700	550	1,000	800	375	300
7/64 0.109	0.120	0.114	0.118	0.113	0.121	0.114	0.106	0.038	0.112	0.108	950	750	1,400	1,125	525	425
1/8 0.125	0.138	0.131	0.136	0.130	0.139	0.131	0.121	0.044	0.129	0.124	1,250	1,000	2,100	1,700	675	550
5/32 0.156	0.171	0.163	0.168	0.161	0.172	0.163	0.152	0.048	0.160	0.155	1,925	1,550	3,000	2,400	1,100	875
3/16 0.188	0.205	0.196	0.202	0.194	0.207	0.196	0.182	0.055	0.192	0.185	2,800	2,250	4,400	3,500	1,500	1,200
7/32 0.219	0.238	0.228	0.235	0.226	0.240	0.228	0.214	0.065	0.224	0.217	3,800	3,000	5,700	4,600	2,100	1,700
1/4 0.250	0.271	0.260	0.268	0.258	0.273	0.260	0.243	0.065	0.256	0.247	5,000	4,000	7,700	6,200	2,700	2,200
5/16 0.312	0.337	0.324	0.334	0.322	0.339	0.324	0.304	0.080	0.319	0.308	7,700	6,200	11,500	9,200	4,440	3,500
3/8 0.375	0.403	0.388	0.400	0.386	0.405	0.388	0.366	0.095	0.383	0.370	11,200	9,000	17,600	14,000	6,000	5,000
7/16 0.438	0.469	0.452	0.466	0.450	0.471	0.452	0.427	0.095	0.446	0.431	15,200	13,000	22,500	18,000	8,400	6,700
1/2 0.500	0.535	0.516	0.532	0.514	0.537	0.516	0.488	0.110	0.510	0.493	20,000	16,000	30,000	24,000	11,000	8,800
5/8 0.625	0.661	0.642	0.658	0.640	...	...	0.613	0.125	0.635	0.618	31,000[3]	25,000	46,000[3]	37,000	...	...
3/4 0.750	0.787	0.768	0.784	0.766	...	...	0.738	0.150	0.760	0.743	45,000[3]	36,000	66,000[3]	53,000	...	...

Table 11.10b Slotted Tubular Spring Pins Dimensional Specifications (All dimensions in inches.)

Nominal Size or Basic Pin Diameter	Average Pin Diameter, *A*		Chamfer Diam., *B*	Chamfer Length, *C*		Stock Thickness, *F*	Recommended Hole Size		Range of Practical Lengths
	Max	Min	Max	Max	Min	Basic	Max	Min	
1/16 0.062	0.069	0.066	0.059	0.028	0.007	0.012	0.065	0.062	3/16–1
5/64 0.078	0.086	0.083	0.075	0.032	0.008	0.018	0.081	0.078	3/16–1 1/2
3/32 0.094	0.103	0.099	0.091	0.038	0.008	0.022	0.097	0.094	3/16–1 1/2
1/8 0.125	0.135	0.131	0.122	0.044	0.008	0.028	0.129	0.125	5/16–2
9/64 0.141	0.149	0.145	0.137	0.044	0.008	0.028	0.144	0.140	3/8–2
5/32 0.156	0.167	0.162	0.151	0.048	0.010	0.032	0.160	0.156	7/16–2 1/2
3/16 0.188	0.199	0.194	0.182	0.055	0.011	0.040	0.192	0.187	1/2–2 1/2
7/32 0.219	0.232	0.226	0.214	0.065	0.011	0.048	0.224	0.219	1/2–3
1/4 0.250	0.264	0.258	0.245	0.065	0.012	0.048	0.256	0.250	1/2–3 1/2
5/16 0.312	0.328	0.321	0.306	0.080	0.014	0.062	0.318	0.312	3/4–4
3/8 0.375	0.392	0.385	0.368	0.095	0.016	0.077	0.382	0.375	3/4, 7/8, 1, 1 1/4, 1 1/2, 1 3/4, 2–4
7/16 0.438	0.456	0.448	0.430	0.095	0.017	0.077	0.445	0.437	1, 1 1/4, 1 1/2, 1 3/4, 2–4
1/2 0.500	0.521	0.513	0.485	0.110	0.025	0.094	0.510	0.500	1 1/4, 1 1/2, 1 3/4, 2–4
5/8 0.625	0.650	0.640	0.608	0.125	0.030	0.125	0.636	0.625	2–6
3/4 0.750	0.780	0.769	0.730	0.150	0.030	0.150	0.764	0.750	. . .

Retaining Rings

A mechanical element that provides a shoulder against which components can be located, retained, or locked is the retaining ring. Also known as *snap rings*, retaining rings are easily installed and removed manually because they are often made out of spring steel. In some cases, these rings are used to eliminate play caused by wear or accumulated tolerances. Shown in Fig. 11-16 are two common types of retaining ring.

There are two general types of retaining ring. The first is *external rings*, which are spread over a shaft. This can be accomplished by using pliers or a special ring tool to set the ring in a circumferential groove. Here the retaining ring provides a protruding shoulder for locating and retaining components on a shaft.

The second type of retaining ring is *internal rings*, which are compressed into a housing. Again, pliers or a special tool can be used to insert the ring manually into a circumferential groove. The shoulder,

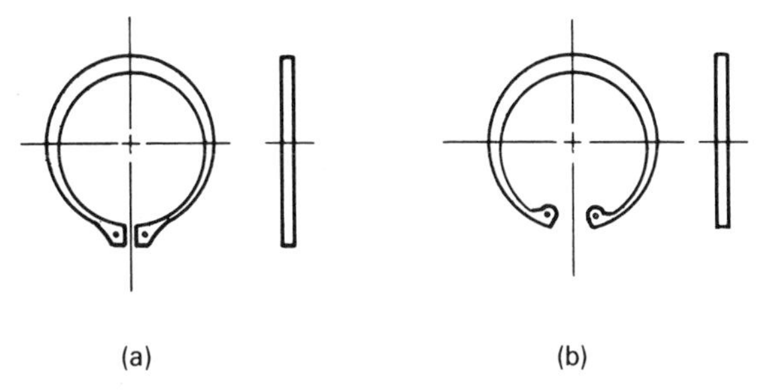

Fig. 11-16. Two types of retaining ring.

however, provides an internal protruding shoulder in the housing for locating and retaining components within it. Both types are shown in Fig. 11-17.

In addition to the two types of retaining ring design are three classifications of ring, according to the material they are made from:

1. *Stamped retaining rings* have a tapered radial width that decreases symmetrically from the center section to the free ends.
2. *Wire-formed retaining rings* have uniform cross-sectional area, and are split rings formed and cut from spring wire material.
3. *Spiral wound retaining rings* are made up of two or more turns of a rectangular metal material. This winding is accomplished on an edge to produce a continuous coil that can be crimped or uncrimped.

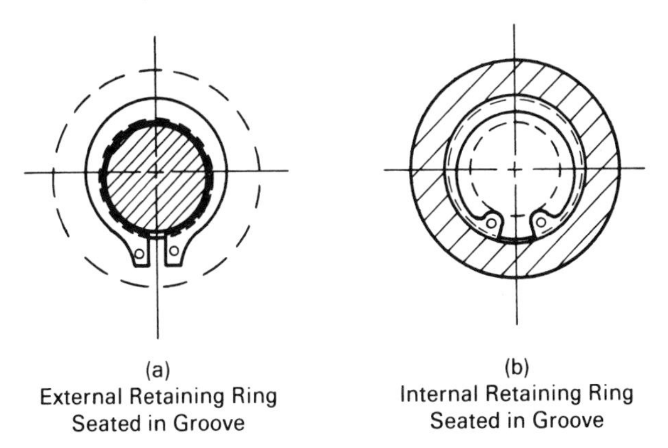

Fig. 11-17. Internal and external retaining rings.

Springs

All materials used in mechanical products have some degree of elasticity, and will deform or bend when a load is applied. If the elastic limit is not exceeded, the material will return to its original shape when the load is removed. In nonspring elements, some degree of deformation is accepted by designers—their function, however, is to remain rigid. By comparison, the underlying function of a spring is to deform, within limits, under applied loads and return to its original size and shape when the load is removed. A more technical definition of its purpose is to store energy elastically as a result of its large displacement.

Types of Spring

Many types of spring are available for commercial use; they come in a variety of shapes, forms, and materials. In mechanical design and engineering they have four major uses. These are:

1. The absorption or storage of energy and mitigation of shock and vibration, as in buffers and the suspension units of vehicles
2. The application of a definite force or torque, as in valves, governors, and pipe supports
3. As indicators or controllers of load or torque, as in weighing machines and dynamometers
4. To act as an elastic pivot or guide, as in balancing machines and expansion bends

For most purposes, springs can be divided into two major groups. The first is *wire springs.* Wire springs are first cold-formed, and then heat-treated to the desired temper (spring temper). These springs include helical and spiral springs that are made of wire with square, round, or special design cross sections. Examples of these springs are compression, extension, and torsion springs.

The second major group of springs is *flat springs.* These springs are made from flat or strip materials and are usually designated as either elliptical or cantilever in design. The materials most frequently used to manufacture flat springs are spring steel, stainless steel, brass, bronze, beryllium, copper, and various nickel alloys.

Representation of Springs

Unlike the other mechanical elements presented thus far, the detailing of springs requires significantly more time and drawing skill. On most drawings, a simplified or schematic drawing of a spring is recommended to save time. In subassembly and assembly drawings, however, springs are often shown in section. When sections are drawn, sectional hatching should be used for large springs, while a solid black shading is recommended for small springs. The distinction between "large" and "small," however, is subjective, and is left to the discretion of the drafter.

COMPRESSION SPRINGS

All compression springs are open-coiled and helical in design. Compression springs are used to offer resistance to compressive loads, and are found in a variety of applications. They also come in round, square, rectangular, and special cross sections. Compression springs are made as either a right- or a left-hand helix, are cylindrical or conical, and come with four types of end: open end not ground, open end ground, closed end not ground, and close end ground (Fig. 11-18).

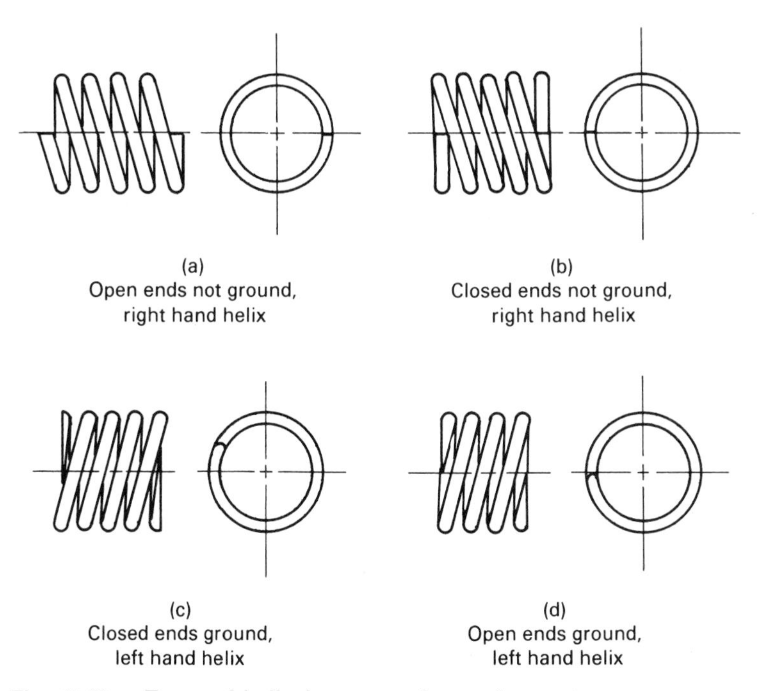

Fig. 11-18. Types of helical compression spring end.

When compression springs are represented in drawings, one of three types of graphic presentation can be used. The first is a plane view, which is a drawing that shows the spring as it would be seen as a whole from a given view. The second is a section of the spring, showing the cross section of the spring material. The last is a simplified, single-line schematic, which is frequently used in detailed drawings. In simplified representations, a notation may be used to indicate if the spring is wound left- or right-hand, and the cross section of the spring material may be indicated either by a symbol or in words. Fig. 11-19 shows the three representations acceptable for compression springs.

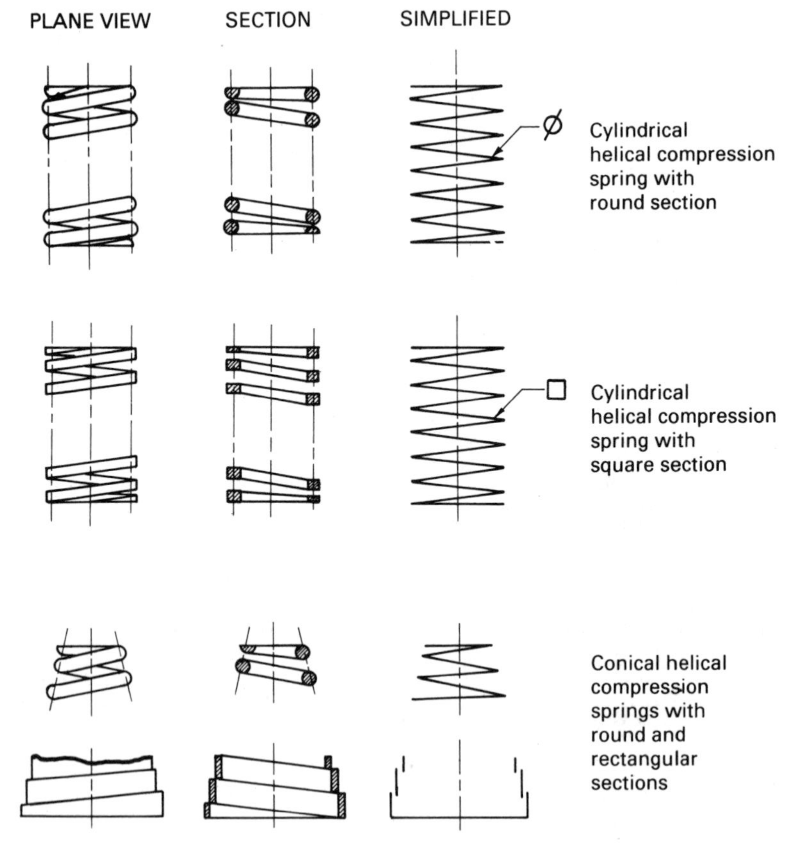

Fig. 11-19. Representation of compression springs.

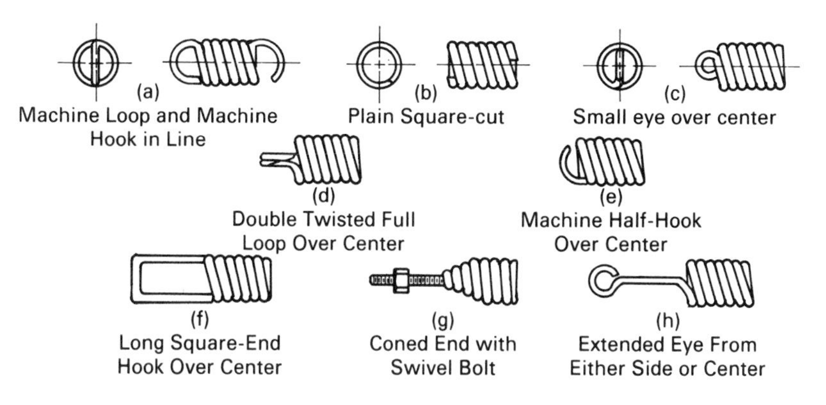

Note: Compression spring types a, b, c, d, and e require no special tool for installation or removal
Compression spring types f, g, and h require special tools for installation and removal

Fig. 11-20. Tension spring ends.

TENSION SPRINGS

Also known as *extension springs*, tension springs are all closed-coil helical springs that offer resistance to pulling forces. They are made from either round or square cross-sectional wire, and are wound so that the coils are usually in contact with each other. Tension springs are available in a wide variety of ends, some requiring special tooling for installation and removal. Fig. 11-20 shows several types of end style used for tension springs.

As in compression spring representation, tension springs can be represented by plane view, section, or simplified presentation. The same notation and specification procedures apply for noting coil wind and cross-sectional material and shape. Fig. 11-21 shows tension spring representations.

TORSION SPRINGS

As their name implies, torsion springs are used to provide a torque, and to exert pressure along a circular arc path. Torsion springs are most commonly made from round, square, or rectangular wire that is wound into a helical spring. Torsion spring end styles are shown in Fig. 11-22.

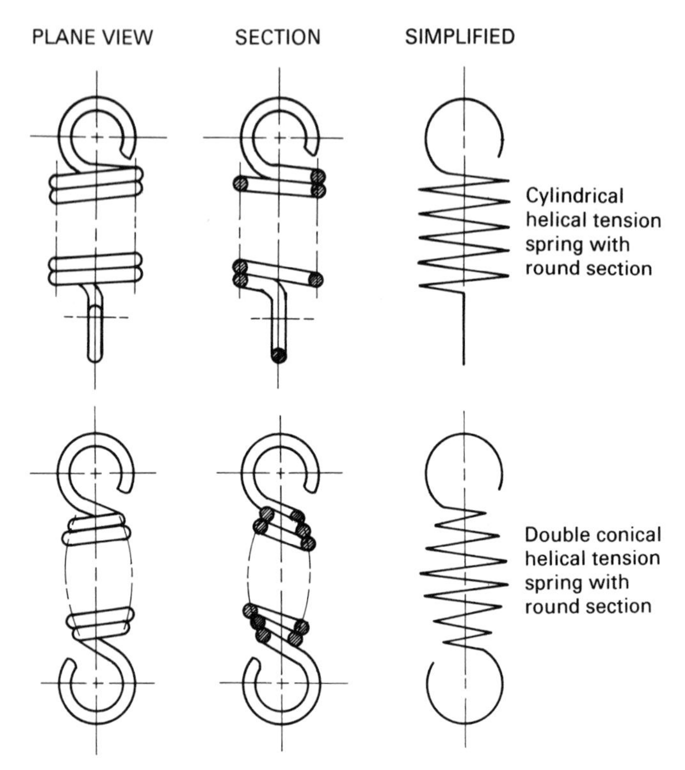

Fig. 11-21. Representation of tension springs.

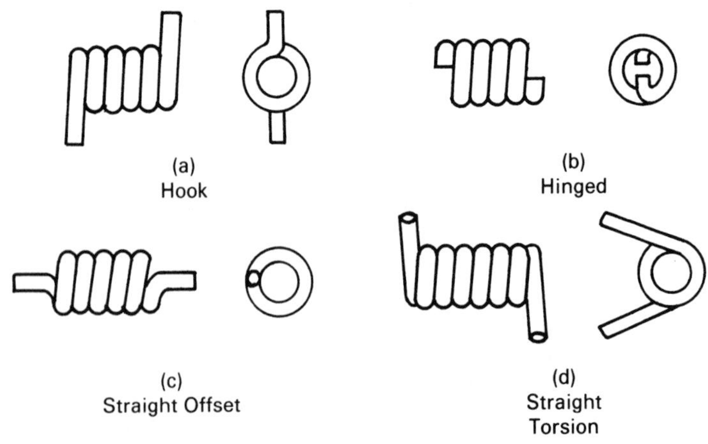

Fig. 11-22. Common types of end for torsion springs.

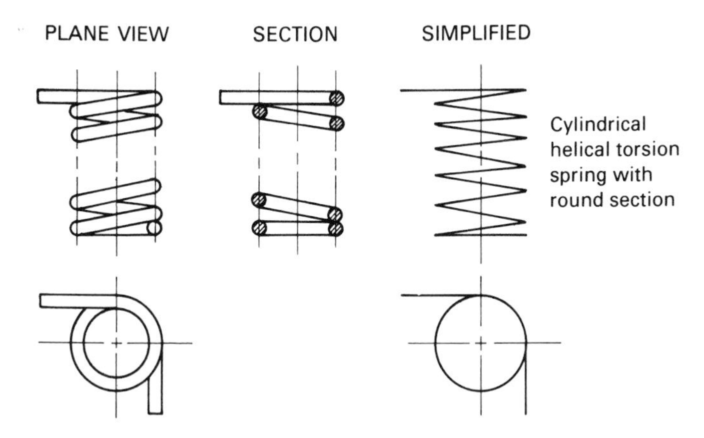

Fig. 11-23. Representation of torsion springs.

Graphic representations for torsion springs fall into the same categories of plane view, section, and simplified drawings. These are shown in Fig. 11-23.

CUP SPRINGS

Sometimes referred to as *conical disk springs* or *Belleville washers*, cup springs are used to sustain large loads with small deflections. Added loads and deflections can be increased, however, by stacking the springs upon one another. Cup springs are normally specified in terms of the ratios of O.D. to I.D. (outside diameter to inside diameter), height to thickness, and other factors generated by mathematical calculations. Fig. 11-24 shows the graphic representations used for cup springs.

SPIRAL SPRINGS

The spiral spring is known by a variety of different names, which include *clock springs* and *motor springs*. They are often used in wind-up motors in toys and other products. Because of design difficulties, they cannot be calculated with any great precision. However, they are used successfully in some precision instruments, such as clocks. Spiral springs are designed to store the energy applied in a winding and deliver a resulting energy as torque to a central shaft or arbor.

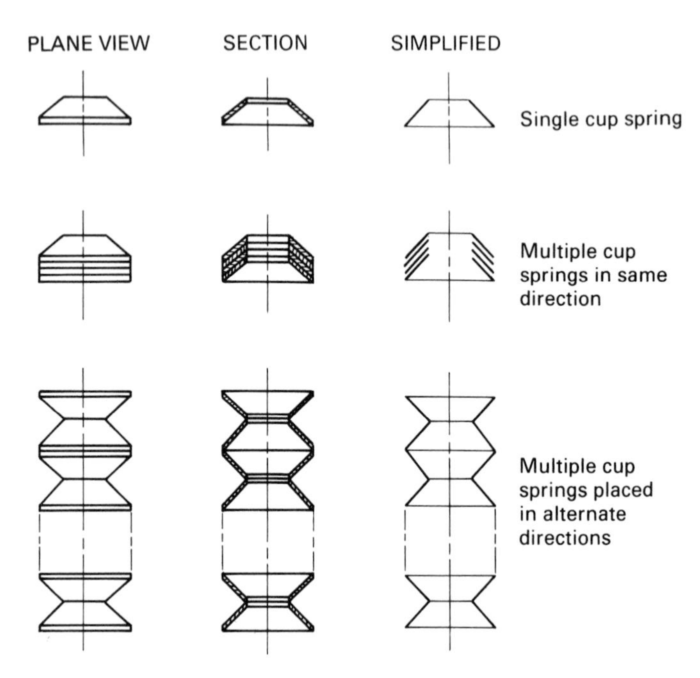

Fig. 11-24. Representation of cup springs.

For spiral springs there are only two recommended representations: plane view and simplified. Fig. 11-25 shows two examples of spiral spring representations.

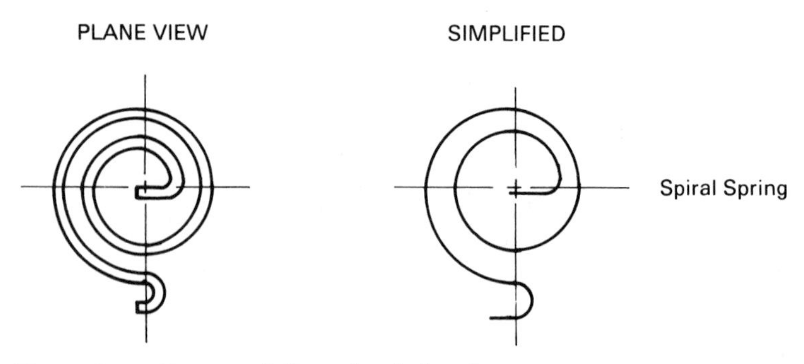

Fig. 11-25. Representation of spiral springs.

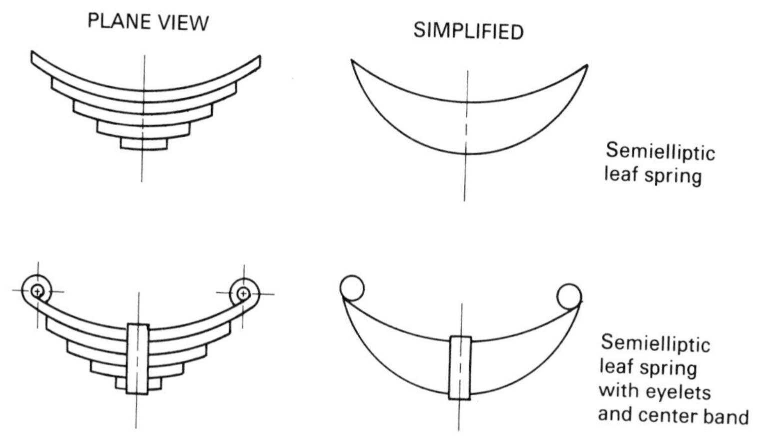

Fig. 11-26. Representation of leaf springs.

LEAF SPRINGS

Unlike helical compression and tension springs, leaf springs are made of material that is not twisted; they are flat springs that absorb energy by a bending action. The primary stresses to which the flat strips of material are exposed are tensile or compressive. As with spiral springs, only plane view and simplified graphic representations are used (see Fig. 11-26).

Rivets and Adhesives

Two permanent types of nonthreaded fastening device are rivets and adhesives. Rivet fastening has been widely used for many years and can be found represented in numerous industrial drawings dating back to the turn of the century. Adhesives, on the other hand, have just recently found acceptance with the development of the "super glues" produced in the 1950s and 1960s. This section will discuss both fastening techniques and their graphic representation.

Rivets

There are two broad categories of rivet. The first are known as large rivets, and are used for structural works on buildings, bridges, and other structures. Most of these rivets, however, have given way to high-strength bolts. Large rivets are still used, though to a lesser degree, in building structures. Rivets that are used at the construction site are known as *field rivets. Shop rivets*, by comparison, are used to assemble a structure or unit (e.g., air-conditioning unit housings) at the shop.

The second type of rivet is *small rivets*, which are used in high-speed, large-volume processes. Small rivets are not considered to be tension fasteners, but are comparable in strength to soldered joints. Because of the speed by which they can be applied, rivets have replaced many soldering operations.

RIVET JOINTS

There are two general types of rivet joint. The first are *butt joints*. Here the sheet material is "butted" together, so that the ends are either in contact or close proximity. To provide material for riveting, two additional sheets are used to sandwich in the sheets. This technique is shown in Fig. 11-27a.

The second type of rivet joint is the *lap joint*. In this configuration, the sheet material is overlapped to provide a riveting surface (Fig. 11-27b). From these two basic designs, more complex joints have been developed.

RIVET REPRESENTATION

Detailing of rivets and rivet joints is not usually required. As a result, a standardized system of rivet symbols has been adopted by a number of countries. These symbols show not only the placement of rivets, but also the installation process and finish. Two general symbols are used here. The first is for shop rivets, and notes the diameter of the rivet head on the drawing. The second identifies field rivets, and notes the diameter of the rivet's shaft. Fig. 11-28 shows these rivet symbols.

In addition to these notations, a series of standardized ISO symbols has been developed for structural notations. Primarily used to identify operations to be accomplished in the shop or field, these symbols are shown in Fig. 11-29. For more information, see ISO 5261-1981 (E) Technical Drawings for Structural Metal Work.

Sometimes rivet dimensions are required for both large and small rivets. When accuracy of size is necessary, the drafter should refer to the standard ANSI B18.1.2 for large rivets, and ANSI B18.1.1 for small rivets. For approximate dimensions, which are acceptable for most drawings, one may use the proportions shown in Fig. 11-30.

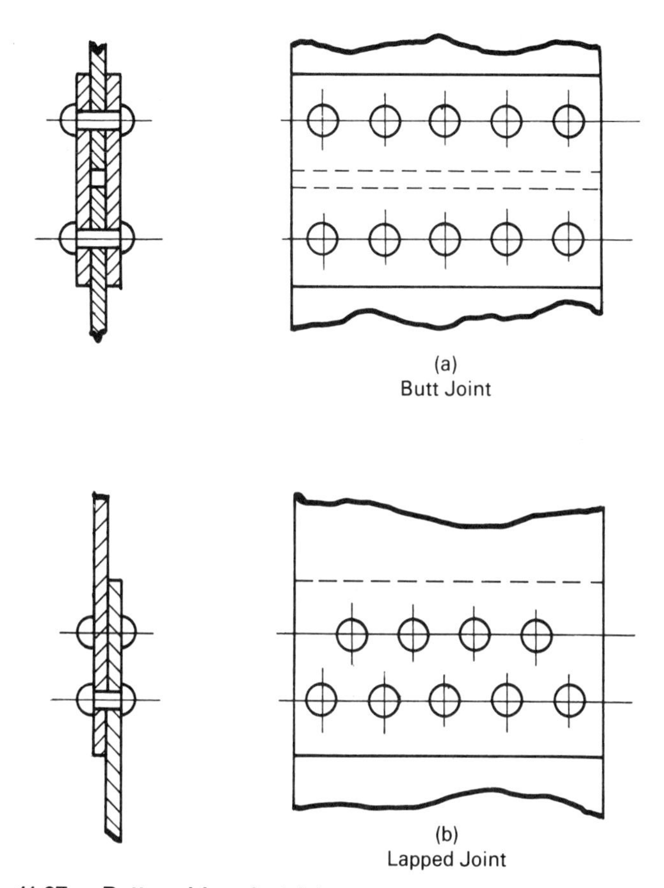

Fig. 11-27. **Butt and lap rivet joints.**

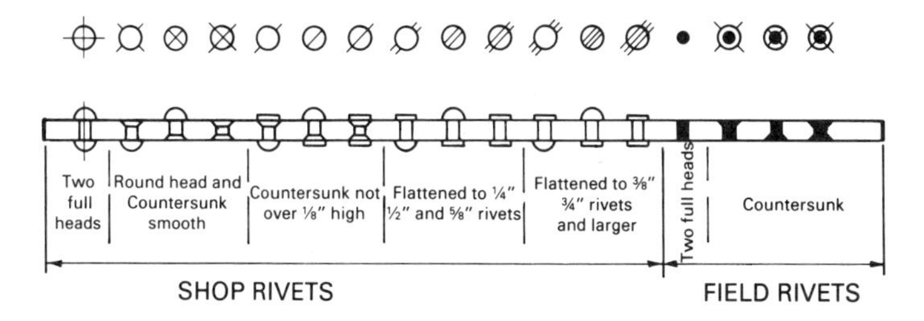

Fig. 11-28. **Rivet symbols.**

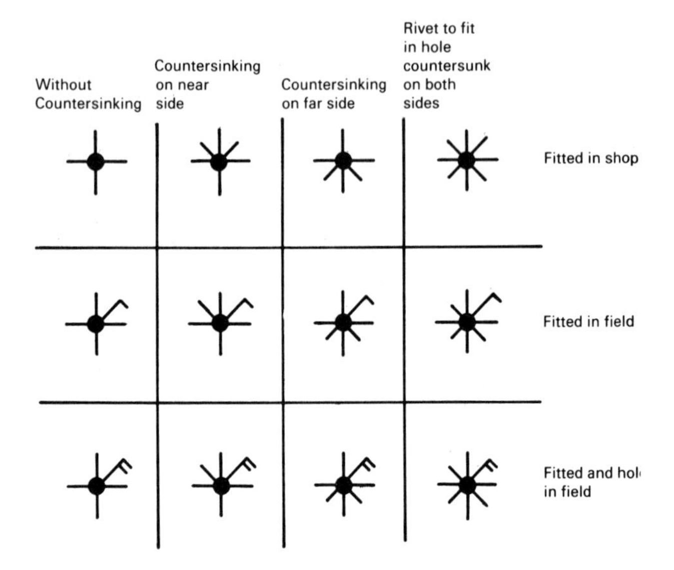

Fig. 11-29. ISO symbols for rivets to fit in hole.

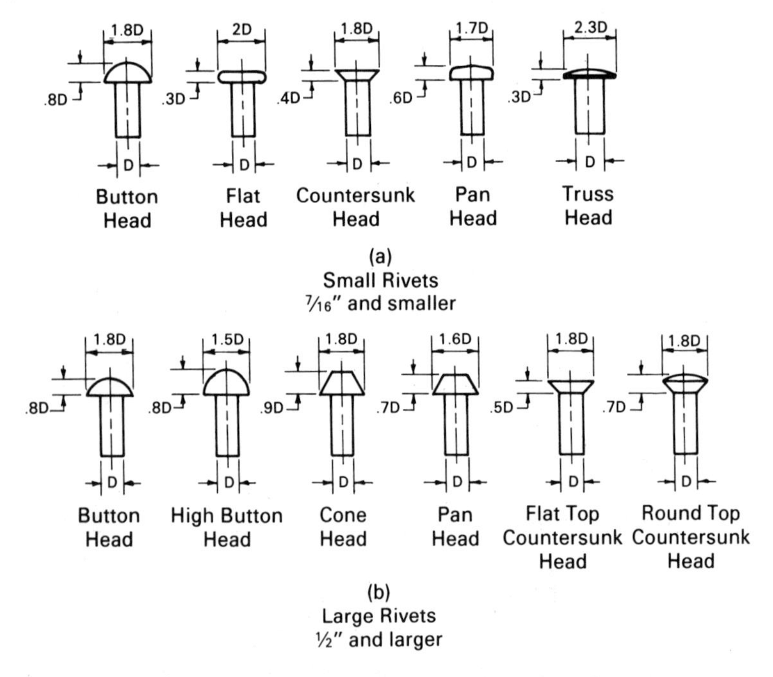

Fig. 11-30. Dimensional proportions for standard rivets.

Adhesives

An adhesive is any material or substance that bonds or fastens materials to be joined by means of surface attachment. In some cases, adhesive bonding has replaced bolting, welding, and riveting. Occasionally an adhesive is used for both bonding and sealing. When it is performing both functions, it will usually be known as an *adhesive sealant.* In addition, some adhesives are also used as insulators against temperature and electrical conductivity.

Perhaps the field in which adhesives have made the largest impact is in the fastening of thin cross sections, where the joint loads are of such unacceptable concentration that adhesives are the only workable alternative. An adhesive is considered to be *structural* when it is able to support heavy loads, and *nonstructural* when it cannot.

Adhesive fastenings, when shown in plane views, are identified by notation only. The only detailing of adhesive bonding should be in section presentations. Because the bonding is thin, crosshatching cannot be used, and the adhesive will be shown as a solid black shading. Both plane view and sectional presentations are shown in Fig. 11-31.

Welding

Welding is a permanent fastening process whereby metal is fused together to form one integral part. There are three types of welding

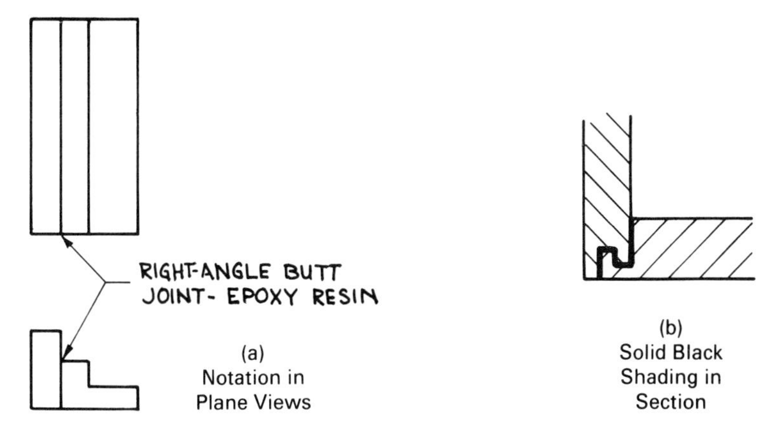

Fig. 11-31. Methods of presenting adhesive joints.

process: pressure, nonpressure, and casting. The most common of the three is nonpressure welding, such as arc and gas welding. Arc welding is the fusion of metal by the application of an electrical arc that is formed between the rod and work. In gas welding, heat is generated by the burning of a mixture of two gases (e.g., oxygen and acetylene). The weld is accomplished by melting a filler rod with a torch along a joint, after the metal joint has been heated to a molten state.

Pressure welding is also known as resistance welding, and is accomplished by applying mechanical pressure and heat. The heat is generated by an electrical current that melts the metal, and the weld occurs when pressure is applied. A casting process is called thermit welding, in which molten metal is poured into a mold that is constructed around the parts and along the joint. Metal is melted in a crucible and poured into the form; fusion occurs when contact is made between the liquid metal and the preheated joining metals.

Because welding is an important fabrication process, a series of graphic procedures and symbols for it have been standardized. Most U.S. welding drawings make use of a basic welding symbol specified by the American Welding Society (AWS) and presented in standard ANSI/AWS A2.4-79. This technique is also recognized by the Canadian Welding Bureau, and the ISO (see ISO 2552-1974 (E), Welds—Symbolic Representation on Drawings). This section briefly discusses basic weld joints and the use of the basic welding symbol.

Welding Joints

The basic types of arc and gas weld are illustrated in Figs. 11-32a and b. Fig. 11-32a shows a cross section of prepared joints that are used in fabrication processes. Square groove and fillet welds cannot be shown as double welds. Single-V, single bevel groove, single-U, and single-J, however, can be double welded by constructing the weld symmetrically about the x——x line. Fig. 11-32b shows joints that are commonly used in sheet and plate fabricated products.

ANSI/AWS Welding Symbol

The use of welding symbols enables the designer and drafter to indicate clearly the type and size of weld desired. The use of the basic welding symbol (Fig. 11-33) is required for contractors and fabricators certified by the American Welding Society and the Canadian Welding

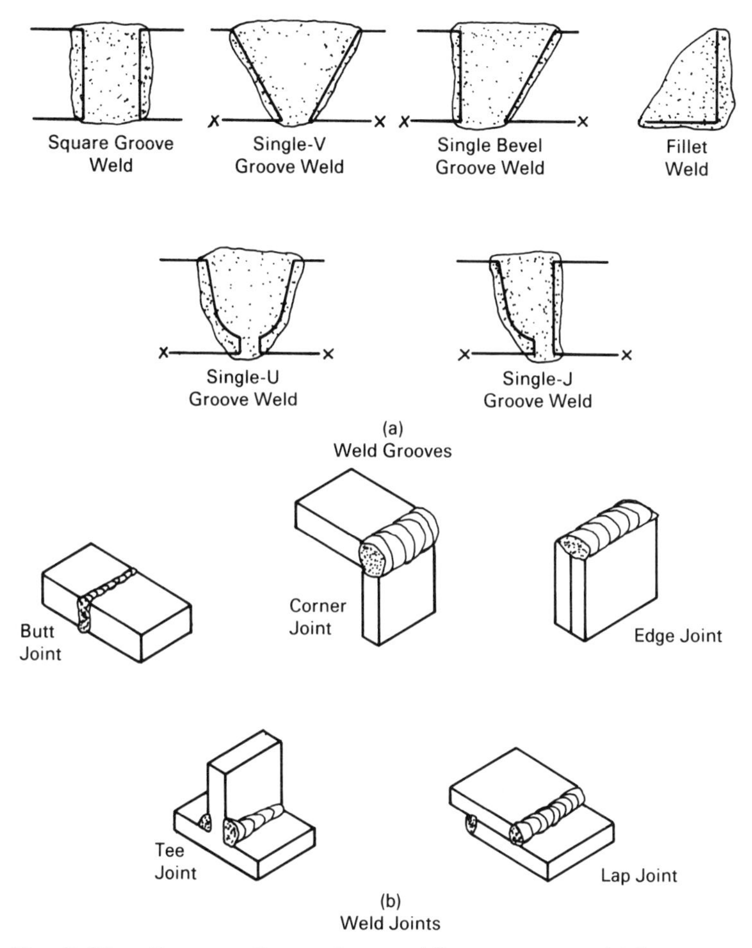

Fig. 11-32. Common Gas and arc welding grooves and joints.

Bureau. Before this standardized symbol was adopted, the terms "far side" and "near side" were used to describe the location of the weld. However, this was confusing, and today, under the present system, with the joint as the point of reference, the joint noted by the symbol is identified with reference to an "arrow side" and the "other side," thereby eliminating any possible confusion.

The arrow is considered the basic element of the welding symbol. It is used to point toward the joint where the weld is to be made.

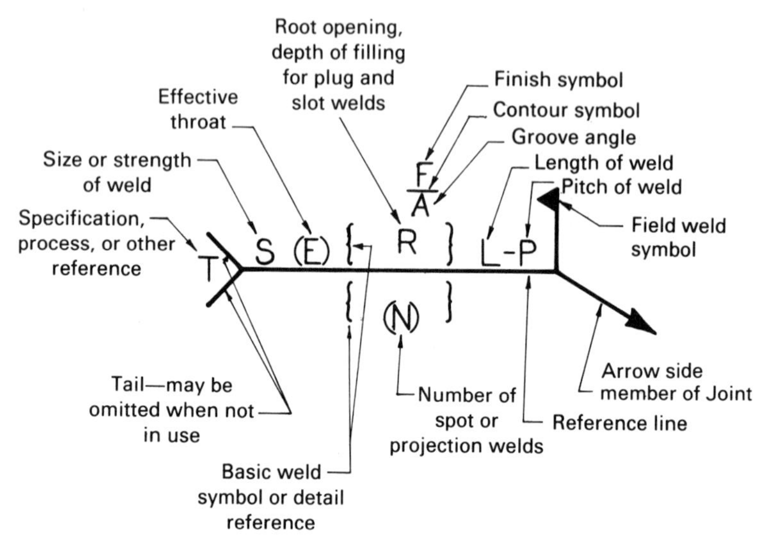

Fig. 11-33. Basic AWS/ANSI/ISO welding symbol.

When the weld is to be executed on the arrow side, the symbol indicating the type of weld is placed below or to the right of the reference line (on the side where the arrow is located). If the weld is located on the other side, the symbol will be located above or to the left of the reference line. If the weld is to be made all around, then one of two notations is drawn. The first is used for round parts, such as tubes and pipes, where a circle is placed at the point where the arrow line and reference line meet. If it is not a circular or tubular piece, then the symbol is placed above and below the line. See Fig. 11-34.

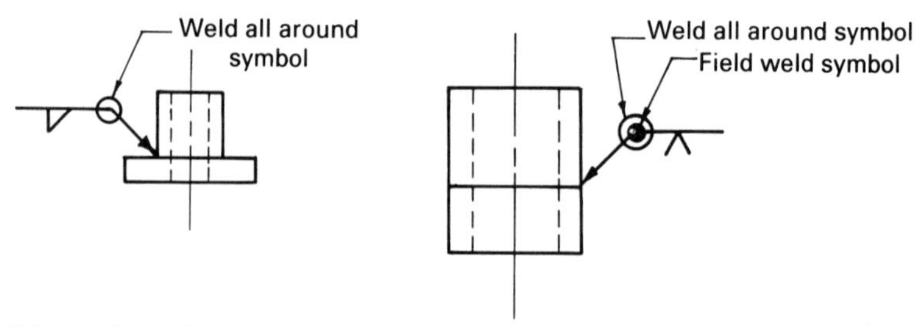

Fig. 11-34. Weld all-around symbol.

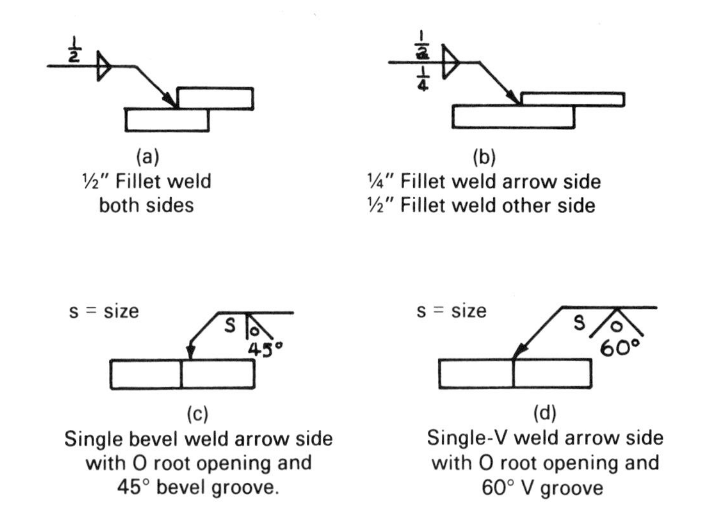

Fig. 11-35. Indicating weld size.

Shown in Fig. 11-35 is the method used for indicating the size of the weld. This dimension should be placed next to the weld symbol. If a weld joint is to be made on the arrow side and the other side, then the appropriate dimension is placed by both symbols. If the dimension is the same, only one notation is necessary.

Weld terms relating to weld specifications are shown in Fig. 11-36a. Figure 11-36b shows the appropriate weld symbols used for single and double groove welds.

To meet the needs for welding contractors and fabricators, a variety of standard welding symbols have been adopted for arc and gas

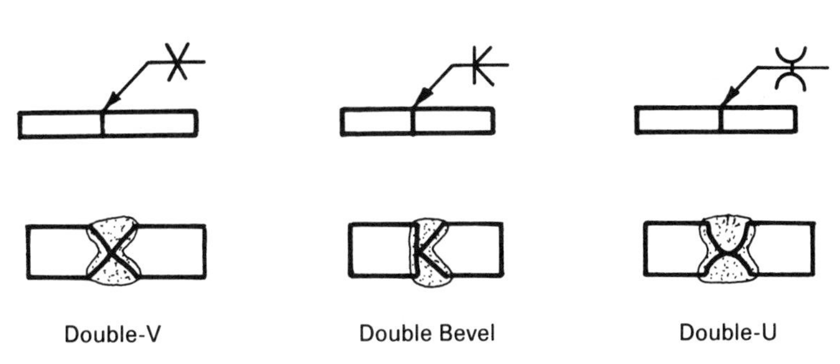

Fig. 11-36. Double-groove weld symbols.

Table 11-11 Welding Abbreviations

Abbreviation	Meaning
AAC	air carbon arc cutting
AAW	air acetylene welding
ABD	adhesive bonding
AB	arc brazing
AC	arc cutting
AHW	atomic hydrogen welding
AOC	oxygen arc cutting
AW	arc welding
B	brazing
BB	block brazing
BMAW	bare metal arc welding
CAC	carbon arc cutting
CAW	carbon arc welding
CAW-G	gas carbon arc welding
CAW-S	shielded carbon arc welding
CAW-T	twin carbon arc welding
CW	cold welding
DB	dip brazing
DFB	diffusion brazing
DFW	diffusion welding
DS	dip soldering
EASP	electric arc spraying
EBC	electron beam cutting
EBW	electron beam welding
ESW	electroslag welding
EXW	explosion welding
FB	furnace brazing
FCAW	flux cored arc welding
FCAW-EG	flux cored arc welding—electrogas
FLB	flow brazing
FLOW	flow welding
FLSP	flame spraying
FOC	chemical flux cutting
FOW	forge welding
FRW	friction welding
FS	furnace soldering
FW	flash welding
GMAC	gas metal arc cutting
GMAW	gas metal arc welding
GMAW-EG	gas metal arc welding—electrogas
GMAW-P	gas metal arc welding—pulsed arc
GMAW-S	gas metal arc welding—short circuiting arc
GTAC	gas tungsten arc cutting
Automatic	AU
Machine	ME

Table 11-11 (continued)

Abbreviation	Meaning
GTAW	gas tungsten arc welding
GTAW-P	gas tungsten arc welding—pulsed arc
HFRW	high frequency resistance welding
HPW	hot pressure welding
IB	induction brazing
INS	iron soldering
IRB	infrared brazing
IRS	infrared soldering
IS	induction soldering
IW	induction welding
LBC	laser beam cutting
LBW	laser beam welding
LOC	oxygen lance cutting
MAC	metal arc cutting
OAW	oxyacetylene welding
OC	oxygen cutting
OFC	oxyfuel gas cutting
OFC-A	oxyacetylene cutting
OFC-H	oxyhydrogen cutting
OFC-N	oxynatural gas cutting
OFC-P	oxypropane cutting
OFW	oxyfuel gas welding
OHW	oxyhydrogen welding
PAC	plasma arc cutting
PAW	plasma arc welding
PEW	percussion welding
PGW	pressure gas welding
POC	metal powder cutting
PSP	plasma spraying
RB	resistance brazing
RPW	projection welding
RS	resistance soldering
RSEW	resistance seam welding
RSW	resistance spot welding
ROW	roll welding
RW	resistance welding
S	soldering
SAW	submerged arc welding
SAW-S	series submerged arc welding
SMAC	shielded metal arc cutting
SMAW	shielded metal arc welding
SSW	solid state welding
SW	stud arc welding
TB	torch brazing
Manual	MA
Semiautomatic	SA

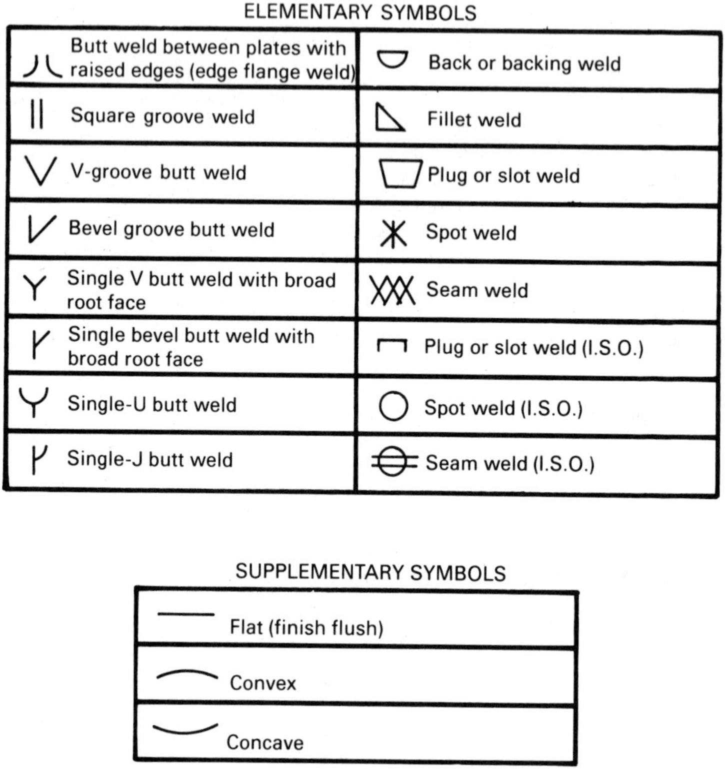

ELEMENTARY SYMBOLS

Butt weld between plates with raised edges (edge flange weld)	Back or backing weld
Square groove weld	Fillet weld
V-groove butt weld	Plug or slot weld
Bevel groove butt weld	Spot weld
Single V butt weld with broad root face	Seam weld
Single bevel butt weld with broad root face	Plug or slot weld (I.S.O.)
Single-U butt weld	Spot weld (I.S.O.)
Single-J butt weld	Seam weld (I.S.O.)

SUPPLEMENTARY SYMBOLS

Flat (finish flush)
Convex
Concave

Fig. 11-37. Basic welding symbols.

welding, as well as for resistance welding. Both of these are given in Fig. 11-37. Common abbreviations used in combination with the welding symbol are given in Table 11-11.

ISO Welding Symbol

The basic ISO welding symbol is the same as that of the ANSI/AWS. The primary difference is not in symbol interpretation, but in the position of the symbol relative to plane view presentation. Since both third and first angle projection are equally acceptable by the ISO, standardized procedures have been developed to meet the unique requirements of each projection method.

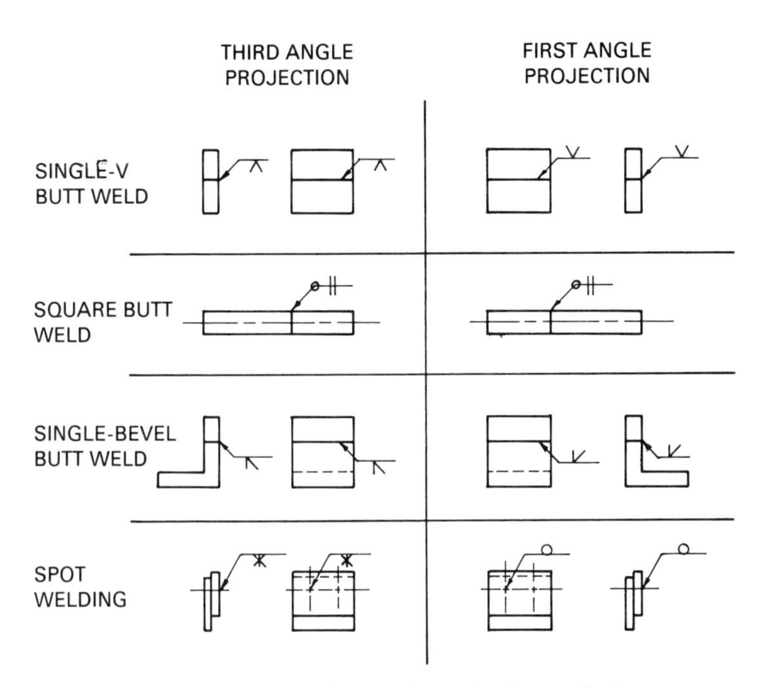

Fig. 11-38. Third angle vs. first angle projection techniques.

Fig. 11-38 compares how the welding symbol is used with third angle and first angle projection. Of particular importance is that the weld symbol face in opposite directions in first angle projection as compared to third angle projection.

Exercises

11.1 Explain the procedure that should be used when drawing and dimensioning standardized mechanical elements, such as dowels, keys, pins, and splines.

11.2 Under what conditions will it be necessary to detail and fully dimension mechanical elements?

11.3 Explain when plane view, section, and simplified presentations of springs should be used.

11.4 Interpret the welding symbol notations presented in Fig. 11-39.

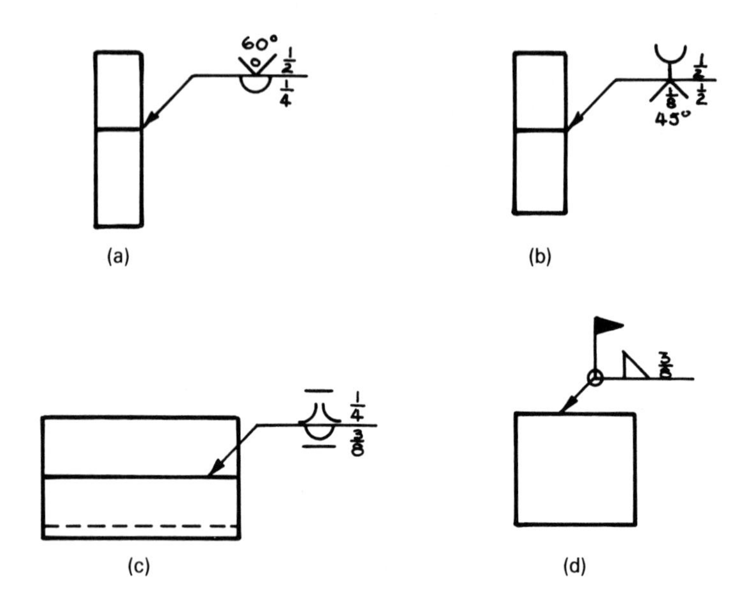

Fig. 11-39. Problem 11.4.

CHAPTER 12

Threaded Devices

- Screw-Thread Systems
- Threaded Machine Fasteners and Components
- Power Transmission and Pipe Thread Forms
- Exercises

One of the most commonly used and easily recognized devices used for fastening mechanical elements is the thread. Threads are used in such standard products as screws, bolts, nuts, and machine screws. Representation of these devices in drawings is common, and must be clearly understood by the drafter. This chapter will discuss the various types of standard threaded device used in industry, and their graphic representation.

Screw-Thread Systems

The work of American, Canadian, British, and other international groups has resulted in the standardization of screw-thread systems that are based on the ISO inch and metric systems. The use of these screw threads has enabled industries to manufacture products that are interchangeable according to internationally agreed standards.

The ISO inch screw-thread system is in close keeping with U.S. unified screw-thread system, because of the economical impact of American chemical, automotive, and aerospace industries in the world economy. Before screw-thread systems can be clearly understood, it is first necessary to define several basic terms. These are shown in Fig. 12-1 and defined as follows:

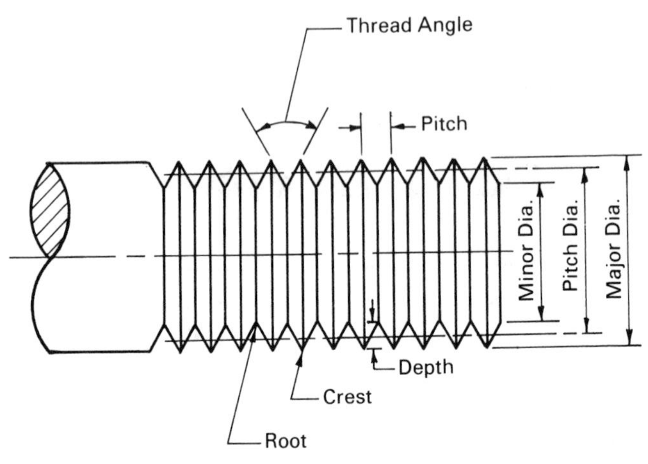

Fig. 12-1. Basic thread nomenclature.

1. *Major diameter* is the largest diameter of a screw thread.
2. *Minor diameter* is the smallest diameter of a screw thread.
3. *Pitch diameter* is the diameter of an imaginary line that is drawn through the thread profile so that the widths of the thread and groove will be equal.
4. *Pitch* (p) is the distance measured parallel with its axis between corresponding points on adjacent thread sides.
5. *Height of thread* (H), also known as *depth* (d), is the distance measured radially between the major and minor diameters.
6. *Root* is that surface of the thread that joins the bottoms of adjacent thread forms.
7. *Crest* is that surface of the thread that joins the tops of adjacent thread forms.
8. *Thread angle* is that angle generated by the sides of adjacent thread forms.
9. *Flat* is a feature that is not found in all thread forms. Flats are the distances across the straight (flat) surface of the thread's crest and/or root.

Unified and American Screw-Thread Standard

Both the Unified and American screw-thread standards are published by the ANSI under the title of American Standard Unified and

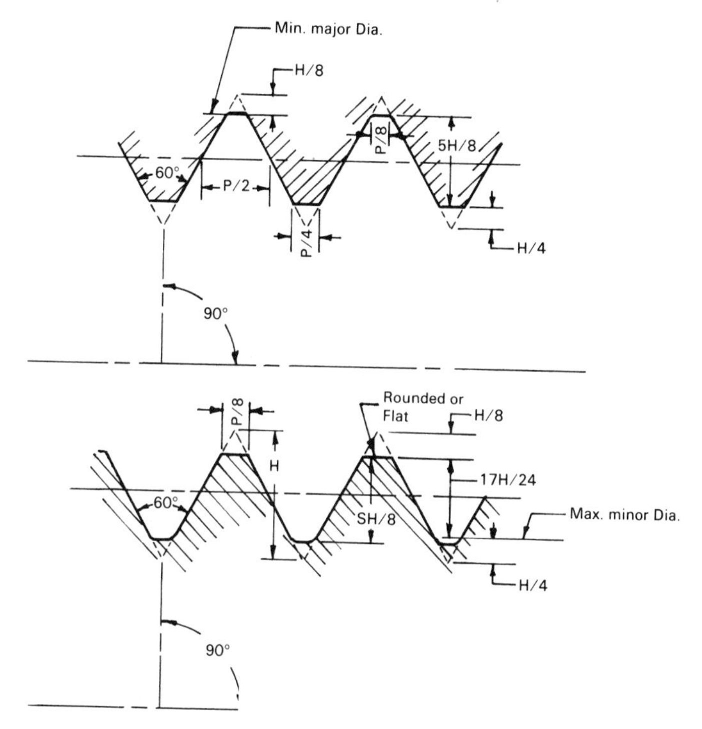

Fig. 12-2. 60° Unified and American screw thread forms.

American Thread Publication B1.1-1974. The thread profile for both types of screw threads is shown in Fig. 12-2. Here the thread series pertains to a standard group of screw threads with diameter-pitch combinations that are specified or grouped according to the number of threads used per diameter size.

All thread designs used today employ the Unified screw-thread standard, unless they are to be used in conjunction with the older American standard. The Unified screw-thread system should therefore be preferred to the American screw-thread system. However, since the American screw-thread standard is still sometimes encountered in industry, it will be discussed along with the Unified system.

There are six standard and three special series of Unified and American standard threads. Special thread series, however, should not be used unless it is impossible to employ one of the standard series in the design. The more common Unified series is similar to the older American standard screw-thread form, and is noted by the prefix letter U in the thread series symbol. If the American standard thread form is specified, then the U prefix is dropped. The standard Unified and American series, and their uses, are as follows:

1. *Coarse-thread series (UNC and NC)* is employed for general use where rapid assembly is required. This series is often used in gray iron, soft metals, and plastic products.
2. *Fine-thread series (UNF and NF)* is used when greater strength is required and/or where the length of thread contact is limited.
3. *Extra-fine-thread series (UNEF and NEF)* is particularly applicable for highly stressed parts. The UNEF and NEF series have also been successfully used in products where internal threads are needed in fastening products with thin cross-sectional walls.
4. *Eight-thread series (8N)* is a coarse thread that is used as a substitute for the UNC and NC series for diameters that exceed 1.5 inches.
5. *Twelve-thread series (12 UN and 12N)* is used as a fine-thread series for diameters that exceed 1.5 inches.
6. *Sixteen-thread series (16 UN and 16N)* is an extra-fine-thread series that is designed for use in products with diameters that exceed two inches.

In addition to the standard thread series, there are three special thread series: 8UN, UNS, and NS.

Both the Unified and American screw-thread standards incorporate a series of thread classes that are used to differentiate threads by the amount of allowance and/or tolerance. Classes with an A designation apply to external threads only, while those with a B designation apply to internal threads. These classes are used as follows:

1. *Classes 1A and 1B* are employed to give liberal allowance for easy assembly. These classes of thread can even be assembled with little difficulty on threads that are slightly damaged or dirty.
2. *Classes 2A and 2B* are used for the production of screws, bolts, nuts, and other threaded commercial fasteners. This design allows external threads (Class 2B) to be plated (e.g., chrome and zinc plating) without effecting performance.

3. *Classes 2 and 3* are used with American standard threads only as a transitional class, for both internal and external threads, to the Unified classes.

The specification of a screw thread is made up of a note or set of numbers and letters that are used to give the nominal diameter, pitch, thread series, class fit, and handedness (i.e., left- or right-handed thread). An example of this would be the following thread specification: 3/8-16 UNC-1A-LH. This specification explains that the screw thread desired has a 3/8-inch nominal size diameter with sixteen threads per inch, is a Unified coarse-thread series with a class 1A tolerance, and is a left-handed thread. Since the majority of threads used are right-handed, if the handedness symbol is not given, it will be assumed to be a right-handed thread. Thus, a right-handed thread would be specified as: 1/4-20 UNC-2A.

Tables 1 through 6 in Appendix B give the dimensional specifications for the six standard series of screw threads.

American and Foreign Standard Screw Threads

Though the Unified screw-thread form has wide national and international acceptance, there are times when one might encounter other American and foreign standard screw threads used for bolts, nuts, and machine screws. To be able to interpret, specify, and convert their dimensional specifications requires comprehension of their forms and proportions.

AMERICAN STANDARD SCREW THREADS

Fig. 12-3 shows the American standard thread form. The basic proportions of this thread are calculated with the following formulas:

p = pitch = 1/(number of threads per inch)
f = flat = p/8

The dimensions of this screw thread are available from the ANSI publication B1.1-1974, and are based on the earlier American National screw-thread standard (American Standard B1.1-1949). The American standard thread form profile was also known as the *Sellers* and *U.S. Standard.* This screw thread is found in four thread series: coarse-thread; fine-thread; special-pitch; and 8-, 12-, and 16-pitch series. In addition, there are four classes of fit for the American National standard: numbers 1, 2, 3, and 4.

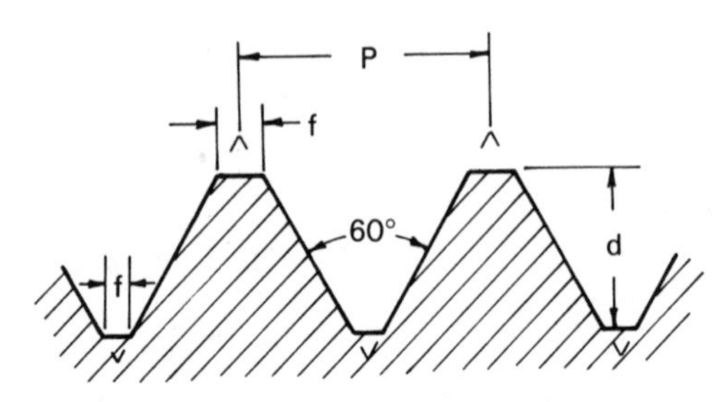

Fig. 12-3. American standard screw thread form.

FOREIGN STANDARD SCREW THREADS

Older thread forms, which were standardized in foreign countries, are still encountered in industries and products found in the international marketplace. For all practical purposes these threads, along with the American National, are obsolete. In new designs, the foreign standard screw threads have been superseded by the ISO inch and metric system.

The *Whitworth Standard Thread* (Fig. 12-4), developed in Britain, was widely used before the standardization of the Unified thread. Today it is primarily used for replacements and spare parts. The formulas used for calculating its proportions are:

d = depth of thread = 0.640327/p
r = radius = 0.137329p

A second type of screw thread sometimes encountered is the *French metric screw thread.* This thread was the predecessor to the

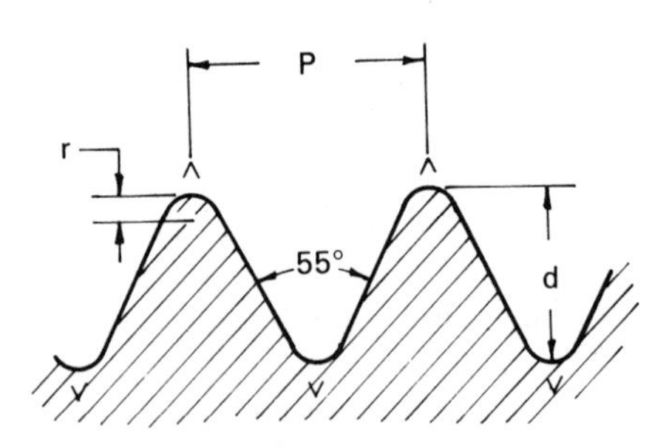

Fig. 12-4. Whitworth screw thread form.

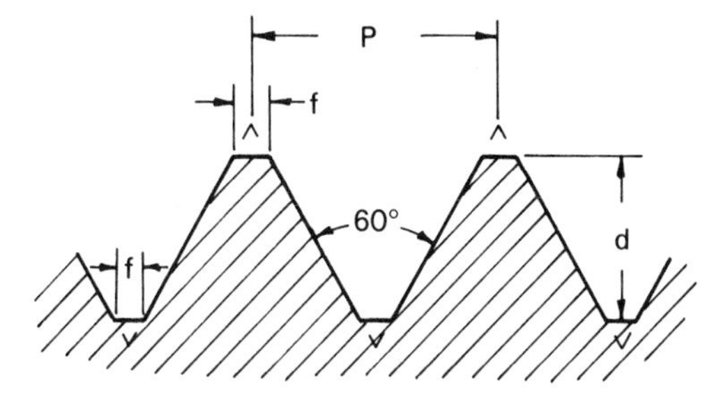

Fig. 12-5. Metric (French) screw thread form.

ISO standard metric screw thread. Shown in Fig. 12-5, its proportions are calculated with the following formulae:

p = pitch in mm
d = 0.6495p
f = p/8

Table 7 in Appendix B compares screw-thread dimensions for several foreign screw thread forms.

ISO Metric Screw Threads

The ISO inch or unified system (Fig. 12-6a) is employed when design requirements dictate interchangeability with products in which the U.S. screw thread designs predominate. There are slight proportional differences between the ISO and the American unified form, especially in thread height (compare Fig. 12-2 and Fig. 12-6a).

Design requirements dictating the use of metric screw threads, however, predominantly incorporate the use of the ISO metric screw thread (Fig. 12-6b). Data for this screw thread form are found in Table 8, Appendix B. Additional information can be obtained from ISO Recommendations R 68 and R 261.

Like the unified screw thread series, ISO metric screw threads are divided into tolerance zones and classes of fit. In the ISO system, tolerances are designated by letters, with capitals used for internal threads and lowercase letters for external threads. The magnitudes or sizes of the tolerance zones are represented by numerical values or tolerance grades. A combination of these forms a tolerance class designation, such as 6g or 5H.

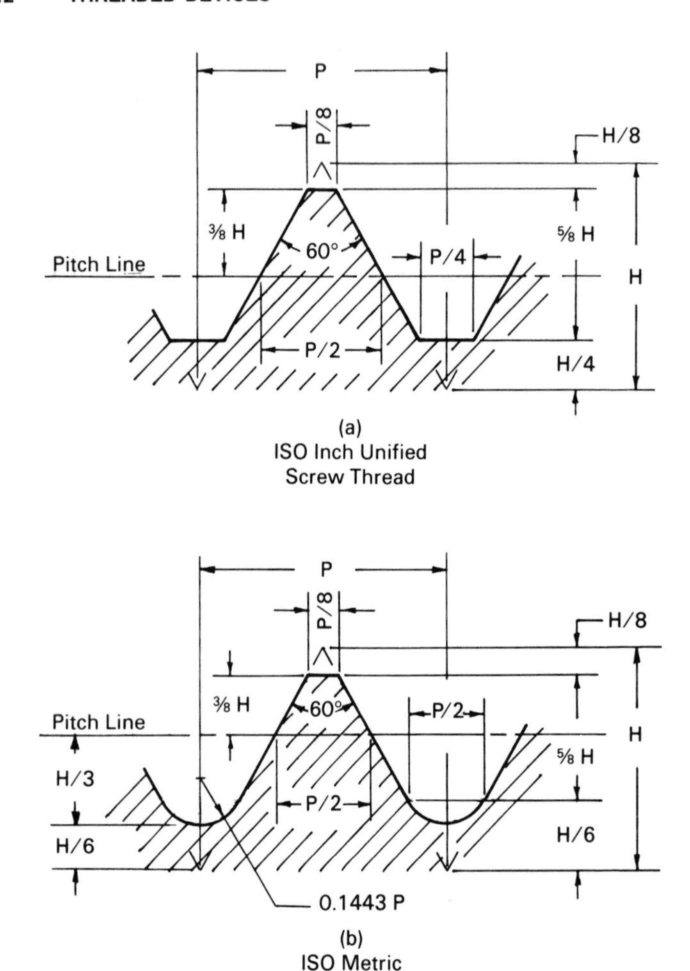

(a)
ISO Inch Unified
Screw Thread

(b)
ISO Metric
Screw Thread

Fig. 12-6. ISO basic thread forms.

ISO defines the degree of fit between external and internal threads as the *tolerance quality*. Thus, we have the following classes of fit for ISO metric screw threads:

1. *Fine tolerance quality* corresponds to a close fit, and is designated by the tolerance classes 5H and 4h. This type of thread should be used where close accuracy of thread form and pitch are necessary.

2. *Medium tolerance quality* is the corresponding designation for a medium fit. It is designated by the tolerance classes of 6H and 6g. The medium class is used for most general purposes, and where rapid and free assembly is necessary.
3. *Coarse tolerance quality* corresponds to a free fit, and is designated by the tolerance classes 7H and 8g. This tolerance quality is used primarily where quick and easy assembly is required, even though threads may become dirty and/or slightly damaged.

The specification method used for these threads is similar to that of the Unified thread system. An example of this is: M5 × 0.5 – 6H. Here the M notes the thread system as ISO metric, the 5 is the nominal diameter size in mm, while the pitch is designated as 0.5 threads per mm. Lastly, the thread tolerance class symbol, 6H, is specified. If the thread was left-handed, an LH designation would be given last. When specifying an ISO inch Unified thread, the same method would be used as for the American Unified screw thread form.

Thread Representation

Three methods are used for representing threaded parts in drawings (Fig. 12-7a). The first is known as a *detailed representation*, where the threads are shown as they would actually appear. When drawn to scale, the thread angle should be drawn as close as possible to the true thread angle, such as 60°. This practice, however, is usually approximated to save time. Of the three methods, this is the most time-consuming and costly, and should be used only as a last resort to clarify the presentation. In general, detailed screw threads are not used in industrial working drawings.

The second, and more common, method of presentation is the *schematic representation* of threads (Fig. 12-7b). Here the cylinder diameter represents the major diameter of the thread form. The longer lines designate the distance between and location of the thread crest, while the short line represents the distance between and location of the thread roots. The distances used to separate the series of lines should be based upon thread pitch proportions. At one time this was the most acceptable method of representation, but it is currently recommended only for complex assembly and subassembly drawings.

The last method of thread representations is a *simplified representation* (Fig. 12-7c). This technique is the international standard (ISO

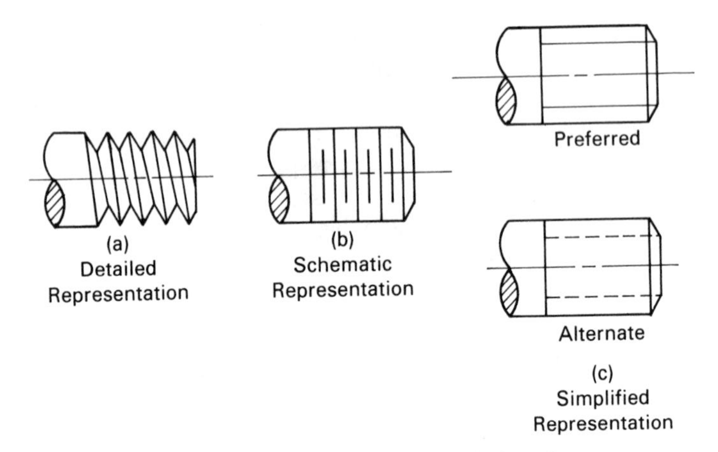

Fig. 12-7. Graphic representations of threaded parts.

6410-1981 (E)) conventional thread representation method. The crests of the threads are defined by a continuous object line, and the roots are shown as a continuous thin line. In some cases, the root line is represented as a hidden line. When ISO standards are followed, and in most American, Canadian, and British drawings, the continuous thin line technique should be used.

Of the three methods used for thread representation, the simplified method should always be the first considered. Again, if drawing interpretation is difficult with this technique, other procedures may be used.

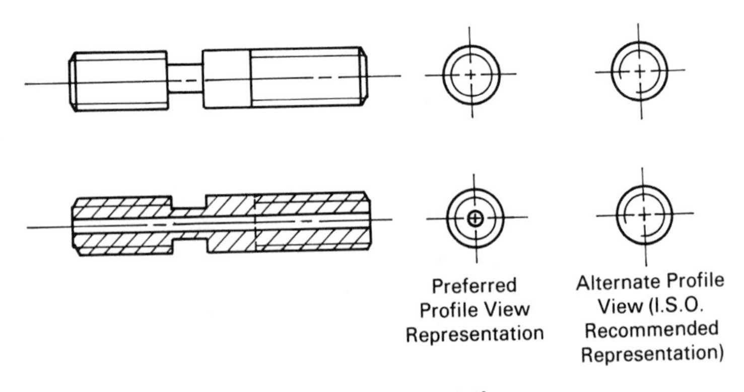

Fig. 12-8. Visible screw thread representation.

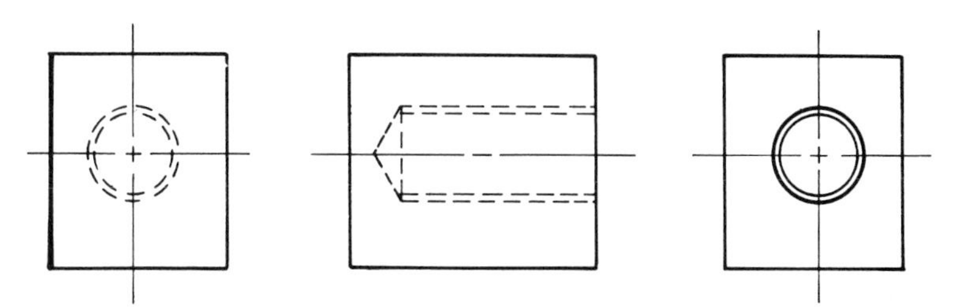

Fig. 12-9. Hidden screw thread representation.

VISIBLE SCREW THREADS

The drawing of visible screw threads should be illustrated by using continuous thick and thin lines to represent the crests and roots of the thread, respectively. The distance between these two lines should approximate the major and minor diameters as close as possible. When a threaded part is shown in section, only two techniques can be used: simplified and detailed. The correct representations for such situations are shown in Fig. 12-8.

HIDDEN SCREW THREADS

For hidden screw threads, the crests and roots should be defined by hidden lines. Again, either simplified or detailed techniques can be used, with the simplified preferred. Fig. 12-9 shows acceptable presentations for hidden screw threads.

RUN-OUTS

A run-out is an incomplete thread or the limits of the usable section of a thread. These presentations should not be used when there is a functional or operational necessity for the unused portion of the thread. Correct graphic representation of a run-out is shown in Fig. 12-10.

Threaded Machine Fasteners and Components

The number of threaded machine fasteners and components is so varied that it would be impossible to cover adequately all the types here; only the more common types will be discussed.

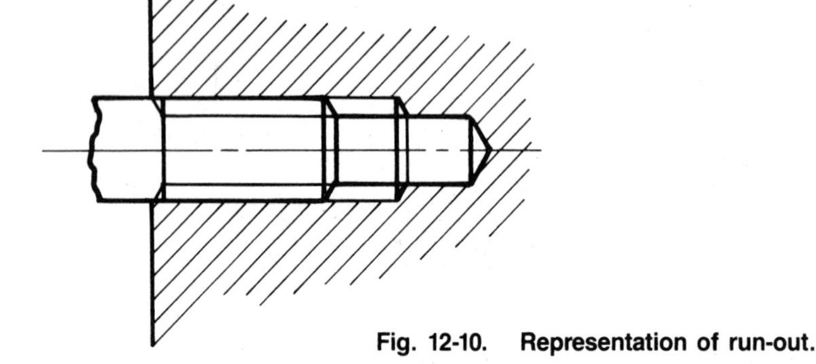

Fig. 12-10. **Representation of run-out.**

Machine Screw Threads and Bolt Heads

A threaded machine fastener that is used in industrial and commercial machines, equipment, and mechanical product design is the *machine screw thread.* Machine screws and bolts will be discussed together, the primary difference between them being in the threaded end of the fastener (screws will have self-tapping pointed ends, while bolts have flat ends). Bolts are designed to be inserted into predrilled holes and held in place by nuts, whereas a screw is held in place by the self-tapping action of the threads in a predrilled hole.

Machine screw and bolt fasteners are categorized according to their head configuration. In all, there are four basic types of machine screw, described as follows:

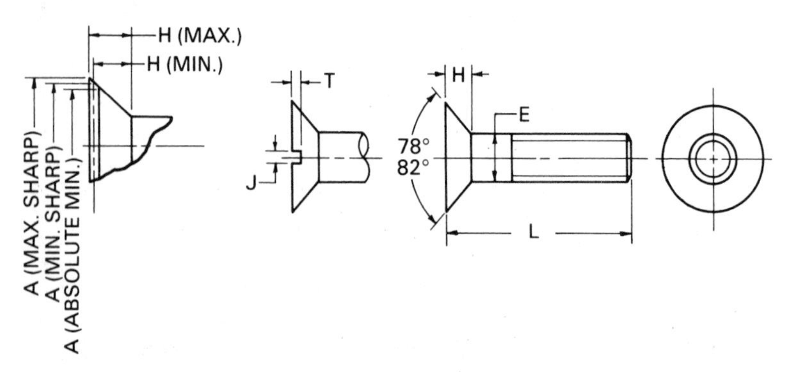

Fig. 12-11. American standard countersunk bolt and slotted countersunk bolt.

1. *Flat head* screws and bolts have a flat surface for the top of the head and a countersink angle of either 78° or 82°. This head type is standard for machine and cap screws and bolts, and wood screws. American standard countersunk and slotted countersunk bolts are shown in Fig. 12-11, and their dimensional specifications are given in Table 12-1.

Table 12.1 American Standard Countersunk Bolts and Slotted Countersunk Bolts Dimensional Specifications (All dimensions in inches.)

Nominal Size or Basic Bolt Diameter	Body Diameter, *E*		Head Diameter, *A*			Flat on Min Diam. Head, *F*
	Max	Min	Max Edge Sharp	Min Edge Sharp	Absolute Min Edge Rounded or Flat	Max
1/4 0.2500	0.260	0.237	0.493	0.477	0.445	0.018
5/16 0.3125	0.324	0.298	0.618	0.598	0.558	0.023
3/8 0.3750	0.388	0.360	0.740	0.715	0.668	0.027
7/16 0.4375	0.452	0.421	0.803	0.778	0.726	0.030
1/2 0.5000	0.515	0.483	0.935	0.905	0.845	0.035
5/8 0.6250	0.642	0.605	1.169	1.132	1.066	0.038
3/4 0.7500	0.768	0.729	1.402	1.357	1.285	0.041
7/8 0.8750	0.895	0.852	1.637	1.584	1.511	0.042
1 1.0000	1.022	0.976	1.869	1.810	1.735	0.043
1 1/8 1.1250	1.149	1.098	2.104	2.037	1.962	0.043
1 1/4 1.2500	1.277	1.223	2.337	2.262	2.187	0.043
1 3/8 1.3750	1.404	1.345	2.571	2.489	2.414	0.043
1 1/2 1.5000	1.531	1.470	2.804	2.715	2.640	0.043
Nom. Size or Basic Bolt Diam.	Head Height, *H*		Slot Width, *J*		Slot Depth, *T*	
	Max	Min	Max	Min	Max	Min
1/4 0.2500	0.150	0.131	0.075	0.064	0.068	0.045
5/16 0.3125	0.189	0.164	0.084	0.072	0.086	0.057
3/8 0.3750	0.225	0.196	0.094	0.081	0.103	0.068
7/16 0.4375	0.226	0.196	0.094	0.081	0.103	0.068
1/2 0.5000	0.269	0.233	0.106	0.091	0.103	0.068
5/8 0.6250	0.336	0.292	0.133	0.116	0.137	0.091
3/4 0.7500	0.403	0.349	0.149	0.131	0.171	0.115
7/8 0.8750	0.470	0.408	0.167	0.147	0.206	0.138
1 1.0000	0.537	0.466	0.188	0.166	0.240	0.162
1 1/8 1.1250	0.604	0.525	0.196	0.178	0.257	0.173
1 1/4 1.2500	0.671	0.582	0.211	0.193	0.291	0.197
1 3/8 1.3750	0.738	0.641	0.226	0.208	0.326	0.220
1 1/2 1.5000	0.805	0.698	0.258	0.240	0.360	0.244

2. *Round head* machine screws and bolts are also known as *button head* machine screws. They have a semielliptical head that is standard in machine and cap bolts and screws, and wood screws. This threaded machine fastener is shown in Fig. 12-12, and Table 12-2 gives its dimensional specifications.

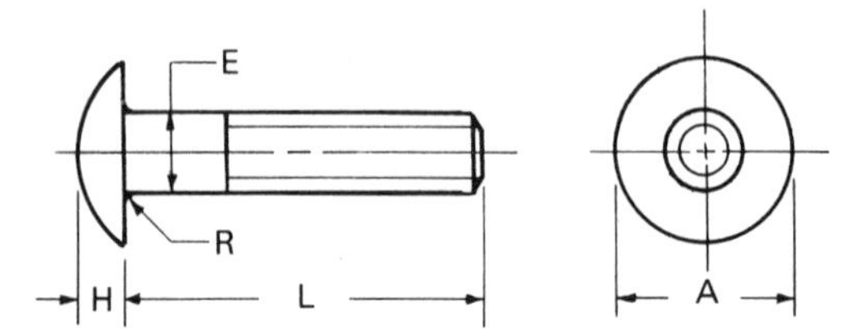

Fig. 12-12. American standard round head bolt.

Table 12.2 American Standard Round Head Bolts Dimensional Specifications (All dimensions in inches.)

Nominal Size	Body Diam., *E*		Diam. of Head, *A*		Height of Head, *H*		Fillet Rad., *R*
	Max.	Min.	Max.	Min.	Max.	Min.	Max.
No. 10	.199	.182	.469	.438	.114	.094	.031
1/4	.260	.237	.594	.563	.145	.125	.031
5/16	.324	.298	.719	.688	.176	.156	.031
3/8	.388	.360	.844	.782	.208	.188	.031
7/16	.452	.421	.969	.907	.239	.219	.031
1/2	.515	.483	1.094	1.032	.270	.250	.031
5/8	.642	.605	1.344	1.219	.344	.313	.062
3/4	.768	.729	1.594	1.469	.406	.375	.062
7/8	.895	.852	1.844	1.719	.469	.438	.062
1	1.022	.976	2.094	1.969	.531	.500	.062

3. *Hexagon head* machine bolts and screws, also referred to as *hex bolts,* have a hexagon-shaped head that is easily installed and disassembled with external wrenches. This design is the standard for machine screws and bolts. It is shown in Fig. 12-13 with dimensional specifications given in Table 12.3.

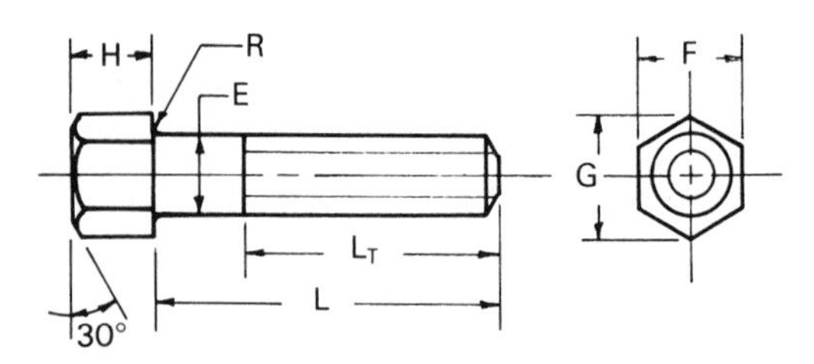

Fig. 12-13. American standard hexagon bolt.

Table 12.3 American Standard Hexagon Bolts Dimensional Specifications (All dimensions in inches.)

Nominal Size or Basic Diam.		Body Diam. *E*	Width Across Flats *F*			Width Across Corners *G*		Height *H*			Thread Length L_T
		Max.	Basic	Max.	Min.	Max.	Min.	Basic	Max.	Min.	Basic
HEX BOLTS											
1/4	0.2500	0.260	7/16	0.438	0.425	0.505	0.484	11/64	0.188	0.150	0.750
5/16	0.3125	0.324	1/2	0.500	0.484	0.577	0.552	7/32	0.235	0.195	0.875
3/8	0.3750	0.388	9/16	0.562	0.544	0.650	0.620	1/4	0.268	0.226	1.000
7/16	0.4375	0.452	5/8	0.625	0.603	0.722	0.687	19/64	0.316	0.272	1.125
1/2	0.5000	0.515	3/4	0.750	0.725	0.866	0.826	11/32	0.364	0.302	1.250
5/8	0.6250	0.642	15/16	0.938	0.906	1.083	1.033	27/64	0.444	0.378	1.500
3/4	0.7500	0.768	1 1/8	1.125	1.088	1.299	1.240	1/2	0.524	0.455	1.750
7/8	0.8750	0.895	1 5/16	1.312	1.269	1.516	1.447	37/64	0.604	0.531	2.000
1	1.0000	1.022	1 1/2	1.500	1.450	1.732	1.653	43/64	0.700	0.591	2.250
1 1/8	1.1250	1.149	1 11/16	1.688	1.631	1.949	1.859	3/4	0.780	0.658	2.500
1 1/4	1.2500	1.277	1 7/8	1.875	1.812	2.165	2.066	27/32	0.876	0.749	2.750
1 3/8	1.3750	1.404	2 1/16	2.062	1.994	2.382	2.273	29/32	0.940	0.810	3.000
1 1/2	1.5000	1.531	2 1/4	2.250	2.175	2.598	2.480	1	1.036	0.902	3.250
1 3/4	1.7500	1.785	2 5/8	2.625	2.538	3.031	2.893	1 5/32	1.196	1.054	3.750
2	2.000	2.039	3	3.000	2.900	3.464	3.306	1 11/32	1.388	1.175	4.250
2 1/4	2.2500	2.305	3 3/8	3.375	3.262	3.897	3.719	1 1/2	1.548	1.327	4.750
2 1/2	2.5000	2.559	3 3/4	3.750	3.625	4.330	4.133	1 21/32	1.708	1.479	5.250
2 3/4	2.7500	2.827	4 1/8	4.125	3.988	4.763	4.546	1 13/16	1.869	1.632	5.750
3	3.0000	3.081	4 1/2	4.500	4.350	5.196	4.959	2	2.060	1.815	6.250
3 1/4	3.2500	3.335	4 7/8	4.875	4.712	5.629	5.372	2 3/16	2.251	1.936	6.750
3 1/2	3.5000	3.589	5 1/4	5.250	5.075	6.062	5.786	2 5/16	2.380	2.057	7.250
3 3/4	3.7500	3.858	5 5/8	5.625	5.437	6.495	6.198	2 1/2	2.572	2.241	7.750
4	4.0000	4.111	6	6.000	5.800	6.928	6.612	2 11/16	2.764	2.424	8.250
HEAVY HEX BOLTS											
1/2	0.5000	0.515	7/8	0.875	0.850	1.010	0.969	11/32	0.364	0.302	1.250
5/8	0.6250	0.642	1 1/16	1.062	1.031	1.227	1.175	27/64	0.444	0.378	1.500
3/4	0.7500	0.768	1 1/4	1.250	1.212	1.443	1.383	1/2	0.524	0.455	1.750
7/8	0.8750	0.895	1 7/16	1.438	1.394	1.660	1.589	37/64	0.604	0.531	2.000
1	1.0000	1.022	1 5/8	1.625	1.575	1.876	1.796	43/64	0.700	0.591	2.250
1 1/8	1.1250	1.149	1 13/16	1.812	1.756	2.093	2.002	3/4	0.780	0.658	2.500
1 1/4	1.2500	1.277	2	2.000	1.938	2.309	2.209	27/32	0.876	0.749	2.750
1 3/8	1.3750	1.404	2 3/16	2.188	2.119	2.526	2.416	29/32	0.940	0.810	3.000
1 1/2	1.5000	1.531	2 3/8	2.375	2.300	2.742	2.622	1	1.036	0.902	3.250
1 3/4	1.7500	1.785	2 3/4	2.750	2.662	3.175	3.035	1 5/32	1.196	1.054	3.750
2	2.0000	2.039	3 1/8	3.125	3.025	3.608	3.449	1 11/32	1.388	1.175	4.250
2 1/4	2.2500	2.305	3 1/2	3.500	3.388	4.041	3.862	1 1/2	1.548	1.327	4.750
2 1/2	2.5000	2.559	3 7/8	3.875	3.750	4.474	4.275	1 21/32	1.708	1.479	5.250
2 3/4	2.7500	2.827	4 1/4	4.250	4.112	4.907	4.688	1 13/16	1.869	1.632	5.750
3	3.0000	3.081	4 5/8	4.625	4.475	5.340	5.102	2	2.060	1.815	6.250

4. *Square head* machine screws and bolts have a square-shaped head that can also be used with external wrenches. They are one of the standard-shaped heads for machine screws and bolts. They are shown in Fig. 12-14, with dimensional specifications given in Table 12-4.

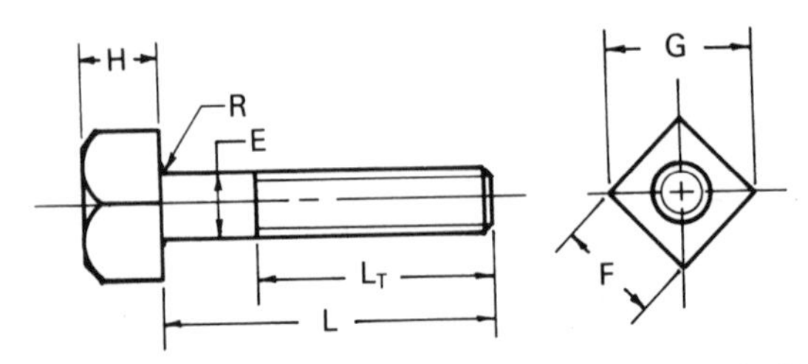

Fig. 12-14. American standard square head machine bolt.

Table 12.4 American Standard Square Head Machine Bolt Dimensional Specifications (All dimensions in inches.)

SQUARE BOLTS											
Nominal Size or Basic Product Diam.		Body Diam. *E*	Width Across Flats *F*			Width Across Corners G		Height *H*			Thread Length L_T
		Max.	Basic	Max.	Min.	Max.	Min.	Basic	Max.	Min.	Basic
1/4	0.2500	0.260	3/8	0.375	0.362	0.530	0.498	11/64	0.188	0.156	0.750
5/16	0.3125	0.324	1/2	0.500	0.484	0.707	0.665	13/64	0.220	0.186	0.875
3/8	0.3750	0.388	9/16	0.562	0.544	0.795	0.747	1/4	0.268	0.232	1.000
7/16	0.4375	0.452	5/8	0.625	0.603	0.884	0.828	19/64	0.316	0.278	1.125
1/2	0.5000	0.515	3/4	0.750	0.725	1.061	0.995	21/64	0.348	0.308	1.250
5/8	0.6250	0.642	15/16	0.938	0.906	1.326	1.244	27/64	0.444	0.400	1.500
3/4	0.7500	0.768	1 1/8	1.125	1.088	1.591	1.494	1/2	0.524	0.476	1.750
7/8	0.8750	0.895	1 5/16	1.312	1.269	1.856	1.742	19/32	0.620	0.568	2.000
1	1.0000	1.022	1 1/2	1.500	1.450	2.121	1.991	21/32	0.684	0.628	2.250
1 1/8	1.1250	1.149	1 11/16	1.688	1.631	2.386	2.239	3/4	0.780	0.720	2.500
1 1/4	1.2500	1.277	1 7/8	1.875	1.812	2.652	2.489	27/32	0.876	0.812	2.750
1 3/8	1.3750	1.404	2 1/16	2.062	1.994	2.917	2.738	29/32	0.940	0.872	3.000
1 1/2	1.5000	1.531	2 1/4	2.250	2.175	3.182	2.986	1	1.036	0.964	3.250

Other machine screw and bolt head configurations used by designers include oval head, socket head, square neck bolts, and eyebolts (Fig. 12-15). For dimensional specifications and sizes, see ANSI 18.2.1-1972 (R1981) and ANSI B18.2.2-1972. Metric sizes can be obtained from ANSI B18.2.3.1M through B18.2.3.8M, ANSI B18.2.4.1M through ANSI B18.2.4.6M, ANSI B18.3.1M, ANSI B18.5.2.2M, and ANSI B18.16.3M.

Specifications given for machine bolts and screws are made by notation, for example:

½-13 × 13 HEX CAP SCREW, SAE GRADE 8 STEEL

¾-16 × 1 SQUARE BOLT, STEEL, ZINC PLATED

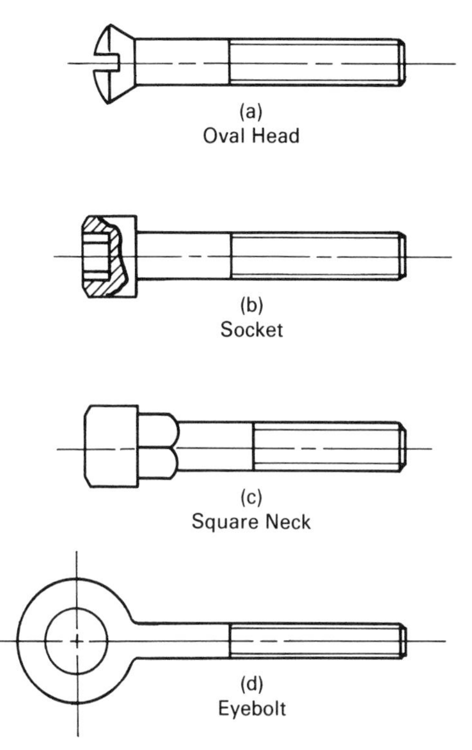

Fig. 12-15. Screw and bolt head configurations.

In these notes, the fraction values designate the nominal diameter of the screw or bolt, followed by the pitch and product length for bolts and screws. The product name follows, along with a specified material and/or protective finish.

Nuts and Washers

Two devices that are often used in combination with bolts are nuts and washers. A typical bolt-washer-nut assembly is illustrated in Fig. 12-16. The washer is used to increase friction, during assembly, between the nut and bolt, as well as to prevent the two from loosening once assembled.

Nuts are internally threaded and used as a torquing device to hold a bolt in place. Nuts will be either hexagonal or square in shape

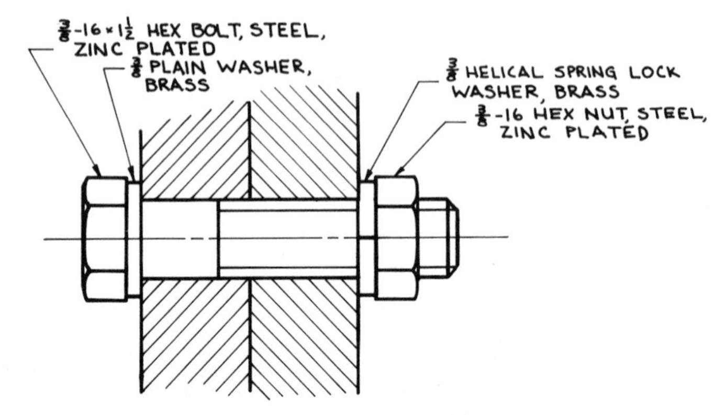

Fig. 12-16. Typical bolt-washer-nut assembly.

so that they can be assembled by using an external wrench. Fig. 12-17 shows various nut types. Tables 12-5 and 12-6 present dimensional specifications for hex and square nuts, respectively.

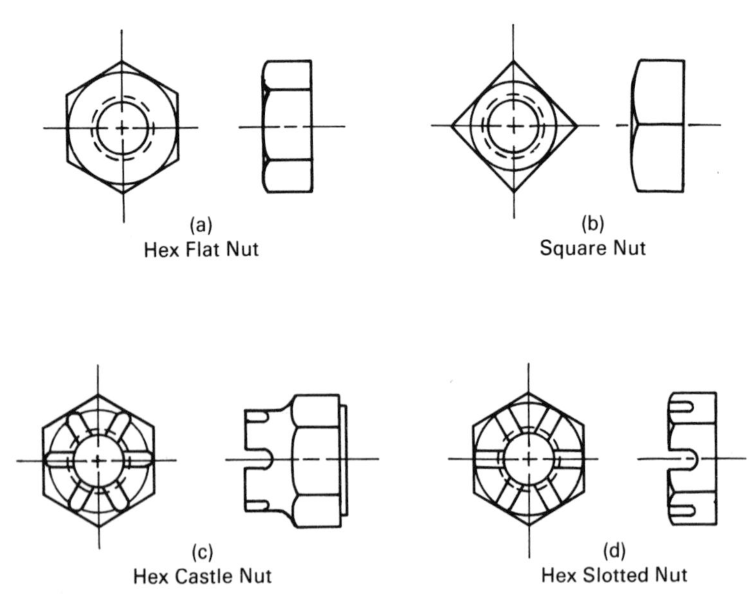

Fig. 12-17. Common types of nut.

Table 12.5 American Standard Hex Nuts Dimensional Specifications (All dimensions in inches.)

Nominal Size or Basic Major Diam. of Thread		Width Across Flats			Width Across Corners		Thickness, Nuts			Thickness, Jam Nuts		
		Basic	Max.	Min.	Max.	Min.	Basic	Max.	Min.	Basic	Max.	Min.
1/4	0.2500	7/16	0.438	0.428	0.505	0.488	7/32	0.226	0.212	5/32	0.163	0.150
5/16	0.3125	1/2	0.500	0.489	0.577	0.557	17/64	0.273	0.258	3/16	0.195	0.180
3/8	0.3750	9/16	0.562	0.551	0.650	0.628	21/64	0.337	0.320	7/32	0.227	0.210
7/16	0.4375	11/16	0.688	0.675	0.794	0.768	3/8	0.385	0.365	1/4	0.260	0.240
1/2	0.5000	3/4	0.750	0.736	0.866	0.840	7/16	0.448	0.427	5/16	0.323	0.302
9/16	0.5625	7/8	0.875	0.861	1.010	0.982	31/64	0.496	0.473	5/16	0.324	0.301
5/8	0.6250	15/16	0.938	0.922	1.083	1.051	35/64	0.559	0.535	3/8	0.387	0.363
3/4	0.7500	1 1/8	1.125	1.088	1.299	1.240	41/64	0.665	0.617	27/64	0.446	0.398
7/8	0.8750	1 5/16	1.312	1.269	1.516	1.447	3/4	0.776	0.724	31/64	0.510	0.458
1	1.0000	1 1/2	1.500	1.450	1.732	1.653	55/64	0.887	0.831	35/64	0.575	0.519
1 1/8	1.1250	1 11/16	1.688	1.631	1.949	1.859	31/32	0.999	0.939	39/64	0.639	0.579
1 1/4	1.2500	1 7/8	1.875	1.812	2.165	2.066	1 1/16	1.094	1.030	23/32	0.751	0.687
1 3/8	1.3750	2 1/16	2.062	1.994	2.382	2.273	1 11/64	1.206	1.138	25/32	0.815	0.747
1 1/2	1.5000	2 1/4	2.250	2.175	2.598	2.480	1 9/32	1.317	1.245	27/32	0.880	0.808

Table 12.6 American Standard Square Nuts Dimensional Specifications (All dimensions in inches.)

Nominal Size or Basic Major Diam. of Thread		Width Across Flats			Width Across Corners		Thickness		
		Basic	Max.	Min.	Max.	Min.	Basic	Max.	Min.
1/4	0.2500	7/16	0.438	0.425	0.619	0.584	7/32	0.235	0.203
5/16	0.3125	9/16	0.562	0.547	0.795	0.751	17/64	0.283	0.249
3/8	0.3750	5/8	0.625	0.606	0.884	0.832	21/64	0.346	0.310
7/16	0.4375	3/4	0.750	0.728	1.061	1.000	3/8	0.394	0.356
1/2	0.5000	13/16	0.812	0.788	1.149	1.082	7/16	0.458	0.418
5/8	0.6250	1	1.000	0.969	1.414	1.330	35/64	0.569	0.525
3/4	0.7500	1 1/8	1.125	1.088	1.591	1.494	21/32	0.680	0.632
7/8	0.8750	1 5/16	1.312	1.269	1.856	1.742	49/64	0.792	0.740
1	1.0000	1 1/2	1.500	1.450	2.121	1.991	7/8	0.903	0.847
1 1/8	1.1250	1 11/16	1.688	1.631	2.386	2.239	1	1.030	0.970
1 1/4	1.2500	1 7/8	1.875	1.812	2.652	2.489	1 3/32	1.126	1.062
1 3/8	1.3750	2 1/16	2.062	1.994	2.917	2.738	1 13/64	1.237	1.169
1 1/2	1.5000	2 1/4	2.250	2.175	3.182	2.986	1 5/16	1.348	1.276

Washers, though not threaded, are used with nuts and bolts as a locking device. Shown in Fig. 12-18 are several types of lock washers used for this purpose. Generally, there are two types of lock washer. The first has internal and/or external teeth, and is known as the *tooth*

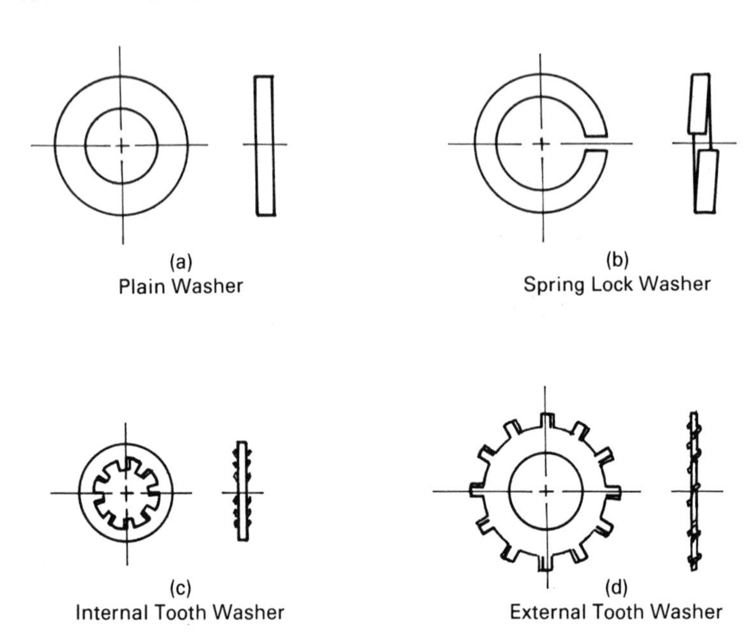

Fig. 12-18. Washers.

lock washer. The second type of washer is the *helical spring lock washer,* which is used to apply a continual spring pressure against the nut and bolt, thereby providing friction for assembly and pressure to prevent loosening.

Self-Tapping and Metal Drive Screws

The various types of self-tapping and metal drive screws are found in the standards ANSI B18.6.1-1972 (R1977) and ANSI 18.6.4-1981. Thread-forming screws are used in materials that allow enough plastic deformation for the thread to form by displacement, without removing any material. Fig. 12-19 shows various types of self-tapping screw thread. Dimensional specifications are given in Table 9 in Appendix B.

When turned into a hole, types A, AB, B, BP, and C will form a thread by a displacement action. Types D, F, G, T, BF, and BT, on the other hand, will form a thread by a cutting action. Type U, when driven into a hole, forms a series of threads by displacement. Applications of these screws are as follows:

Type A was formerly used for light sheet metal, resin-impregnated plywood, and asbestos compositions. This type, however, is no longer recommended, and has been replaced by Type AB.

Type AB is a spaced-thread screw that is now used in the same applications Type A was designed for. All new designs now use Type AB in place of Type A.

Type B has the greatest application for fastening thin metal, nonferrous castings, plastics, resin-impregnated plywood, and asbestos compositions.

Type BP has a conical point design that is used for fastening fabrics and in assemblies in which holes are improperly aligned.

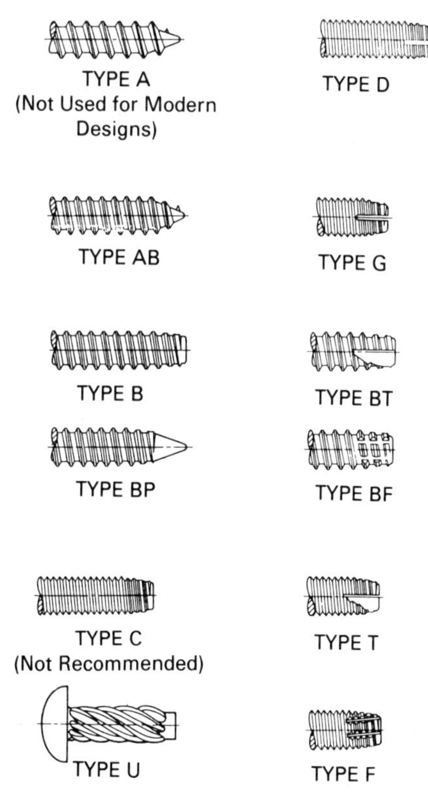

Fig. 12-19. Self-tapping screws.

Type C are used for high-driving torques, where machine screw thread is preferable to spaced threads of thread-forming screws. This type of screw, however, is not widely used in newer products.

Types D, F, G, and T are used in a variety of materials such as aluminum, zinc, lead die castings, sheet steel, steel shapes, cast iron, brass, and plastics.

Types BF and BT have found wide application in plastics and asbestos compositions.

Types BF and BT have found wide application in plastics and asbestos compositions.

Type U is used for permanent fastening only.

The most common form of screw fastener is the wood screw. Wood screws are made in lengths ranging from ¼ to 5 inches for steel screws, and from ¼ to 3½ inches for brass screws. American standard sizes are given in Table 12-7. Wood screws are made with flat, round, or oval heads (Fig. 12-20).

When specifying screws in drawings, the following sequence of information is given: nominal size; threads per inch; nominal length;

Table 12-7 American Standard Wood Screw Sizes (All dimensions in inches.)

Screw Number	Threads per Inch	Diam.
0	32	0.060
1	28	0.073
2	26	0.086
3	24	0.099
4	22	0.112
5	20	0.125
6	18	0.138
7	16	0.151
8	15	0.164
9	14	0.177
10	13	0.190
11	12	0.203
12	11	0.216
14	10	0.242
16	9	0.268
18	8	0.294
20	8	0.320
24	7	0.372

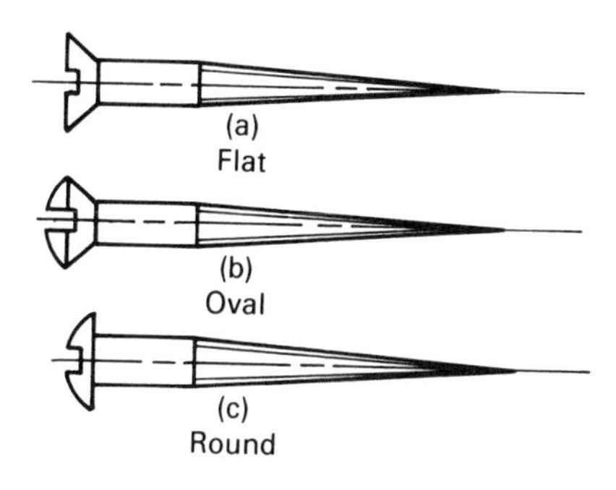

Fig. 12-20. Wood screws.

point type; product name, including head type and driving provision; material; and any necessary protective finish or coating. Examples of this type of specification are:

¼-14 × 1 TYPE AB, SLOTTED HEAD TAPPING SCREW, STEEL

6-32 × ¾ TYPE G, WASHER HEAD TAPPING SCREW, STEEL, NICKEL PLATED

Power Transmission and Pipe Thread Forms

Two other thread forms used in industrial drawings are those involving the transmission of power and the use of piping or tubing materials. Though they are not as common as other thread forms, drafters working with machine and industrial equipment design should be aware of these thread forms.

Acme Screw Threads

The acme screw thread is standardized in ANSI B1.5-1977, and accounts for two general applications for this thread form: general purpose and centralization. Both types of screw thread (Fig. 12-21) are used for providing rapid relative traverse motion and power transmission when needed. The basic dimensions for both types of acme screw thread are given in Table 10 of Appendix B.

There are three classes of general-purpose acme thread: 2G, 3G, and 4G. General-purpose threads are used in assemblies that require internal threads to be rigidly stationary, yet have sufficient clearance to allow for free and easy movement. It is common design practice to have the same class for both internal and external threads in assemblies designed for general use. The most preferred thread class

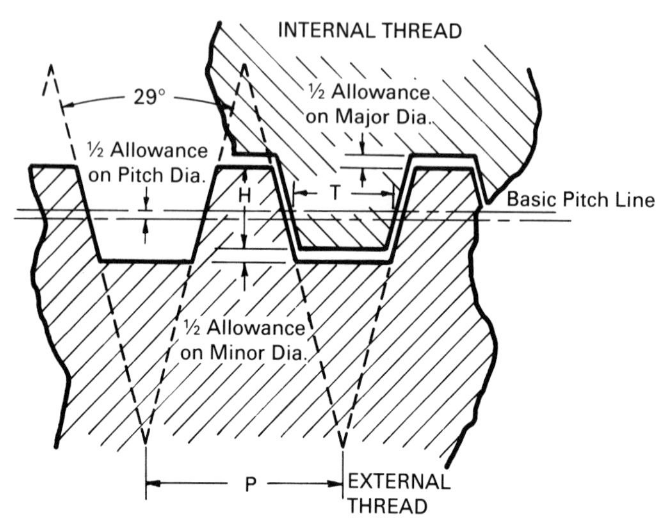

Fig. 12-21. Acme screw thread form.

is 2G, with classes 3G and 4G being incorporated where less backlash is desired. When minimum backlash is required, class 5G should be used.

Centralizing acme threads are divided into five classes, designated as 2C, 3C, 4C, 5C, and 6C. Class 2C is used to provide maximum end play or backlash. If less backlash is desired, the other classes can be used. Class 2C external threads have a larger pitch diameter allowance than classes 3C or 4C, and can be used interchangeably with classes 2C, 3C, or 4C internal threads. Likewise, class 3C external threads can be used interchangeably with classes 3C or 4C internal threads.

Screw Threads for Pipes

Pipe threads are usually categorized according to their application. In all, there are four general usage categories:

1. Threads that are assembled with a sealer, so as to produce a pressure-tight joint
2. Threads that do not use a sealer to produce a pressure-tight joint
3. Threads that produce free and loose-fitting mechanical joints with no pressure tightness

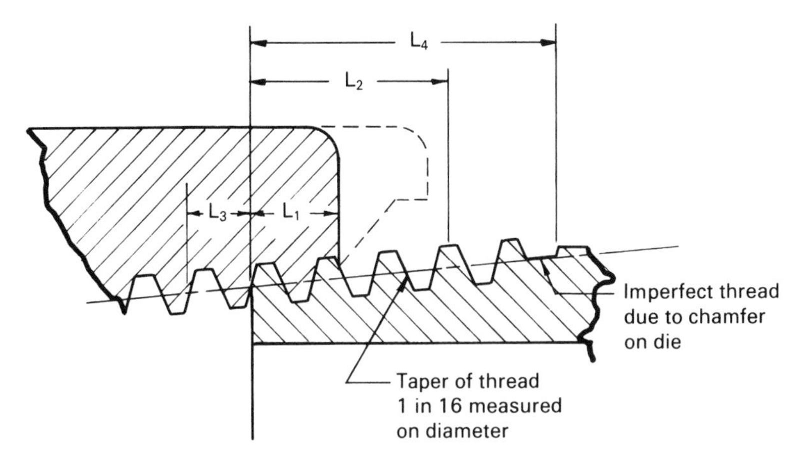

Fig. 12-22. Standard taper pipe threads.

4. Threads that produce rigid-fitting mechanical joints with no pressure tightness

There are three common screw threads used for pipes. The first is the *American standard taper pipe thread*, whose specifications are given in ANSI B2.1-1968. As shown in Fig. 12-22, it is manufactured to the following specifications:

taper	= 1:16	= 0.75 inch/foot
D	= basic outside diameter	
L_2	$= p(0.8D + 6.8)$	= basic length of the effective external thread

Basic dimensions for the standard taper pipe thread are given in Table 11 in Appendix B.

The second common pipe screw thread form is the *American standard straight pipe thread*. ANSI B2.1 provides specifications for this screw-thread form. The straight pipe thread is perhaps the most versatile, and is found in a variety of pipe, tube, and hose couplings. Straight pipe threads for mechanical joints are used with and without sealants, and include:

1. *NPSM, free-fitting mechanical joints for fixtures*, which are used for standard iron, steel, and brass pipe where there is no internal pressure

2. *NPSL, loose-fitting mechanical joints with locknuts,* designed for a pipe thread to have the largest diameter possible on a standard pipe
3. *NPSH, loose-fitting mechanical joints for hose couplings,* designed for hose coupling joints made with straight internal and external loose-fitting joints.

The last pipe screw thread to be considered is the *American Dryseal Pipe Thread.* These threads are specified in ANSI B1.20.3-1973, and are used when lubricants or sealers are undesirable. There are four basic types of Dryseal pipe thread:

1. *NPTF threads* are used for pipe joints in most general service functions. These threads are considered to be the best type for strength and seal because they have the longest length of thread.
2. *PTF-SAE SHORT threads* are used where the amount of clearance is insufficient for the full length of the NPTF thread. Thus, the NPTF thread should be considered first.
3. *NPSF threads* are internal straight threads only. They are used with soft materials that tend to adjust at assembly to the taper of the external threads. NPSF threads can be used with harder materials if the section provided is thick.
4. *NPSI threads* are also straight internal threads only. Their applications are similar to the NPSF thread, though they have wider use with harder materials having thick sections.

Exercises

12.1 Explain the difference between the American National Standard and Unified thread forms.

12.2 What are the basic differences between the American Unified and ISO Metric Unified thread forms?

12.3 Using the various thread representation techniques, illustrate and explain the use of each.

12.4 Explain and describe the methods used to dimension and specify various types of threaded devices.

12.5 What is the difference between a screw and bolt?

12.6 What is meant by the expression "power transmission thread"?

12.7 Describe the basic types of pipe thread form and their uses.

CHAPTER 13

Machine Mechanisms and Elements

- Linkages and Cams
- Gears
- Bearings
- Exercises

Discussion of the graphic representation of machine mechanisms and elements involves two different, yet closely allied, concepts. A mechanism is defined as that portion of a machine having two or more pieces or parts that are so arranged that the movement or motion of one causes the other to move. An element, by comparison, is that part of a machine that is used for fastening, securing, anchoring, coupling, or transmission of motion.

Fastening machine elements, such as screw and rivet fasteners, splines, pins, and springs have already been discussed. The topics covered here will therefore deal primarily with those mechanisms employed in the kinematics and motion-producing phase of machines. A wide variety of machine mechanisms and elements are encountered in industrial drawings; therefore, only a brief description of each will be given, emphasizing their graphic representation and dimensional specifications.

Linkages and Cams

Two types of mechanism that are closely related and are often used in combination are linkages and cams. Linkages are used by machine designers and drafters as a means of mechanizing various interrelated motions. These motions are usually critical to the successful operation of parts within machinery such as printing presses, valves and valve lifters, packaging equipment, and textile machinery. Cams, on the other hand, are machine elements that are used to generate a desired motion in a follower by direct contact.

Linkages

The physical design of a link may be of any form, as long as it does not change or interfere with the desired motion. It is therefore not uncommon to find different types of links used to generate the same motion. When drawing linkages, it is customary practice to use the term "crank" to denote a rotating link, a "lever" or "beam" to denote an oscillating link, and "connecting rods" to denote a connecting link.

Most links belong to one of six general categories. These are shown in Fig. 13-1, and described as follows:

1. *Beam and crank mechanisms* are perhaps one of the simplest forms of linkage. As shown in Fig. 13-1a, if point D is fixed, A can rotate while C oscillates; while if B is stationary, the same motion will be generated. This type of linkage is found in machines such as side-wheel steamers.
2. *Drag link mechanisms* are employed to feather the floats on paddle wheels. Here (Fig. 13-1b), if A is fixed, B and D will be able to rotate.
3. *Rocker mechanisms* are often used in straight-line motion machines. In Fig. 13-1c, C is fixed so as to produce an oscillating motion for D and B.
4. *Sliding block linkages* are used to generate motion within a given path. This path is usually defined by a slide that is housed in a fixed frame, and can take two basic forms (see Fig. 13-1d).
5. *Swing block linkages* are variations of the sliding linkage, except in this case the frame moves. As shown in Fig. 13-1e, A rotates and E oscillates, while C and D are defined as infinite links.
6. *Turning block linkages* are illustrated in Fig. 13-1f. Here the short line (A) is fixed, while B and E rotate. This type of linkage is typically referred to as *Whitworth quick-return motion.*

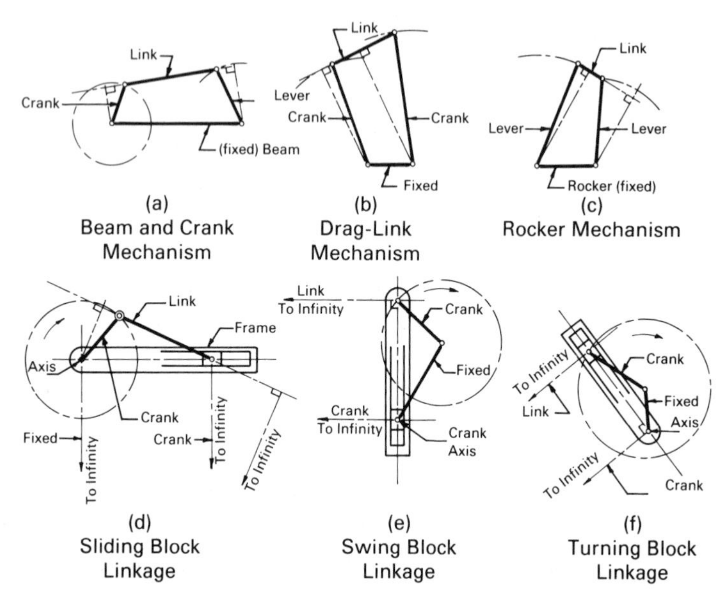

Fig. 13-1. Types of linkage.

Many linkage designs are prepared with accompanying *kinematic diagrams*. These diagrams are schematics of the overall design of the linkage system. Shown in Fig. 13-2 are examples of a kinematic diagram.

To draw and interpret these schematics properly, the drafter must be familiar with the symbols used and their interpretation. These symbols have been standardized on the international level (ISO 3952/1-1981 (E/F/R)), and are divided into four major categories:

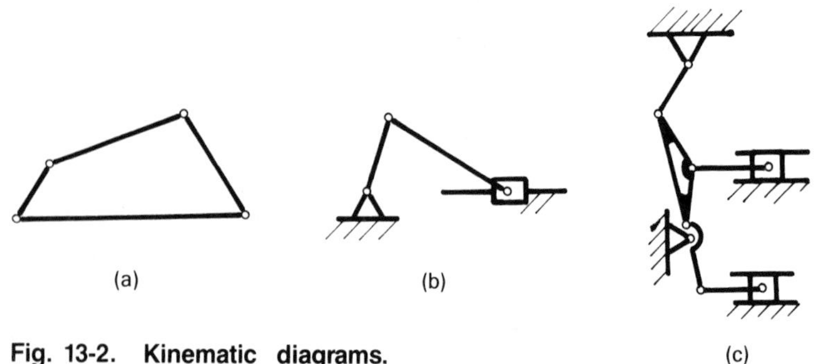

Fig. 13-2. Kinematic diagrams.

1. Motion of links of mechanisms
2. Kinematic pairs
3. Links and connections of their components
4. N-bar linkages and their components

Figs. 13-3 through 13-6 show these schematic symbols.

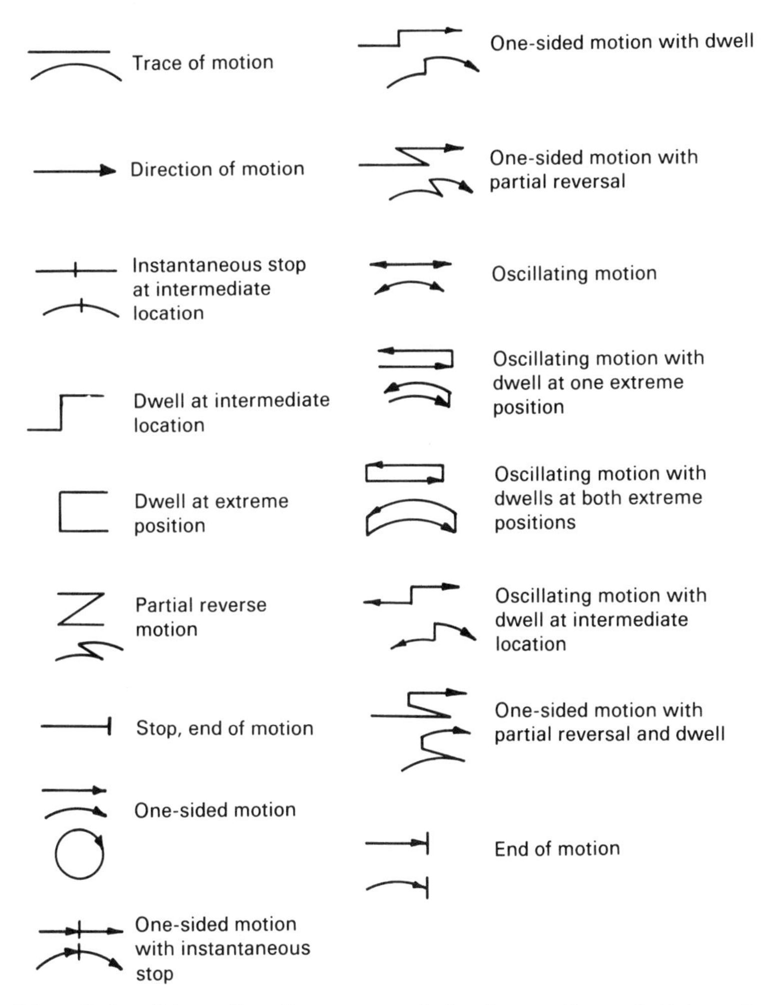

Fig. 13-3. Kinematic diagram symbols for motion of links of mechanisms.

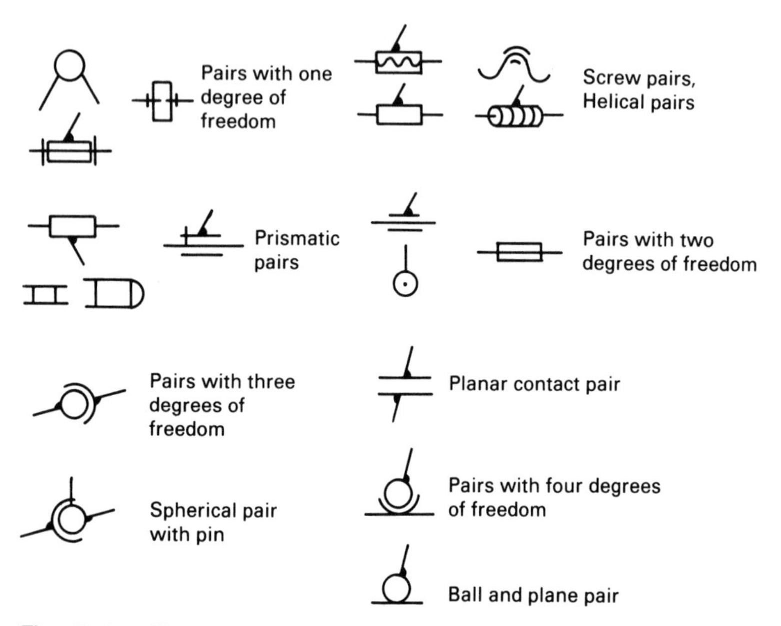

Fig. 13-4. Kinematic pairs symbols.

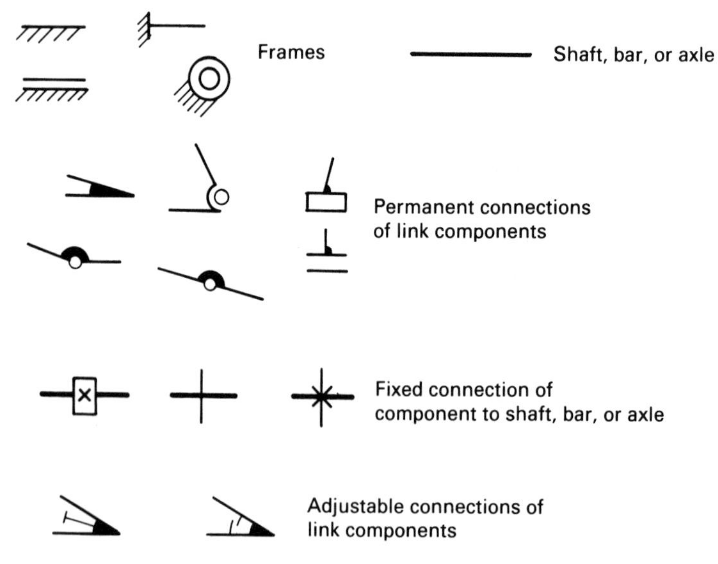

Fig. 13-5. Kinematic symbols for links and connections of components.

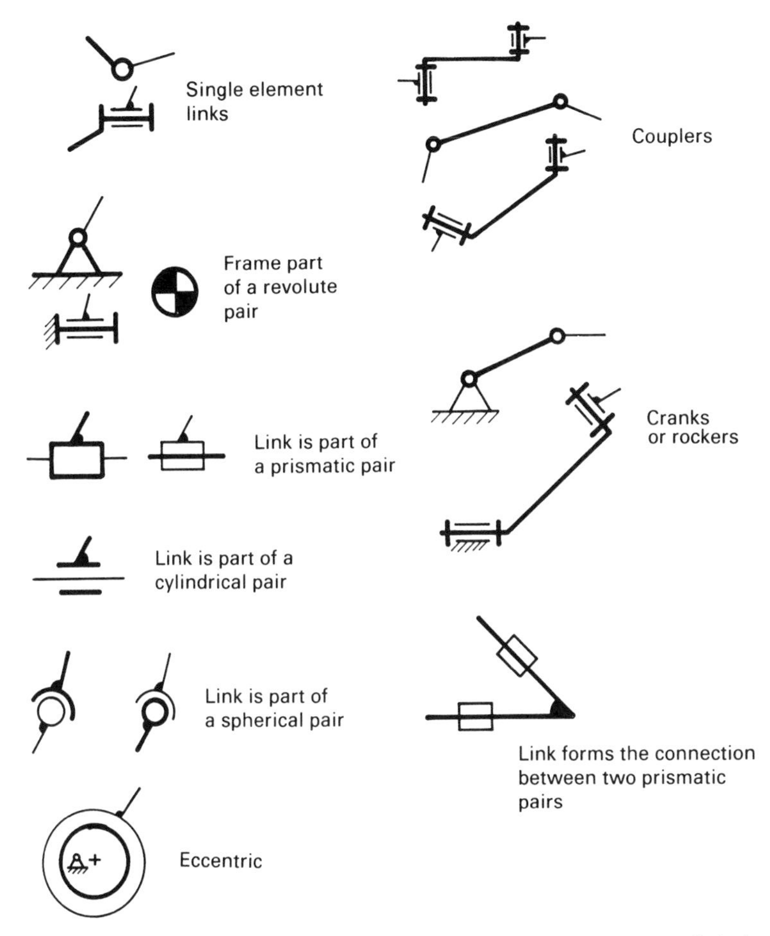

Fig. 13-6. Kinematic diagram symbols for N-bar linkages and their components.

Cams

Cams are used in machines and other devices to communicate motion via a follower that traces its edge or a groove cut into its surface. Cams are usually designed as a plate or cylinder. As shown in Fig. 13-7, there are several basic terms that must be understood to gain competence in cam drawings. These are as follows:

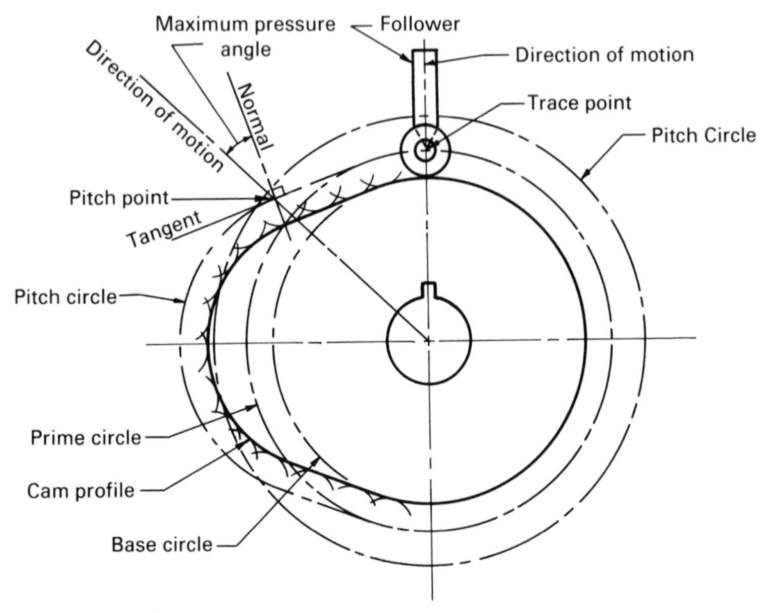

Fig. 13-7. Cam nomenclature.

1. *Follower displacement* is the position of the follower from a rest position; also known as the zero position, in relationship to time or some other fraction of a measurement unit (e.g., degrees, inches, or millimeters).
2. *Cam displacement* is specified in units such as degrees, inches, or millimeters that relate to follower displacement, and is the cam motion as measured from a given zero or rest position.
3. *Cam profile* is the machined or formed contour of the cam.
4. *Trace point* is the center line of the follower.
5. *Base circle* is the smallest-diameter circle to which any portion of the cam profile fits.
6. *Pitch curve* is the locus of each position that the trace point occupies as it travels about the cam.
7. *Prime circle* is the smallest-diameter circle in the pitch curve from the center of the cam.
8. *Pressure angle* is that angle located between the normal to the pitch curve and the direction of motion of the follower.

9. *Pitch point* is that position on the pitch curve of the maximum pressure angle.
10. *Pitch circle* is a circle that is defined by a radius that passes through the pitch point.
11. *Transition point* is the location of maximum velocity where the acceleration of the follower changes from plus to minus. In some cam designs (e.g., closed cams) this is referred to as the crossover point.

Cams are generally divided into two classes: uniform-motion cams and accelerated-motion cams. Uniform motion pertains to cams that move followers at a constant rate of speed from start to finish. However, as the movement starts from the zero point to full speed and stops at the end, there will be a noticeable and abrupt "shock." Accelerated-motion cams are best suited for moderate speeds, and have sudden changes in acceleration at various points of the stroke.

Three general types of cam follower systems are used: radial, offset, and swinging roller followers (Fig. 13-8). In some cases, design requirements call for the roller follower to work within an enclosed or closed path (Fig. 13-9). These types of cam are referred to as closed track (as opposed to open track) cams.

DISPLACEMENT DIAGRAMS

The designs of cams are made with displacement diagrams. Shown in Fig. 13-10 is a simple displacement diagram. Here one cycle is equal to one complete revolution of the cam, which is equal to 360°. The

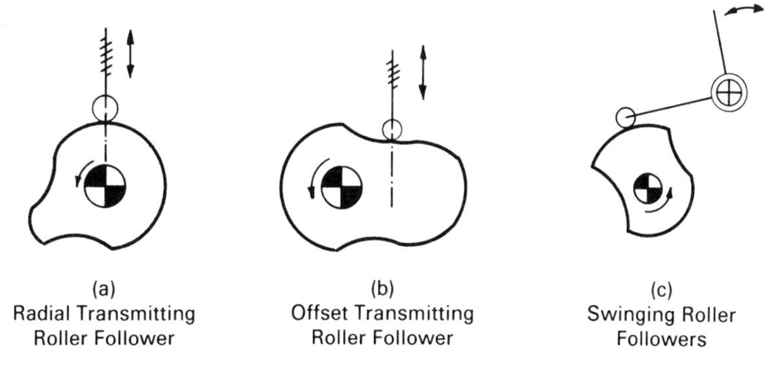

Fig. 13-8. Cam follower systems.

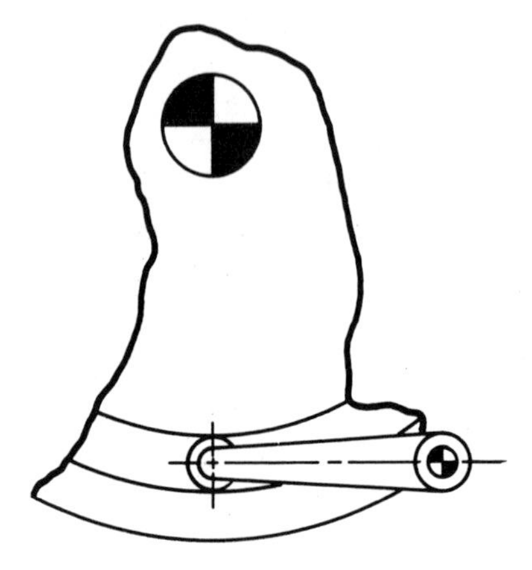

Fig. 13-9. Closed-track cam.

horizontal distances t_1, t_2, etc., are used to represent standard units of measure, such as time, radians, or degrees. The vertical distance h denotes the "rise" or stroke of the follower.

A variety of cam curves are used to move followers. The vast majority of cam displacement diagrams, however, employ one of four displacement curves. These are:

1. *Constant velocity motion* (Fig. 13-11). These displacement curves are used in only the most crude type of device. They are not very popular because the motion from rest to constant velocity occurs instantly, so that accelerations are quite great at transition points, resulting in substantial "jerks" and "shocks" in the equipment. Nonetheless, this type of motion can be, and often is, successfully used by modifying the start and finish of the follower stroke.

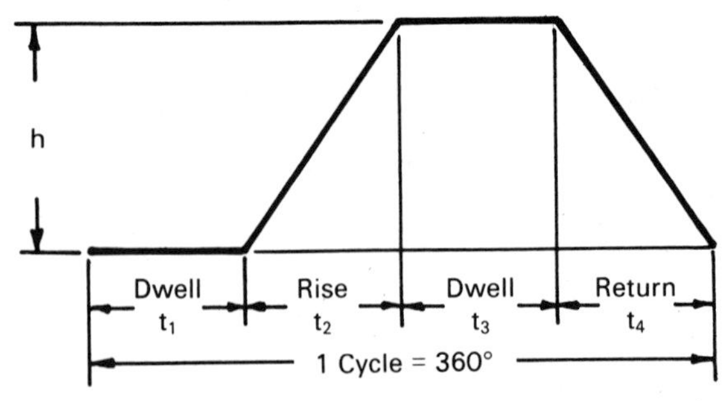

Fig. 13-10. Simple displacement diagram.

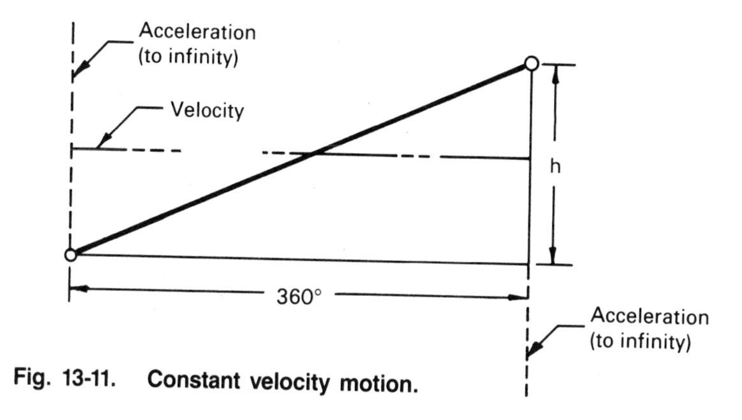

Fig. 13-11. Constant velocity motion.

2. *Parabolic motion* (Fig. 13-12). This design is used when the smallest amount of acceleration for a given angle of rotation and rise is desired. There are, however, sudden changes in acceleration at the beginning, middle, and end of the stroke, resulting in follower and cam shocks.
3. *Simple harmonic motion* (Fig. 13-13). The primary advantage of this displacement curve is the smoothness in velocity and acceleration during the stroke. Vibration, noise, and wear do occur here, owing to the instantaneous changes in acceleration at the beginning and end of each stroke. If the follower is subjected to inertia loads, it must be so designed as to overcome related stresses to the members.

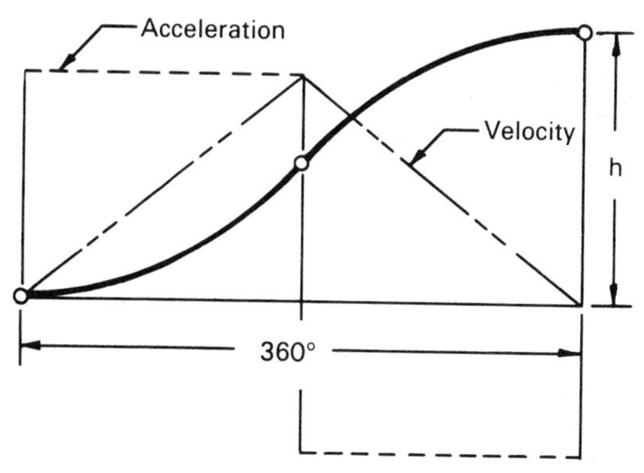

Fig. 13-12. Parabolic motion.

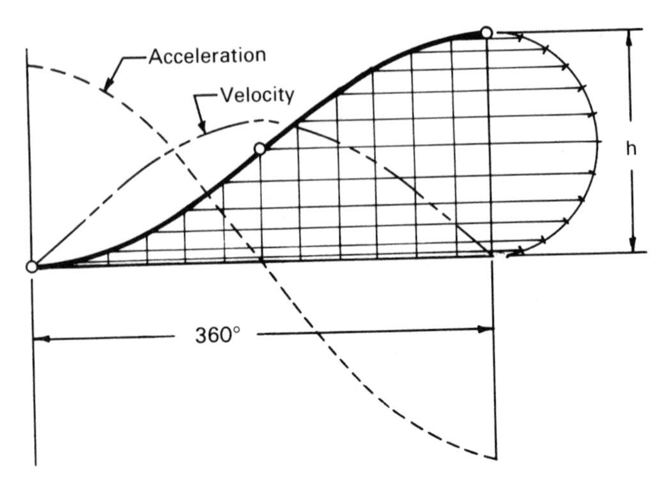

Fig. 13-13. Simple harmonic motion.

4. *Cycloidal motion* (Fig. 13-14). This curve is used where good acceleration characteristics are desired; there are no abrupt or sharp changes in the acceleration curve. Even though the maximum value of the acceleration of the follower is higher than the simple harmonic motion curve, cycloidal curves are frequently used as the basis for cams in high-speed equipment because of the lower levels of noise, wear, and vibration.

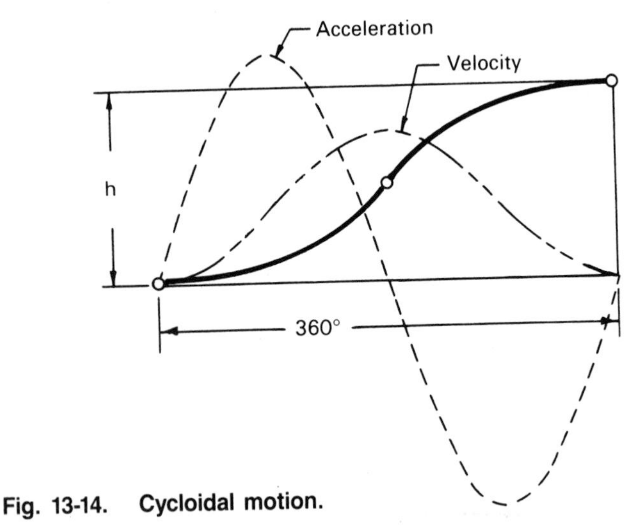

Fig. 13-14. Cycloidal motion.

DRAWING CAM PROFILES

The machining of a cam requires transforming the abstract displacement diagram into a tangible product, which was formerly accomplished by preparing a detailed drawing of the cam profile itself, from which direct measurements were made. This is not widely practiced today, except for crudely toleranced mechanisms. However, there are times when the drafter must be able to develop a cam profile from the displacement diagram so as to show how the cam should appear, and provide a means of giving manufacturing measurements and tolerances.

Although such is not the case in the actual operation of the cam, cam profile drawings are made as if the cam were fixed and the follower rotated about it. Thus, the profile requires the drawing of many follower positions. Since the cam design is first made via the displacement diagram, this process is quite simple.

Fig. 13-15 shows how the displacement diagram is converted into the cam profile. Here each horizontal distance is converted into degree measures, so that the base circle can be appropriately divided. The rise per degree increment is measured directly from the diagram to the profile until the entire 360° cycle has been completed. As can be surmised, two additional givens must be known before this process can be accomplished; namely, follower design and pitch circle.

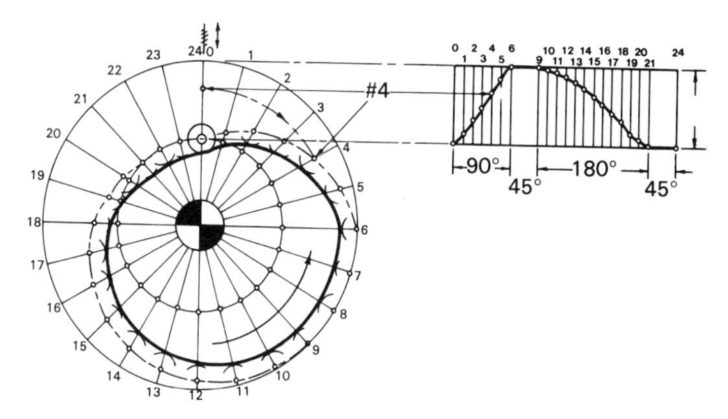

Fig. 13-15. Constructing cam profile from a displacement diagram.

CAM DIMENSIONS AND TOLERANCES

One of the most difficult processes for a manufacturer and machinist is the construction of a cam from a drawing. Today these profiles are primarily used as references to the general appearance of the cam. Accompanying these cam drawings must be data pertaining to the dimensional specifications and tolerances of the contour.

When specifications are given for the cam profile, dimensions must *always* be made to the pitch curve. Follower location is then determined by the radial and angular displacements. Here radial displacement is given in terms of distance from the center of the cam or displacement from the prime circle. Angular displacement is given in degrees from the zero reference.

Rather than dimensioning directly on the cam profile itself, design information is usually given in a table provided on the drawing. It

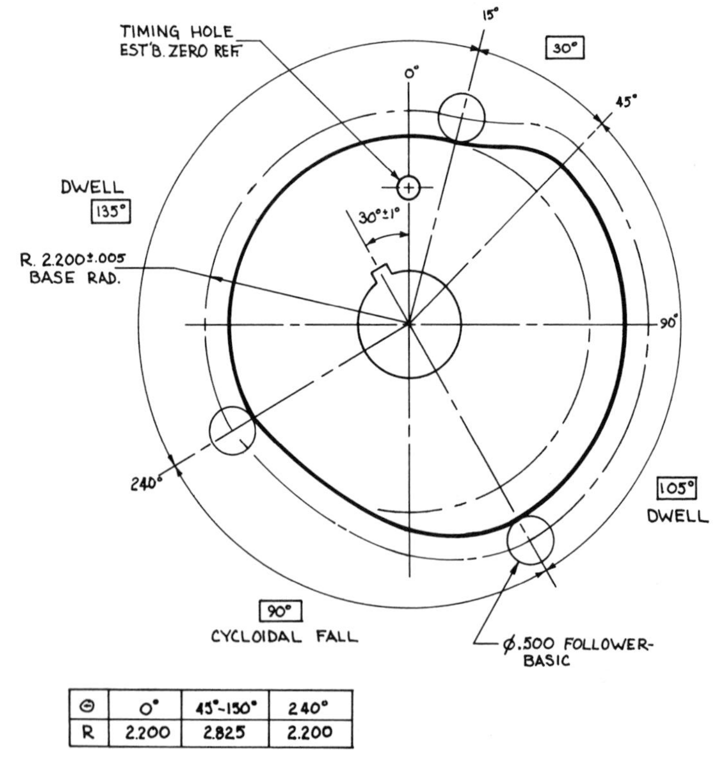

Θ	0°	45°-150°	240°
R	2.200	2.825	2.200

Fig. 13-16. Detailed cam profile.

is recommended that specifications be given in increments of 30′ (1/2 degree), since it gives the machinist the choice in selecting intervals for the finished cam.

Tolerancing should be made on displacement values, rather than angular data. The only angle that should be toleranced is the one that pertains to zero references to another point on the cam (e.g., keyways, slots, dowel holes, etc.). A typical detailed cam drawing, including dimensional and tolerance specifications, is shown in Fig. 13-16.

KINEMATIC DIAGRAMS

Cam mechanisms are often used as elements in kinematic diagrams, and there is a series of standardized graphic representations used for different cam mechanisms. Fig. 13-17 shows those graphic symbols for cam mechanisms used in kinematic diagrams.

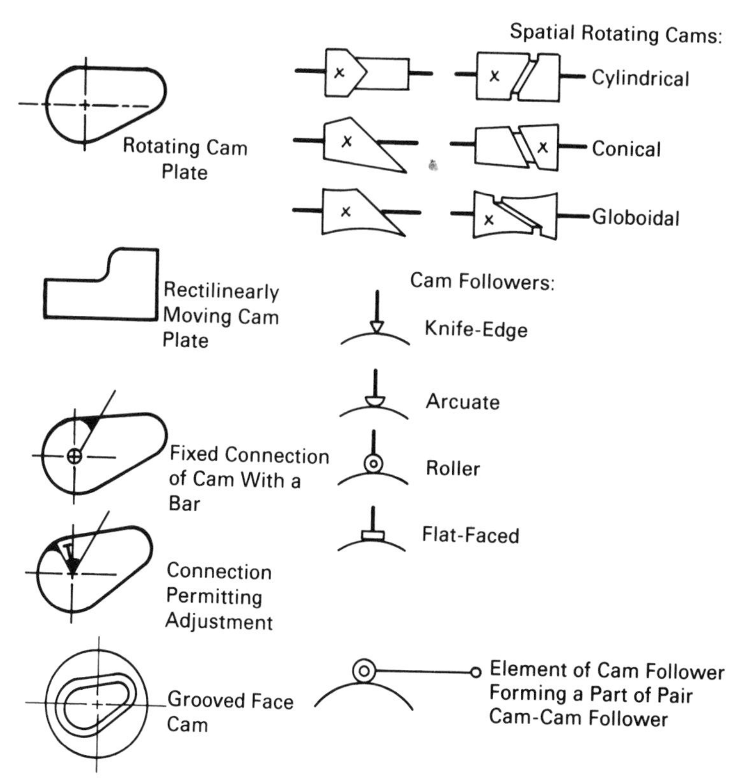

Fig. 13-17. Cam mechanism symbols for kinematic diagrams.

Gears

The primary function of gears is to transmit motion and/or power from one location to another, or to change the rpm of a shaft. Gears are classified according to tooth form, shaft arrangement, pitch, and quality. The basic tool forms and shaft arrangements are given in five major categories:

Tooth Form	Shaft Arrangement
Spur	Parallel
Helical	Parallel or skew
Worm	Skew
Bevel	Intersecting
Hypoid	Skew

Pitch classifications are coarse (less than twenty teeth to diametral pitch) and fine (twenty teeth to diametral pitch, and finer). Quality categories are: commercial, precision, and ultraprecision.

Fig. 13-18 illustrates basic gear-tooth nomenclature. These terms are defined as follows:

1. *Addendum* is the height of the tooth above the pitch circle, or the radial distance from the pitch circle to the crest of the tooth.
2. *Backlash* pertains to the width of a tooth space that exceeds the thickness of engagement.
3. *Chordal addendum* is the height from the crest of the tooth to the chord drawn along the circular thickness arc.
4. *Chordal thickness* is the length of the chord drawn subtending the circular thickness arc.
5. *Circular thickness* is the length of the arc between two sides of a gear tooth along the pitch circle.
6. *Clearance* pertains to the amount that the dedendum exceeds the addendum of its mating gear.
7. *Dedendum* is the depth of the tooth area (space) from the pitch circle to the bottom of the tooth space.
8. *Diametral pitch* is the ratio of the number of teeth to the number of inches of the pitch diameter.
9. *Pitch circle* has a radius equal to the distance from the gear axis to the pitch point.

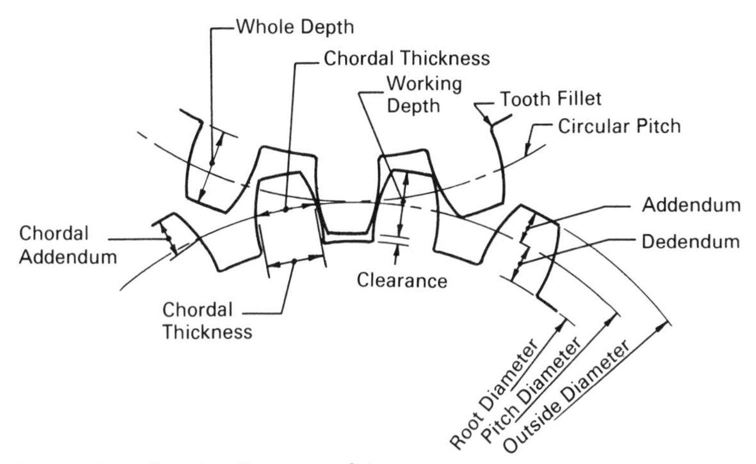

Fig. 13-18. Gear-tooth nomenclature.

10. *Pitch point* is the point of tangency of two pitch circles.
11. *Pressure angle* is the angle formed between the gear tooth profile and the radial line at its pitch point.
12. *Root circle* is the circle that coincides with (or is tangent to) the bottoms of the tooth space.
13. *Working depth* refers to the depth of the tooth space.

Spur Gears

One of the most basic and recognizable gear tooth forms is the spur gear. Tooth proportions have been standardized in ANSI B6.1-1968, R1974. In this gear system, the 14.5° pressure angle was, and still is, useful for duplicate or replacement gearing, and where backlash control is important. The 20° pressure angle spur gear has since become the standard for newer designs because of its quieter and smoother operating characteristics, its ability to carry greater loads, and the smaller number of teeth affected by undercutting.

The specification of gear data is calculated by the use of standard formulas. The formulas used for the 14.5° and 20° full teeth spur gear are the same, the primary difference being that the first will have between sixteen and forty teeth in a mating pair, while the latter will have thirteen to twenty six teeth per mating pair. The formulas used for dimensions of standard spur gears are given in Table 13.1.

Table 13-1 Standard Spur Gear Formulas

Notations

$\emptyset$	= Pressure Angle	D_O	= Outside Diameter
a	= Addendum	D_R	= Root Diameter
a_G	= Addendum of Gear	F	= Face Width
a_P	= Addendum of Pinion	h_k	= Working Depth of Tooth
b	= Dedendum	h_t	= Whole Depth of Tooth
c	= Clearance	m_G	= Gear Ratio
C	= Center Distance	N	= Number of Teeth
D	= Pitch Diameter	N_G	= Number of Teeth in Gear
D_G	= Pitch Diameter of Gear	N_P	= Number of Teeth in Pinion
D_P	= Pitch Diameter of Pinion	P	= Diametral Pitch
D_B	= Base Circle Diameter		
p	= Circular Pitch		

General Formulas

Unknown	Formulas
Base Circle Diameter	$D_B = D \cos \emptyset$
Circular Pitch	$p = \frac{\pi D}{N} = \frac{\pi}{P}$
Center Distance	$C = \frac{N_P(m_G + 1)}{2P} = \frac{D_P + D_G}{2} = \frac{N_G + N_P}{2P} = \frac{(N_G + N_P)p}{2\pi}$
Diametral Pitch	$P = \frac{\pi}{p} = \frac{N}{D} = \frac{N_P(m_G + 1)}{2C}$
Gear Ratio	$m_G = \frac{N_G}{N_P}$
Number of Teeth	$N = PD = \frac{\pi D}{p}$
Outside Diameter (Full Depth Teeth)	$D_O = \frac{N + 2}{P} = \frac{(N + 2)p}{\pi}$
Outside Diameter (Stub Teeth)	$D_O = \frac{N + 1.6}{\pi} = \frac{N + 1.6}{P}$
Outside Diameter	$D_O = D + 2a$
Pitch Diameter	$D = \frac{Np}{\pi}$

General Formulas (continued)

Unknown	Formulas
Root Diameter	$D_R = D - 2b$
Whole Depth	$h_t = a + b$
Working Depth	$h_k = a_G + a_p$

The details of a gear are not normally provided on a set of working drawings. Instead, a simplified representation of gears is recommended. A gear is usually represented (except in axial section views) as a solid part without teeth. ANSI standards recommend that one or more gear teeth be drawn with phantom lines used to represent the outside and root diameters, while center lines are used to indicate the pitch circle. ISO standards, by comparison, recommend that a solid line be used to represent the outside root circles, while a center line should indicate the pitch circle. Fig. 13-19 shows examples of each method.

Helical Gears

When a spur gear's teeth are machined at an angle to the gear shafts, the gear is known as a helical gear. Helical gears are used for the same

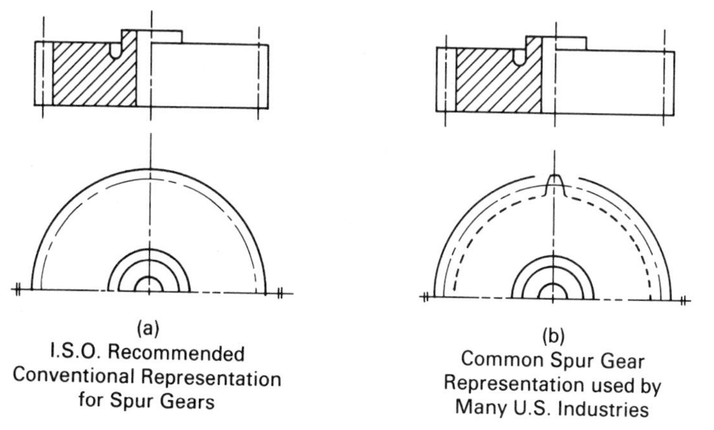

(a)
I.S.O. Recommended Conventional Representation for Spur Gears

(b)
Common Spur Gear Representation used by Many U.S. Industries

Fig. 13-19. Spur gear representation.

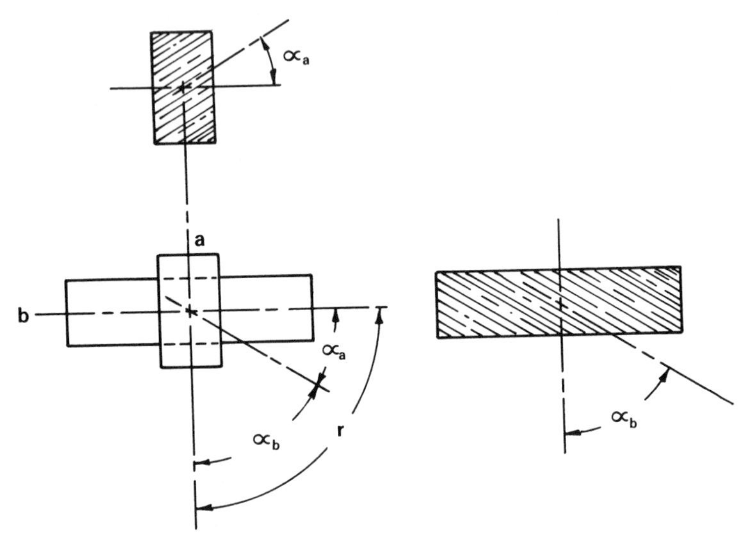

Fig. 13-20. Helical gear dimensions for formulas used in calculations.

general purposes as the common spur gear, but cannot be used as sliding gears. Thus, the mating gears are required to have the same helix angle in opposite directions. As a general rule, helical gears will run quieter than those with straight-cut teeth.

Fig. 13-20 shows the dimensions for helical gears and Table 13-2 gives the formulas.

It is important to show the direction of the teeth in helical gear drawings. A gear or rack representation in a parallel view of the gear should illustrate the direction of the gear teeth by using three thin, continuous lines. If additional lines are needed, they should be drawn. This technique is shown in Fig. 13-21.

Table 13-2 Basic Formulas for Helical Gears

Notations			
P_n	= Normal Diametral Pitch	L	= Lead of Tooth Helix
D	= Pitch Diameter	S	= Addendum
N	= Number of Teeth	W	= Whole Tooth Depth
α	= Helix Angle	T_n	= Normal Tooth Thickness at Pitch Line
γ	= Angle between Shafts		
C	= Center Distance	O	= Outside Diameter

General Formulas

Unknown	Formulas
Pitch Diameter	$D = \frac{N}{P_n \cos\alpha}$
Center Distance	$C = \frac{D_a + D_b}{2}$
Lead of Tooth Helix	$L = \pi D \cot\alpha$
Addendum	$S = \frac{1}{P_n}$
Whole Depth of Tooth	$W = \frac{2.157}{P_n}$
Normal Tooth Thickness at Pitch Line	$T_n = \frac{1.571}{P_n}$
Outside Diameter	$O = D + 2S$

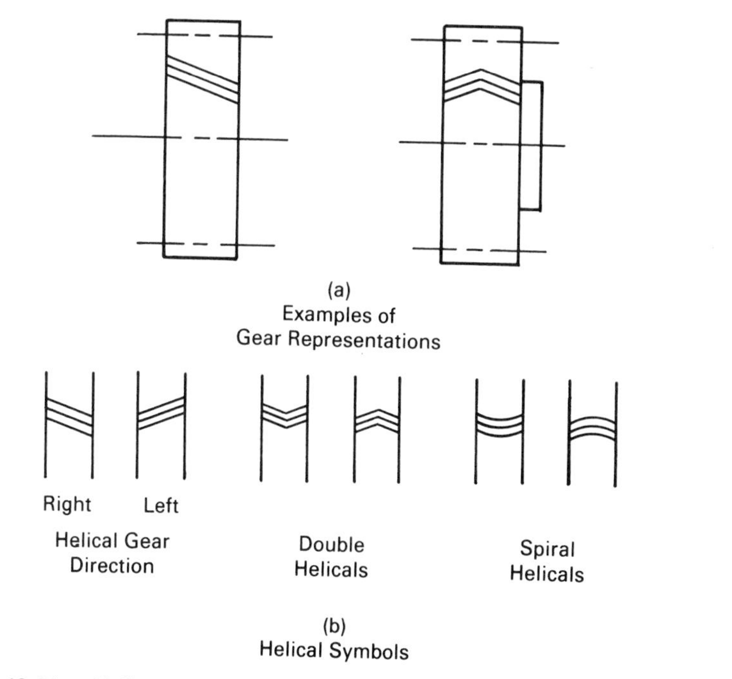

Fig. 13-21. Helical gear conventional representations.

Bevel Gears

Bevel gears are conical in shape, and are used to connect shafts that are not in line, but have intersecting axes. There are four general types of bevel gear. The first, and most common, is known as *straight bevel gears.* Their gear teeth are straight with tapered sides, and are used for peripheral speeds up to 1,000 fpm, where maximum smoothness and quietness are not critical. *Zerol bevel gears* are a second type, in which the gear teeth are curved but lie in the same general direction as the straight bevel gear. These gears are used for similar applications as straight bevel gears, but are more advantageous where hardened gears of high precision are required.

A third type of bevel gear is *spiral bevel gears.* These gears are designed with curved oblique teeth so that contact starts gradually and continues smoothly from end to end. The primary advantage of this gear design is that of complete control of the localized tooth contact. Spiral bevel gears are used where peripheral speeds exceed 1,000 fpm or rpm. The fourth gear type is *hypoid bevel gears.* These gears are similar in appearance to spiral bevel gears except that the pinion's

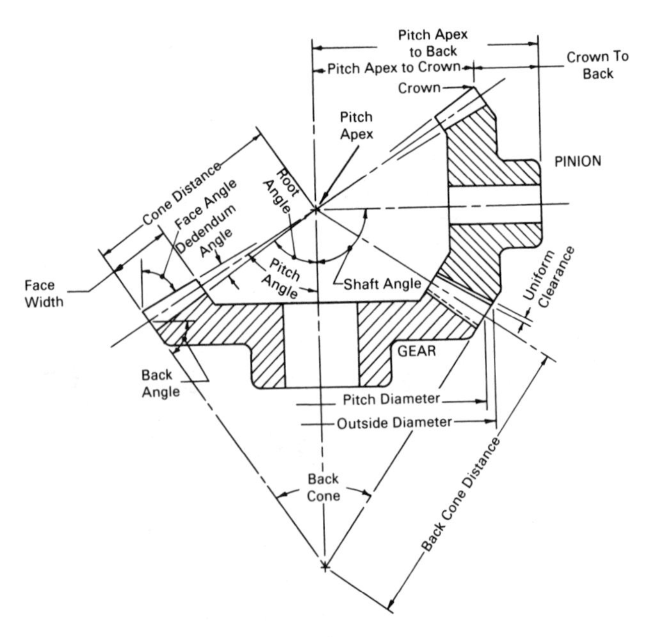

Fig. 13-22. Bevel gear nomenclature.

axis is offset from the gear axis. Hypoid gears are used under the same general conditions as the spiral type, as well as where maximum smoothness is desired, and for nonintersecting shafts.

Fig. 13-22 shows the angle and dimensional nomenclature associated with bevel gear drawings. It should be noted that relative to the face angles of these gear teeth, the face cones are constructed parallel to the root cones of the mating gears. This is necessary to provide for sufficient and uniform clearance along the length of the gear teeth.

Straight bevel gear standards are given in ANSI 208.03 for both the American Standard and Gleason System straight bevel gears. The Gleason System meets all requirements of the universal system for straight bevel gears specified by ANSI standards. The most common type of bevel gear arrangement encountered is the 20° straight bevel gear with a 90° shaft angle. Formulas used for the Gleason System are found in Table 13.3. For other bevel gear forms, the designer and drafter is referred to ANSI and American Gear Manufacturers Association (AGMA) standards (i.e. ANSI/AGMA 208.03 and AGMA 202.03).

Specific guidelines used to draw bevel gearing are the same as for other gear forms, and are shown in Fig. 13-23. The drawing itself should give only the dimensions of the gear blank. Cutting data and gear specifications should be given in a table or a note. In some cases, only the gear will be drawn; however, many designs call for the presentation of both gear and pinion. Seldom should the drafter display actual gear teeth in the drawing, for their detailing will add substantial time and cost to the drawing.

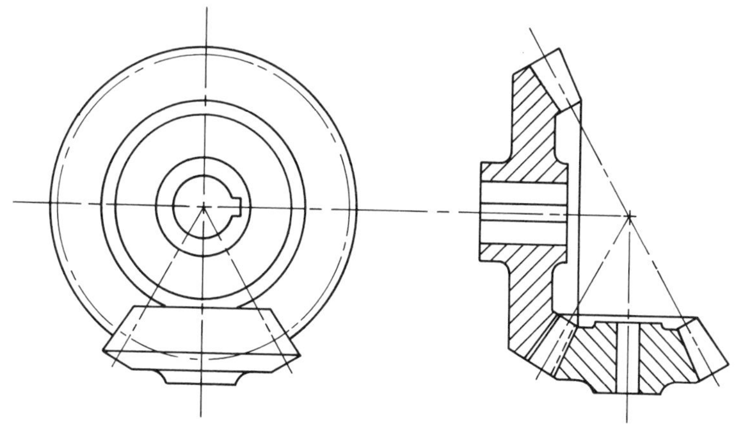

Fig. 13-23. Bevel gear conventional graphic representation.

Table 13-3 Basic Formulas for the Gleason System 20-Degree Straight Bevel Gear with 90-Degree Shaft Angle

Given

n	= Number of Pinion Teeth	F	= Face Width
N	= Number of Gear Teeth	ϕ	= Pressure Angle (20°)
P	= Diametral Pitch	Σ	= Shaft Angle (90°)

General Formulas

Unknown	Formulas: *Pinion*	Formulas: *Gear*
Working Depth	$h_k = \frac{2.000}{P}$	Same as pinion
Whole Depth	$h_t = \frac{2.188}{P} + 0.002$	Same as pinion
Pitch Diameter	$d = \frac{n}{P}$	$D = \frac{N}{P}$
Pitch Angle	$\gamma = \tan^{-1}\frac{n}{N}$	$\Gamma = 90° - \gamma$
Cone Distance	$A_O = \frac{D}{2 \sin\Gamma}$	Same as pinion
Circular Pitch	$p = \frac{\pi}{P}$	Same as pinion
Addendum	$a_P = h_k - a_G$	$a_G = \frac{0.540}{P} + \frac{0.460}{P(N/n)^2}$
Dedendum	$b_P = \frac{2.188}{P} - a_P$	$b_G = \frac{2.188}{P} - a_G$
Clearance	$c = h_t - h_k$	Same as pinion
Dedendum Angle	$\delta_P = \tan^{-1}\frac{b_P}{A_O}$	$\delta_G = \tan^{-1}\frac{b_G}{A_O}$
Face Angle of Blank	$\gamma_O = \gamma + \delta_G$	$\Gamma_O = \Gamma + \delta_P$
Root Angle	$\gamma_R = \gamma - \delta_P$	$\Gamma_R = \Gamma - \delta_G$
Outside Diameter	$d_O = d + 2a_P \cos\gamma$	$D_O = D + 2a_G \cos\Gamma$
Pitch Apex to Crown	$x_O = \frac{D}{2} - a_P \sin\gamma$	$X_O = \frac{d}{2} - a_G \sin\Gamma$

General Formulas (continued)

Unknown	Formulas: *Pinion*	Formulas: *Gear*
Circular Thickness	$t = p - T$	$T = \frac{p}{2} - (a_p - a_G)\tan\emptyset - \frac{K}{P}$
Chordal Thickness	$t_C = t - \frac{t^3}{6d^2} - \frac{B}{2}$	$T_C = T - \frac{T^3}{6D^2} - \frac{B}{2}$
Chordal Addendum	$a_{CP} = a_P + \frac{t^2 \cos\gamma}{4d}$	$a_{CG} = a_G + \frac{T^2 \cos\Gamma}{4D}$
Tooth Angle	min. $(3438/A_O)(t/2 + b_P \tan\emptyset)$	min. $(3438/A_O)(T/2 + b_G \tan\emptyset)$
Limit Point Width	$\frac{A_O - F}{A_O}(T - 2b_P \tan\emptyset) - 0.0015$	$\frac{A_O - F}{A_O}(t - 2b_G \tan\emptyset) - 0.0015$

Worm Gears

Worm gears are used for the transmission of motion and power between two shafts that are at 90° to each other. These gear forms are standardized in ANSI B6.9, and are divided into the two general classes of fine-pitch and coarse-pitch worm gearing. This differentiation is important for the following reasons:

1. Fine-pitch worms and worm gears are used to transmit motion, while coarse-pitch are used for transmitting power.
2. Gear housings are constructed differently for each class of worm gear.
3. Profile deviations cannot be accurately determined for fine-pitch worm gearing, owing to its small size. These deviations, however, can be accurately found for coarse-pitch worms and worm gears.
4. Significant differences exist in the product and inspection methods used for each class of gear.

A worm gear that is considered to be *single-thread* is one in which one complete revolution will advance the worm gear one tooth and space, which results in a significant reduction in velocity. The ratio of worm speed to worm-gear speed will functionally range from about 1.5 to over one hundred. The higher the ratios, the less efficient will be the transmission of power. This ratio is calculated by the number of worm-gear teeth divided by the number of threads (starts) on the worm. As a general design rule, a ratio of fifty should be the maximum for single-thread worms.

Multithread worms are incorporated where the transmission of power is important. The lead angle used here is relatively high, and will range between 25° and 45°. Typical worm-gear teeth to worm thread ratios are commonly in the range of four to seven.

Tables 13.4 and 13.5 present the basic dimension formulas for fine-pitch and coarse-pitch worm gearing. It should be noted that in many industries, the coarse-pitch class is also referred to as *industrial worm gearing.*

Table 13-4 Basic Formulas for Fine-Pitch Worm Gearing

Notations

P	= Circular Pitch of Worm Gear	Ψ	= Helix Angle of Worm Gear
P	= Axial Pitch of the Worm, (P_x), in the Central Plane	n	= Number of Threads in Worm
P_n	= Normal Circular Pitch of Worm	N	= Number of Teeth in Worm Gear
P_n	= P_x cos = P cos	N	= nm_G
		m_G	= Ratio of Gearing (N:n)
		λ	= Lead Angle of Worm

General Formulas

Unknown	Formulas
Worm Dimensions	
Lead	$l = nP_x$
Pitch Diameter	$d = \dfrac{l}{\pi \tan\lambda}$
Outside Diameter	$d_O = d + 2a$
Safe Minimum Length of Threaded Portion of Worm	$F_W = \sqrt{D_O{}^2 - D^2}$

General Formulas (continued)	
Unknown	**Formulas**
Worm Gear Dimensions	
Pitch Diameter	$D = \frac{NP}{\pi} = \frac{NP_x}{\pi}$
Outside Diameter	$D_O = 2C - d + 2a$
Face Width	$F_{Gmin} = 1.125\sqrt{(d_O + 2C)^2 - (d_O - 4a)^2}$
Dimensions for both Worm and Worm Gear	
Addendum	$a = 0.3183P_n$
Whole Depth	$h_t = 0.7003P_n + 0.002$
Working Depth	$h_k = 0.6366P_n$
Clearance	$c = h_t - h_k$
Tooth Thickness	$t_n = 0.5P_n$
Center Distance	$C = 0.5(d + D)$
Approximate Normal Pressure Angle	$\emptyset_N = 20°$

Table 13-5 Basic Formulas for Coarse-Pitch Worm Gearing (Industrial Worm Gearing)

Notations

A	= Addendum	T	= Threads or Starts
C	= Center Distance	R	= Ratio
F	= Gear Face Width	h_t	= Whole Depth
FL	= Worm Face Length	OD_W	= Worm Outside Diameter
$\emptyset$	= Lead Angle		
N	= Number of Gear Teeth	OD_G	= Gear Outside Diameter
L	= Lead		
D_W	= Worm Pitch Diameter	D_T	= Throat Diameter
		D_R	= Radius of Throat
		R_R	= Rim Radius
		D_G	= Gear Pitch Diameter
		P	= Pitch

General Formulas

Unknown	Formulas
Addendum	$A = 0.318P$ for single and double threads $= 0.286P$ for triple and quadruple threads
Center Distance	$C = \frac{D_W + D_G}{2}$
Face Width of Gear	$F = 2.38P + 0.25$ for single and double threads $= 2.15P + 0.2$ for triple and quadruple threads
Face Length of Worm	$FL = (0.02N + 4.5)P$
Gear Teeth	$N = \frac{\pi D_G}{P}$
Lead	$L = \frac{D_G}{R} = PT = \text{Tan}\emptyset \pi D_W$
Lead Angle	$\tan\emptyset = \frac{L}{D_W \pi}$
Outside Diameter of Gear	$OD_G = D_T + 0.4775P$ for single and double threads $= D_T + 0.3183P$ for triple and quadruple threads
Outside Diameter of Worm	$OD_W = D_W + 2A$
Pitch	$P = \frac{L}{T} = \frac{(2C - D_W)}{N}$
Pitch Diameter of Gear	$D_G = 2C - D_W = \frac{NP}{\ }$
Pitch Diameter of Worm	$D_W = 2C - D_G$
Ratio	$R = \frac{N}{T}$
Rim Radius	$R_R = \frac{D_W}{2} + P$
Throat Diameter	$D_T = D_G + 2A$
Throat Radius	$R_T = \frac{D_W}{2} - A$
Whole Depth	$h_t = 0.686P$ for single and double threads $= 0.623P$ for triple and quadruple threads

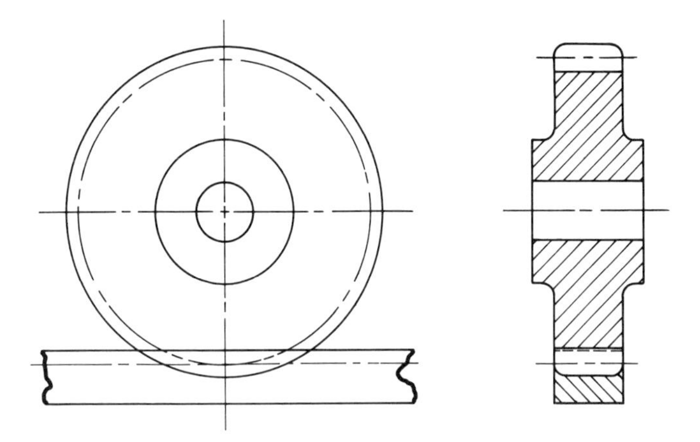

Fig. 13-24. Worm gear conventional graphic representation.

The working drawings used for worm gearing are the same as for other gears. Normally, a single section view drawing is used, with a second view provided when necessary. Gear dimension data will be presented in either a table or a note. See Fig. 13-24.

Kinematic Diagrams

Like other machine elements and mechanisms, gearing is also used in kinematic diagrams. In addition to gearing, close associations are made with other friction mechanisms, such as are found in friction wheels and transmission. To avoid confusion between the proper representations of gearing and friction mechanisms, graphic representations for each are provided in Fig. 13-25.

Bearings

Bearings are elements that are used to promote easy movement of machine parts. There are three general classifications of plane bearing, each defined by its function. The first is *journal bearings,* which are cylindrically shaped and are used to carry rotating shafts. *Thrust bearings* are used to encourage the free rotation of shafts, but prevent lengthwise motion. Finally, *guide bearings* are used to guide machine

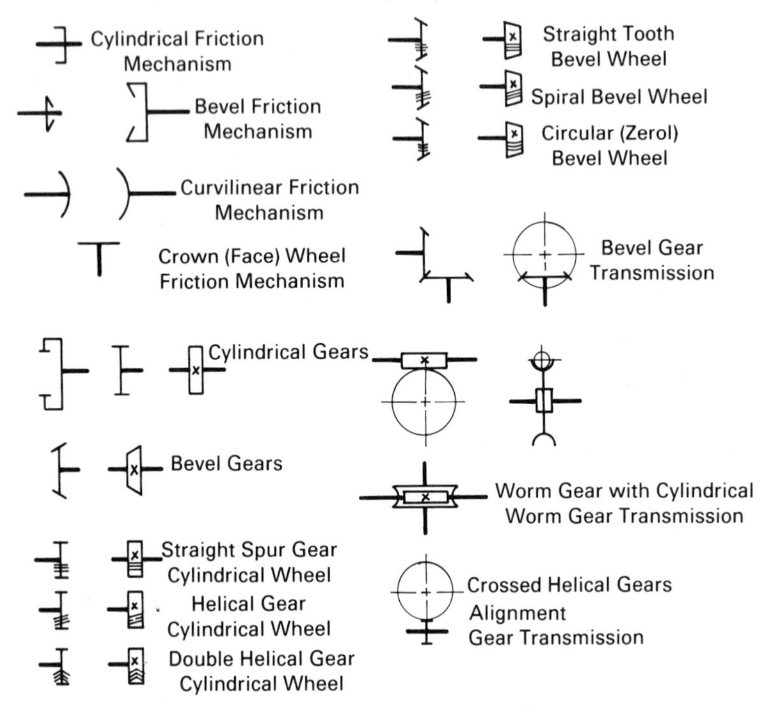

Fig. 13-25. Friction wheel and gear symbols for kinematic diagrams.

parts in a lengthwise motion, and are usually applied where rotation is undesired. These bearings are often referred to as *fluid-film bearings* because they require the use of liquid, semifluid, or mixed lubrication.

One of the most common forms of bearing employs rolling contacts. These machine elements are designed to support and locate rotating shafts and parts between rotating and stationary members. Generally, these types of bearings are divided into two major design formats: ball and roller bearings. Examples of each are shown in Fig. 13-26, and are described as follows:

BALL BEARINGS.

1. *Single-row radial* bearings are also known as *deep groove* or *conrad bearings*, and are primarily used for radial and moderate thrust loads.

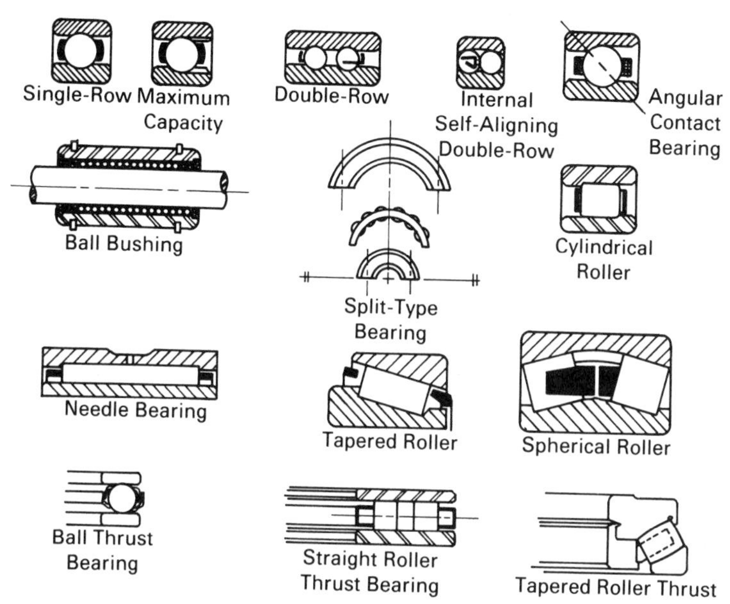

Fig. 13-26. Common types of bearing.

2. *Maximum capacity* bearings are used for carrying heavier loads than the single row radial bearing.
3. *Double-row* bearings are used to provide for heavy radial and light thrust loads, while not increasing the outside diameter of the bearing.
4. *Internal self-aligned double-row* bearings are designed primarily for radial loads that require self-alignment between 0.003 and 0.005 inch.
5. *Angular contact* bearings are designed to support combined radial and thrust loads.
6. *Ball bushings* are a type of bearing used to promote linear motions for hardened shafts.
7. *Split-type ball bearings* have a split inner ring, outer ring, and cage that are useful for assemblies where solid bearings are difficult to install.

ROLLER BEARINGS.

1. *Cylindrical roller* bearings are used for free axial movement of shafts with high speed limits.

2. *Needle bearings* are used where space is at a premium and rotational motion is desired.
3. *Tapered roller* bearings are used for heavy radial and thrust loads.
4. *Spherical roller* bearings are used for heavy radial loads and moderate thrust.

THRUST BEARINGS.

1. *Ball thrust bearings* are employed for low-speed situations where other bearings are used to carry radial loads.
2. *Straight roller thrust bearings* are designed for moderate speeds and loads to minimize skidding.
3. *Tapered roller thrust* bearings eliminate all skidding actions.

Exercises

13.1 Draw a cam displacement diagram and cam profile for a cam that has a follower stroke of 1.5 inches and a pitch diameter of six inches with the following specifications: 0°–90° rise interval of 1.5 inches with parabolic motion, 90°–180° dwell interval, 180°–250° fall interval of 0.75 inch with constant velocity motion, and a final 250°–360° fall interval with simple harmonic motion.

13.2 Develop a displacement diagram requiring an equal rise and fall of five inches, with cycloidal motion where one complete rotation of the cam will equal twenty seconds.

13.3 Calculate the following dimensional requirements for a standard spur with twenty eight teeth, having a pitch of 0.250 inch and a circular pitch of four inches: outside diameter, diametral pitch, center distance, pitch diameter, addendum and dedendum, and whole depth.

13.4 Given a bevel gear with a diametral pitch of 1.5 inches with twelve pinion teeth and twenty four gear teeth, with a face width of 0.125 inch, a pressure angle of 20°, and a perpendicular shaft angle, calculate the following for the pinion and gear: addendum, dedendum, working depth, circular pitch clearance, chordal thickness, circular thickness, outside diameter, and cone distance.

CHAPTER 14

Industrial Working Drawings

Industrial working drawings are prepared to provide sufficient information, specifications, and directions for the manufacturing, fabrication, construction, installation, and/or servicing of products. These drawings are generally grouped into three categories: layout, detailed, and assembly drawings. The drafter should not only be familiar with the characteristics of these drawings, but should also know when each type should be used, and the procedures required for its production. Presented in this chapter will be a description of layout, detailed, and assembly drawings, and the practices used in their presentation.

Layout Drawings

An important transition drawing, which is prepared between product design and detailing, is the layout. Original designs are usually expressed by engineers and designers as sketches. From these initial sketches, additional sketches and drawings are prepared for developing an idea in more detail. In some industries this second drawing process is referred to as *computation sketching*. Computation sketches are usually drawn by project engineers and designers for the purpose of working out major design problems. If layout drafters have had substantial training and experience, they may also be responsible for the further development of a design.

These sketches are then followed by further study and research into the kinematic and mechanical problems involved in a device's production, installation, and application. Hence, it is not uncommon first to draw kinematic diagrams, showing the function and movement of the device, to see if the design concept will function as anticipated. Involved here are such problems as:

1. The need for power sources, such as motors and drive units
2. The number and placement of moving parts (e.g., links, cranks, levers, etc.)
3. Types of motion required (e.g., oscillating, rotating, and linear)
4. Application of various principles of mechanics and thermodynamics
5. Selection of proper mechanisms, such as cams, gears, threaded devices, and friction wheels

From these preliminary designs and studies, a *design layout* is made. The layout is usually an instrument drawing that is prepared by the layout drafter or designer. The function of this drawing is accurately to show the full-sized representation of the design so that engineers can clearly view and interpret sizes, proportions, and interrelationships of parts. At this time, decisions are made pertaining to production processes and costs, as well as whether or not the design will function as desired.

As illustrated in Fig. 14-1, the layout drawing shows the product as it would appear fully assembled. Here one can see how parts fit together and their relative sizes and proportions. However, not all information and specifications (e.g., dimensions, parts lists, materials, etc.) are given, but only such information as is needed to make decisions relative to the design of the product. Further product specifications are worked out in detailed and assembly drawings.

Detail Drawings

After the layout drawing has been reviewed and approved, the next step is to prepare a complete set of working drawings. Drafters or detailers are now required to study the layout drawing and identify

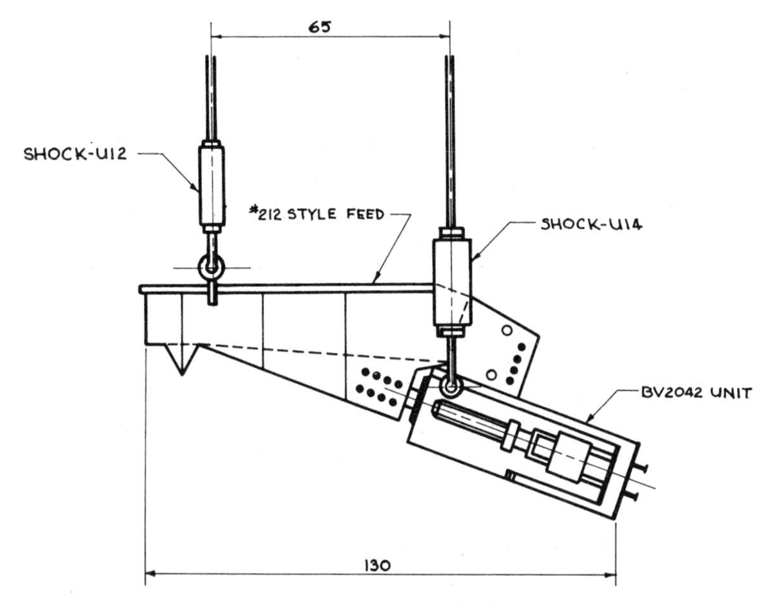

Fig. 14-1. Layout drawing for a vibrating feeder.

"details" or component parts for which detail drawings must be prepared. The detail drawing provides all the information needed for the making of a part. The information given in these drawings includes:

1. *Geometric shape.* This is a graphic description of how the object appears in the necessary orthographic and auxiliary views.
2. *Size.* Descriptions pertaining to dimensions, both shape and location, are then shown in dimensional or notation form.
3. *Specifications.* Here information is given that is used for the object's manufacture or fabrication, or for servicing. Examples of these data are type of material to be used, machining technique, heat treatment, surface finish, and the number required.

Depending upon company policy and the nature of the detail, part drawings may be prepared on individual sheets of drawing paper, or grouped together on a single sheet. This latter technique is usually employed when the parts are small enough to fit together on a single sheet of paper. As a general rule of thumb, when metal parts must be machined or fabricated, they will usually be drawn on separate sheets of paper.

In some cases, two or more drawings may be prepared for a single part. This technique is used when specific information is needed for a particular operation or process. For example, a part may have one drawing pertaining to a casting operation that is used by the pattern or die maker, and another for machining operations to be performed once the part is cast.

The first decision to be made by the drafter is the choice of parts to be detailed. This decision is based upon the concepts of design standardization and interchangeability. If the part is to be custom-made for a particular product, then a detailed drawing must be prepared. On the other hand, if the part is standardized, within either the company or the industry, then detailing is not required. Examples of parts that are not detailed are standard types and sizes of bolts, nuts, screws, keys and keyways, springs, bearings, bushings, die blocks, and punches. These parts are specified in notations in terms of name, type, size, and material. A typical detailed drawing is shown in Fig. 14-2.

Assembly and Subassembly Drawings

After the designated parts have been detailed, an assembly drawing is made to show how all parts are to fit or be assembled in the completed product. The function of assembly drawings is twofold. First,

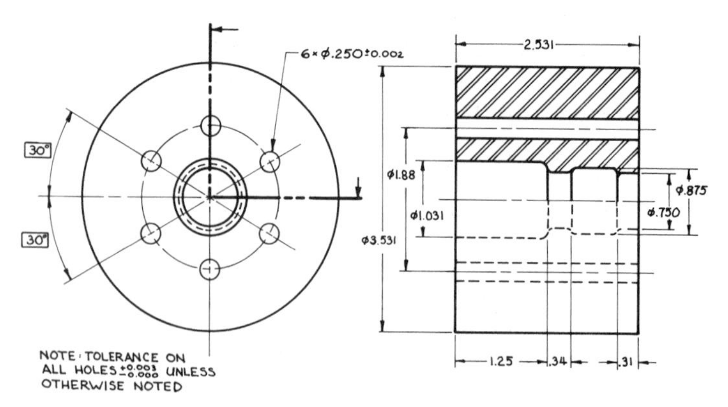

Fig. 14-2. Detail drawing for a circular forming tool blank.

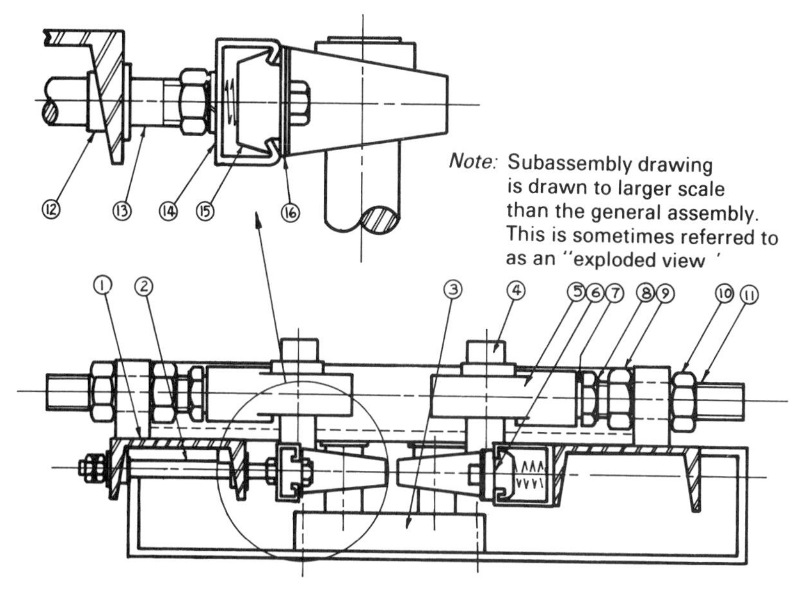

Fig. 14-3. Assembly and subassembly drawings for automatic nozzle wash unit.

they are used to show how the product is to be constructed and how its parts fit together. Second, they illustrate how the product functions or operates. The assembly, installation, and servicing of products are often accomplished after careful review of the assembly drawing.

A subassembly drawing is similar to the assembly drawing, except that it is used to show how a section or portion of the entire product is assembled. In other words, it is a portion of a general assembly drawing that further explains the assembly of a component. Subassemblies should be used when the general assembly drawing cannot provide the necessary clarity for product assembly and operational interpretation. Fig. 14-3 shows an assembly drawing and accompanying subassembly.

Drawing Procedures

The procedures used in preparing an assembly drawing are quite simple and straightforward. Assembly and subassembly drawings differ from detailed drawings in the way specifications are given and parts noted. As a general rule, detailed dimensioning should not be given. Instead, each part is referenced in tabular form.

Both national and international standards require that each part in the assembly be identified by name on the drawing. The name may be referenced to an identification number or letter and listed in tabular form, or may be identified directly on the drawing itself. These two techniques are shown in Fig. 14-4.

The general notation requirements for assembly and subassembly drawings are:

1. Part references should be given in sequential order to each item shown in the assembly and/or detailed part on the drawing.
2. All identical parts should carry the same item reference.
3. When a subassembly is used in the main assembly drawing, a single item reference can be made to it.
4. Every part that is referenced must be shown in a part list that presents the appropriate information on the parts referenced.

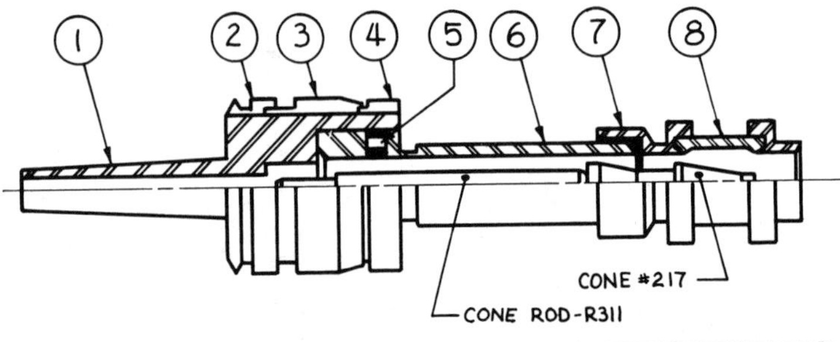

NO.	DESCRIPTION	QTY	SPEC.
1	DRIVER	1	SAE 2030
2	ADJ. LOCK	1	AL-20
3	ADJ. HEAD	1	AH-30
4	ADJ. BASE	1	AB-20
5	SPRING. LOCK	1	#1414
6	DRIVE SHAFT	1	SAE 2030
7	BODY	1	SAE 1045
8	EXPANDER	1	SAE 1080

FIX. HONING TOOL-340		
SCALE: FULL	DATE: 3-2-87	
DR:	CK.	E-1
APPR.	REF. 34087-1	

Fig. 14-4. **Assembly drawing using part identification numbers and part name call-outs.**

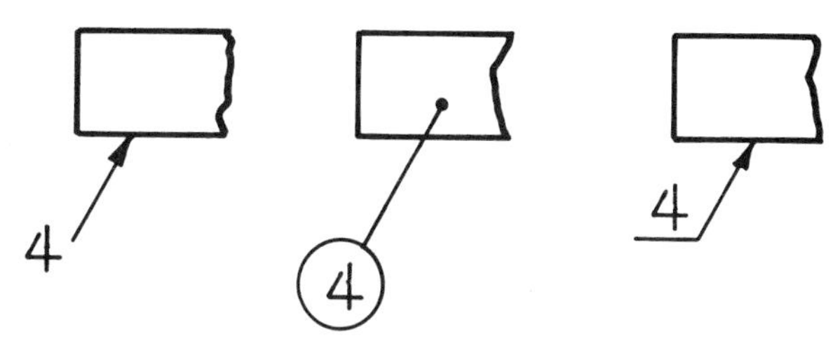

Fig. 14-5. Correct methods for referencing parts.

When referencing a part, the reference symbols should in general use arabic numerals only. In some cases, it may be necessary to augment this system by the use of capital letters. Part references should be placed outside the general outline of the assembly and connected by a leader line, as shown in Fig. 14-5. The leader lines should not intersect, and must be kept as short as possible.

When referencing parts, it is good practice to arrange the items in vertical columns and horizontal rows. Part references of related items, such as nut-bolt-washer assemblies may be indicated by the same leader line. When numbering parts, three recommended criteria are used for sequencing:

1. Number parts according to the order of which they will be assembled.
2. Number them in order of importance, such as subassemblies, major parts, and minor parts.
3. Number them in accordance with any other logical sequence that is pertinent to the product.

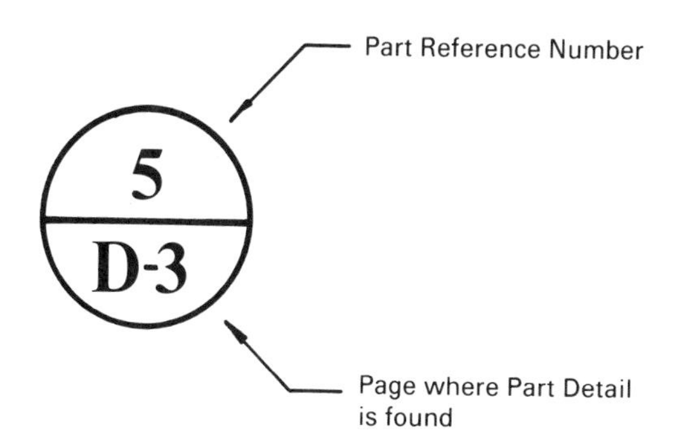

Fig. 14-6. Technique for identifying location of a part's detail.

As already mentioned, detailed drawings will be prepared for individual parts that are specific to the design of the product. When there are numerous details on a number of different drawings, additional information pertaining to the location of the details may be needed; the sheet or page number should then accompany the reference number. The technique usually employed is shown in Fig. 14-6. Here the numerator pertains to the part number, while the denominator identifies the page on which the detail can be found.

Presentations

In addition to the information already presented, the drawing conventions used in the preparation of assembly drawings should be considered. Presentation decisions consist basically of the following:

1. *View selection.* When selecting views for assembly drawings, the main consideration should be which view(s) will best show how the parts fit together. The views should also illustrate how the product will function and operate. The primary purpose of view selection, then, is based not upon describing shapes of individual parts, but their functional interrelationships.

   If the manufacturing engineer and/or machinist needs information about a given part that cannot be obtained from the assembly drawing, they should refer to the detail drawings. Since assembly drawings are designed to show relationships between parts, the number of views or partial views selected should be kept to a minimum. In some cases, this may mean drawing only one view.
2. *Sections.* Because assemblies show the fitting together of parts, the reliance on hidden lines is usually not acceptable. Therefore, sectioning is widely used to show how internal components fit. The drafter is encouraged to use any type of section that will enhance understanding of the drawing.
3. *Hidden lines.* Since sectioning is used extensively in assembly drawings, hidden lines should be kept to a minimum. Again, the incorporation of hidden lines in assembly drawings is left to the discretion of the drafter; they should be used whenever it helps to clarify the drawing.
4. *Dimensions.* Dimensioning is not recommended in assembly drawings. Dimensional specifications, where needed, can be obtained from the parts list and/or accompanying detailed

drawings. There are times, however, when there might be a need for dimensions, for instance, indicating the functional limits of the product (e.g., maximum height of a lift, or swing of a rotating head).

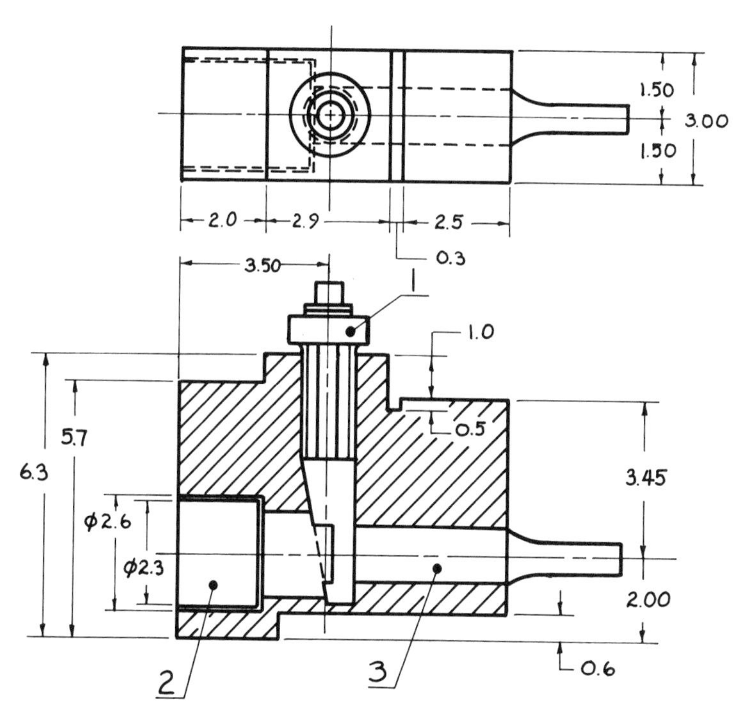

NOTES:
1. ALL DIMENSIONS IN mm.
2. ALL SIZE TOLERANCES:
+0.10, −0.20.
3. ALL LOCATION TOLERANCES:
+0.002, −0.003.

NO.	DESCRIPTION
1	C-10WEDGE HOLDER ADJ.
2	PUNCH SEAT #SC412
3	EXTRUSION PUNCH

PUNCH HOLDER

SCALE: FULL DWG. acm

Fig. 14-7. Working drawing assembly.

Alternative Assembly Drawings

In addition to the general assembly and subassembly drawing, other types of specialty assemblies are used in industry. In reality, these assembly drawings are combinations of assembly and other drawing techniques. The selection of an alternate type of assembly drawing should be based upon how the drawing is to be used. The three most common assembly alternatives are:

1. *Working drawing assemblies.* When details must be shown in combination with the assembly, then a working drawing assembly should be used. These drawings are used when separate detail drawings are not wanted. This type of drawing, however, can only be effective with simple devices that have relatively few parts. Its most obvious difference from a general assembly drawing is that it presents a complete set of dimensions and notations, along with part identification numbers. An example of this is shown in Fig. 14-7.
2. *Installation assemblies.* As their name implies, installation assembly drawings are specifically used to show the proper installation of a machine, structure, or device. Also known as

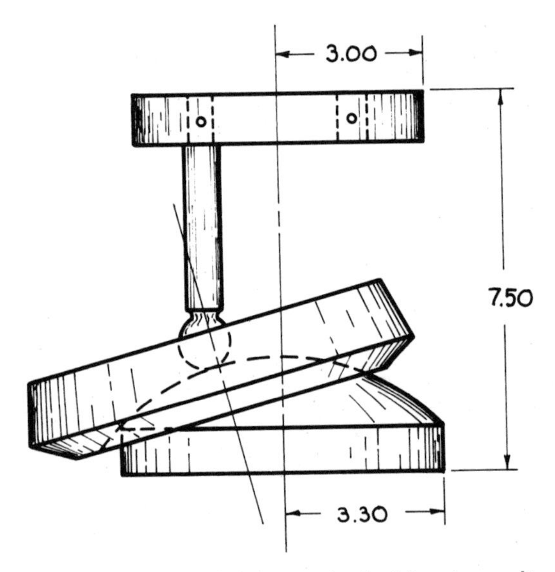

Fig. 14-8. Installation assembly for spherical lapping unit.

an *outline assembly*, this drawing shows the outline of the product and the dimensional features that are critical for installation. An example of an installation assembly is shown in Fig. 14-8.

3. *Check assemblies.* A check assembly is used during the design phase to ensure that all parts can fit together correctly. It is frequently used when a number of changes have been made in the design. Here drawing accuracy is critical to evaluate the details and their relationship in the assembly. Check assemblies are often converted into general assembly drawings.

Review Questions

14.1 Explain the function of detail and assembly drawings. What relationship do they have to one another?

14.2 Give several examples of situations in which a subassembly drawing might be used.

14.3 How should one reference a subassembly in a general assembly drawing?

14.4 How many views should be used in an assembly drawing?

14.5 Explain why dimensioning should not be included in assembly drawings. Are there exceptions to this? If so, explain.

CHAPTER 15

Jig, Fixture, and Die Drawings

- **Jigs and Fixtures**
- **Die Drawings**
- **Review Questions**

Working drawings, in particular assembly drawings, are extensively used in the machining industry. Machining is not limited here to the traditional metal machining, but also includes work with materials such as polymers, ceramics, and woods. The drawing techniques used in jig, fixture, and die design are similar in appearance, but differ in concept and application. This chapter discusses jig and fixture and die drawing practices.

Jigs and Fixtures

An underlying requirement of mass production is the interchangeability of parts. For parts to be interchangeable requires that they be sized and machined to identical specifications. To accomplish this, devices known as jigs and fixtures are used to hold or locate the workpiece securely and guide the cutting tools during processing. Through proper design, jigs and fixtures can increase machining accuracy, reduce production time, and lower product costs.

Jigs

A jig is any device that is used to hold a workpiece and locate the path of the cutting tool. They are usually designed so that they can be easily moved about a particular machine. For example, a milling jig may be positioned in several locations for each hole that is milled. Here the cutting tool is located by a bushing that is positioned within the jig.

Jigs are primarily used in such production operations as drilling, tapping, boring, and reaming. Proper jig design must not only allow for holding the part, but must also help guide the cutting tool. The size of the jig will depend on the measurements of the machine and the operational characteristics of the jig itself. To a lesser degree, jigs are also used for assembly operations. When used for assembly, jigs locate the separate components of the product and hold them in their proper position while they are being fastened. Assembly jigs are larger than machining jigs.

Regardless of the jig's application, there are several factors in its design that are constant throughout the industry. These are as follows:

1. The nature of the machining operation (e.g., drilling, boring, tapping, or reaming)
2. The total number of parts to be machined
3. The level of accuracy required for each part
4. The stage of product completeness when machining occurs

In addition to these four factors, jig and tool designers must also concern themselves with production questions such as:

1. The size, type, and operating characteristics of the machine on which the jig is to be used
2. How the part is to be mounted and unloaded before and after machining
3. Clamping time requirements (i.e., time allowed for clamping between operations)
4. Use of cutting fluids and other coolants
5. Methods used to remove chips
6. Quality control measures
7. Requirements for observing the machining operation during processing
8. Safety requirements for machine operators and other personnel in the area

TYPES OF JIG

There is no one way of designing jigs. They often differ significantly in appearance according to the shape and design of the workpiece,

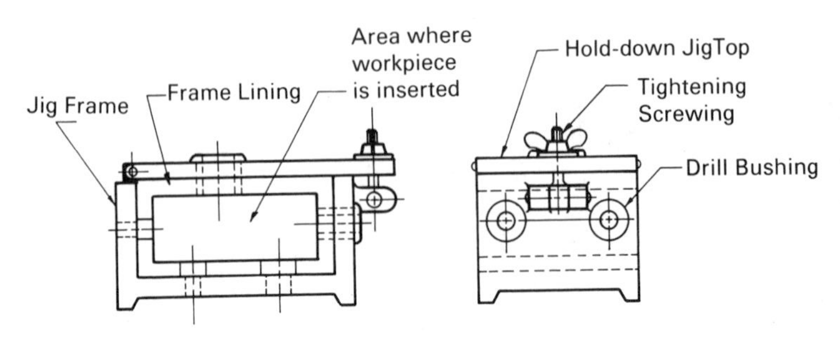

Fig. 15-1. Box-type jig.

even though the same machining operations are executed on each one. There are several classifications of jig based upon general appearance and construction. Two of the most common are *box-type* and *table-type jigs.*

As shown in Fig. 15-1, box-type jigs are used for drilling two or more sides of a part. In this example, the work is placed inside a frame that is box-shaped, and held in place by the action of a tightening screw. Adjusting screws or some other similar device are used to ensure accurate placement of the part.

Table-type jigs (Fig. 15-2) are primarily used for drilling holes into one side of a part. This jig is a simpler design, and is made to be mounted onto the machine's table or machining surface. Other types of jig used for production purposes are templet, closed, open, indexing, diameter, and universal jigs.

JIG COMPONENTS

A number of different devices can be used in the design and construction of jigs. In fact, it is possible to use almost any component that will meet manufacturing requirements. To minimize manufacturing costs, tool and manufacturing engineers will attempt to use standard jig components whenever possible, but it may sometimes be necessary to design and make a specialized component.

The most basic jig component is known as the *jig body*. Jig bodies are employed to hold the workpiece. The primary requirement here is that the body be rigidly constructed to withstand the stress and strains of machining, and still be light enough for easy handling. Jig bodies are available in standard shapes and sizes, which if possible

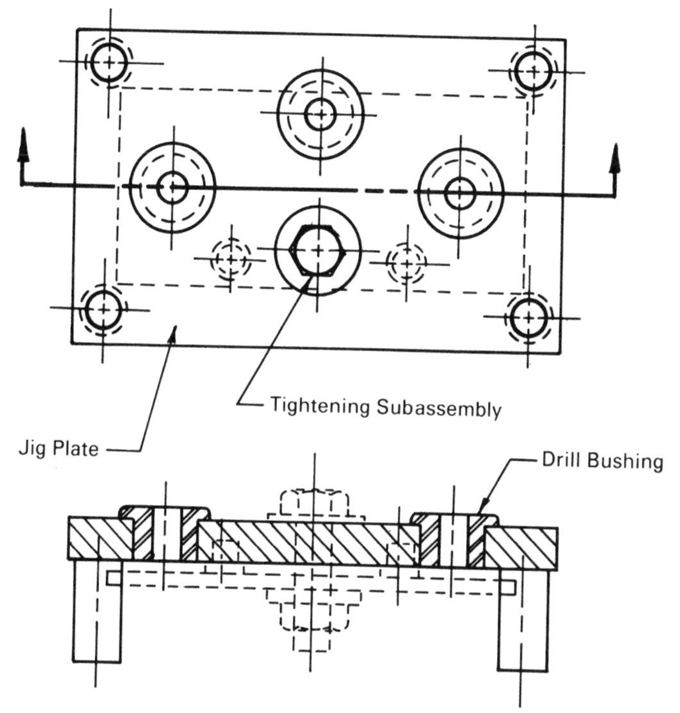

Fig. 15-2. Table-type jig.

should be preferred to customed designed and fabricated bodies. The most basic part of the jig body is the *jig plate,* which can function as the jig itself. These plates come in standard thicknesses of 5/16, 3/8, 1/2, 1, 1 3/8, 1 3/4, 2 1/8, 2 1/2, and 3 inches.

Components used to hold a product's parts together are *cap screws* and *dowel pins.* Shown in Fig. 15-3, dowel pins make possible accurate alignment of the parts, while the cap screw is used for tightening. In some cases, the cap screw is also used for alignment. In situations that allow for greater flexibility in tolerances, flat-head machine screws can be used. Tapered dowel pins are used more often than straight dowels. In both cases, press fits are employed to maximize alignment accuracy.

An important component in jigs is the *drill bushing.* Drill bushings are made out of a high-grade tool steel, and are used for locating holes within extremely close tolerances and for guiding the

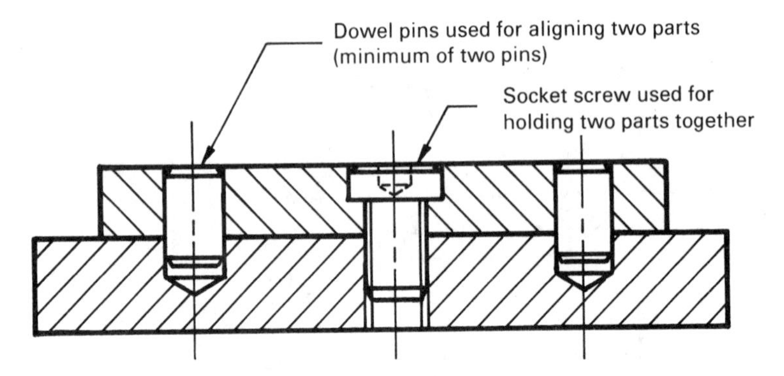

Fig. 15-3. Use of cap screws and dowel pins in jig construction.

cutting tool to the workpiece. *Liners* are used in combination with bushings to prevent wear of the jig plate and provide for precision mounting of the bushing itself. Fig. 15-4 shows examples of liners and bushings.

Generally, there are four types of jig bushings:

1. *Renewable bushings* are employed in jigs where the bushing will wear out, break, or become obsolete before the jig itself. They can also be used where two or more bushings are interchanged for one hole (e.g., drilling and reaming). *Fixed* renewable bushings are installed in liners, and left until there

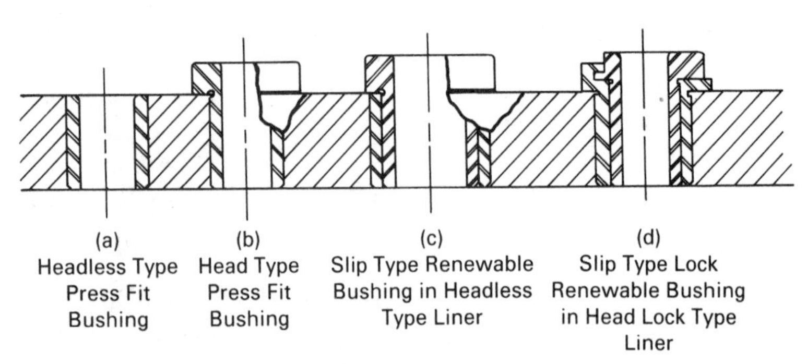

Fig. 15-4. Bushing and liners.

is significant wear. *Slip* renewable bushings are interchangeable with all standard size of lines, and are often made with a knurled head for easy removal.

2. *Press fit bushings* are installed in liners and are employed for short production runs so as not to require replacement or servicing. They are also used for closely spaced holes.
3. *Liner bushings*, also known as *master bushings*, are permanently mounted in the jig without a liner, and are available with and without heads.
4. *Embedment bushings* are designed to "grip" onto soft metals. They are primarily used in jigs that are made out of materials such as fiberglass, castable metals, and potting compounds.

A fourth component used in jigs includes several types of *locating device*. These devices are used to locate the workpiece accurately in or on the jig itself. There are five classifications of locating devices; examples of each are shown in Fig. 15-5, and described as follows:

1. *Internal locating devices* are located within the jig body, and may be machined into the jig body itself, or attached with pins, set screws, bushings, and machine screws.
2. *External locating devices* can be the same type of device as is used for internal location, except that they are mounted to the external surfaces of the jig.
3. *Stops* can be either fixed or adjustable, and are used as a surface against which the work is fitted. They are used when recessed or projected devices cannot be employed.
4. *Centralizers* are used with spherical, circular, cylindrical, or other curved surface parts. Common types of centralizers are V-blocks and dowel pins.
5. *Part supports* provide a surface or base upon which the workpiece will set or rest. Frequently used here are hardened tool steel *rest buttons*, which offer exceptional resistance to wear.

Another type of jig component employed for manufacturing includes various *clamping devices*. They are used to hold the workpiece securely in place while the machining operation is executed. Most clamping devices are threaded, but when speed is critical, cam-action and spring-loaded clamps prove most useful. See Fig. 15-6.

The last jig component to be discussed is the *locking pin*, which is designed to be inserted into an existing hole in the workpiece. They

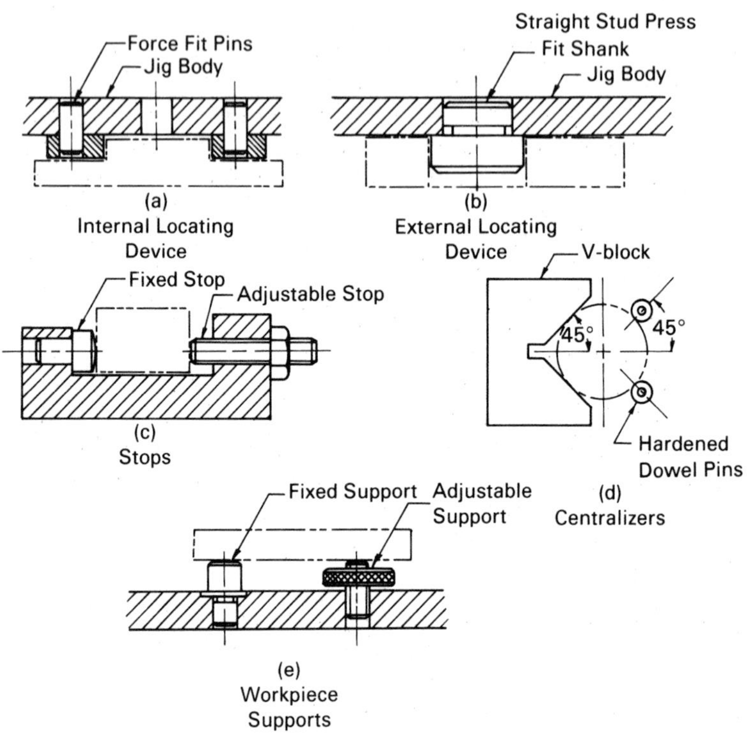

Fig. 15-5. Types of locating device.

are used for multiple drilling and machining operations. Unlike other clamping devices, locking pins are usually held in place by cam-action and spring-loaded clamps.

Fixtures

The primary functions of a fixture are the support, location, and fastening of a workpiece during machining operations. The function of fixtures is to hold and support the workpiece, and unlike jigs, they do not guide the tool. Properly designed fixtures will reduce machining costs, increase production, and stabilize complex-shaped and heavy parts.

Fixtures are most often used on lathes, turret lathes, milling machines, boring equipment, shapers, and planers. They are usually

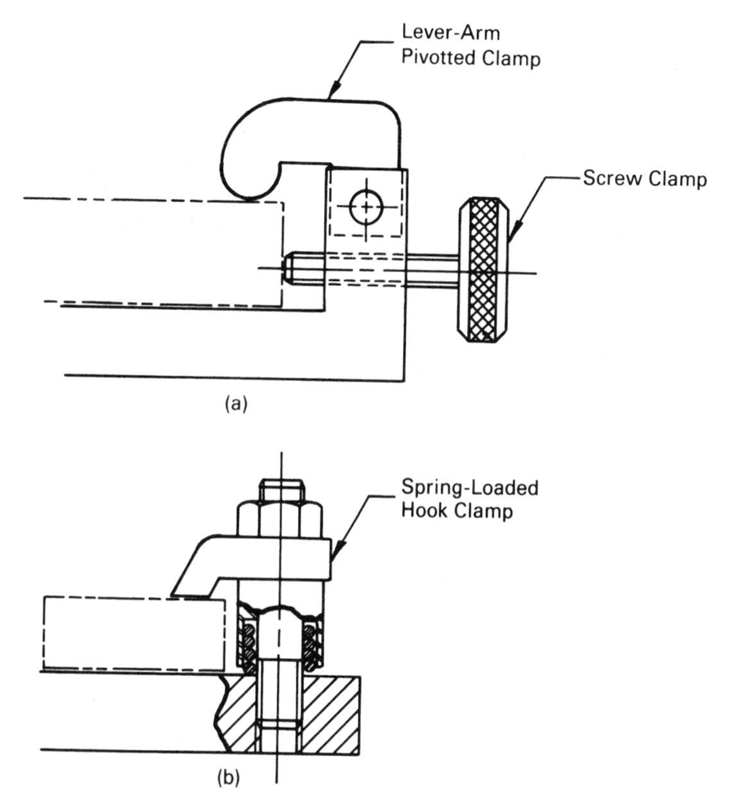

Fig. 15-6. Clamping devices.

made from gray cast iron or steel plate, and can be "fixed" to machines by bolting or "set" with low-melting alloys via welding. High-production fixtures will often be extremely large because they need to withstand heavy dynamic forces of the machining operation.

All fixtures will include locating pins and/or machined blocks against which the workpiece is fitted. Fastening is then accomplished by using either a clamping device or bolting. The major components of fixtures consists of:

1. *Fixture bases* are used to provide the clamping lugs or slots to which the workpiece and other components are attached. The size availability of the lug openings will correspond to the T-slot with openings found on milling machine tables,

which can receive bolt sizes of 1/4, 5/16, 3/8, 1/2, 5/8, 3/4, and 1 inch. Fixture bases are available in a variety of standard sizes.

2. *Clamps* used for fixtures are of the same nature as jigs. The primary difference is that fixture clamps must be of heavier design to withstand machining stresses.
3. *Set blocks* are used for proper positioning of cutting tools (e.g., milling cutters) in relation to the workpiece.

Jig and Fixture Drawings

The preparation of jig and fixture drawings incorporates the use of assembly drawing concepts. The reason for this is that most, if not all, components are made up of standard-sized devices; their specification will thus be found in a parts list. The procedures used for drawing jigs and fixtures are the same. In some cases, a fully dimensioned detailed drawing of the workpiece will also be provided with the jig and fixture assembly. The use of section views is often helpful as well.

The procedures used to prepare the jig and fixture drawings in Figs. 15-7 and 15-8 are as follows:

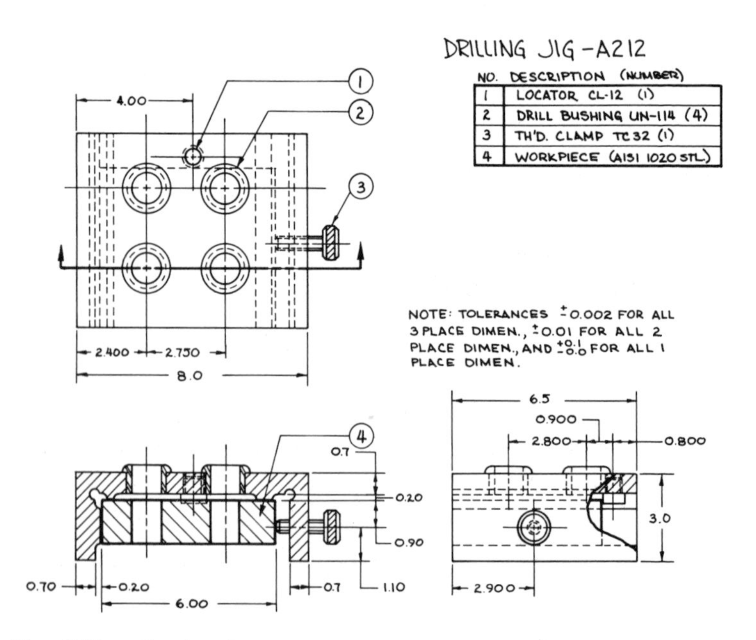

Fig. 15-7. Jig drawing.

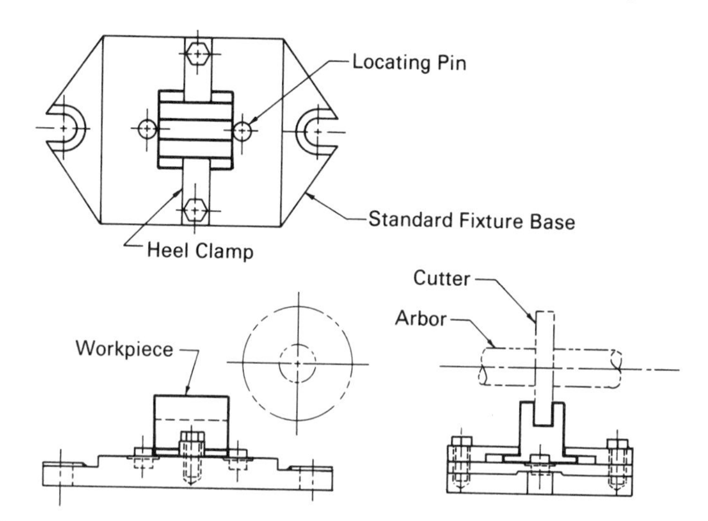

Fig. 15-8. Fixture drawing designed for simple milling operation.

1. Lay out and draw the required number of views necessary to illustrate the *workpiece*. Many companies require that the workpiece be drawn in with extra-thick object lines. Some use some form of colored pencil or ink for this purpose (note that if a colored pencil or ink is used, be sure that it will appear as a black object line when reproduced as a print).
2. Locate and draw in all locating devices used for workpiece placement.
3. For *jig drawings*, locate and draw in all devices used for guiding the cutting tool (e.g., bushings). For *fixture drawings*, locate and draw with phantom lining, cutters, arbors, shafts, and machine parts directly related to the operation.
4. Locate and draw in all clamping and fastening devices.
5. For *fixture drawings*, locate and draw in set blocks, if required.
6. For *jig drawings*, lay out and draw in jig plate and body. For *fixture drawings*, lay out and draw the fixture base.
7. Identify all parts, and specify all dimensions that are not given in the parts list.

Die Drawings

The extensive use of sheet and pliable materials has made the use of dies critical to industrial processes. Die-forming operations have made die design a critical function of tool and manufacturing engineering, and technicians, technologists, and drafters are required to interpret and prepare die drawings.

Before proceeding with a discussion of die drawings, it is first necessary to understand the basic terms used in the industry. These basic terms are as follows:

1. *Die.* The term "die" has two meanings in the engineering field. First, it is a complete operational tool that is employed to make piece parts to within given specifications. Second, it is the female or cavity part of the die assembly. Dies can be simple punch dies that make flat products, forming dies that generate contoured products, or a combination of the two.
2. *Piece part.* The result or product that is produced by a die is known as the piece part. It can be either a finished product or a part of a product; hence, dies are used to produce both finished and unfinished piece parts.
3. *Stock material.* Any material that is used to make the piece part is known as the stock material.
4. *Punch.* The male segment of the die is known as the punch. It functions in unison with the die to produce the piece part.

Presswork

Dies are used in combination with power presses to work the material. Generally, presswork is divided into two broad categories: cold and hot working. As their name implies, cold-working presses work the material without the addition of heat, while hot-working presses raise the temperature of the material to make it more pliable.

COLD WORKING

Cold-working press dies can perform cutting and noncutting operations. All die-cutting operations are a result of applied stresses that produce a fracture in the material beyond its ultimate strength. The major classifications of die-cutting operation are described as follows:

1. *Blanking and punching* are the most basic types of die-cutting operation. Blanking is the process whereby the stock material is cut around the perimeter of the shape to produce a piece part blank. Punching, by comparison, produces slugs from the stock and makes holes in the piece part.

2. *Perforating* is the process whereby a pattern of holes is punched in continuous or patterned rows.
3. *Slotting* is a punching process that produces rectangular or elongated holes.
4. *Piercing* is sometimes used synonymously with punching operations. Technically, the term refers to a specific hole-generating process in which no scrap (i.e., no slug) is produced from the hole.
5. *Notching* removes material from the edge of a strip to allow for further processing, provide for relief, or produce partial cuts on contour blanks too complex for blanking operations.
6. *Seminotching* is used only to remove or cut material for further processing.
7. *Lancing* includes all operations that cut a single line across a part of the material strip or coil.
8. *Trimming* is the removal of excess material.
9. *Slugging* is a punching operation that produces a slug. In some products, slugging is used to make reference points for progressive dies.
10. *Parting* is used to separate or "part" blanks by cutting the material from the strip.

As with cutting dies, a number of different noncutting dies are used for forming contours and providing for material bending. The basic operations performed by these dies are described as follows:

1. *Bending and flanging* are the two most basic forming operations that produce shapes by stressing the material beyond its yield strength. Bending can be applied to most forming processes, while flanging is used where a bent edge or "flange" is produced. The two types of formed flange generated are stretch and shrink flanges.
2. *Embossing* is the producing of a shallow design in the piece part without modifying the thickness of the material. Here the piece part will have the same design on both parts, with one side raised and the other depressed.
3. *Coining* is related to embossing, except that it is used to produce different designs on each side of the material, and is considered to be the most drastic form of metal-squeezing operation.
4. *Drawing* pertains to those die-forming operations in which flat blanks are formed into shapes such as cups and cylinders.

5. *Bulging* includes all pressworking used to expand a preformed tubular or cylindrical blank.
6. *Sizing* is similar to coining, except that it is performed in an opened face die and is primarily used to finish a product by sharpening its edge or corners and flattening material around pierced holes.
7. *Swaging* is a compression operation that is also related to coining. The two terms are sometimes used synonymously; however, in swaging the material flow is not confined, and can produce flashing along the edge of the piece part.

HOT WORKING

Though the majority of die drawings are prepared for cold working processes, hot working still plays a significant part in the manufacturing of products. Hot working pertains to those processes whereby material is plastically deformed by working it above the recrystallization temperature. The work of these materials often require the use of dies.

The most common forms of hot working that employ dies are described as follows:

1. *Hot rolling* employs rolling dies to make continuously formed shapes. This process is frequently used for shaping solid metal into useable forms such as plates, sheets, bars, and structural forms. In many cases, hot rolling is followed by cold rolling to form the metal into more complex shapes and incorporate desired properties into its internal metallic structure.
2. *Forging* is a controlled process where metals are formed into desired shapes. In all, there are four basic types of forging operation: open-die, blocker-type impression, conventional-type impression, and precision-type impression forgings.
3. *Extrusion* is a process that forces heated material through die openings to form continuous and highly controlled shaped products. This process is used in a wide variety of products, including structural members, aircraft parts, electrical conduits, evaporator tubing, and sporting goods.
4. *Swaging* is a metalworking process in which the cross-sectional portions of the workpiece are reduced to the desired shape by repeated blows of radially acting hammers. These hammers, in effect, are the forming dies.

5. *Creep forming* is primarily used in the aerospace industry to produce large, thin, and shallow contoured parts. This forming technique deforms the metal below its yield point with the application of heat for a period of time that permits metallurgical creep to occur. The relaxing of the metal's elastic stress then makes it possible to deform the material into a die.

Dies

Numerous die designs are found in industry, and their complete description is beyond the scope of this book. It is, however, essential to understand basic die components and classifications. This section will present an overview of both topics.

DIE COMPONENTS

There are a number of different die components that make up a complete die assembly. The first is known as the *die set* and is shown in Fig. 15-9. This component is a subassembly or subpress unit that is made up of seven parts: 1) *die shoe*, also referred to as the die holder; 2) *guidepost*, sometimes termed the guide pin or leader pin; 3) *guidepost bushing;* 4) *punch shoe*, also known as the punch holder; 5) *shank;* 6) *flange;* and 7) *bolt slot.* The function of the die set is to hold the die and provide power to it from the press. Die sets are available in standard sizes.

The second component is the *punch* or male segment of a die. Punches are available in standard shapes and sizes, and either *segregated* or *integrated.* The first type are self-mounting in that they have their own mounting screws and dowels, while the latter depend on some other component for mounting. The punch is then retained and/or positioned in a *punch plate.* An example of a punch subassembly is shown in Fig. 15-10.

The next components employed in die assemblies are *mounting screws and dowels.* These elements are used to hold the die securely together. Most designers use both screws and dowels so as to enable the diemaker to construct a stable tool. Rarely would one use either doweling or screwing devices alone.

The *die block* component is a generic term used to describe the complete subassembly that mates the male and female portions of the die, so as to produce the piece part. The construction of the die block

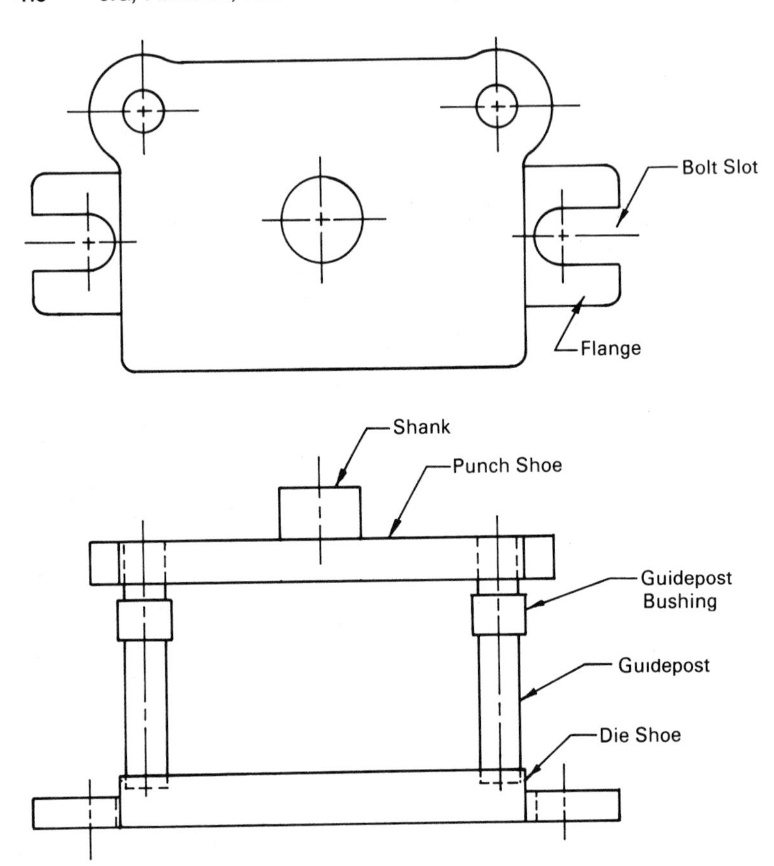

Fig. 15-9. Basic die set components.

may consist of either a single piece or an assembly of numerous pieces. Sometimes the term "sectional die" is used; in most cases, it can refer to either type of die block construction. Die blocks are usually made out of a tool steel, the exact type depending upon the material being cut or formed, and whether the process is cold or hot. An example of a die block detail is shown in Fig. 15-11.

The purpose of die *pilots* is to position the stock accurately for working. When located in its position by pilots, it is said to be *registered*. An example of pilot registry is shown in Fig. 15-12.

The next component is the *stripper*. Strippers are used to remove the metal from the punch after the hole has been made. During punch-

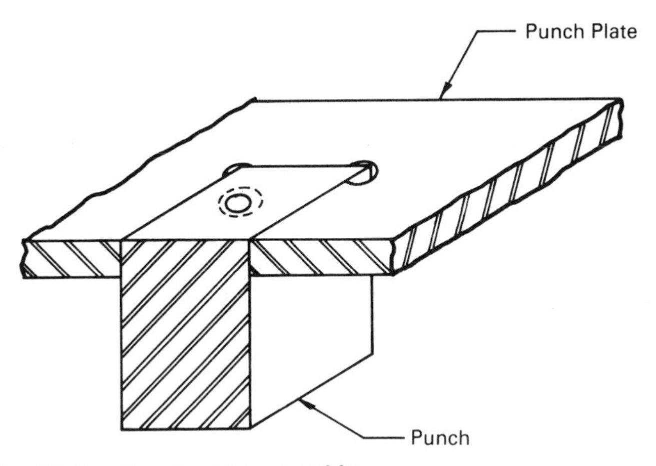

Fig. 15-10. Punch plate assembly.

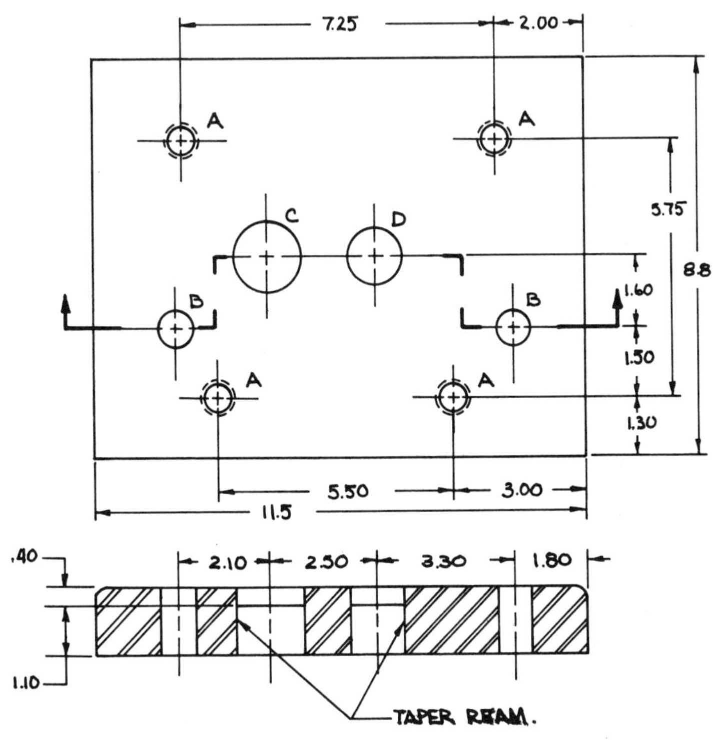

Fig. 15-11. Detail for die block.

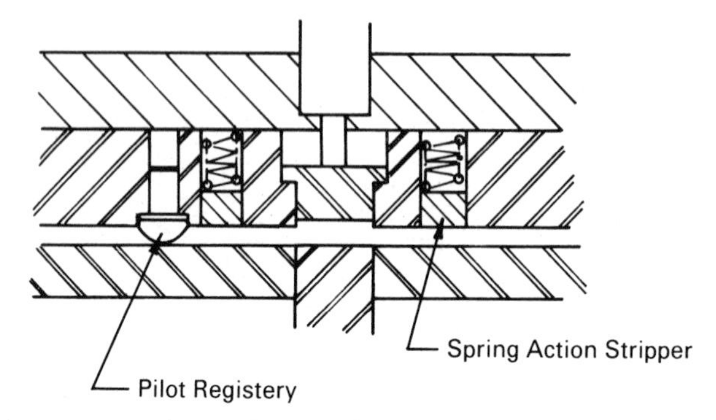

Fig. 15-12. Registry and stripping devices.

ing and forming, it is not uncommon for the material to hold onto the male portion of the die. To prevent this, and enable processing to continue, strippers are built into the die assembly. The exact type of stripper used can vary from simple stationary units to pneumatic devices. See Fig. 15-12.

A component that is used to stop the movement of the stock strip is known as the *die stop.* In reference to the die stop, designers must note the *stop* and *registry* positions. The first is the actual stopping point or area against which the stock strip material is arrested, while the latter pertains to the exact position that the stock strip must be in to maintain dimensional tolerance. Stops may be blocks, ponts, or any other type of device.

The last component to be discussed is the *nest gauge.* These are usually employed in dies that perform more than one operation. The function of nest gauges is to locate properly the position of the workpiece so that further processing can be executed. In die design it is not uncommon to use several pins, dowels, or blocks for this purpose.

CLASSIFICATION OF DIES

Dies are often classified according to the type of operation that they perform. For example, cutting operations are performed on cutting dies, drawing on drawing dies, and so on. Complex die designs, such as combination and progressive, execute more than one operation. Combination dies perform both cutting and noncutting operations, while progressive dies take the stock material through several opera-

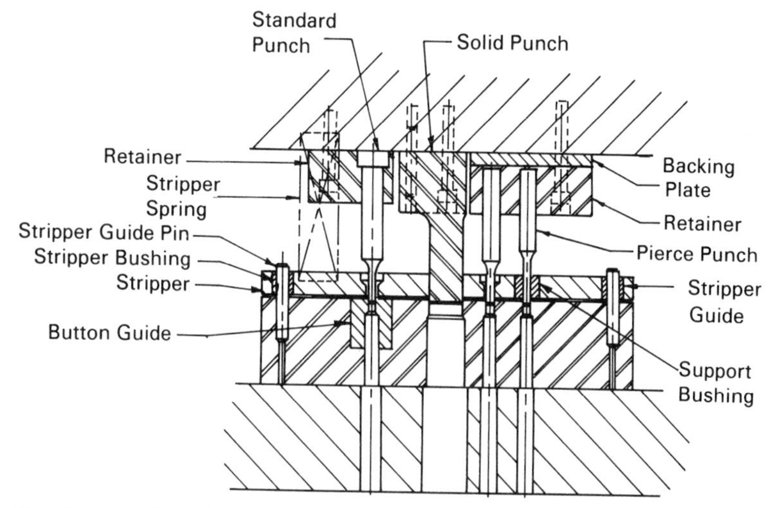

Fig. 15-13. Punch die.

tions that usually result in a finished or semifinished product.

The most basic type of die used is the punching die. Shown in Fig. 15-13 is a simple punching die that illustrates the elements necessary to the design of such dies.

Drawing dies require more complex design considerations. Shown in Fig. 15-14 is an example of a single-action drawing die that incorporated an air-cushion device for holding the blank. This design is typically used for shallow draws used for pans and deep draws required for washer and drier tubs. By comparison, double-action drawing dies are able to execute two drawing actions in sequence.

In some industries *multiple-station dies* are used to describe combination, progressive, and transfer operation dies. Critical in the design of these dies is the effective utilization of pilots, guides, and nest gauges. In these types of die, the sheet material is taken from a continuous coil of stock. The piece part remains attached to this strip until the final operation, at which time it is cut off. An example of this is shown in Fig. 15-15.

Preparing Die Drawings

The procedures used to prepare die drawings are similar to those used for jig and fixture drawings. Since most die assembly parts are stan-

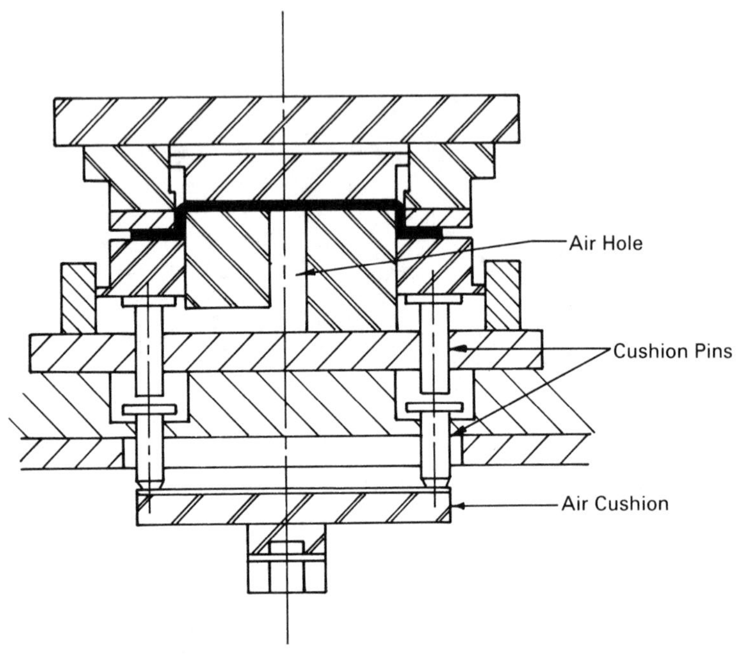

Fig. 15-14. Single-action drawing die with air cushion device.

dard, there is infrequent need for part detailing; hence the extensive use of assembly drawings.

There is one significant difference in view arrangements for die assemblies: the placement of the bottom view. In orthographic projection, the bottom view is placed beneath the frontal view. In the

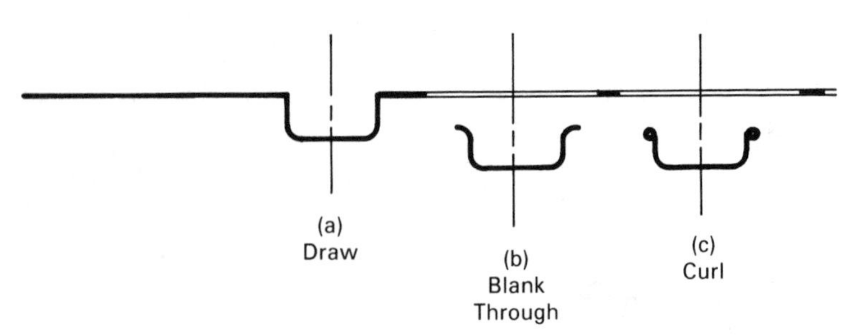

Fig. 15-15. Material stages in simple progressive die.

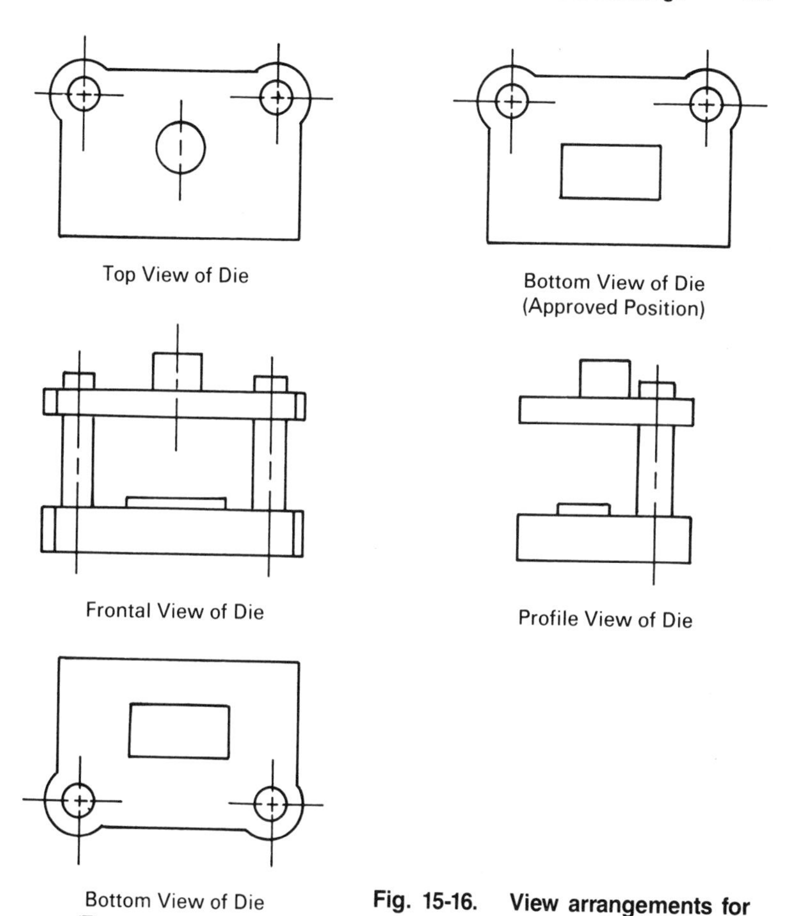

Fig. 15-16. View arrangements for die drawing.

field of die design, an acceptable and frequent practice is to place the bottom view over the profile view. These two are compared in Fig. 15-16.

The reason for drawing bottom views is their importance in the construction of the die assembly. Many die assemblies will incorporate the following four views: frontal, horizontal, profile, and bottom. The specific selection of views, however, is left to the discretion of the drafter and designer. Drawings should be as simple as possible, while presenting all the information necessary for die construction.

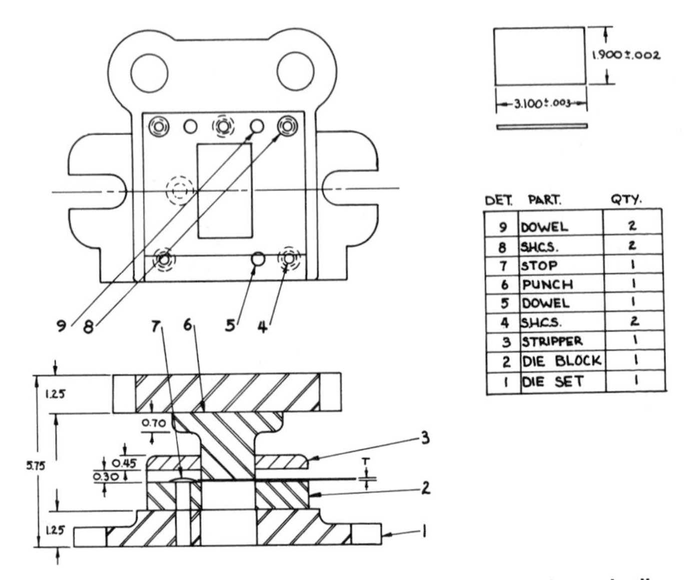

DET.	PART.	QTY.
9	DOWEL	2
8	SHCS.	2
7	STOP	1
6	PUNCH	1
5	DOWEL	1
4	SHCS.	2
3	STRIPPER	1
2	DIE BLOCK	1
1	DIE SET	1

Fig. 15-17. Die assembly drawing for simple blank-through die.

Fig. 15-17 shows a typical die assembly drawing. A common practice is the drawing of the workpiece (again, colored pencils or inking may be used as long as they reproduce well on prints). Depending on the nature of the die, the workpiece may be shown as either a part piece or a stock material strip. If they are important to the function of the die, the width and thickness of the workpiece may be dimensioned. In most cases, the drawing will include a detail of the piece part itself. When progressive dies are used, the progression of the stock material strip is sometimes illustrated between the frontal and horizontal views.

Owing to the standardization of die assembly parts, details are kept to a minimum. In most cases, the only details provided are for the die block (Fig. 15-11), which is used to form or cut the stock material into the desired shape. Since it is not possible to anticipate all product designs, die blocks must be machined into finished form—hence the need for detailing.

Like other assembly drawings, dimensioning is not usually shown on the drawing, since size specifications can be obtained from the parts list. There are times, however, when assembly dimensions will be necessary. An example of typically dimensioned features are stock material width and thicknesses, punch and material clearances, gauge locations, and material advancement distances.

Review Questions

15.1 Briefly explain the differences between a jig and fixture.

15.2 Identify the major components and functions of jigs and fixtures.

15.3 Describe the essential characteristics of jig and fixture drawings.

15.4 What is the difference between hot and cold presswork? Identify and describe common operations used in each.

15.5 Briefly describe the features of the major die types.

15.6 Describe the characteristics of a well-prepared die drawing assembly.

15.7 What are the requirements for detailed drawings in jigs and fixtures and die drawings?

CHAPTER 16

Developments

- **Sheet Material Developments**
- **Parallel-Line Developments**
- **Radial-Line Developments**
- **Triangulation Development**
- **Exercises**

The drawing of *surface developments* has traditionally been associated with the sheet metal trade. This has changed in recent years, with advances in sheet materials and products, and the principles employed in development drawings are now applicable to plastics, fabrics, and wood veneer, as well as sheet metals.

Developments are still widely used in sheet metal work, pattern making, and related industries. The term "development" refers to a process whereby a surface is laid out on a flat plane, so as to produce a *stretch-out* or pattern. This chapter will discuss the basic surface development procedures used in industrial drafting.

Sheet Material Developments

The development of patterns is one of the more important types of drawing procedure used with sheet materials. A development is a drafting term that refers to the actual layout of a pattern. This pattern can be either drawn on the material itself, traced, or transferred from the master drawing. Patterns are used to produce sheet products such as boxes, cartons, duct systems, and chassis.

Not all surfaces can be developed precisely. Flat and single-curved surfaces, such as the surfaces of polyhedra, can be accurately

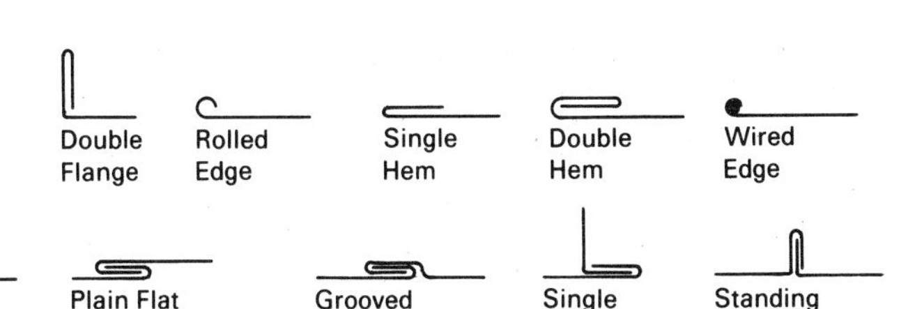

Fig. 16-1. Common sheet metal seams, hems, and joints.

developed. Double-curved, or warped surfaces, such as spheres and cones, can only be approximated—but with sufficient accuracy for most products.

Most sheet materials used for fabricating purposes are sufficiently pliable to permit stretching, pressing, stamping, and other forming techniques. In sheet material layout, additional material is allowed for seams and hems. If the material is thick, allowance must be made for bending it. This is commonly referred to as a *bend allowance.*

Within the industry, there is a definite difference between a seam and hem. A seam is used for attaching or fastening one surface to another, while a hem is used for providing additional strength or giving a more finished appearance to an edge. A variety of hems and joints are used, of which some of the more common are shown in Fig. 16-1.

Parallel-Line Developments

Sheet material products that have parallel sides, edges, or elements can employ parallel-line development. Examples of shapes that employ parallelism are prisms and cylinders. Directly related to parallelism is the concept of perpendicularity. Here, a straight line(s) is drawn perpendicular to the object's elements, along which the pattern or stretch-out is developed.

When drawing parallel line developments, several givens must be known. The first is a true size and shape view of a right section, or perpendicular plane view, of the object (the distance around this view will define the length of the stretch-out). Next, a true size view of the parallel elements of the object must also be given. This view will provide height measures along the stretch-out. This section will briefly describe and discuss how parallel-line development is used in prism and cylindrical developments.

Prism Development

Fig. 16-2 shows the correct development of a prism. The horizontal and frontal views of the prism are shown in section (a) of the drawing, while the stretch-out development is seen in section (b). In the two views the corners have been appropriately labeled for easy identification. Once this is accomplished, these steps are followed:

1. Using light construction lines, project all the height dimensions into the stretch-out.
2. In a clockwise manner, beginning with corner 1 in the top view, transfer the prism's periphery measures to the stretch-out. Be sure that you begin and end with the first (number 1) element or corner.
3. Lightly draw these elements vertically until they intersect their height measure, and label appropriately. Thus, corner 1 will have a height of 1-A, corner 2 will have a height of 2-B, and so on.

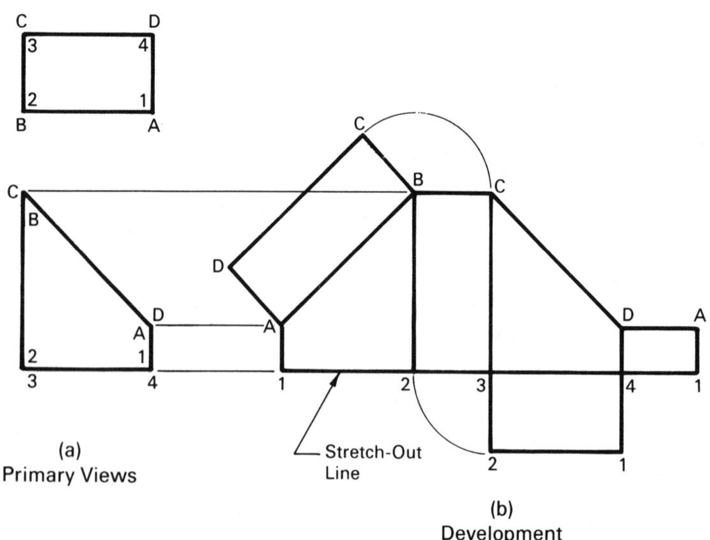

Fig. 16-2. Prism development.

4. Connect the points of each plane surface of the stretch-out.
5. To construct a top and bottom, project 90° lines to form the bevel and bottom surfaces, and lay out their length with a divider or compass.
6. Join the points to form the prism's top and bottom.
7. Add any material needed for hems and seams.

To check to see if the development is correct, you may cut out the pattern and fold it together. This procedure is usually accomplished with a heavier stock paper. The pattern should be transferred from the master drawing.

Right Cylinder Development

Another parallel-line development problem commonly encountered is to construct a stretch-out for a right cylinder. This shape is used in a number of different products, such as containers, ducts, and supports. The procedures used here are shown in Fig. 16-3, and described as follows:

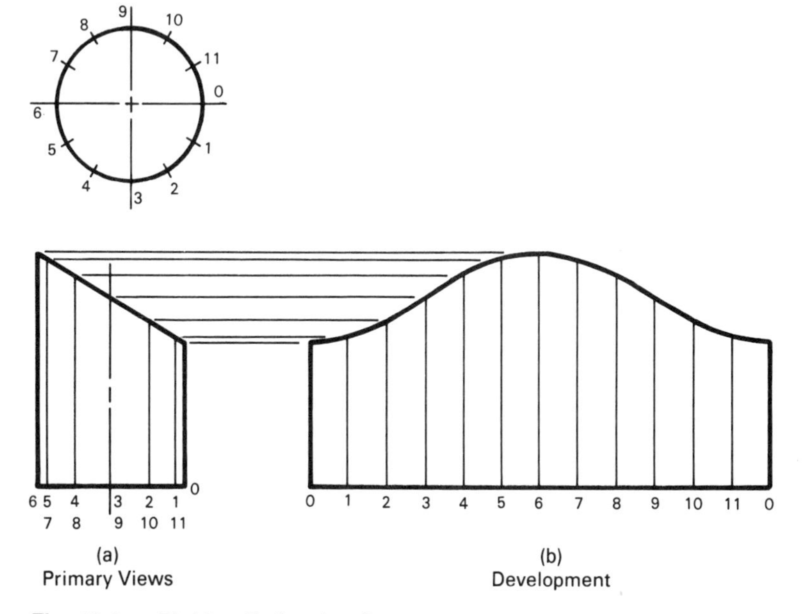

Fig. 16-3. Right cylinder development.

1. With the drawn frontal and horizontal views in place, divide the top view into a given number of equal parts and number them accordingly. Each division then represents an element along the side of the cylinder.
2. The length of the stretch-out will be equal to the circumference of the complete right section (top view). This can be calculated with the formula $C = \pi D$, or measured off by the distance between each division.
3. Draw the stretch-out to the appropriate length and divide it into the same number of divisions used in the top view, and number each element.
4. Determine the length of each element by projecting its height from the frontal view, and label accordingly.
5. Allow sufficient material for required seam.

Oblique Cylinder Development

The development of an oblique cylinder presents a unique problem as compared to a right cylinder, the primary difference being that of the two primary views provided, none shows the true size of the

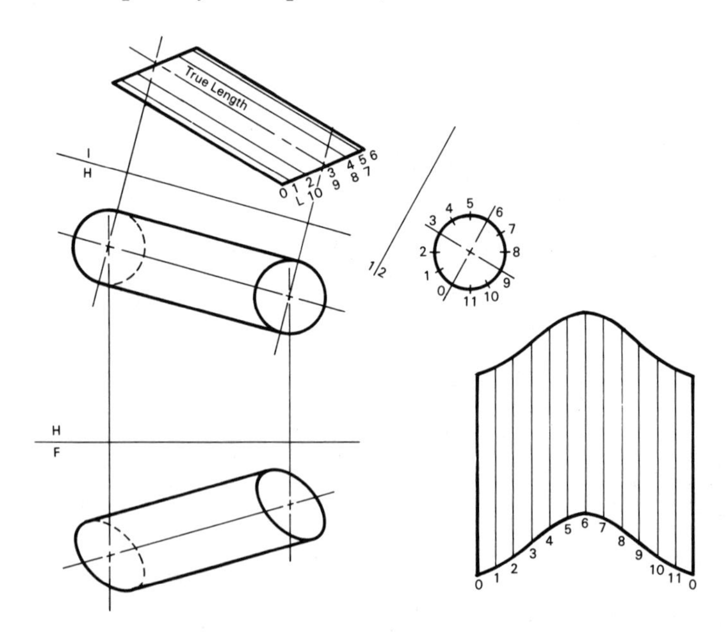

Fig. 16-4. Oblique cylinder development.

right section, nor do the elements appear as true length. To develop an oblique cylinder, auxiliary views must be drawn so that the elements will appear true length and the right section true size.

Fig. 16-4 shows how this is accomplished. Note that the second auxiliary view is used to identify and number each element, while the length of the elements is taken from the first auxiliary view. The length of the stretch-out is calculated by the circumference of the cylinder's circle.

Radial-Line Developments

Objects such as pyramids and cones do not have elements that are parallel to each other, but radiate to a common point or vertex; a pattern for these shapes could therefore not be constructed by development techniques. The drawing technique used for these types of object is known as radial-line development, which generally employs the construction of a series of triangles.

Pyramid Development

Fig. 16-5 shows the procedures used for pyramid development by the radial-line method. Note that this process involves finding the true length of the edges of the pyramid. The procedures are as follows:

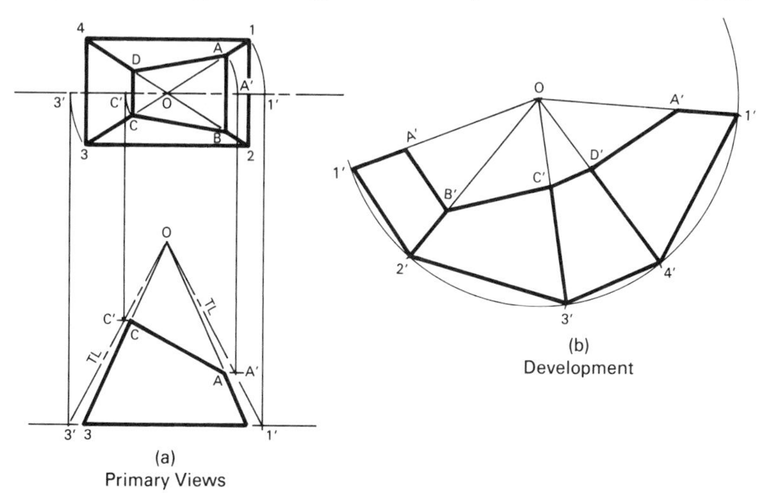

Fig. 16-5. Pyramid development.

1. Label the vertex and all corners of the pyramid.
2. Since the edges of the pyramid do not appear true length in the frontal view (they are not parallel to the frontal plane), their true length must be found. This is accomplished by rotating, from the vertex, each edge until it is parallel to the frontal plane and projecting it to the front view. The true length lines are then labeled as 1′, 2′, etc.
3. With the true length of the sides known, strike an arc whose radius is equal to that length.
4. Along the arc, step off the base of the pyramid.
5. The length of each radial element is then transferred to the stretch-out (see A′, B′, C′, and D′).

Right-Circular Cone Development

The development of a right-circular cone can be accomplished by dividing the base into a given number of parts and stepping it off along an arc, or by calculating the angle of the sector. This formula is $360° r/s$, where r is the radius of the cone's base and s is its slant height. Thus, if a cone has a base diameter of 14 inches and a slant height of 10 inches, the stretch-out's angle will be $360°(7)/10 = 252°$.

An example of the development of a right-circular cone is shown in Fig. 16-6, and described as follows:

1. Given the frontal and horizontal views of a truncated cone, divide the top view into a given number of equal parts.
2. Strike an arc for the development of the cone whose radius will be equal to the slant height. Calculate the stretch-out's angle either by formulation, or by graphic procedures.
3. Find the true length of each element by rotating it parallel to the frontal plane, and transfer to the stretch-out.
4. Connect all points, and allow for sufficient seam material.

Oblique Cone Development

The development technique used for an oblique cone is similar to that used in the oblique cylinder, except that the elements radiate from a vertex. Fig. 16-7 shows how the true length of individual radial elements are determined. The true length elements are noted by the use of the prime (′) symbol.

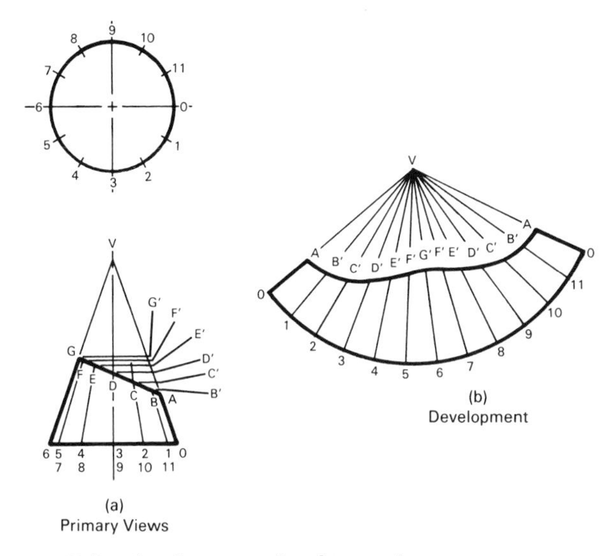

Fig. 16-6. Right-circular cone development.

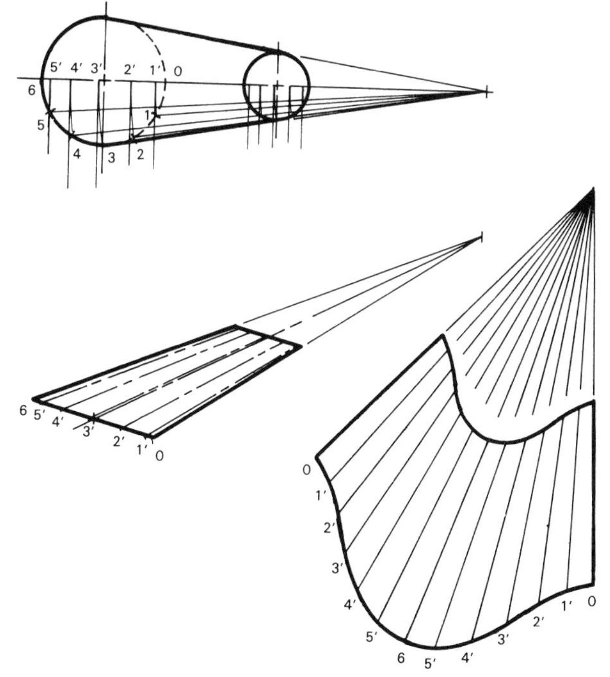

Fig. 16-7. Oblique cone development.

Triangulation Development

The use of triangulation development simply means that the surface of a plane is divided into a number of triangles, and transferred to a development. Radial line development is a form of triangulation, but is used for the construction of developments in which the elements radiate to a common vertex. Within the sheet material industry, there are a number of nonradial forms that employ triangulation.

The most common application of triangulation development involves *transition* pieces. A transition piece is defined as a segment that connects two or more differently shaped or sized openings, as in a duct system connecting a rectangular opening to a round opening. Transition pieces are also used in other fabricated products.

Transition of Two Rectangular Openings

A simple transition problem is shown in Fig. 16-8, where a transition piece is made for two rectangular openings of different sizes. The procedures used here are:

1. Where convenient, extend the sides of the transition piece to form triangles. Label each corner with the appropriate notation.

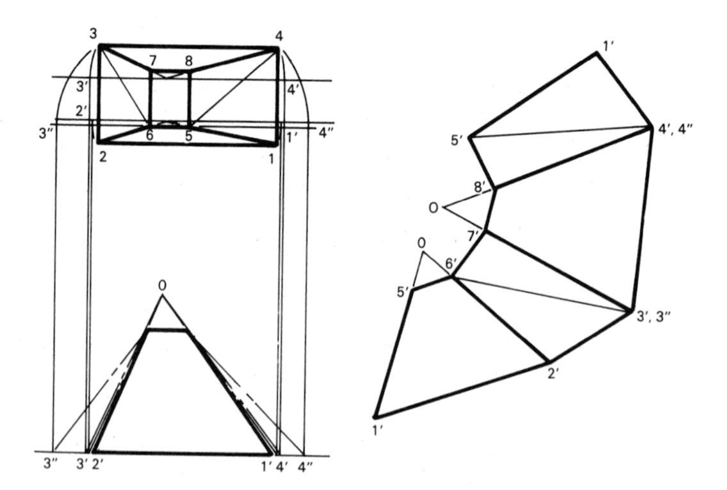

Fig. 16-8. Transition of two rectangular openings.

2. Construct the true length of each side of the transition piece, and label.
3. Beginning with the first surface, construct its stretch-out based upon the true length measures obtained. Repeat this process for each surface.
4. In this particular problem, accuracy can be checked by ensuring that the top and bottom edges of each plane are parallel to each other.

Transition of Square to Round Openings

A typical arrangement of many transition pieces is square to round (Fig. 16-9). This transition piece is made up of four flat surfaces and four warped or curved (conical) surfaces. The procedures used here are the same as for radial and other triangulation procedures.

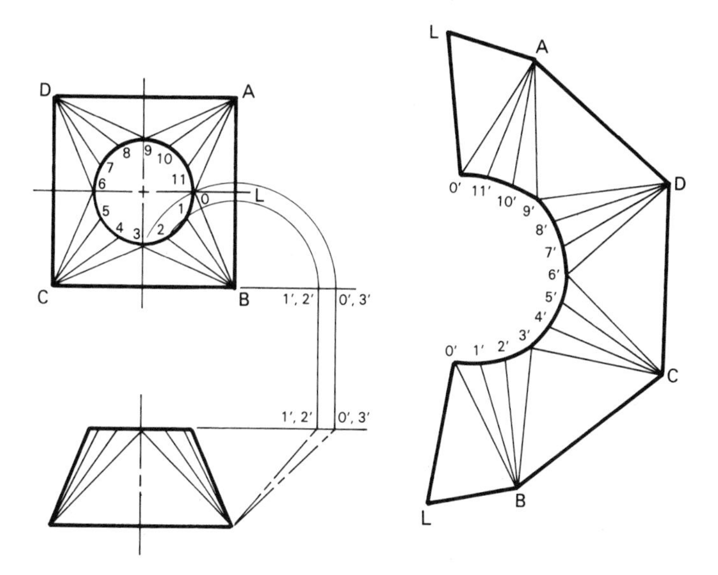

Note: For ease of assembly, plane AOB was split in half, thus, planes LOB and LOA

Fig. 16-9. Transition of square to round opening.

Exercises

16.1 Given the objects in Fig. 16-10, use parallel-line development techniques to construct their stretch-out.

16.2 Use radial-line development techniques for the development of patterns for the surfaces shown in Fig. 16-11.

16.3 Develop stretch-outs for the transition pieces shown in Fig. 16-12.

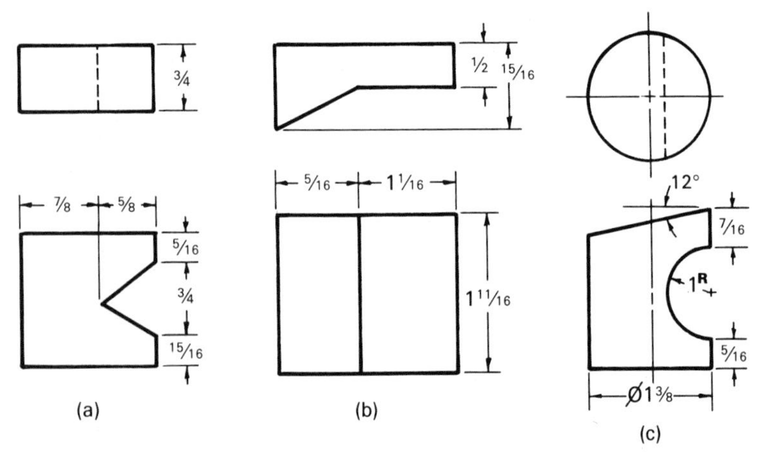

Fig. 16-10. Problem 16.1.

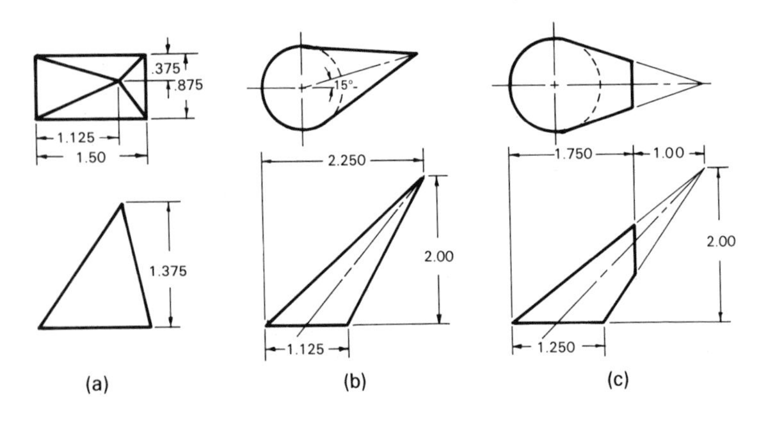

Fig. 16-11. Problem 16.2.

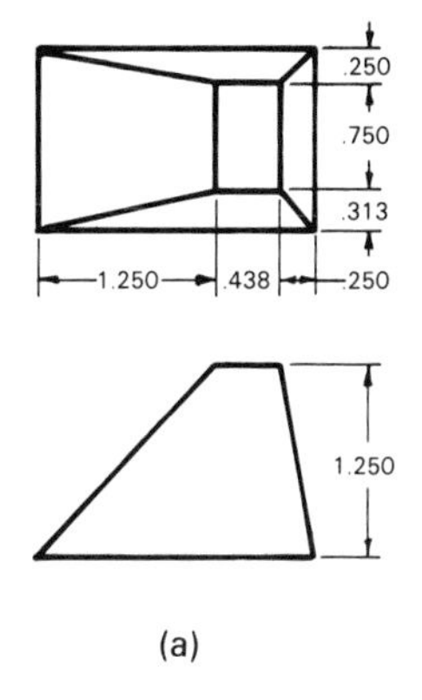

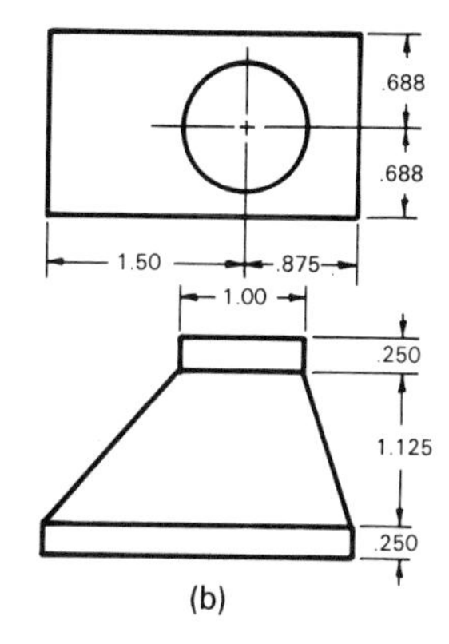

Fig. 16-12. Problem 16.3.

CHAPTER 17

Pictorial Drawings

The oldest form of graphic communication between human beings is the pictorial drawing. Early cave drawings and designs on pottery incorporated pictorial drawings to illustrate events and beliefs. As the human race grew more technically sophisticated, pictorial drawing procedures became more formalized and specialized, to the point where we now have several broad fields of graphics, such as fine art, commercial art, technical illustration, and industrial design.

Of particular interest to industrial drafters are the pictorial drawing procedures categorized as technical illustration and industrial design. Pictorial drawings are not often used by drafters, whose primary method of graphic representation is the multiview, or orthographic projection, drawing.

There are times, however, when drafters find it advantageous to use pictorial drawing techniques. In those cases, they will select one of three types of pictorial drawing projection: axonometric, oblique, and perspective. It is important that the industrial drafter not only be familiar with these types of drawing, but also have some knowledge and competence in the preparation and interpretation of pictorial drawings. This chapter will discuss the basic concepts and drawing procedures used in axonometric, oblique, and perspective drawings.

Axonometric Projection

Axonometric projection is similar to orthographic projection, except that the object being viewed is "tilted" or rotated in such a way as

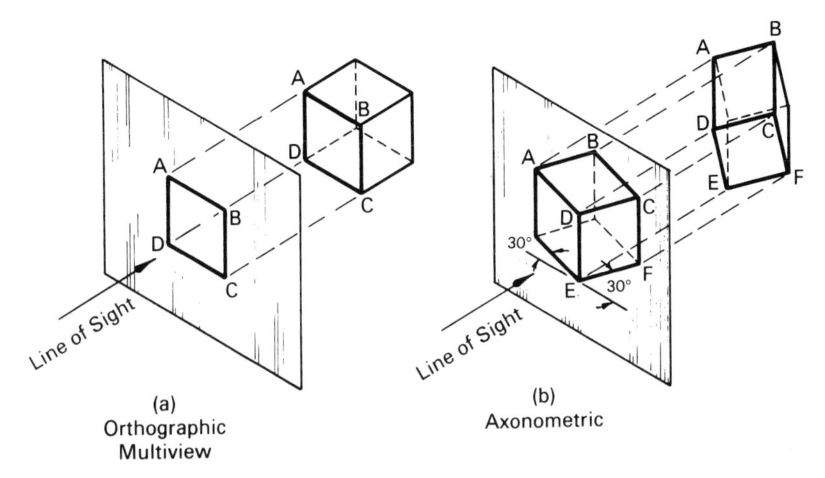

Fig. 17-1. Projection types.

to expose several surfaces in that view to the observer and the projection plane. As shown in Fig. 17-1, both orthographic and axonometric projections use lines of sight that are parallel to each other (projected to infinity), to produce the desired result.

The difference between axonometric and orthographic projection is the inclined position of the projected object. Because the lines, edges, and surfaces of the object are not parallel to the plane of projection, their lengths, sizes, and proportions can have an infinite number of variations. Three types of axonometric projection techniques have been standardized in the drafting field: isometric, dimetric, and trimetric projection.

Isometric Projection

The form of axonometric projection most frequently used by industrial drafters is isometric projection. The term "isometric" is derived from the Greek word *isometros*, which means to have equality of measure. To draw an isometric projection, the object must be positioned so that its principal edges (axes) form equal angles with the plane of projection. Thus, each side or edge will be equally proportioned. As shown in Fig. 17-2, the axes of the object form three equal angles of 120°, the proportion of all edges being the same.

The scale used in drawing an isometric can be of two types. The first is full size, where all measures are made as if the object were

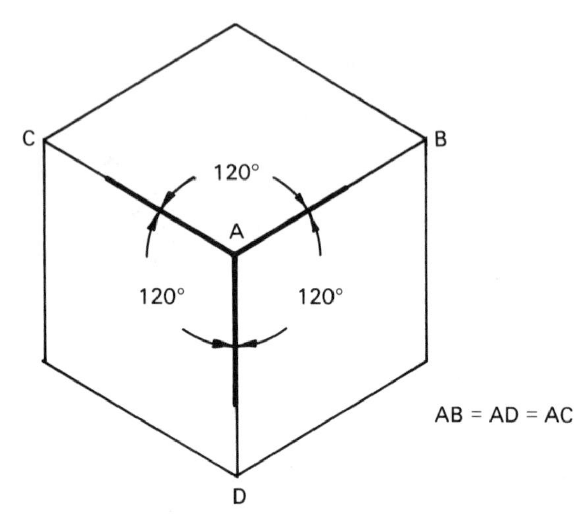

Fig. 17-2. Isometric projection proportions.

projected true size. For example, if a 2-inch isometric cube were drawn, all sides would be measured to the length of 2 inches. In reality, the length of the sides of an isometric object is actually $\sqrt{2/3}$ of true size, equal to 81.65 percent of actual size. Thus, the 2-inch cube would be drawn with sides equal to 1.633 inches. To approximate this proportion, drafters frequently use a 3/4 size or 9″ = 1′-0″ scale. See Fig. 17-3.

PROCEDURES FOR PRODUCING ISOMETRIC DRAWINGS

The basic procedures used for drawing an isometric projection are not difficult or complex. Shown in Fig. 17-4 are the steps employed in drawing an isometric projection for a block. Note that all the measurements are parallel to the three axes (i.e., length, width, and height). Measurements should not be made along nonisometric paths, that is, not parallel to one of the three axes. As shown in our example, this does not mean, however, that an inclined or oblique surface cannot be drawn.

The basic procedure recommended here is sometimes referred to as *box construction* technique. The reason for this is quite obvious; first, an isometric box is constructed representing the overall length, width, and height measurements of the object. Using this procedure,

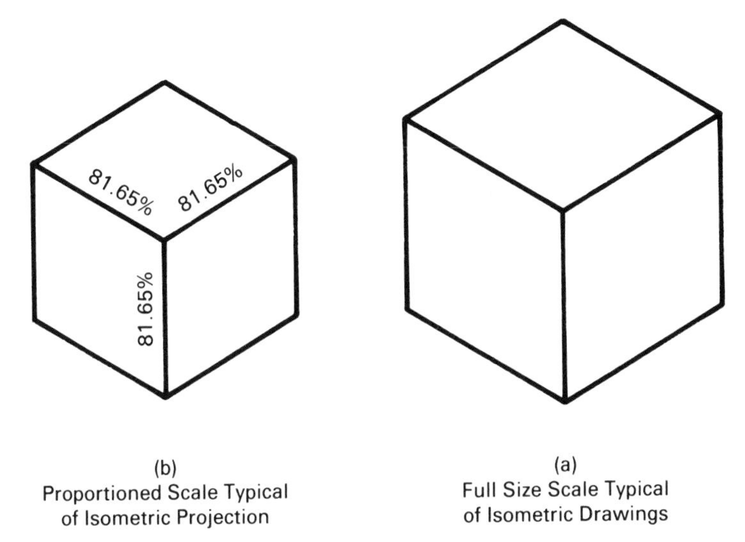

(b)
Proportioned Scale Typical of Isometric Projection

(a)
Full Size Scale Typical of Isometric Drawings

Fig. 17-3. Isometric scales.

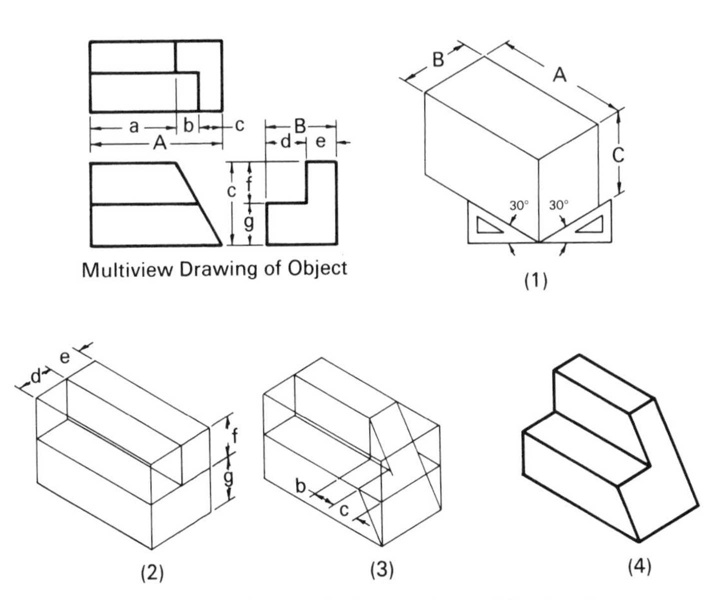

Fig. 17-4. Procedures for producing an isometric drawing.

a surface or object of any shape can be drawn by locating its boundaries (e.g., edges, corners, or circumference) within the box. Because the isometric box is used for construction purposes only, it should be drawn in lightly, and erased when the object is completed. The basic steps recommended for producing an isometric drawing are as follows:

1. Lightly draw the three major axes by using a 30°-60° triangle.
2. Draw an isometric box whose dimensions are equal to the overall length, width, and height of the object.
3. Carefully measure and mark off the distances along one of the axes, and draw a line along the appropriate axis.
4. Repeat step 3 for each axis.
5. When all corners and edges have been identified and drawn, carefully compare the multiview and isometric drawings to make sure that the pictorial is a correct representation of the object.
6. Darken in the completed isometric drawing and erase all construction lines.

ISOMETRIC DRAWING PRACTICES

Several practices should be followed when preparing an isometric drawing. The first pertains to the drawing of hidden surfaces or edges by means of hidden lines. The basic rule of thumb is that hidden lines should be omitted, unless they are needed to clarify the drawing. In the majority of drawings there should be no need for hidden lines, but if there is any doubt, they should be included, for it is better to "overdescribe" an object than to have insufficient information.

The drawing of center lines is another concern of the drafter. Center lines should be used only if they are necessary to indicate symmetry or for dimensioning. Generally center lines are kept to a minimum to prevent confusion and overcluttering.

The last major practice to be considered is isometric dimensioning. Isometric dimensioning employs the same principals found in multiview dimensioning, with the exception that isometric dimensions are presented in pictorial form. Both ANSI and ISO standards recognize two dimensioning techniques for isometric drawings: aligned and unidirectional. Of the two techniques, the unidirectional is preferred because it is the easier and less time-consuming.

When using aligned (Fig. 17-5a) dimensions, the lettering and numbers are made within guide lines that are drawn parallel to the

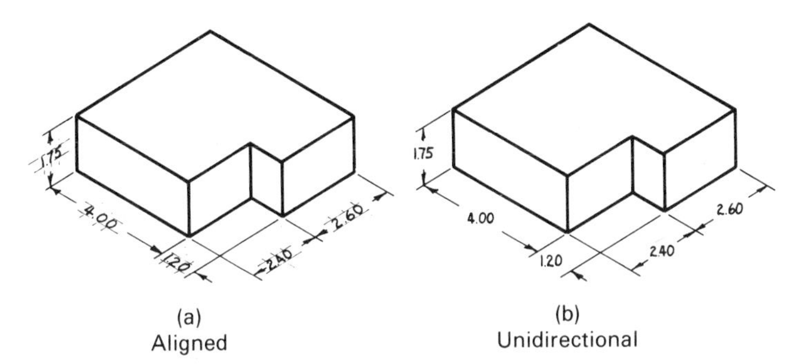

Fig. 17-5. Dimensioning isometric figures.

given axis and the vertical. All dimension lines are drawn parallel to their respective axes, while extension lines are drawn parallel to one of the other axes. When a dimension is written in place, it must align itself with the axis to which the dimensioned feature is parallel.

In unidirectional dimensioning (Fig. 17-5b), all numbers and letters are presented in alignment with the vertical axis. As can be seen, the unidirectional technique does not have the inherent problem associated with lettering at an angle that is found in the aligned technique. The drawing of extension and dimension lines follows the same procedure as used for aligned dimensioning.

ISOMETRIC ELLIPSES

Shapes that appear as circles in conventional orthographic views are drawn as ellipses in isometric drawings. The two most commonly used techniques for drawing isometric ellipses are geometric construction and the isometric ellipse template. Whenever possible, drafters will use the template over the construction method. There may be times, however, when one must construct an isometric ellipse because a template is not available, or the ellipse's size is not found in the template.

The construction method recommended for drawing an ellipse is the approximate ellipse method. This and other construction methods are outlined and illustrated in Chapter 5 of this book. See also Fig. 17-6, and follow these procedures:

1. Draw an isometric square whose sides are equal to the diameter of the circle.

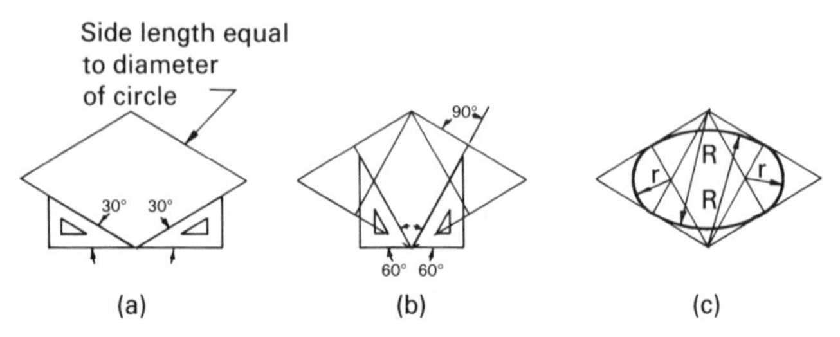

Fig. 17-6. Approximate ellipse.

2. Using a 30°-60° triangle, draw perpendicular bisectors to each side. The four points where the bisectors intersect will be the centers for the circular arcs.
3. Draw the large (R) and small (r) circular arcs as shown in Fig. 17-6.

Dimetric Projection

Another form of axonometric projection is known as dimetric projection. Though not widely used, it is sometimes preferable to isometric projection. As its name implies, this projection method views an object in such a way that two of its axes will be equal. Similarly, two of the axes will be equally foreshortened, while the third axis will be measured with a completely different scale.

The angles of the axes selected for dimetric drawings will be determined by how the object is to be presented for greatest clarity. Unlike isometric projection, in which linear proportions and axes angles are always equal, dimetric specifications can vary greatly. To simplify matters, six approximate axis angles and proportions are recommended for dimetric drawings. See Fig. 17-7.

The procedures used to prepare a dimetric drawing are the same as for isometric drawings, except that the axis angles will conform to those just presented. An example of this is shown in Fig. 17-8.

Trimetric Projection

The third type of axonometric projection technique available to drafters is trimetric projection. Compared to isometric drawings, trimetric drawings are rarely used. This form of axonometric drawing should

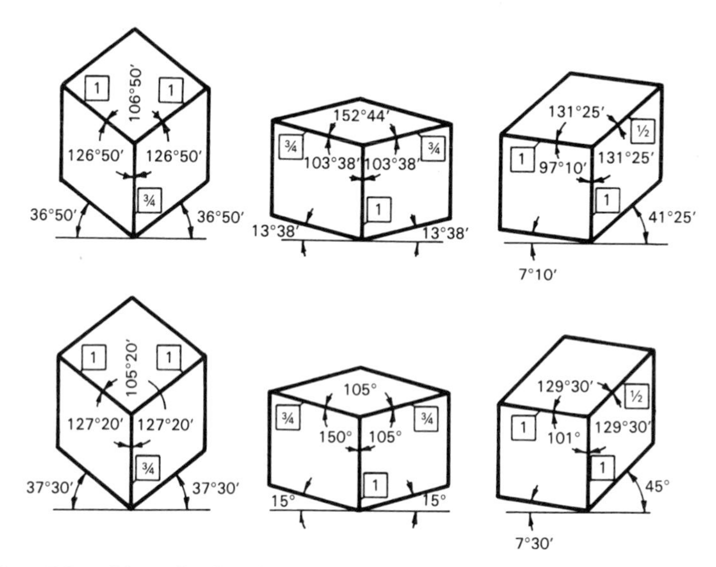

Fig. 17-7. **Dimetric drawing angles and scales.**

be employed only when all other methods fail to present the object adequately. In a trimetric projection set-up, the object is so aligned that no two axes will have the same angle. Respectively, the scales used to measure along each axis will also be different.

The nature of trimetric projection dictates that three different scales be used along each axis. To ensure proper proportioning, special trimetric scales made of clear plastic are available and can be ordered

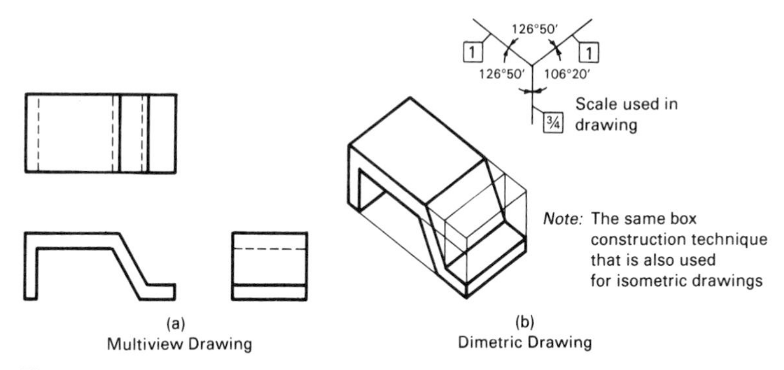

Fig. 17-8. **Application of dimetric drawing procedures.**

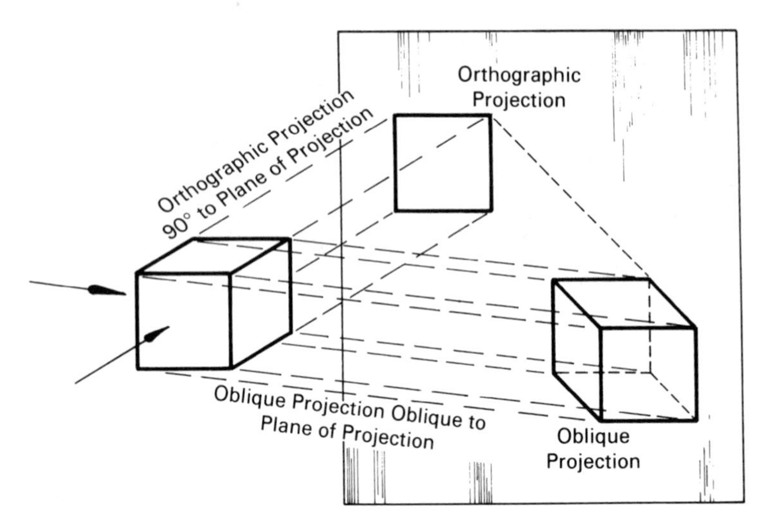

Fig. 17-9. Oblique projection.

with standard angles and scales. Again, the procedures used for preparing a trimetric drawing will be the same for other axonometric drawings, except for axes angles and scales.

Oblique Projection

A projection technique that is used when one particular surface is of critical importance is oblique projection. In this technique, the object is viewed so that the projectors are parallel to each other (go to infinity) and are *oblique* to the viewing plane. This results in one surface of the object appearing true size and shape (Fig. 17-9). This drawing technique is particularly useful for drawing contours that would be extremely difficult to represent on an inclined axis.

Axis Angles and Scales

Oblique projection axes can be made at any convenient angle. For the majority of drawings, however, an inclination of 45°, 30°, or 60° are used because they are easily drawn with standard triangles. How these angles are used to illustrate various objects is shown in Fig. 17-10. As can be seen, the angle and direction of the axis not only deter-

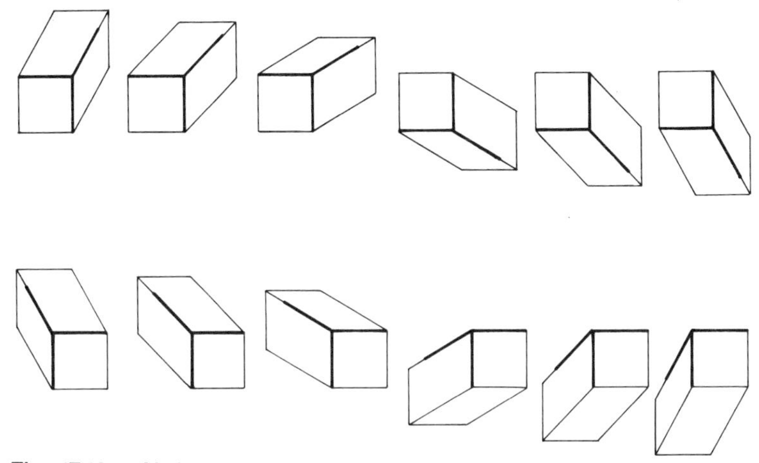

Fig. 17-10. Various positions of oblique axes.

mines which surfaces are seen, but also the amount of detailing that can be clearly shown: large angles show better views of an object's recess on the top, while small angles are better for showing details on the side.

The human eye observes objects in perspective; that is, all parallel lines that recede from view appear to converge at a distant point. Owing to the nature of oblique projection, these receding lines are parallel and tend to exhibit more distortion than in axonometric or perspective drawings. To eliminate or reduce this distortion, receding lines may be foreshortened.

When the receding lines of an oblique drawing are full size, it is known as a *cavalier* oblique. Obliques that are drawn half size are said to be a *cabinet* projection. Cavalier obliques were first used in medieval times for drawing fortifications, when the term "cavalier" was used to describe an elevated command position. Cabinet drawings were first widely used in the furniture industry.

Though cavalier and cabinet obliques are the two most common forms of scaling, other foreshortening proportions can be used. Fig. 17-11 shows several examples of how the foreshortening of receding lines can dramatically affect the appearance of an object. In this example, the cavalier projection makes the receding line of the cube appear too long, though all sides are drawn to the same scale. For ease of drawing, it is recommended that the drafter use one of the scales available on the mechanical scale.

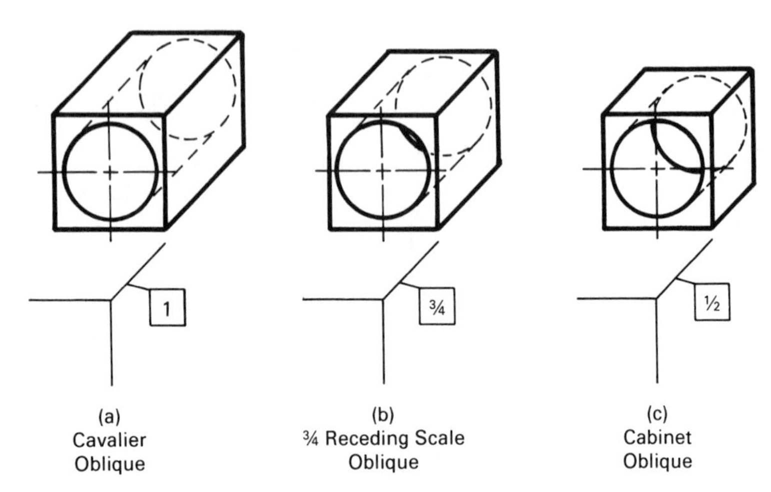

Fig. 17-11. Various scales for receding lines.

Procedures Used for Preparing Oblique Drawings

Before beginning work on an oblique drawing, the optimal positioning of the object must be selected first. As a general rule, the surface that shows the important contours and shapes should be parallel to the plane of projection. The selection of this surface is often similar to selecting the frontal view of an object. Thus, it is not unusual to find that the face drawn parallel to the plane of projection is the same one shown in the frontal view of a multiview drawing.

When no one view has a significant advantage over the other, then the surface with the greatest number of curves and circles should be selected. The reason for this is that the curves and circles can be more easily drawn true size and shape than at a projected angle.

The procedures used in finding locations and drawing shapes in oblique projection are similar to those used in axonometric projection. That is, all relative lines are drawn parallel to their respective axis. The steps used in oblique drawing are shown in Fig. 17-12 and described as follows:

1. Select the surface to be drawn parallel to the plane of projection and the scale to be used for the receding lines.
2. Draw an oblique box whose dimensions are equal to the overall width, length, and height of the object.

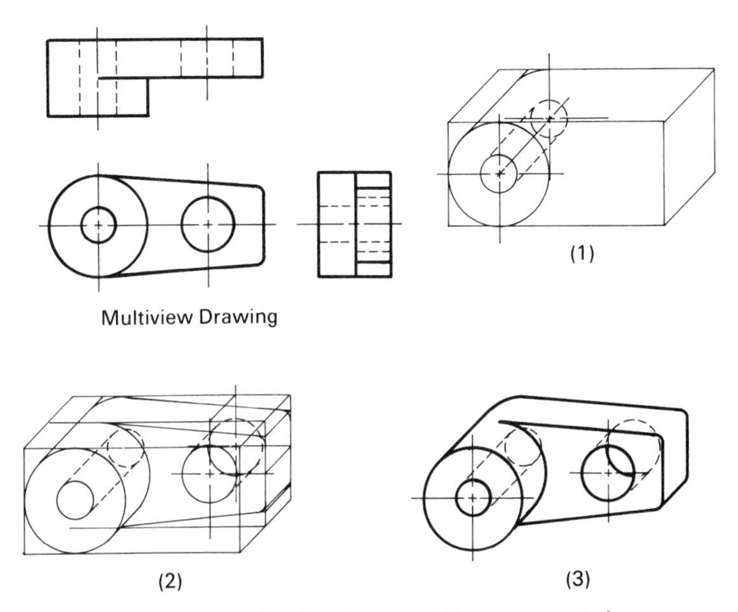

Fig. 17-12. Procedures for drawing an oblique presentation.

3. Draw in all details on the surface parallel to the plane of projection.
4. Locate and draw all other features.
5. Darken in all visible lines and erase all construction lines.

The axonometric rules used for hidden and center lines also apply to oblique drawings. In addition, oblique drawings may be dimensioned by either aligned or unidirectional methods. Again, the easiest and most recommended dimensioning method is the unidirectional technique. When lettering notes and other specifications, drafters should use vertical lettering only, unless otherwise specified.

Perspective Drawings

Drawings that are made according to perspective or central projection principles are usually superior to all other forms of pictorial drawings. The reason for this is that perspectives more closely represent what the human eye would view. Even with this advantage, perspective drawings are infrequently used by industrial drafters, although they are widely used by industrial designers and illustrators. Because from time to time drafters are concerned with perspective presentations, they should understand the basic principles employed in these drawings.

Principles of Perspective Drawings

Perspective drawings incorporate the use of four major factors:

The first is the position of the viewer. This is important to identify, for it will dictate what surfaces will be visible and how they will be shown.

Second is the object being viewed relative to its position and complexity.

Third is the plane of projection, and in particular where it is placed, in that its location will control how all perspective features of the object will appear.

Finally, the projectors are important in that they are used to "project" all features of the object onto the projection plane. Fig. 17-13 shows these four factors.

In Fig. 17-13, the observer's eye is looking through a theoretical plane or projection that is referred to as a *picture plane*. On drawings, the picture plane is usually labeled as "PP." The *station point*, or "SP," is the location of the observer's eye. When lines are drawn from the station point to various points on the object, they become the projectors or *visual rays*. The point at which each visual ray intersects the picture plane is a perspective point. Together these points will form a perspective of the object.

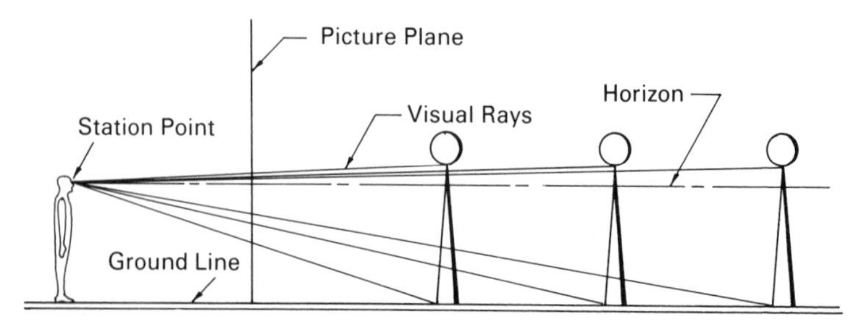

Fig. 17-13. Factors influencing perspective drawings.

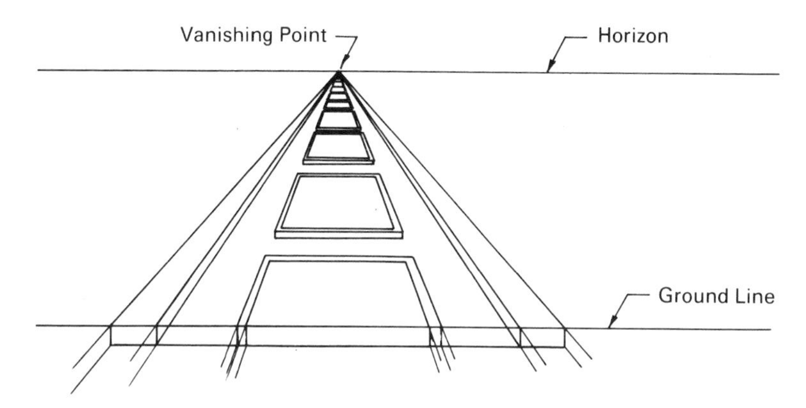

Fig. 17-14. Parallel lines in perspective appearing as lines converging to the vanishing point.

The line known as the *horizon* is parallel to the *ground line* and is at the same height as the station point. It represents the eye level of the observer, and is the edge view of the horizontal plane. In Fig. 17-14 the receding lines, though parallel, do not appear so. In fact, they converge toward a point on the horizon known as the *vanishing point*, which is labeled "VP." Parallel lines that do not appear parallel in the picture plane will converge to a vanishing point.

One-Point Perspective

A simple perspective drawing technique involves the use of one vanishing point, hence the name "one-point perspective." In this technique, the object is arranged so that two of its principal axes will be parallel to the picture plane, while the third is perpendicular to it. With this type of arrangement, the third axis will have all parallel lines converging toward the vanishing point.

To draw a one-point perspective, two views of the object are required. The views selected will depend upon how the object is to be viewed. An example of how one-point perspectives are drawn is shown in Fig. 17-15. Here the frontal surface of the object is placed in the picture plane. The station point should be the location from which the object is to be viewed. In this example, it is located to the right of the object. A horizontal line is drawn and one vanishing point is located on it.

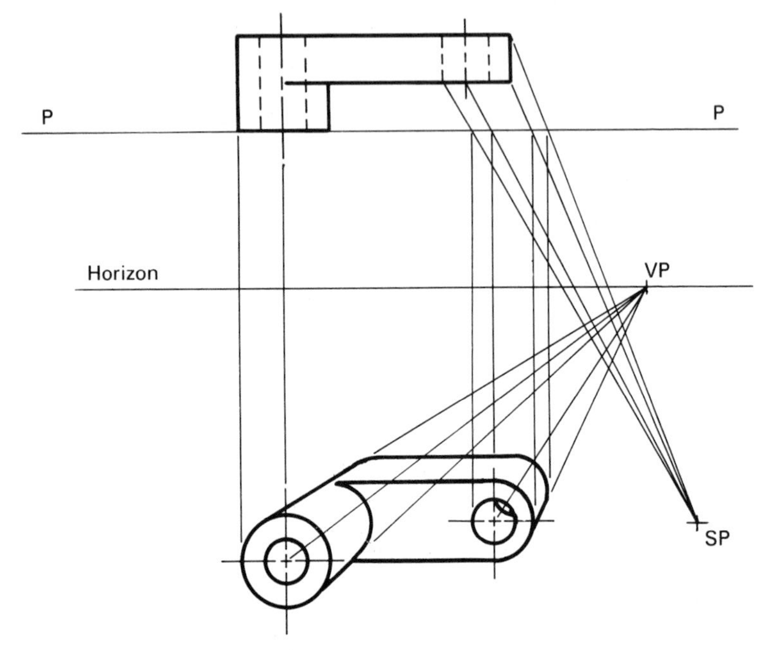

Fig. 17-15. One-point perspective.

The procedures used in one-point perspective are as follows:

1. To locate the position of the various features on the picture plane, visual rays are drawn from the top view of the object to SP. The point where the visual rays intersect the PP will be its location.
2. From their location on the PP, the rays are projected downward to the perspective. If the feature is behind the PP, they are located by projecting backward toward the VP. Only those surfaces in the picture plane will appear true size; all others will be foreshortened relative to their distance behind the picture plane.
3. Once all object lines are located, the visible lines are darkened in.

The top view is usually not drawn on the drawing material, but is placed above it on the drawing surface, so that when the drawing is completed, all that appears on the drawing sheet is the perspective.

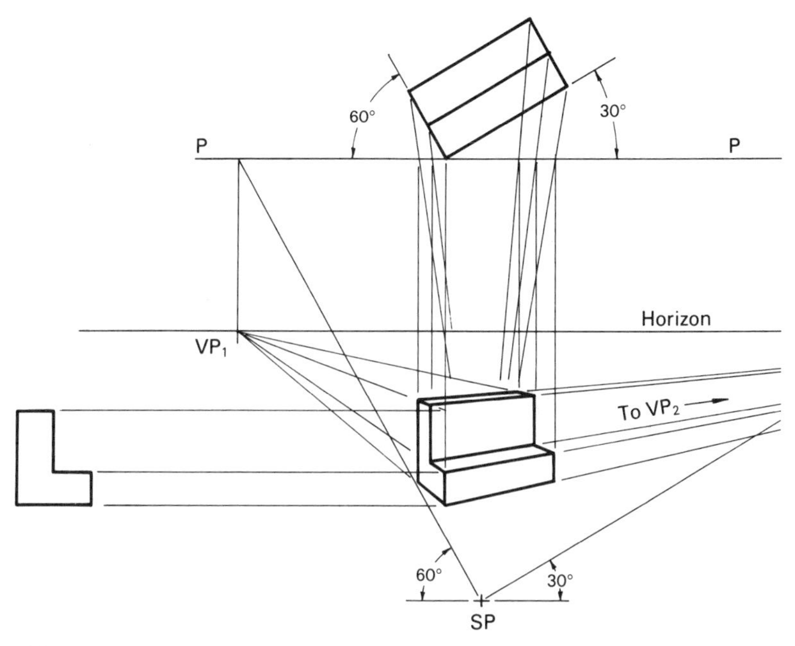

Fig. 17-16. Two-point perspective.

Two-Point Perspective

A second, and more complex, form of perspective drawing is the two-point perspective. Here the object is positioned so that the vertical axis does not project to a vanishing point (i.e., all vertical lines are parallel), while the other two axes have lines that converge to a vanishing point. This form of perspective is frequently used in architectural and civil engineering drawings of buildings and large structures, but is infrequently used in industrial drawing presentations.

An example of this perspective technique is given in Fig. 17-16. The procedures used here are:

1. As shown, one edge of the object is placed in the picture plane so that direct height measures can be made to that line.
2. The top view of the object is placed at a convenient angle, such as 30° or 45°.
3. As in one-point perspective, the horizontal and frontal views of the object are placed off the drawing sheet, but are posi-

tioned so that they can be accurately projected to the appropriate reference lines.

4. The only line that will appear true length is the edge that is touching the picture.
5. To locate the position of all points and features on PP, visual rays are drawn to SP so that they intersect PP in the horizontal view. From PP, they are projected 90° downward.
6. The heights and vertical locations of these points and features are then determined by projecting all height measures to the true-length line in the perspective and drawing a line to the appropriate vanishing point. Where the converging lines intersect the appropriate downward projected lines will be the location of the point or feature.
7. Once all features are located and drawn, darken in the perspective and erase all construction lines.

Three-Point Perspective

In three-point perspective, the object is located so that none of its edges are parallel to the picture plane. As a result, all axis lines will converge to a vanishing point. Of the three perspective techniques, this is the most realistic form of pictorial drawing. However, because of the complexity of this projection technique, it is rarely used by industrial drafters. Three-point perspective drawings, for all practical purposes, are restricted to commercial artists and technical illustrators.

Exercises

17.1 Prepare an isometric drawing for the objects shown in Fig. 17-17.

17.2 Prepare a dimetric and trimetric drawing for one of the objects illustrated in Fig. 17-17.

17.3 Draw an oblique projection for the objects shown in Fig. 17-18. Determine the correct foreshortening scale (i.e., cavalier or cabinet) for each figure.

17.4 Draw a one-point perspective for the parts shown in Fig. 17-19.

17.5 Prepare a two-point perspective for the object shown in Fig. 17-20.

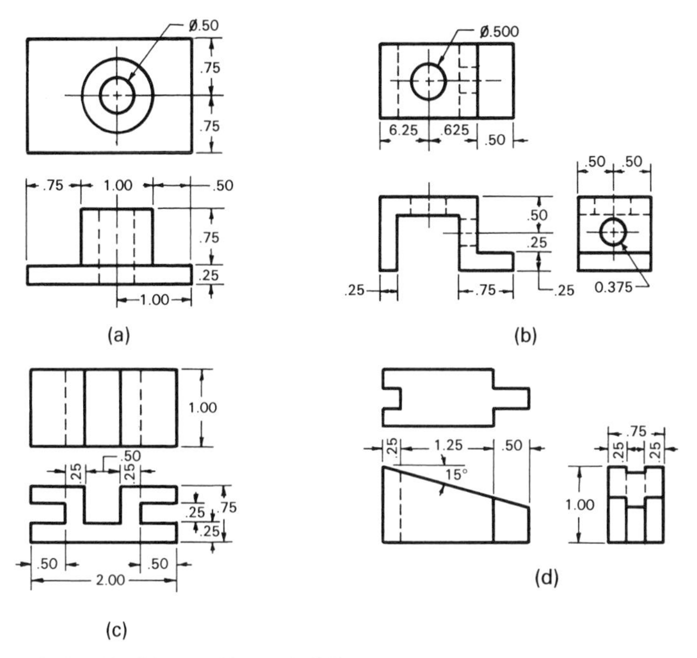

Fig. 17-17. **Problems 17.1 and 17.2.**

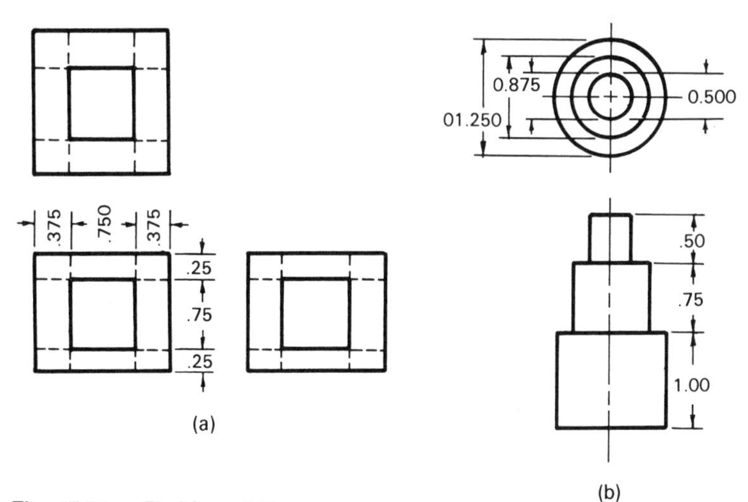

Fig. 17-18. **Problem 17.3.**

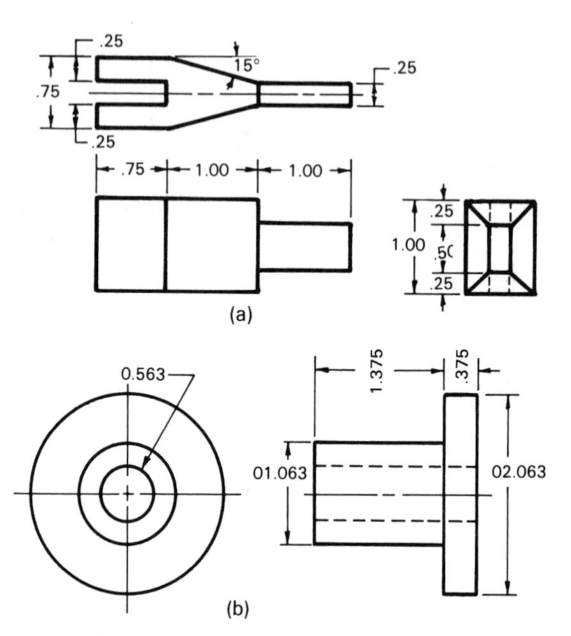

Fig. 17-19. **Problem 17.4.**

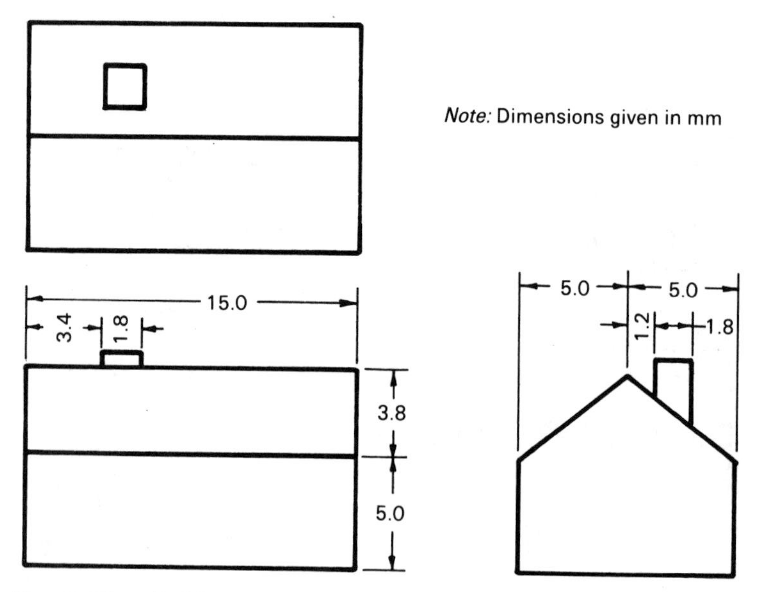

Fig. 17-20. **Problem 17.5.**

CHAPTER 18

Computer-Aided Graphics

- Definitions in Computer Graphics
- The CADD Environment
- Interactive Computer Graphics
- Review Questions

We are experiencing a computer revolution, we are often told. Computers are frequently seen as overtaking all aspects of business and industry by increasing productivity and manufacturing efficiency. Managers and executives are constantly hearing that they must have state-of-the-art computer systems if they want to remain competitive in today's marketplace. Drafters and technicians too are being told that they must become "computer literate" and have some understanding of how to use computers, or their employability will be limited.

In the field of drafting and design, businesses are taking a serious look at the potentials of computer systems to aid in product development and the generation of working drawings. These systems are referred to as CAD or CADD, which are abbreviations for "Computer-Aided Drafting," "Computer-Aided Design," or "Computer-Aided Drafting and Design."

Before rushing out and studying to become a computer programmer, or deciding to eliminate drafting and design staff in favor of a CADD system, it is important to take a sane and logical look at just what computers and computer graphics are and what they are capable of doing. Before this is possible, however, some "common ground" needs to be established in order to understand what is being presented. This chapter discusses the special terms used in the CADD field, and what is really involved in a CADD system.

Definitions in Computer Graphics

One of the first things encountered when learning about computer systems and CADD is an array of new terminology. Talking with sales, marketing, technical, and professional personnel involved with CADD systems can be somewhat intimidating when they use the nomenclature associated with computer systems and graphics. It is therefore necessary to give a brief overview of terminology in the field.

Computer Systems

Computers are devices consisting of a series of miniature electronic circuits that are used to move or transmit electrical signals, or store them. They are designed to receive information, store it, retrieve it, and manipulate it upon command. The ability of computers to retrieve data and manipulate that information in numerous ways is what makes them so valuable. For example, a computer can "search" a business's product inventory of hundreds of thousands of products and very quickly determine the exact number and specifications of any one product.

The two major elements of a computer system are the *hardware* and the *software*. Hardware refers to any physical equipment that makes up the system. Examples of hardware are keyboards, disc and tape drives, printers, plotters (drum and flatbed), CRTs or display screens, and central processing units. Software consists of the programs or instructions that make the computer execute different tasks. These programs are very detailed and give a step-by-step sequence through which the computer must work. Together, hardware and software provide the basic design by which jobs are performed.

As might be expected, an entire vocabulary of terms has evolved to describe the various elements of computer system hardware and software. Presented here is some of the common terminology used in the computer field, accompanied by a brief definition.

Access time. The amount of time required to obtain data from storage or memory. This will vary according to the amount of data to be retrieved, and can range from fractions of a second to tens of minutes.

Application program. Computer software that is specifically designed to meet the needs of the user.

Architecture. The organization or design of the computer's central processing unit.

ASCII (American Standard Code for Information Interchange). A binary code that is used to control printers or plotters and communication devices. The particular binary number given for each letter or number (alphanumeric character) is called its ASCII code.

Backup. A process whereby data is copied into storage for safekeeping, in case the original data storage medium is lost or damaged.

BASIC. A programming language that is widely used in interactive computer systems, especially in personal computers. This stands for Beginners' All-purpose Symbolic Instruction Code.

Batch processing. A process by which a set or series of computer programs is implemented without human direction or interaction.

Binary. There are two definitions. The first is a numbering system made up of the two digits 1 and 0, in which each symbol designates a decimal power of two. Second, any system that has only two levels or states, such as "on" and "off," where 1 would represent on or a presence of current, and 0 would represent off or an absence of current. All computer programs are processed entirely in binary form.

Bit. Related to the binary form, bit is short for *b*inary dig*it*. It is the smallest unit of information that is read by the computer. All alphanumeric data processed by computers are expressed solely as a combination of bits (i.e., 0s and 1s).

Byte. The total number of bits used to represent an alphanumeric character.

Cathode-ray tube (CRT). A fluorescent screen used to produce images and characters. The CRT is a vacuum tube of the same type as that used in television sets.

Central processing unit (CPU). The components of a computer that control the transfer of data and execute arithmetic and logic calculations.

Character. An individual letter, number, or symbol.

Chip. Semiconductor material that contains integrated circuits. These circuits are usually microscopic in character.

COBOL. A computer language that stands for Common Business-Oriented Language. It is a sophisticated language that is designed for business applications where large and complex data records are used.

Command. An instruction given to the computer.

Compatibility. This term is used in two senses. The first is to denote that a given software or component can be used with two or more different computers. The second refers to the ability of a computer to interact with other computers that are different in design and capability.

Computer network. The interconnection of different computer systems, terminals, and facilities.

Configuration. The assortment of hardware and software used in a system.

Cursor. A blinking, movable light or marker (usually a rectangular box or a line) that is used to designate the position for the next data entry point or change.

Database. A large quantity of organized information needed to perform a task or calculation.

Data processing. The use of a computer primarily for working with numerical information (as opposed to word processing).

Diagnostic. A software program that checks for errors and malfunctions.

Disk. A plate that is flat and circular, and made out of a rigid material that has a magnetic coating for storing data. The physical size and storage capacity of disks vary.

Disk or diskette drive. A piece of hardware that is designed to place data in storage on or read information from one or more diskettes.

Diskette. A plate that is flexible, flat, circular, and coated with magnetic material for storing data.

Display screen. Any device used to provide a visual representation of data, such as a cathode-ray tube or liquid crystal displays.

Downtime. The length of time a device is not operational.

Electronic file cabinet. Any storage unit that stores information in the same manner as an office file cabinet.

File. A collection of related information or data.

Filename. Any title or name given to a file.

Floppy or flexible disk. A diskette.

FORTRAN. A technical and widely used programming language that stands for *For*mula *Tran*slation. This language is primarily used for mathematical and scientific applications.

Hard copy. Data presented in permanent form such as paper or tape.

Instruction. A command given to the computer that tells it to do a particular task and when to do it.

Interactive. A computer system that is capable of carrying out a dialogue with the user.

Interface. Electronic connections that hook the computer up to external devices such as terminals, printers, and plotters.

Keyboard. A set of keys, similar to a typewriter, that is on a terminal for the transmission of data.

Kbyte. Kilobytes, or 1,024 bytes (2^{10}).

Magnetic tape or magtape. A magnetic tape that is used for storing large quantities of data. Information access on magtapes can only be obtained serially, and they are therefore practical only for larger computer systems.

Mainframe. A computer that is physically large and powerful enough to perform computations and information processing that require large amounts of data (e.g., large payrolls, finite element analysis, stress/strain resolution, and multicolor pictorial presentations).

Mbyte. Megabyte, or 1,048,576 bytes (10^{20}).

Medium. Any physical material upon which data is recorded and/or stored.

Memory. The primary storage area within the computer. It can also be a device used to store and retrieve data.

Menu. A displayed list of options or tasks that can be selected and implemented by certain command statements.

Microcomputer. A physically small computer that can fit on or beneath a desk.

Microprocessor. A CPU that is contained on a single chip. This is the basis of LSI technology.

Minicomputer. A computer that is smaller than a mainframe and generally more powerful than a microcomputer.

Modem. A device, whose name is derived from *Mo*dulator/*Dem*odulator, that converts electrical signals sent by computers into high-frequency signals that can be sent and received over telephone lines.

Multiprocessing. The execution of more than one program by a computer having more than one central processor.

Multitasking. The ability of a computer to perform more than one task at a time, without having to complete one before beginning another.

Network. A group of computers that are connected and can communicate with each other.

On line. Anything that is directly under the control of the computer.

Operating system. A series of programs for the purpose of controlling the overall operation of the computer.

Peripheral. Any device that is external to the CPU.

Power supply. A transistor that converts ac to dc power, energizing components, and in some cases steps down the power to some electrical components.

Program. Software that gives a complete sequence of instructions to the computer.

RAM. Random Access Memory, the amount of memory that can be read and written into the computer during normal operation.

Realtime. The time used to perform a computation while it is actually happening, and does not include the time involved for any other operation.

ROM. Read Only Memory; the amount of fixed data or instructions that are permanently fixed into the computer during manufacture. The ROM of a computer cannot be changed.

Soft copy. Displays or information presented in nonpermanent form (e.g., video screens).

Sort. The rearranging of records in a file into a more practical and convenient form for the user.

Storage unit. Place where information can be stored for later use.

System. Any combination of software and hardware used to perform processing and computational operations.

Tape. A recording medium, usually magnetic or perforated tape.

Terminal. Any device used for data input and output as hard copy and/or soft copy.

Timesharing. Providing computer service to many users by having the computer execute each user's task on a sharing basis (part of the time).

Turnkey system. A completely packaged computer system that is ready to use without the addition of any hardware or software.

Word processing system. Any system that processes text material and performs functions such as sentencing, paragraphing, margin justification, indenting, line and word rearranging, page numbering, and text printing.

Computer Graphics

The development of computer systems and the improvements made in them have greatly influenced the way we live and the way business is conducted. The first computers were made of vacuum tubes and filled entire rooms. They were temperamental and difficult to maintain, and could accomplish little more than today's hand-held calculators.

Eventually the vacuum tube gave way to the transistor, making miniaturization possible. Since then the development of chips and "super chips" has put computer systems within reach of everyone. It is interesting to note that the typical personal computer (PC) is now capable of executing operations that just ten years ago required a mainframe. It is predicted that by the turn of the century PCs will have the ability to store several Mbytes of information, and be affordable for the average consumer.

Until recently, computer-aided drafting consisted of little more than a machine that was programmed to generate drawings. Today graphic systems are essential to the entire design and manufacturing process, of which drawings are but one segment. Computerization of the drawing and design process has greatly affected all industries. Because CADD employs computer systems, the terminology used here is the same as for computers in general. There are, however, several terms with unique application to the CADD field; they include:

Automated drafting. Preparing the drawings by use of automatic equipment such as plotters.

B-rep. Solid model programs used for irregular geometries.

Bed. The surface upon which the drawing paper is placed on a plotter.

Bit-map graphics. A method in which individual pixels on a display screen are controlled to generate graphic elements of high resolution.

CAD. Denotes two different concepts: computer-aided drafting and computer-aided design. The first applies only to the generation of drawings traditionally produced by drafters, designers, and illustrators. The second involves an entire array of tasks that include finite element analysis, solid modeling, simulations, and drawings.

CADD. Computer-aided drafting and design.

CAM. Computer-aided manufacturing. Usually closely linked to CADD systems.

Character printer. Any printer that prints one character at a time (similar to a typewriter).

Computer-generated synthetic imagery (CGSI). A photograph-based system used in visual simulation.

Constructive solid geometry (CSG). Programs that construct models in a building block procedure by combining elements or forms such as cubes, pyramids, and cones.

Coordinate system. A method of locating points, lines, and planes within a three-dimensional framework by the use of axes. The most common is the Cartesian coordinate system.

Cross-section analysis. The calculation of the section properties in any specific cross section of a part.

Daisywheel. A print head with individual "spokes" for each character.

Dot-matrix printer. A printer that forms characters and symbols with a series of dots.

Draft-quality printer. A printer that produces characters that are easily read, but of inferior quality to those produced on typewriters.

Dragmesh. Modeling programs developed for copying nodes and elements (see finite element analysis).

Finite element analysis. A mathematical technique used for calculating stresses, whereby parts are drawn by connecting nodes along an object's surface, so that the finished drawing resembles element meshes.

Geometric modeling. The creation, usually in color, of a solid-looking part or product. This is usually accomplished on a CRT, and printed later.

Graphics. The use of lines, points, and figures for displaying data.

Impact printer. A printer that forms characters and symbols by striking an inked ribbon.

Interactive computer graphics (ICG). CAD systems that permit dialogue between the user and computer.

Kinematic programs. Software used to mimic the paths of various mechanical elements such as cams, gears, links, hinges, and human appendages.

Laser printer. A printer that creates images with a laser beam scanning rapidly across a charged photoconductive drum so that a toner or ink with the opposite charge adheres to the photoconductor and transfers to the paper.

Letter-quality printer. A printer that produces characters comparable to those of typewriters.

Light pencil. Special pointing device used to locate graphic images.

Lineprinter. A high-speed printer that prints an entire line at one time.

Liquid crystal display. Liquid crystals that produce images by selectively reflecting ambient light.

Modeling. A simulation of a real-world product or part to determine how it would appear or function under application.

Mouse. A device used for inputting information for drawing.

Multicopy form. Paper sets with several carbon papers between each page.

Node. A point, usually on the surface of a part.

Numerical control (NC). The use of a numerical code, usually on perforated or magnetic tape, to move a stylus or tool.

Pixels. Picture elements that generate images on display screens.

Plotter. A device used to produce hard copies of drawings. They are usually of either flat-bed or barrel design.

Printer. A device used to produce hard copies of documents.

Printout. Any printed material or hard copy.

Resolution. The clarity and detailing shown on a display screen.

Simulation. Provides designers with a preview of how a product will behave and function in service.

Stylus. Pencil or ink heads used for drawing on plotters.

Visual simulation. The use of computer-controlled television (CCTV) systems to reproduce exact replicas of an environment. Used in flight simulators to train pilots, in highway traffic analysis, and in mapping.

Wire frame model. See finite element analysis.

The CADD Environment

Individual organizations make use of CADD systems in various ways, the exact procedure and system used depending on the nature of the business. For example, an aeronautical design firm and a civil engineering firm will be concerned with different kinds of analysis and modeling. There are, however, basic kinds of analysis and procedure that are available in most CADD systems. This section will briefly review each of them.

CADD Systems

The manufacturers of products such as shoes, automobiles, aircraft, artificial limbs, gears, robots, and water pumps have found it advantageous to use modeling and drawing procedures available through CADD systems. Originally, CADD was nothing more than an extension of the drafter's drawing board, but now it offers designers and manufacturers the ability to build solid models, field-test products, and determine stress areas and levels, all on a display screen.

In reality, CADD is only half of a total discipline known as CAD/CAM or Computer-Aided Design/Computer Aided Manufacturing. In fact, advances in computer systems will soon make it economically feasible for many companies to handle almost all aspects of the evolution of a product, from conception to design to manufacturing.

CADD systems are significantly different from the original systems available in the early 1960s. At that time CADD (or CAD) was nothing more than a computer-controlled automatic drawing machine. From this beginning, the computer was linked with the CRT for display, so that the model could be made into a working drawing.

Today's CADD systems come with a wide range of software that enables them to perform such functions as geometric modeling, simulation, stress/strain analysis, field testing, automated drafting, and manufacturing set-up. What makes these systems even more desirable is that most are turnkey systems, so that with little training, most drafters and designers can quickly learn to use the system even though they may have absolutely no experience with computer language or hardware. In fact, these systems are so user-friendly (easy to use) that no computer language or programming skills are required.

The potentials of CADD systems are only beginning to be realized. Computer modeling, for example, makes it possible to create

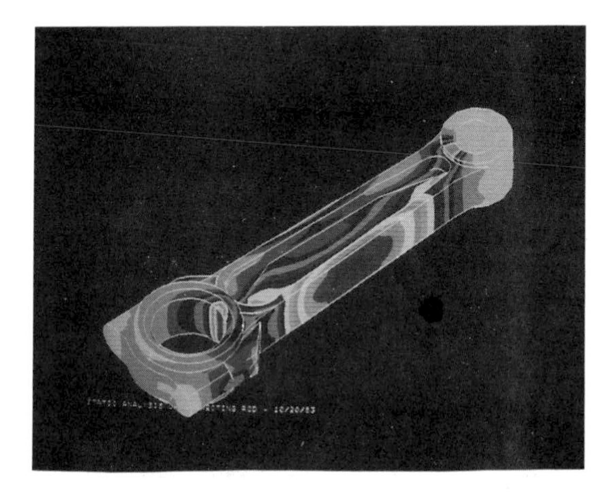

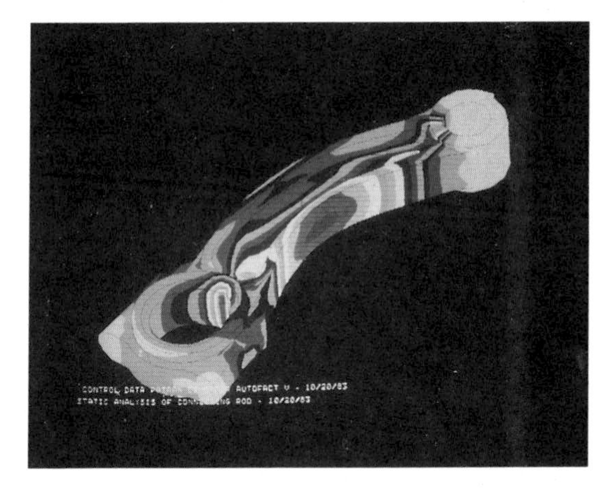

Fig. 18-1. Computer simulations. *(Courtesy of Control Data)*

a multicolor solid part appear on a display screen. It can then be rotated about any axes and cut into cross sections to reveal both external and internal surface details. Analysis can also be performed to check how the part fits with other devices or systems. In manufacturing, tooling and machining paths can be automatically designed and checked against the model.

Computer simulations have also greatly enhanced the capacities of designers and engineers. It is now possible to obtain a visual display,

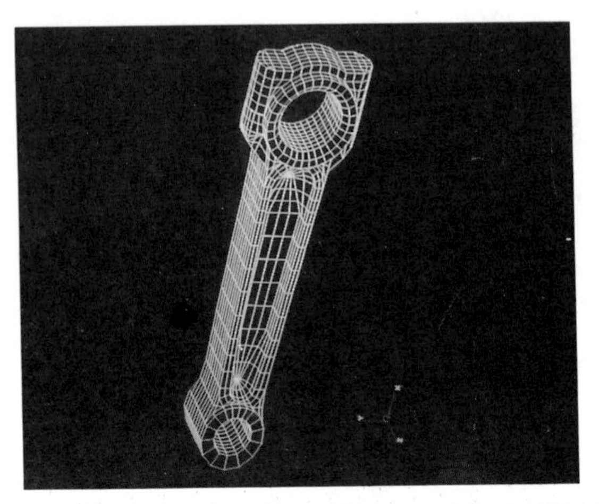

Fig. 18-2. Finite element analysis. *(Courtesy of Control Data)*

in color, of the location of stress and strains on a product as it is placed under different working conditions (Fig. 18-1). For example, simulations are used to determine the location and levels of stress/strain on airplane wing spans during various flight conditions, and on automobile panels that are exposed to collisions.

A third element in CADD systems is their ability to conduct finite element analysis for calculating stress levels. At one time all mesh drawings were prepared by hand, and frequently took hours to accomplish even with the use of a minimum number of points or nodes to locate and connect. Automatic mesh displays (Fig. 18-2) can produce these same crosshatched sectors for any model in a matter of minutes. In addition, the number of nodes used can be increased significantly.

Mesh drawings are derived by producing the crosshatched sectors or meshes on the surface of the model prior to analysis. Within the CADD program, complex sets of simulation equations are used to calculate the levels of stress/strain along the mesh. From this, variations can be made in the product's design to see how stress/strain can be reduced or transferred to other parts of the product.

The last aspect of CADD to consider is data field testing. These systems enable the designer to generate display screen videos where motion and movement can be observed, so that the part can be seen under working conditions. Potential problem areas, such as vibration and deflections, can then be pinpointed and eliminated.

CAD/CAM

As yet, relatively few companies have implemented a fully integrated CAD/CAM system that is capable of designing and manufacturing a product through interactive computer graphics. The sequence of such a system is quite straightforward:

1. Geometric modeling
2. Simulation
3. Analysis
4. Testing
5. Drafting
6. Tooling
7. Machine movement configuration
8. Manufacturing

Throughout this sequence, there is continual reworking of previous procedures for the improvement of design. Each step in the process is built upon the information for design improvement obtained from the previous step. In each case, all steps will have access to the same product design and manufacturing information, which is stored in a shared database.

In aircraft design, for example, the geometric model is first used to define the general shape of the aircraft. The initial models created will be rough and show only the overall appearance of the structure. Later, more detailing will be added.

Using the initial geometric model as a beginning, a *dynamic* computer model is derived to show all structural joints and connections. This model is also used to simulate the aircraft under various flight conditions. Whenever the aircraft behaves in an unexpected and undesirable manner, changes are made in the design.

Such simulations enable aircraft designers to perform stress analyses, determining the weights and loads on individual aircraft sections. A finite element model is displayed to show the exact location and level of stress. Again, the model can be changed or modified and reanalyzed until stress behavior is as desired.

After the aircraft is designed, automated drafting is used to produce accurate and high-speed drawings. These drawings are usually produced on plotters that generate hard copy documentation. Then a prototype model is tested showing how the aircraft will bend, twist, and behave under various levels of stress and vibration.

Fig. 18-3. Design geometry used for numerical control machining. *(Courtesy of Control Data)*

Tooling is then designed for the machining of specialized parts. Robotic units or automatic machines can be programmed to move these tools in a particular path so that the aircraft material can be machined or fabricated with great accuracy. Fig. 18-3 shows how design geometry is used directly with numerical control machining.

Image Synthesis Graphics

Though not directly associated with CADD systems, image synthesis graphics is a newly developed technology that has unique potential for graphic presentation. Image synthesis graphics is a technology that makes it possible to produce pictures of nearly photographic quality by computer.

This computer graphics breakthrough was made possible by advances in supercomputers and high-performance dedicated processors that give extremely high computing power, together with a variety of new types of computational procedure (algorithms). It is now possible for programmers to generate complex and detailed images effectively and efficiently.

Perhaps the most recognizable products of this imaging technique have been seen in the motion picture industry. Movies such as *Tron*, *The Last Starfighter*, and *Star Trek* all used computer graphics imagery for backgrounds. It is anticipated that synthetic imaging will increasingly replace the use of costly movie sets and scale modeling.

Interactive Computer Graphics

Some kinds of drawing can be done better by human beings, while others can be more economically made by machines. By combining the best qualities and functions of each, engineers and drafters can produce higher-quality drawings at a lower cost.

One area of CADD that makes use of both human and machine functions is known as interactive computer graphics (ICG). ICG incorporates both human and computer functions so that they are able to communicate with each other for design, production, and servicing. The use of ICG systems, in turn, has spawned a new design process known as *interactive problem solving* (IPS).

ICG is the continual and instant communication between a person and a CADD system. When integrated with manufacturing processes, the total system is then known as CAD-CAM (computer-aided design–computer-aided manufacturing). Here visual displays are used so that the engineer, designer, and drafter can review a potential product and see how it will function under given conditions. A typical sequence of events in a ICG system is as follows:

Employing communication between the CADD system and design, a product is presented as a “solid” model. Fig. 18-4 shows a solid model of a connecting rod. This model was produced by a designer who instructed a CADD system to construct the product in true three-dimensional perspective by combining fourteen basic geometric forms, such as spheres and cones. In addition, color and shading may be added.

Once the solid model has been produced, the ICG can create a wire frame model from the solid geometry. Fig. 18-5 is such a drawing generated from the solid model of the connecting rod. Here, hidden lines are not shown, unless commanded.

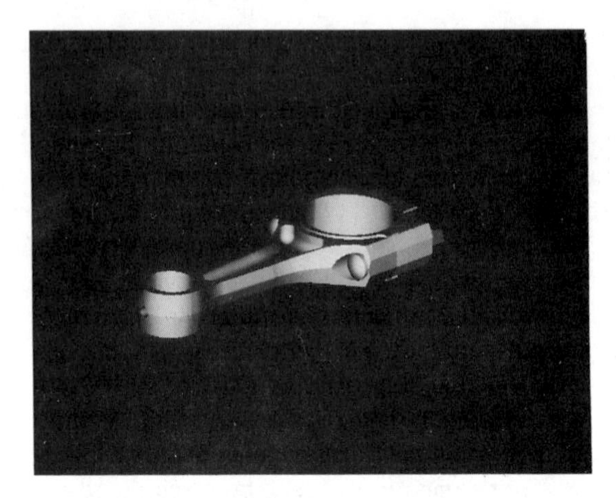

Fig. 18-4. Three-dimensional modeling used in the design of a connecting rod. *(Courtesy of Control Data)*

Next, exploded views of the product are generated to see how it would "fit" into an assembly. Fig. 18-6 is an exploded wire model view of the connecting rod, as it would be assembled into its subassembly. The geometric data presented here will also tell the designer and drafter the weight, volume, surface area, moment of inertia, and radius of gyration specifications for the part. All this can

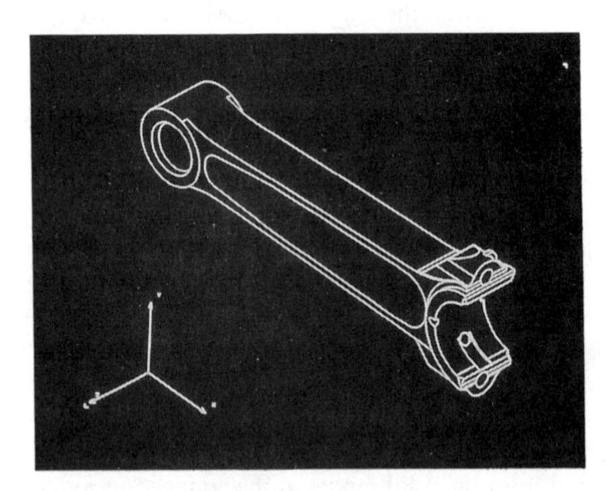

Fig. 18-5. Wire frame model. *(Courtesy of Control Data)*

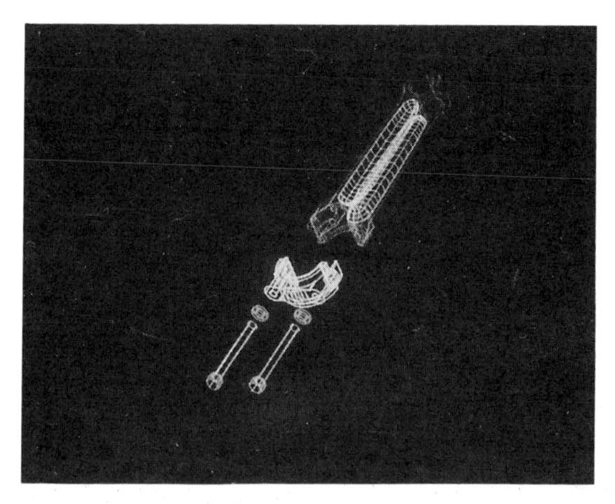

Fig. 18-6. Exploded view of wire model. *(Courtesy of Control Data)*

be accomplished with a given command statement, and produced in a matter of seconds.

Variations in dimensional and proportional designs can be examined easily. Multiple dimension parameters for a product are automatically generated by one command statement. If a drafter were to prepare such a presentation manually, the design costs would increase dramatically. Various designs can be displayed on a CRT in a matter of seconds or minutes, as opposed to days or even weeks for manual drawings.

Once all specifications are finalized, an ICG system will produce accurate and detailed drawings of a design. These drawings may be produced in one color or many. When they are displayed on a CRT, it is not uncommon to use multicolorations, from which a photographic copy can be made. When a hard-copy drawing is produced, a single color is used. Here the working drawings will usually be drawn with an inking stylus.

It might be supposed that an ICG system would complete all its functions with the final working drawings. The flexibility of such systems, however, enables engineers and designers to become more directly involved with manufacturing parameters and specifications. An example of how an ICG system is merged with manufacturing is shown in Fig. 18-7. Here design engineers have been able to identify the most efficient cutter path for the part. This path was precisely specified to ensure the most accurate machining of the connecting

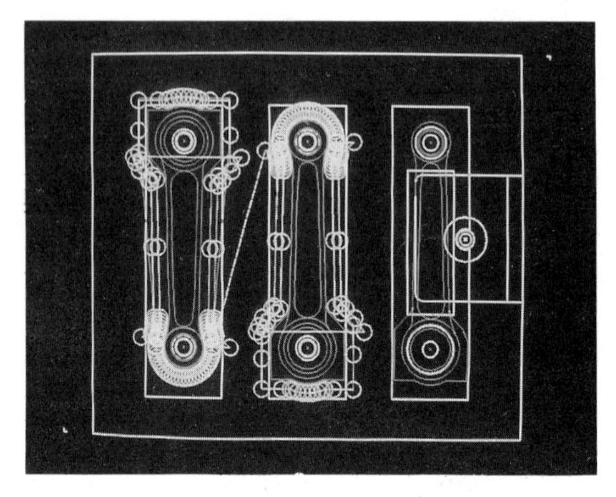

Fig. 18-7. Cutter path design. *(Courtesy of Control Data)*

rod. Within the context of manufacturing processes, this significantly shortens the time of manually readjusting cutter paths to obtain the optimum cutter path design.

ICG systems are of particular importance in automated machining processes. Here, ICG aids engineers and designers in determining numerical control machining for complex contours or surfaces of dies,

Fig. 18-8. Creating numerical control machining for complex contours. *(Courtesy of Control Data)*

molds, and finished parts (Fig. 18-8). Examples of factors considered here are point-to-point construction, pocketing, profiling, three-axis machining, five-axis swarf cutting, and dynamic tool pathing.

Review Questions

18.1 Explain the difference between hardware and software.

18.2 Is there any difference between hard copy and soft copy? Explain.

18.3 Define the following terms:

Architecture	Binary	Bit	Byte
CPU	Chip	Command	Cursor
Disk	Magtape	Medium	Memory

18.4 Within the context of drafting and graphics, explain the meaning of the following terms:

Bit-map graphics	CAD	CADD
CSG	Dragmesh	ICG
Simulation Modeling	Finite Element Analysis	Kinematic Programs

18.5 Explain the function of geometric modeling in product design. How is simulation incorporated?

18.6 Explain what image synthesis graphics is, and give examples.

18.7 Explain the relationship between ICG and IPS.

18.8 Describe the working relationship between ICG and manufacturing processes.

CHAPTER 19

Drawing Reproduction and Storage

- Reproduction Systems
- Storage Systems
- Review Questions

One of the most important aspects of any engineering project is cost. Costs in industrial design and drafting are most obviously associated with board time. There are, however, other factors that can increase the cost of a project and the operation of a drafting department. Two of these are reproduction and storage costs.

A wide variety of systems can be used in reproducing industrial drawings. The proper selection of a reproduction technique will not only influence project costs, but its appearance can also affect whether or not a project is accepted or rejected. Another factor that is often overlooked in assessing drafting costs is drawing storage. This chapter will discuss systems of reproduction and storage commonly used within the field of industrial drafting.

Reproduction Systems

When most inexperienced people are asked to give an example of how drawings are reproduced, they will probably first think of *blueprints*. In practice, however, the blueprint is seldom used today as a reproduction method, because of the time and cost involved in the process.

As recently as the 1950s, the blueprint was the primary method of reproduction. Blueprints can easily be identified by their negative appearance, white lines on a deep blue background. The blueprint image is actually a negative image of a drawing that is made by a wet process. The original drawing must be prepared on translucent paper, which is placed on top of a special copy paper. This overlay is exposed to an intense light, and then developed by immersion or soaking in a series of baths or sprays. The print is then hung up to dry. It is a quite time-consuming and expensive process, and of little value when prints are needed in short order. In addition, as a result of the baths and the drying, these reproductions tend to be less dimensionally stable than those made by other systems.

Significant advances and improvements have been made in drawing reproduction since the 1950s. Though there are a large number of techniques that can be used for copying drawings, most can be categorized in one of four classifications: intermediate prints, ammonia-sensitized prints, xerographic systems, and photographic systems.

Intermediate Prints

Another term for intermediates is *vanDyke prints.* As the name suggests, intermediates are reproductions that fall between the master drawing and typical reproductions. Technically, they are reproductions that are made from masters on translucent material, from which other reproductions can be made; thus, they function as a secondary master.

The primary use of intermediates is in drawings that must be taken to the work site for modifications in project designs, and where environmental exposure can lead to potential damage and fading. Rather than using the master drawing, which would result in excessive wear and tear from exposure to the work environment and from modifications on the drawing itself, it is easier and more cost-effective to make the changes directly on the intermediate, from which other prints can be generated. Once all the changes have been made and finalized, the drafter can then go back and incorporate them in the master drawing.

Intermediates are considered to be high-quality prints on material with good dimensional stability. The major difference between intermediates and other print systems is the line color. The image produced is considered to be positive, since the lines will be dark and the background white. In many cases, the lines will appear brown or sepia, which is the reason why some drafters refer to this type of reproduction as *sepias*. Intermediates, however, can also be made with black and dark blue lines.

Ammonia-Sensitized Prints

By far the most common form of reproduction used in industrial drafting is the ammonia-sensitized print. Prints made by this technique are also referred to as *ozalid* or *diazode* prints, or as *blue lines* or *blue line prints*. Here the positive-image print will have blue lines on a white background. Reproduction paper can also be purchased that produces prints with black and sepia (brown) lines.

The process of producing an ozalid print is accomplished by placing the translucent master or intermediate drawing on top of a presensitized paper. The sensitized side can easily be identified by its color, which is usually yellow (some sensitized materials are pink and blue). The overlay sheets are then inserted into a *copying machine*, where they are exposed to ultraviolet light. Where there are no lines or markings on the master drawing, the ultraviolet light will pass through the translucent material and "burnoff" the sensitized material. The only sensitized material remaining will be those areas not exposed to the light source, so the darker the lines that are drawn on the master sheet, the darker and more pronounced its image in the print.

After exposure to the ultraviolet light, the sensitized paper is then exposed to an ammonia vapor or mist; ammonia vapor is usually preferred, since it requires no drying time. The ammonia "develops" the remaining sensitized coating and produces the print. The print is considered to be permanent, though it will eventually fade if exposed to excessive sunlight.

Two types of print machines are used, the first providing only the ultraviolet light source. Once removed from the copying machine, the print paper is usually inserted in a tube that holds the ammonia vapor. A more popular and less time-consuming system is shown in Fig. 19-1. In this design, the ultraviolet light source and the ammonia vapors are both contained within the machine, so that when the print paper leaves the printer, a finished reproduction is generated.

Fig. 19-1. Ammonia-sensitized copying machine. *(Courtesy of Bruning)*

Xerographic Systems

Over the past fifteen to twenty years, substantial advances have been made in the field of xerography. Xerography is a form of photocopying that employs the principles of electrostatics and heat processing to generate an image on paper. Such systems are available with dry powder or liquid ink media. Either way, copies can be made as rapidly as one per second, and in a variety of sizes.

A recent application of xerography is its merging with CADD systems. Laser printers make one of the highest quality hard copies, comparable to inked line drawings found in periodicals and books. In some CADD systems, prints are made by a xerographic unit that is coupled to a laser printer.

When xerographic copiers were first introduced, they were limited to reproducing full-sized drawings on paper no larger than 8.5″ × 11″ and 8.5″ × 14″. Today copying machines can produce prints to a variety of scales on paper the equivalent size of most master drawings. More expensive units are capable of reproducing full-color prints and halftone pictures on papers and films.

The advantage of xerographic systems is that copies can be made from master drawings prepared on any type of material (i.e., translucent and opaque). In addition, it is now possible to produce prints on a variety of media, including bond papers, vellums, and acrylics. Since advances in xerographic technology have been so rapid, few engineering and drafting departments have realized its full potential. It is expected, however, that in coming years more and more firms will make effective use of xerographic systems.

Photographic Systems

Photographic systems have not been widely used as a reproductive method in industrial drafting. Their primary application has been in legal documents, in which precise duplication is necessary. The most common use of photographic systems has been for storing large quantities of drawings, or for reducing exceptionally large drawings to a more manageable size. Shown in Fig. 19-2 is an engineering drawing that has been photographically reduced to one-eighteenth actual size, as seen in an engineering data reader.

Fig. 19-2. Photographically reduced engineering drawing displayed on a data reader. *(Courtesy of Bruning)*

In addition to use in documents, there are times when photographic processes are employed to meet high quantity requirements. That is, when a hundred or more copies of a drawing are necessary, photographic techniques are used to make offset (printing) plates. These plates are mounted on a printing press, from which the desired number of copies are made. Drawings are frequently reduced to fit on standard 8.5″ × 11″ paper, but can be printed in sizes up to 58″ × 77″.

The offset printing technique is used, for example, for products sold in large quantities that require assembly and/or servicing by the purchaser. Here drawings are used to show the purchaser how to assemble the product and service it (e.g., lubrication and cleaning).

Another photographic process used to produce high-quality copies is the *PMT*, which is a direct print made from the drawing. PMTs can be made full size or to any given scale, and are considered to be of extremely high quality. As might be expected, photographic reproductions are one of the more expensive techniques available. Before selecting any one of these processes, therefore, careful analysis must be made of the cost-benefit ratio of the copy produced.

Other Reproduction Systems

There are a number of other reproduction systems available to the drafter and engineer, those just described being the techniques more commonly used in drafting and engineering departments. In addition to these, there are two other techniques for drawing reproduction, but they are infrequently used and tend to be more applicable to the reproduction of specifications and text materials.

The first are *stencils*, which are "cut" by using a stylus or typewriter, where the stencil serves as the master. Color selection is available via inks, with typical sheet sizes being 5″ × 3″, 9″ × 15″, and 14″ × 18″. The second technique is known as the *spirit process*, in which the image is transferred by a pressure procedure. Here the ink image is on the back of the master and is transferred to paper by chemical and contact pressure. Color selection is also available with this technique.

Storage Systems

Working practices and legal requirements frequently dictate that project drawings be accessible for a given length of time. The system used for this purpose will be determined by the storage requirements. There are two broad categories of storage systems employed by industry: short and long term. Short-term storage refers to a length of time equivalent to the run of the project, while long-term storage usually follows the end of the project and lasts for a number of years.

Short-term Storage

The most visible and recognizable form of storage is the short-term storage system. The physical location of such a system will be either in the drafting area or in a nearby room. Short-term storage provides a place where drawings and their prints are kept for ready reference by the engineering and technical staff. If these drawings are kept handy, it will be easy to make additional prints or drawing changes as needed, particularly when prints must be taken to the job site and are subject to damage and loss.

A number of different types of storage unit are available to the drafting department. The first is *drawer storage units* (Fig. 19-3). Here drawers are used for storing drawings and prints in such a way that they will not be folded or rolled. These units are of sufficient size so that they will be capable of holding most drawings, and are typically designed to hold a number of drawings that total 1 inch to 3 inches in thickness.

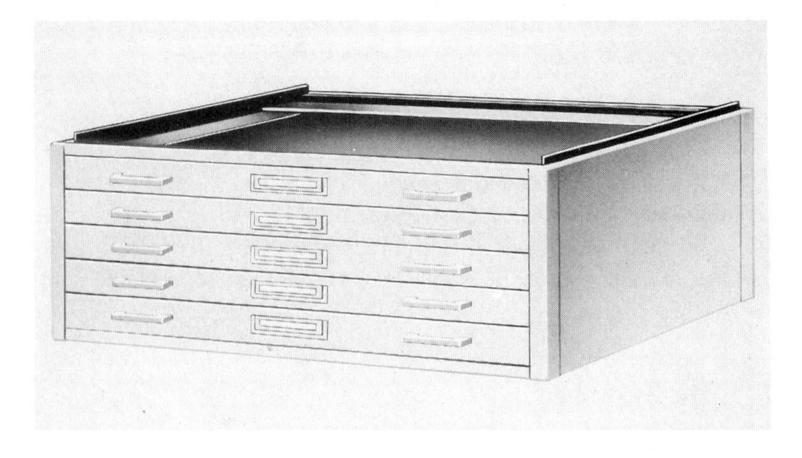

Fig. 19-3. Drawer storage unit. *(Courtesy of Bruning)*

Another common type of storage unit used for short periods of time is *rack storage*. In this design, hinges (Fig. 19-4a) are attached to the top or side of the drawings and act as a binding so that a set of working drawings can be opened like a book. The hinges are then used to hang on racks that are either permanently fixed to walls (Fig. 19-4b) or are portable units. The chief advantages of rack storage are that it makes the drawings more accessible to the project staff, and also reduces the amount of required storage space.

A third type of short-term storage unit is the *roll file*. These units are used when drawings are in frequent use, or when modifications are still being made on them. As shown in Fig. 19-5, the roll file can be easily placed at any location, and makes it easy for drafters to take a drawing and mount it directly to their drawing table, or for engineers to refer to.

Fig. 19-4a. Hinged drawing for hanging in rack storage unit. *(Courtesy of Bruning)*

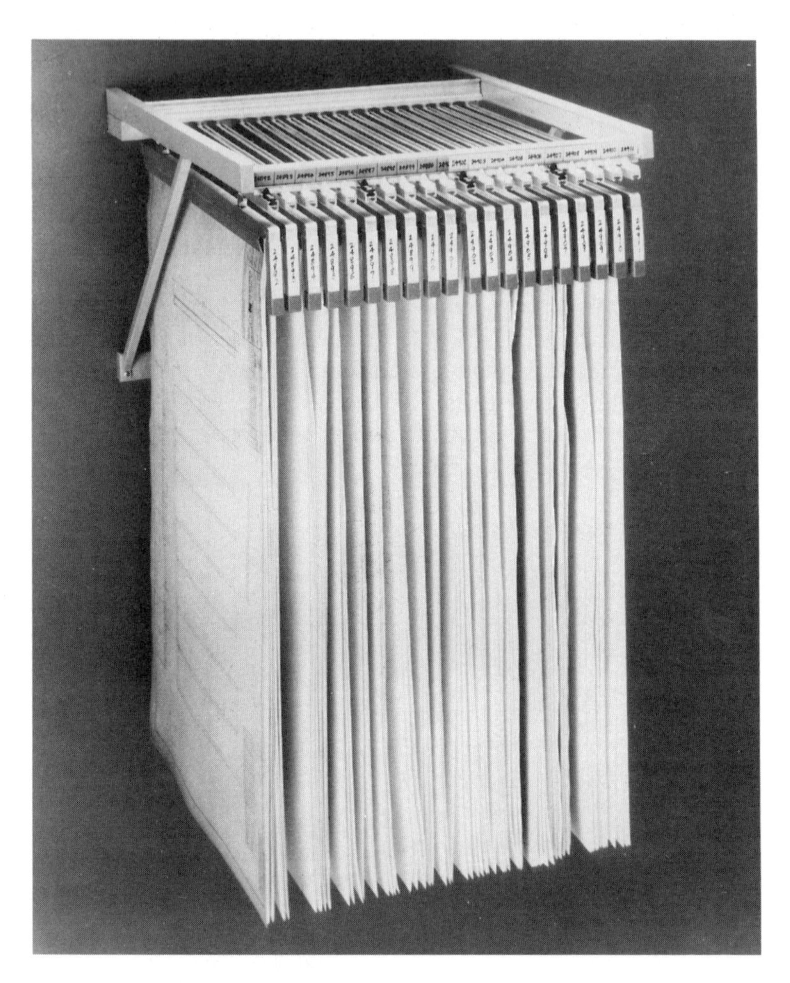

Fig. 19-4b. Rack storage unit. *(Courtesy of Bruning)*

Long-term Storage

Many companies keep their drawings, or copies, for seven years or more; some firms have drawings that date back to the 1800s. The reason for keeping working drawings for this length of time is twofold. The first is that in some cases (e.g., contract agreements) the firm has a legal responsibility to make available these plans to their customers for a given number of years. Another and more common reason is

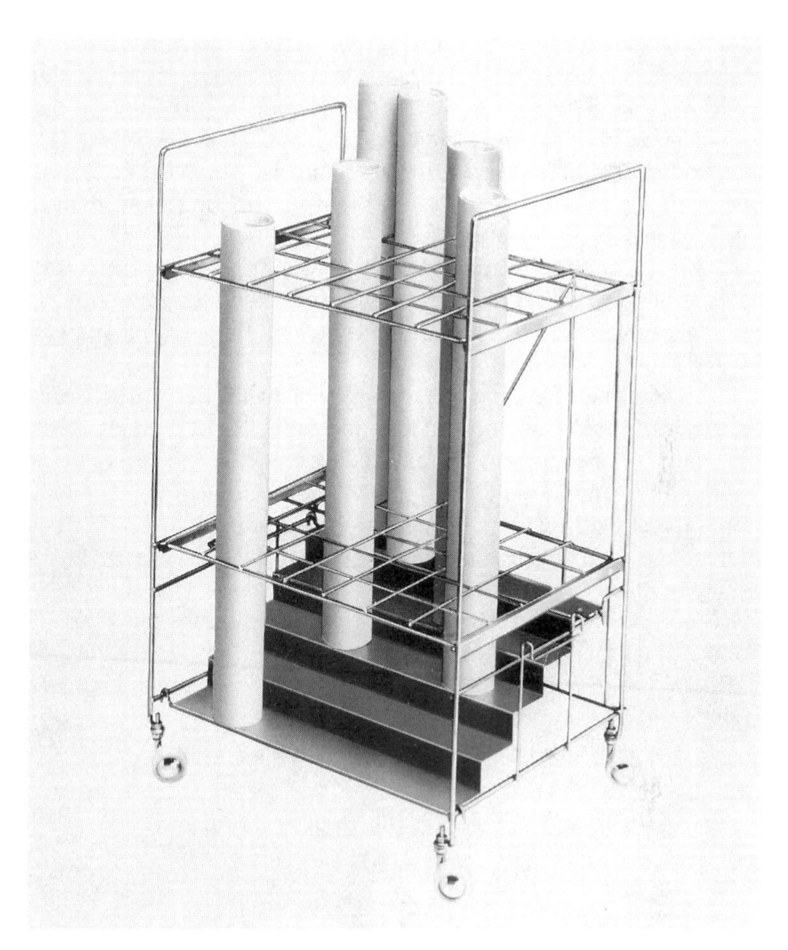

Fig. 19-5. Roll file unit. *(Courtesy of Bruning)*

that companies may use these drawings for identifying and/or manufacturing replacement parts. This latter reason is typical for heavy industries, especially those incorporating such processes as forging, casting, die forming, and machining. In addition, it is not uncommon for a company to make a product for thirty or more years, especially when there is a continuing demand for it.

One of the more traditional methods of long-term storage is the placement of drawings in *storage vaults.* Before being placed in these vaults, the drawings will be put in a *storage tube*, which is made out of cardboard, plastic, or metal, after which they are placed in the vault on racks. Their exact location in the vault will be kept on file for reference. In many cases, vaults are environmentally controlled (i.e., temperature and humidity controlled) to minimize drawing deterioration. Other companies have "vaults" that are nothing more than a storage room or screened-in area.

A number of industries are moving away from the storage vault concept because of the costs involved in maintaining such a system, particularly when a large number of drawings are prepared annually, and require a large storage area.

An alternate long-storage system, which has been briefly discussed, is the use of photoreduction procedures. As its name implies, the drawings are photographically reduced in size, most frequently to one-third original size. These photographs are held on tape and stored in cabinets for later viewing on film-reading screens. The actual image is a negative of the drawing, white lines with black background. Since the film is made of an acetate or acrylic film, there is less concern for environmental controls, and the size reduction of the drawings greatly reduces the amount of storage space needed. If a copy of the drawing is desired, a photographic print can be made from the roll of film. Shown in Fig. 19-6 is a photograph of a typical roll film and film adapter used for scanning the film.

Fig. 19-6. Roll film and adapter. *(Courtesy of Bruning)*

Another form of photographic reduction is the *aperture card system*. Instead of having a series of drawings on an entire roll of film, individual photographs are mounted on aperture cards. The cards are then coded for easy identification by a card reader or manually. Both photographic reduction procedures are often subcontracted out to firms specializing in engineering micrographic systems.

Review Questions

19.1 What is a blueprint, and how does it compare to a blue line print?

19.2 Explain the purpose and advantages of intermediate prints.

19.3 Describe how ammonia-sensitized prints are made. What are the two terms used to describe this process?

19.4 Briefly identify and describe three other reproduction systems.

19.5 Give examples of long- and short-term storage applications.

APPENDIX A

Allowance and Fits Tables

Table 1 American National Standard Running and Sliding Fits

Nominal Size Range, Inches	Class RC 1			Class RC 2			Class RC 3			Class RC 4		
		Standard Tolerance Limits			Standard Tolerance Limits			Standard Tolerance Limits			Standard Tolerance Limits	
	Clear-ance	Hole H5	Shaft g4	Clear-ance	Hole H6	Shaft g5	Clear-ance	Hole H7	Shaft f6	Clear-ance	Hole H8	Shaft f7
Over To	Values shown below are in thousandths of an inch											
0– 0.12	0.1 0.45	+0.2 0	−0.1 −0.25	0.1 0.55	+0.25 0	−0.1 −0.3	0.3 0.95	+0.4 0	−0.3 −0.55	0.3 1.3	+0.6 0	−0.3 −0.7
0.12– 0.24	0.15 0.5	+0.2 0	−0.15 −0.3	0.15 0.65	+0.3 0	−0.15 −0.35	0.4 1.12	+0.5 0	−0.4 −0.7	0.4 1.6	+0.7 0	−0.4 −0.9
0.24– 0.40	0.2 0.6	+0.25 0	−0.2 −0.35	0.2 0.85	+0.4 0	−0.2 −0.45	0.5 1.5	+0.6 0	−0.5 −0.9	0.5 2.0	+0.9 0	−0.5 −1.1
0.40– 0.71	0.25 0.75	+0.3 0	−0.25 −0.45	0.25 0.95	+0.4 0	−0.25 −0.55	0.6 1.7	+0.7 0	−0.6 −1.0	0.6 2.3	+1.0 0	−0.6 −1.3
0.71– 1.19	0.3 0.95	+0.4 0	−0.3 −0.55	0.3 1.2	+0.5 0	−0.3 −0.7	0.8 2.1	+0.8 0	−0.8 −1.3	0.8 2.8	+1.2 0	−0.8 −1.6
1.19– 1.97	0.4 1.1	+0.4 0	−0.4 −0.7	0.4 1.4	+0.6 0	−0.4 −0.8	1.0 2.6	+1.0 0	−1.0 −1.6	1.0 3.6	+1.6 0	−1.0 −2.0
1.97 – 3.15	0.4 1.2	+0.5 0	−0.4 −0.7	0.4 1.6	+0.7 0	−0.4 −0.9	1.2 3.1	+1.2 0	−1.2 −1.9	1.2 4.2	+1.8 0	−1.2 −2.4
3.15– 4.73	0.5 1.5	+0.6 0	−0.5 −0.9	0.5 2.0	+0.9 0	−0.5 −1.1	1.4 3.7	+1.4 0	−1.4 −2.3	1.4 5.0	+2.2 0	−1.4 −2.8
4.73– 7.09	0.6 1.8	+0.7 0	−0.6 −1.1	0.6 2.3	+1.0 0	−0.6 −1.3	1.6 4.2	+1.6 0	−1.6 −2.6	1.6 5.7	+2.5 0	−1.6 −3.2
7.09– 9.85	0.6 2.0	+0.8 0	−0.6 −1.2	0.6 2.6	+1.2 0	−0.6 −1.4	2.0 5.0	+1.8 0	−2.0 −3.2	2.0 6.6	+2.8 0	−2.0 −3.8
9.85–12.41	0.8 2.3	+0.9 0	−0.8 −1.4	0.8 2.9	+1.2 0	−0.8 −1.7	2.5 5.7	+2.0 0	−2.5 −3.7	2.5 7.5	+3.0 0	−2.5 −4.5
12.41–15.75	1.0 2.7	+1.0 0	−1.0 −1.7	1.0 3.4	+1.4 0	−1.0 −2.0	3.0 6.6	+2.2 0	−3.0 −4.4	3.0 8.7	+3.5 0	−3.0 −5.2
15.75–19.69	1.2 3.0	+1.0 0	−1.2 −2.0	1.2 3.8	+1.6 0	−1.2 −2.2	4.0 8.1	+2.5 0	−4.0 −5.6	4.0 10.5	+4.0 0	−4.0 −6.5

Table 1 (continued)

Nominal Size Range, Inches	Class RC 5			Class RC 6			Class RC 7			Class RC 8			Class RC 9		
		Standard Tolerance Limits			Standard Tolerance Limits			Standard Tolerance Limits			Standard Tolerance Limits			Standard Tolerance Limits	
Over To	Clearance	Hole H8	Shaft e7	Clearance	Hole H9	Shaft e8	Clearance	Hole H9	Shaft d8	Clearance	Hole H10	Shaft c9	Clearance	Hole H11	Shaft
	Values shown below are in thousandths of an inch														
0– 0.12	0.6 1.6	+0.6 0	− 0.6 − 1.0	0.6 2.2	+1.0 0	− 0.6 − 1.2	1.0 2.6	+1.0 0	− 1.0 − 1.6	2.5 5.1	+ 1.6 0	− 2.5 − 3.5	4.0 8.1	+ 2.5 0	− 4.0 − 5.6
0.12– 0.24	0.8 2.0	+0.7 0	− 0.8 − 1.3	0.8 2.7	+1.2 0	− 0.8 − 1.5	1.2 3.1	+1.2 0	− 1.2 − 1.9	2.8 5.8	+ 1.8 0	− 2.8 − 4.0	4.5 9.0	+ 3.0 0	− 4.5 − 6.0
0.24– 0.40	1.0 2.5	+0.9 0	− 1.0 − 1.6	1.0 3.3	+1.4 0	− 1.0 − 1.9	1.6 3.9	+1.4 0	− 1.6 − 2.5	3.0 6.6	+ 2.2 0	− 3.0 − 4.4	5.0 10.7	+ 3.5 0	− 5.0 − 7.2
0.40– 0.71	1.2 2.9	+1.0 0	− 1.2 − 1.9	1.2 3.8	+1.6 0	− 1.2 − 2.2	2.0 4.6	+1.6 0	− 2.0 − 3.0	3.5 7.9	+ 2.8 0	− 3.5 − 5.1	6.0 12.8	+ 4.0 0	− 6.0 − 8.8
0.71– 1.19	1.6 3.6	+1.2 0	− 1.6 − 2.4	1.6 4.8	+2.0 0	− 1.6 − 2.8	2.5 5.7	+2.0 0	− 2.5 − 3.7	4.5 10.0	+ 3.5 0	− 4.5 − 6.5	7.0 15.5	+ 5.0 0	− 7.0 − 10.5
1.19– 1.97	2.0 4.6	+1.6 0	− 2.0 − 3.0	2.0 6.1	+2.5 0	− 2.0 − 3.6	3.0 7.1	+2.5 0	− 3.0 − 4.6	5.0 11.5	+ 4.0 0	− 5.0 − 7.5	8.0 18.0	+ 6.0 0	− 8.0 − 12.0
1.97– 3.15	2.5 5.5	+1.8 0	− 2.5 − 3.7	2.5 7.3	+3.0 0	− 2.5 − 4.3	4.0 8.8	+3.0 0	− 4.0 − 5.8	6.0 13.5	+ 4.5 0	− 6.0 − 9.0	9.0 20.5	+ 7.0 0	− 9.0 − 13.5
3.15– 4.73	3.0 6.6	+2.2 0	− 3.0 − 4.4	3.0 8.7	+3.5 0	− 3.0 − 5.2	5.0 10.7	+3.5 0	− 5.0 − 7.2	7.0 15.5	+ 5.0 0	− 7.0 − 10.5	10.0 24.0	+ 9.0 0	− 10.0 − 15.0
4.73– 7.09	3.5 7.6	+2.5 0	− 3.5 − 5.1	3.5 10.0	+4.0 0	− 3.5 − 6.0	6.0 12.5	+4.0 0	− 6.0 − 8.5	8.0 18.0	+ 6.0 0	− 8.0 − 12.0	12.0 28.0	+10.0 0	− 12.0 − 18.0
7.09– 9.85	4.0 8.6	+2.8 0	− 4.0 − 5.8	4.0 11.3	+4.5 0	− 4.0 − 6.8	7.0 14.3	+4.5 0	− 7.0 − 9.8	10.0 21.5	+ 7.0 0	− 10.0 − 14.5	15.0 34.0	+12.0 0	− 15.0 − 22.0
9.85–12.41	5.0 10.0	+3.0 0	− 5.0 − 7.0	5.0 13.0	+5.0 0	− 5.0 − 8.0	8.0 16.0	+5.0 0	− 8.0 − 11.0	12.0 25.0	+ 8.0 0	− 12.0 − 17.0	18.0 38.0	+12.0 0	− 18.0 − 26.0
12.41–15.75	6.0 11.7	+3.5 0	− 6.0 − 8.2	6.0 15.5	+6.0 0	− 6.0 − 9.5	10.0 19.5	+6.0 0	− 10.0 − 13.5	14.0 29.0	+ 9.0 0	− 14.0 − 20.0	22.0 45.0	+14.0 0	− 22.0 − 31.0
15.75–19.69	8.0 14.5	+4.0 0	− 8.0 − 10.5	8.0 18.0	+6.0 0	− 8.0 − 12.0	12.0 22.0	+6.0 0	− 12.0 − 16.0	16.0 32.0	+10.0 0	− 16.0 − 22.0	25.0 51.0	+16.0 0	− 25.0 − 35.0

Table 2 American National Standard Clearance Locational Fits

Nominal Size Range, Inches	Class LC 1			Class LC 2			Class LC 3			Class LC 4			Class LC 5		
		Standard Tolerance Limits			Standard Tolerance Limits			Standard Tolerance Limits			Standard Tolerance Limits			Standard Tolerance Limits	
Over To	Clear-ance	Hole H6	Shaft h5	Clear-ance	Hole H7	Shaft h6	Clear-ance	Hole H8	Shaft h7	Clear-ance	Hole H10	Shaft h9	Clear-ance	Hole H7	Shaft g6
	Values shown below are in thousandths of an inch														
0– 0.12	0 0.45	+0.25 0	0 −0.2	0 0.65	+0.4 0	0 −0.25	0 1	+0.6 0	0 −0.4	0 2.6	+1.6 0	0 −1.0	0.1 0.75	+0.4 0	−0.1 −0.35
0.12– 0.24	0 0.5	+0.3 0	0 −0.2	0 0.8	+0.5 0	0 −0.3	0 1.2	+0.7 0	0 −0.5	0 3.0	+1.8 0	0 −1.2	0.15 0.95	+0.5 0	−0.15 −0.45
0.24– 0.40	0 0.65	+0.4 0	0 −0.25	0 1.0	+0.6 0	0 −0.4	0 1.5	+0.9 0	0 −0.6	0 3.6	+2.2 0	0 −1.4	0.2 1.2	+0.6 0	−0.2 −0.6
0.40– 0.71	0 0.7	+0.4 0	0 −0.3	0 1.1	+0.7 0	0 −0.4	0 1.7	+1.0 0	0 −0.7	0 4.4	+2.8 0	0 −1.6	0.25 1.35	+0.7 0	−0.25 −0.65
0.71– 1.19	0 0.9	+0.5 0	0 −0.4	0 1.3	+0.8 0	0 −0.5	0 2	+1.2 0	0 −0.8	0 5.5	+3.5 0	0 −2.0	0.3 1.6	+0.8 0	−0.3 −0.8
1.19– 1.97	0 1.0	+0.6 0	0 −0.4	0 1.6	+1.0 0	0 −0.6	0 2.6	+1.6 0	0 −1	0 6.5	+4.0 0	0 −2.5	0.4 2.0	+1.0 0	−0.4 −1.0
1.97– 3.15	0 1.2	+0.7 0	0 −0.5	0 1.9	+1.2 0	0 −0.7	0 3	+1.8 0	0 −1.2	0 7.5	+4.5 0	0 −3	0.4 2.3	+1.2 0	−0.4 −1.1
3.15– 4.73	0 1.5	+0.9 0	0 −0.6	0 2.3	+1.4 0	0 −0.9	0 3.6	+2.2 0	0 −1.4	0 8.5	+5.0 0	0 −3.5	0.5 2.8	+1.4 0	−0.5 −1.4
4.73– 7.09	0 1.7	+1.0 0	0 −0.7	0 2.6	+1.6 0	0 −1.0	0 4.1	+2.5 0	0 −1.6	0 10.0	+6.0 0	0 −4	0.6 3.2	+1.6 0	−0.6 −1.6
7.09– 9.85	0 2.0	+1.2 0	0 −0.8	0 3.0	+1.8 0	0 −1.2	0 4.6	+2.8 0	0 −1.8	0 11.5	+7.0 0	0 −4.5	0.6 3.6	+1.8 0	−0.6 −1.8
9.85–12.41	0 2.1	+1.2 0	0 −0.9	0 3.2	+2.0 0	0 −1.2	0 5	+3.0 0	0 −2.0	0 13.0	+8.0 0	0 −5	0.7 3.9	+2.0 0	−0.7 −1.9
12.41–15.75	0 2.4	+1.4 0	0 −1.0	0 3.6	+2.2 0	0 −1.4	0 5.7	+3.5 0	0 −2.2	0 15.0	+9.0 0	0 −6	0.7 4.3	+2.2 0	−0.7 −2.1
15.75–19.69	0 2.6	+1.6 0	0 −1.0	0 4.1	+2.5 0	0 −1.6	0 6.5	+4 0	0 −2.5	0 16.0	+10.0 0	0 −6	0.8 4.9	+2.5 0	−0.8 −2.4

Table 2 (continued)

Nominal Size Range, Inches	Class LC 6			Class LC 7			Class LC 8			Class LC 9			Class LC 10			Class LC 11		
	Clearance	Std. Tolerance Limits		Clearance	Std. Tolerance Limits		Clearance	Std. Tolerance Limits		Clearance	Std. Tolerance Limits		Clearance	Std. Tolerance Limits		Clearance	Std. Tolerance Limits	
Over To		Hole H9	Shaft f8		Hole H10	Shaft e9		Hole H10	Shaft d9		Hole H11	Shaft c10		Hole H12	Shaft		Hole H13	Shaft
	Values shown below are in thousandths of an inch																	
0– 0.12	0.3 1.9	+1.0 0	−0.3 −0.9	0.6 3.2	+1.6 0	−0.6 −1.6	1.0 2.0	+1.6 0	−1.0 −2.0	2.5 6.6	+2.5 0	−2.5 −4.1	4 12	+4 0	−4 −8	5 17	+6 0	−5 −11
0.12– 0.24	0.4 2.3	+1.2 0	−0.4 −1.1	0.8 3.8	+1.8 0	−0.8 −2.0	1.2 4.2	+1.8 0	−1.2 −2.4	2.8 7.6	+3.0 0	−2.8 −4.6	4.5 14.5	+5 0	−4.5 −9.5	6 20	+7 0	−6 −13
0.24– 0.40	0.5 2.8	+1.4 0	−0.5 −1.4	1.0 4.6	+2.2 0	−1.0 −2.4	1.6 5.2	+2.2 0	−1.6 −3.0	3.0 8.7	+3.5 0	−3.0 −5.2	5 17	+6 0	−5 −11	7 25	+9 0	−7 −16
0.40– 0.71	0.6 3.2	+1.6 0	−0.6 −1.6	1.2 5.6	+2.8 0	−1.2 −2.8	2.0 6.4	+2.8 0	−2.0 −3.6	3.5 10.3	+4.0 0	−3.5 −6.3	6 20	+7 0	−6 −13	8 28	+10 0	−8 −18
0.71– 1.19	0.8 4.0	+2.0 0	−0.8 −2.0	1.6 7.1	+3.5 0	−1.6 −3.6	2.5 8.0	+3.5 0	−2.5 −4.5	4.5 13.0	+5.0 0	−4.5 −8.0	7 23	+8 0	−7 −15	10 34	+12 0	−10 −22
1.19– 1.97	1.0 5.1	+2.5 0	−1.0 −2.6	2.0 8.5	+4.0 0	−2.0 −4.5	3.6 9.5	+4.0 0	−3.0 −5.5	5.0 15.0	+6 0	−5.0 −9.0	8 28	+10 0	−8 −18	12 44	+16 0	−12 −28
1.97– 3.15	1.2 6.0	+3.0 0	−1.0 −3.0	2.5 10.0	+4.5 0	−2.5 −5.5	4.0 11.5	+4.5 0	−4.0 −7.0	6.0 17.5	+7 0	−6.0 −10.5	10 34	+12 0	−10 −22	14 50	+18 0	−14 −32
3.15– 4.73	1.4 7.1	+3.5 0	−1.4 −3.6	3.0 11.5	+5.0 0	−3.0 −6.5	5.0 13.5	+5.0 0	−5.0 −8.5	7 21	+9 0	−7 −12	11 39	+14 0	−11 −25	16 60	+22 0	−16 −38
4.73– 7.09	1.6 8.1	+4.0 0	−1.6 −4.1	3.5 13.5	+6.0 0	−3.5 −7.5	6 16	+6 0	−6 −10	8 24	+10 0	−8 −14	12 44	+16 0	−12 −28	18 68	+25 0	−18 −43
7.09– 9.85	2.0 9.3	+4.5 0	−2.0 −4.8	4.0 15.5	+7.0 0	−4.0 −8.5	7 18.5	+7 0	−7 −11.5	10 29	+12 0	−10 −17	16 52	+18 0	−16 −34	22 78	+28 0	−22 −50
9.85–12.41	2.2 10.2	+5.0 0	−2.2 −5.2	4.5 17.5	+8.0 0	−4.5 −9.5	7 20	+8 0	−7 −12	12 32	+12 0	−12 −20	20 60	+20 0	−20 −40	28 88	+30 0	−28 −58
12.41–15.75	2.5 12.0	+6.0 0	−2.5 −6.0	5.0 20.0	+9.0 0	−5 −11	8 23	+9 0	−8 −14	14 37	+14 0	−14 −23	22 66	+22 0	−22 −44	30 100	+35 0	−30 −65
15.75–19.69	2.8 12.8	+6.0 0	−2.8 −6.8	5.0 21.0	+10.0 0	−5 −11	9 25	+10 0	−9 −15	16 42	+1.6 0	−16 −26	25 75	+25 0	−25 −50	35 115	+40 0	−35 −75

Table 3 ANSI Standard Transition Location Fits

Nominal Size Range, Inches	Class LT 1			Class LT 2			Class LT 3			Class LT 4			Class LT 5			Class LT 6		
	Fit	Std. Tolerance Limits		Fit	Std. Tolerance Limits		Fit	Std. Tolerance Limits		Fit	Std. Tolerance Limits		Fit	Std. Tolerance Limits		Fit	Std. Tolerance Limits	
Over To		Hole H7	Shaft js6		Hole H8	Shaft js7		Hole H7	Shaft k6		Hole H8	Shaft k7		Hole H7	Shaft n6		Hole H7	Shaft n7
	Values shown below are in thousandths of an inch																	
0– 0.12	−0.12 +0.52	+0.4 0	+0.12 −0.12	−0.2 +0.8	+0.6 0	+0.2 −0.2							−0.5 +0.15	+0.4 0	+0.5 +0.25	−0.65 +0.15	+0.4 0	+0.65 +0.25
0.12– 0.24	−0.15 +0.65	+0.5 0	+0.15 −0.15	−0.25 +0.95	+0.7 0	+0.25 −0.25							−0.6 +0.2	+0.5 0	+0.6 +0.3	−0.8 +0.2	+0.5 0	+0.8 +0.3
0.24– 0.40	−0.2 +0.8	+0.6 0	+0.2 −0.2	−0.3 +1.2	+0.9 0	+0.3 −0.3	−0.5 +0.5	+0.6 0	+0.5 +0.1	−0.7 +0.8	+0.9 0	+0.7 +0.1	−0.8 +0.2	+0.6 0	+0.8 +0.4	−1.0 +0.2	+0.6 0	+1.0 +0.4
0.40– 0.71	−0.2 +0.9	+0.7 0	+0.2 −0.2	−0.35 +1.35	+1.0 0	+0.35 −0.35	−0.5 +0.6	+0.7 0	+0.5 +0.1	−0.8 +0.9	+1.0 0	+0.8 +0.1	−0.9 +0.2	+0.7 0	+0.9 +0.5	−1.2 +0.2	+0.7 0	+1.2 +0.5
0.71– 1.19	−0.25 +1.05	+0.8 0	+0.25 −0.25	−0.4 +1.6	+1.2 0	+0.4 −0.4	−0.6 +0.7	+0.8 0	+0.6 +0.1	−0.9 +1.1	+1.2 0	+0.9 +0.1	−1.1 +0.2	+0.8 0	+1.1 +0.6	−1.4 +0.2	+0.8 0	+1.4 +0.6
1.19– 1.97	−0.3 +1.3	+1.0 0	+0.3 −0.3	−0.5 +2.1	+1.6 0	+0.5 −0.5	−0.7 +0.9	+1.0 0	+0.7 +0.1	−1.1 +1.5	+1.6 0	+1.1 +0.1	−1.3 +0.3	+1.0 0	+1.3 +0.7	−1.7 +0.3	+1.0 0	+1.7 +0.7
1.97– 3.15	−0.3 +1.5	+1.2 0	+0.3 −0.3	−0.6 +2.4	+1.8 0	+0.6 −0.6	−0.8 +1.1	+1.2 0	+0.8 +0.1	−1.3 +1.7	+1.8 0	+1.3 +0.1	−1.5 +0.4	+1.2 0	+1.5 +0.8	−2.0 +0.4	+1.2 0	+2.0 +0.8
3.15– 4.73	−0.4 +1.8	+1.4 0	+0.4 −0.4	−0.7 +2.9	+2.2 0	+0.7 −0.7	−1.0 +1.3	+1.4 0	+1.0 +0.1	−1.5 +2.1	+2.2 0	+1.5 +0.1	−1.9 +0.4	+1.4 0	+1.9 +1.0	−2.4 +0.4	+1.4 0	+2.4 +1.0
4.73– 7.09	−0.5 +2.1	+1.6 0	+0.5 −0.5	−0.8 +3.3	+2.5 0	+0.8 −0.8	−1.1 +1.5	+1.6 0	+1.1 +0.1	−1.7 +2.4	+2.5 0	+1.7 +0.1	−2.2 +0.4	+1.6 0	+2.2 +1.2	−2.8 +0.4	+1.6 0	+2.8 +1.2
7.09– 9.85	−0.6 +2.4	+1.8 0	+0.6 −0.6	−0.9 +3.7	+2.8 0	+0.9 −0.9	−1.4 +1.6	+1.8 0	+1.4 +0.2	−2.0 +2.6	+2.8 0	+2.0 +0.2	−2.6 +0.4	+1.8 0	+2.6 +1.4	−3.2 +0.4	+1.8 0	+3.2 +1.4
9.85–12.41	−0.6 +2.6	+2.0 0	+0.6 −0.6	−1.0 +4.0	+3.0 0	+1.0 −1.0	−1.4 +1.8	+2.0 0	+1.4 +0.2	−2.2 +2.8	+3.0 0	+2.2 +0.2	−2.6 +0.6	+2.0 0	+2.6 +1.4	−3.4 +0.6	+2.0 0	+3.4 +1.4
12.41–15.75	−0.7 +2.9	+2.2 0	+0.7 −0.7	−1.0 +4.5	+3.5 0	+1.0 −1.0	−1.6 +2.0	+2.2 0	+1.6 +0.2	−2.4 +3.3	+3.5 0	+2.4 +0.2	−3.0 +0.6	+2.2 0	+3.0 +1.6	−3.8 +0.6	+2.2 0	+3.8 +1.6
15.75–19.69	−0.8 +3.3	+2.5 0	+0.8 −0.8	−1.2 +5.2	+4.0 0	+1.2 −1.2	−1.8 +2.3	+2.5 0	+1.8 +0.2	−2.7 +3.8	+4.0 0	+2.7 +0.2	−3.4 +0.7	+2.5 0	+3.4 +1.8	−4.3 +0.7	+2.5 0	+4.3 +1.8

Table 4 ANSI Standard Interference Locational Fits

Nominal Size Range, Inches	Class LN 1			Class LN 2			Class LN 3		
	Limits of Interference	Standard Limits		Limits of Interference	Standard Limits		Limits of Interference	Standard Limits	
		Hole H6	Shaft n5		Hole H7	Shaft p6		Hole H7	Shaft r6
Over To	Values shown below are given in thousandths of an inch								
0– 0.12	0 0.45	+0.25 0	+0.45 +0.25	0 0.65	+0.4 0	+0.65 +0.4	0.1 0.75	+0.4 0	+0.75 +0.5
0.12– 0.24	0 0.5	+0.3 0	+0.5 +0.3	0 0.8	+0.5 0	+0.8 +0.5	0.1 0.9	+0.5 0	+0.9 +0.6
0.24– 0.40	0 0.65	+0.4 0	+0.65 +0.4	0 1.0	+0.6 0	+1.0 +0.6	0.2 1.2	+0.6 0	+1.2 +0.8
0.40– 0.71	0 0.8	+0.4 0	+0.8 +0.4	0 1.1	+0.7 0	+1.1 +0.7	0.3 1.4	+0.7 0	+1.4 +1.0
0.71– 1.19	0 1.0	+0.5 0	+1.0 +0.5	0 1.3	+0.8 0	+1.3 +0.8	0.4 1.7	+0.8 0	+1.7 +1.2
1.19– 1.97	0 1.1	+0.6 0	+1.1 +0.6	0 1.6	+1.0 0	+1.6 +1.0	0.4 2.0	+1.0 0	+2.0 +1.4
1.97– 3.15	0.1 1.3	+0.7 0	+1.3 +0.8	0.2 2.1	+1.2 0	+2.1 +1.4	0.4 2.3	+1.2 0	+2.3 +1.6
3.15– 4.73	0.1 1.6	+0.9 0	+1.6 +1.0	0.2 2.5	+1.4 0	+2.5 +1.6	0.6 2.9	+1.4 0	+2.9 +2.0
4.73– 7.09	0.2 1.9	+1.0 0	+1.9 +1.2	0.2 2.8	+1.6 0	+2.8 +1.8	0.9 3.5	+1.6 0	+3.5 +2.5
7.09– 9.85	0.2 2.2	+1.2 0	+2.2 +1.4	0.2 3.2	+1.8 0	+3.2 +2.0	1.2 4.2	+1.8 0	+4.2 +3.0
9.85–12.41	0.2 2.3	+1.2 0	+2.3 +1.4	0.2 3.4	+2.0 0	+3.4 +2.2	1.5 4.7	+2.0 0	+4.7 +3.5
12.41–15.75	0.2 2.6	+1.4 0	+2.6 +1.6	0.3 3.9	+2.2 0	+3.9 +2.5	2.3 5.9	+2.2 0	+5.9 +4.5
15.75–19.69	0.2 2.8	+1.6 0	+2.8 +1.8	0.3 4.4	+2.5 0	+4.4 +2.8	2.5 6.6	+2.5 0	+6.6 +5.0

Table 5 ANSI Standard Force and Shrink Fits

Nominal Size Range, Inches	Class FN 1			Class FN 2			Class FN 3			Class FN 4			Class FN 5		
		Standard Tolerance Limits			Standard Tolerance Limits			Standard Tolerance Limits			Standard Tolerance Limits			Standard Tolerance Limits	
Over To	Interference	Hole H6	Shaft	Interference	Hole H7	Shaft s6	Interference	Hole H7	Shaft t6	Interference	Hole H7	Shaft u6	Interference	Hole H8	Shaft x7
	Values shown below are in thousandths of an inch														
0–0.12	0.05 0.5	+0.25 0	+0.5 +0.3	0.2 0.85	+0.4 0	+0.85 +0.6				0.3 0.95	+0.4 0	+0.95 +0.7	0.3 1.3	+0.6 0	+1.3 +0.9
0.12–0.24	0.1 0.6	+0.3 0	+0.6 +0.4	0.2 1.0	+0.5 0	+1.0 +0.7				0.4 1.2	+0.5 0	+1.2 +0.9	0.5 1.7	+0.7 0	+1.7 +1.2
0.24–0.40	0.1 0.75	+0.4 0	+0.75 +0.5	0.4 1.4	+0.6 0	+1.4 +1.0				0.6 1.6	+0.6 0	+1.6 +1.2	0.5 2.0	+0.9 0	+2.0 +1.4
0.40–0.56	0.1 0.8	+0.4 0	+0.8 +0.5	0.5 1.6	+0.7 0	+1.6 +1.2				0.7 1.8	+0.7 0	+1.8 +1.4	0.6 2.3	+1.0 0	+2.3 +1.6
0.56–0.71	0.2 0.9	+0.4 0	+0.9 +0.6	0.5 1.6	+0.7 0	+1.6 +1.2				0.7 1.8	+0.7 0	+1.8 +1.4	0.8 2.5	+1.0 0	+2.5 +1.8
0.71–0.95	0.2 1.1	+0.5 0	+1.1 +0.7	0.6 1.9	+0.8 0	+1.9 +1.4				0.8 2.1	+0.8 0	+2.1 +1.6	1.0 3.0	+1.2 0	+3.0 +2.2
0.95–1.19	0.3 1.2	+0.5 0	+1.2 +0.8	0.6 1.9	+0.8 0	+1.9 +1.4	0.8 2.1	+0.8 0	+2.1 +1.6	1.0 2.3	+0.8 0	+2.3 +1.8	1.3 3.3	+1.2 0	+3.3 +2.5
1.19–1.58	0.3 1.3	+0.6 0	+1.3 +0.9	0.8 2.4	+1.0 0	+2.4 +1.8	1.0 2.6	+1.0 0	+2.6 +2.0	1.5 3.1	+1.0 0	+3.1 +2.5	1.4 4.0	+1.6 0	+4.0 +3.0
1.58–1.97	0.4 1.4	+0.6 0	+1.4 +1.0	0.8 2.4	+1.0 0	+2.4 +1.8	1.2 2.8	+1.0 0	+2.8 +2.2	1.8 3.4	+1.0 0	+3.4 +2.8	2.4 5.0	+1.6 0	+5.0 +4.0
1.97–2.56	0.6 1.8	+0.7 0	+1.8 +1.3	0.8 2.7	+1.2 0	+2.7 +2.0	1.3 3.2	+1.2 0	+3.2 +2.5	2.3 4.2	+1.2 0	+4.2 +3.5	3.2 6.2	+1.8 0	+6.2 +5.0
2.56–3.15	0.7 1.9	+0.7 0	+1.9 +1.4	1.0 2.9	+1.2 0	+2.9 +2.2	1.8 3.7	+1.2 0	+3.7 +3.0	2.8 4.7	+1.2 0	+4.7 +4.0	4.2 7.2	+1.8 0	+7.2 +6.0
3.15–3.94	0.9 2.4	+0.9 0	+2.4 +1.8	1.4 3.7	+1.4 0	+3.7 +2.8	2.1 4.4	+1.4 0	+4.4 +3.5	3.6 5.9	+1.4 0	+5.9 +5.0	4.8 8.4	+2.2 0	+8.4 +7.0
3.94–4.73	1.1 2.6	+0.9 0	+2.6 +2.0	1.6 3.9	+1.4 0	+3.9 +3.0	2.6 4.9	+1.4 0	+4.9 +4.0	4.6 6.9	+1.4 0	+6.9 +6.0	5.8 9.4	+2.2 0	+9.4 +8.0

Table 5 (continued)

Nominal Size Range, Inches	Class FN 1			Class FN 2			Class FN 3			Class FN 4			Class FN 5		
		Standard Tolerance Limits			Standard Tolerance Limits			Standard Tolerance Limits			Standard Tolerance Limits			Standard Tolerance Limits	
Over To	Interference	Hole H6	Shaft	Interference	Hole H7	Shaft s6	Interference	Hole H7	Shaft t6	Interference	Hole H7	Shaft u6	Interference	Hole H8	Shaft x7
	Values shown below are in thousandths of an inch														
4.73– 5.52	1.2 2.9	+ 1.0 0	+ 2.9 + 2.2	1.9 4.5	+ 1.6 0	+ 4.5 + 3.5	3.4 6.0	+ 1.6 0	+ 6.0 + 5.0	5.4 8.0	+ 1.6 0	+ 8.0 + 7.0	7.5 11.6	+ 2.5 0	+ 11.6 + 10.0
5.52– 6.30	1.5 3.2	+ 1.0 0	+ 3.2 + 2.5	2.4 5.0	+ 1.6 0	+ 5.0 + 4.0	3.4 6.0	+ 1.6 0	+ 6.0 + 5.0	5.4 8.0	+ 1.6 0	+ 8.0 + 7.0	9.5 13.6	+ 2.5 0	+ 13.6 + 12.0
6.30– 7.09	1.8 3.5	+ 1.0 0	+ 3.5 + 2.8	2.9 5.5	+ 1.6 0	+ 5.5 + 4.5	4.4 7.0	+ 1.6 0	+ 7.0 + 6.0	6.4 9.0	+ 1.6 0	+ 9.0 + 8.0	9.5 13.6	+ 2.5 0	+ 13.6 + 12.0
7.09– 7.88	1.8 3.8	+ 1.2 0	+ 3.8 + 3.0	3.2 6.2	+ 1.8 0	+ 6.2 + 5.0	5.2 8.2	+ 1.8 0	+ 8.2 + 7.0	7.2 10.2	+ 1.8 0	+ 10.2 + 9.0	11.2 15.8	+ 2.8 0	+ 15.8 + 14.0
7.88– 8.86	2.3 4.3	+ 1.2 0	+ 4.3 + 3.5	3.2 6.2	+ 1.8 0	+ 6.2 + 5.0	5.2 8.2	+ 1.8 0	+ 8.2 + 7.0	8.2 11.2	+ 1.8 0	+ 11.2 + 10.0	13.2 17.8	+ 2.8 0	+ 17.8 + 16.0
8.86– 9.85	2.3 4.3	+ 1.2 0	+ 4.3 + 3.5	4.2 7.2	+ 1.8 0	+ 7.2 + 6.0	6.2 9.2	+ 1.8 0	+ 9.2 + 8.0	10.2 13.2	+ 1.8 0	+ 13.2 + 12.0	13.2 17.8	+ 2.8 0	+ 17.8 + 16.0
9.85–11.03	2.8 4.9	+ 1.2 0	+ 4.9 + 4.0	4.0 7.2	+ 2.0 0	+ 7.2 + 6.0	7.0 10.2	+ 2.0 0	+ 10.2 + 9.0	10.0 13.2	+ 2.0 0	+ 13.2 + 12.0	15.0 20.0	+ 3.0 0	+ 20.0 + 18.0
11.03–12.41	2.8 4.9	+ 1.2 0	+ 4.9 + 4.0	5.0 8.2	+ 2.0 0	+ 8.2 + 7.0	7.0 10.2	+ 2.0 0	+ 10.2 + 9.0	12.0 15.2	+ 2.0 0	+ 15.2 + 14.0	17.0 22.0	+ 3.0 0	+ 22.0 + 20.0
12.41–13.98	3.1 5.5	+ 1.4 0	+ 5.5 + 4.5	5.8 9.4	+ 2.2 0	+ 9.4 + 8.0	7.8 11.4	+ 2.2 0	+ 11.4 + 10.0	13.8 17.4	+ 2.2 0	+ 17.4 + 16.0	18.5 24.2	+ 3.5 0	+ 24.2 + 22.0
13.98–15.75	3.6 6.1	+ 1.4 0	+ 6.1 + 5.0	5.8 9.4	+ 2.2 0	+ 9.4 + 8.0	9.8 13.4	+ 2.2 0	+ 13.4 + 12.0	15.8 19.4	+ 2.2 0	+ 19.4 + 18.0	21.5 27.2	+ 3.5 0	+ 27.2 + 25.0
15.75–17.72	4.4 7.0	+ 1.6 0	+ 7.0 + 6.0	6.5 10.6	+ 2.5 0	+ 10.6 + 9.0	9.5 13.6	+ 2.5 0	+ 13.6 + 12.0	17.5 21.6	+ 2.5 0	+ 21.6 + 20.0	24.0 30.5	+ 4.0 0	+ 30.5 + 28.0
17.72–19.69	4.4 7.0	+ 1.6 0	+ 7.0 + 6.0	7.5 11.6	+ 2.5 0	+ 11.6 + 10.0	11.5 15.6	+ 2.5 0	+ 15.6 + 14.0	19.5 23.6	+ 2.5 0	+ 23.6 + 22.0	26.0 32.5	+ 4.0 0	+ 32.5 + 30.0

Table 6 ISO Standard Limits and Fits

Nominal Sizes, mm		Tolerance Grades									
Over	To	IT 01	IT 0	IT 1	IT 2	IT 3	IT 4	IT 5	IT 6	IT 7	IT 8
...	3	0.3	0.5	0.8	1.2	2	3	4	6	10	14
3	6	0.4	0.6	1	1.5	2.5	4	5	8	12	18
6	10	0.4	0.6	1	1.5	2.5	4	6	9	15	22
10	18	0.5	0.8	1.2	2	3	5	8	11	18	27
18	30	0.6	1	1.5	2.5	4	6	9	13	21	33
30	50	0.6	1	1.5	2.5	4	7	11	16	25	39
50	80	0.8	1.2	2	3	5	8	13	19	30	46
80	120	1	1.5	2.5	4	6	10	15	22	35	54
120	180	1.2	2	3.5	5	8	12	18	25	40	63
180	250	2	3	4.5	7	10	14	20	29	46	72
250	315	2.5	4	6	8	12	16	23	32	52	81
315	400	3	5	7	9	13	18	25	36	57	89
400	500	4	6	8	10	15	20	27	40	63	97

Nominal Sizes, mm		Tolerance Grades							
Over	To	IT 9	IT 10	IT 11	IT 12	IT 13	IT 14†	IT 15†	IT 16†
...	3	25	40	60	100	140	250	400	600
3	6	30	48	75	120	180	300	480	750
6	10	36	58	90	150	220	360	580	900
10	18	43	70	110	180	270	430	700	1100
18	30	52	84	130	210	330	520	840	1300
30	50	62	100	160	250	390	620	1000	1600
50	80	74	120	190	300	460	740	1200	1900
80	120	87	140	220	350	540	870	1400	2200
120	180	100	160	250	400	630	1000	1600	2500
180	250	115	185	290	460	720	1150	1850	2900
250	315	130	210	320	520	810	1300	2100	3200
315	400	140	230	360	570	890	1400	2300	3600
400	500	155	250	400	630	970	1550	2500	4000

† Not applicable to sizes below 1 mm.

The dimensions are given in 0.001 mm, except for the nominal sizes which are in millimeters.

Table 7 ISO Standard Fundamental Deviations for Shafts

Nominal Size, mm		Grade																
		01 to 16												5–6	7	8	4–7	≦3 >7
		Fundamental (Upper) Deviation es												Fundamental (Lower) Dev'n ei				
Over	To	a*	b*	c	cd	d	e	ef	f	fg	g	h	js†	j			k	
...	3	−270	−140	−60	−34	−20	−14	−10	−6	−4	−2	0		−2	−4	−6	0	0
3	6	−270	−140	−70	−46	−30	−20	−14	−10	−6	−4	0		−2	−4	...	+1	0
6	10	−280	−150	−80	−56	−40	−25	−18	−13	−8	−5	0		−2	−5	...	+1	0
10	14	−290	−150	−95	...	−50	−32	...	−16	...	−6	0		−3	−6	...	+1	0
14	18	−290	−150	−95	...	−50	−32	...	−16	...	−6	0		−3	−6	...	+1	0
18	24	−300	−160	−110	...	−65	−40	...	−20	...	−7	0		−4	−8	...	+2	0
24	30	−300	−160	−110	...	−65	−40	...	−20	...	−7	0		−4	−8	...	+2	0
30	40	−310	−170	−120	...	−80	−50	...	−25	...	−9	0		−5	−10	...	+2	0
40	50	−320	−180	−130	...	−80	−50	...	−25	...	−9	0		−5	−10	...	+2	0
50	65	−340	−190	−140	...	−100	−60	...	−30	...	−10	0		−7	−12	...	+2	0
65	80	−360	−200	−150	...	−100	−60	...	−30	...	−10	0		−7	−12	...	+2	0
80	100	−380	−220	−170	...	−120	−72	...	−36	...	−12	0		−9	−15	...	+3	0
100	120	−410	−240	−180	...	−120	−72	...	−36	...	−12	0	±IT/2	−9	−15	...	+3	0
120	140	−460	−260	−200	...	−145	−85	...	−43	...	−14	0		−11	−18	...	+3	0
140	160	−520	−280	−210	...	−145	−85	...	−43	...	−14	0		−11	−18	...	+3	0
160	180	−580	−310	−230	...	−145	−85	...	−43	...	−14	0		−11	−18	...	+3	0
180	200	−660	−340	−240	...	−170	−100	...	−50	...	−15	0		−13	−21	...	+4	0
200	225	−740	−380	−260	...	−170	−100	...	−50	...	−15	0		−13	−21	...	+4	0
225	250	−820	−420	−280	...	−170	−100	...	−50	...	−15	0		−13	−21	...	+4	0
250	280	−920	−480	−300	...	−190	−110	...	−56	...	−17	0		−16	−26	...	+4	0
280	315	−1050	−540	−330	...	−190	−110	...	−56	...	−17	0		−16	−26	...	+4	0
315	355	−1200	−600	−360	...	−210	−125	...	−62	...	−18	0		−18	−28	...	+4	0
355	400	−1350	−680	−400	...	−210	−125	...	−62	...	−18	0		−18	−28	...	+4	0
400	450	−1500	−760	−440	...	−230	−135	...	−68	...	−20	0		−20	−32	...	+5	0
450	500	−1650	−840	−480	...	−230	−135	...	−68	...	−20	0		−20	−32	...	+5	0

The dimensions are in 0.001 mm, except the nominal sizes, which are in millimeters.

* Not applicable to sizes up to 1 mm. † In grades 7 to 11, the two symmetrical deviations ±IT/2 should be rounded if the IT value in micro-meters is an odd value by replacing it with the even value immediately below. For example, if IT = 175, replace it by 174.

Table 7 (continued)

Nominal Size, mm		Grade													
		01 to 16													
		Fundamental (Lower) Deviation ei													
Over	To	m	n	p	r	s	t	u	v	x	y	z	za	zb	zc
...	3	+2	+4	+6	+10	+14	...	+18	...	+20	...	+26	+32	+40	+60
3	6	+4	+8	+12	+15	+19	...	+23	...	+28	...	+35	+42	+50	+80
6	10	+6	+10	+15	+19	+23	...	+28	...	+34	...	+42	+52	+67	+97
10	14	+7	+12	+18	+23	+28	...	+33	...	+40	...	+50	+64	+90	+130
14	18	+7	+12	+18	+23	+28	...	+33	+39	+45	...	+60	+77	+108	+150
18	24	+8	+15	+22	+28	+35	...	+41	+47	+54	+63	+73	+98	+136	+188
24	30	+8	+15	+22	+28	+35	+41	+48	+55	+64	+75	+88	+118	+160	+218
30	40	+9	+17	+26	+34	+43	+48	+60	+68	+80	+94	+112	+148	+200	+274
40	50	+9	+17	+26	+34	+43	+54	+70	+81	+97	+114	+136	+180	+242	+325
50	65	+11	+20	+32	+41	+53	+66	+87	+102	+122	+144	+172	+226	+300	+405
65	80	+11	+20	+32	+43	+59	+75	+102	+120	+146	+174	+210	+274	+360	+480
80	100	+13	+23	+37	+51	+71	+91	+124	+146	+178	+214	+258	+335	+445	+585
100	120	+13	+23	+37	+54	+79	+104	+144	+172	+210	+254	+310	+400	+525	+690
120	140	+15	+27	+43	+63	+92	+122	+170	+202	+248	+300	+365	+470	+620	+800
140	160	+15	+27	+43	+65	+100	+134	+190	+228	+280	+340	+415	+535	+700	+900
160	180	+15	+27	+43	+68	+108	+146	+210	+252	+310	+380	+465	+600	+780	+1000
180	200	+17	+31	+50	+77	+122	+166	+236	+284	+350	+425	+520	+670	+880	+1150
200	225	+17	+31	+50	+80	+130	+180	+258	+310	+385	+470	+575	+740	+960	+1250
225	250	+17	+31	+50	+84	+140	+196	+284	+340	+425	+520	+640	+820	+1050	+1350
250	280	+20	+34	+56	+94	+158	+218	+315	+385	+475	+580	+710	+920	+1200	+1550
280	315	+20	+34	+56	+98	+170	+240	+350	+425	+525	+650	+790	+1000	+1300	+1700
315	355	+21	+37	+62	+108	+190	+268	+390	+475	+590	+730	+900	+1150	+1500	+1900
355	400	+21	+37	+62	+114	+208	+294	+435	+530	+660	+820	+1000	+1300	+1650	+2100
400	450	+23	+40	+68	+126	+232	+330	+490	+595	+740	+920	+1100	+1450	+1850	+2400
450	500	+23	+40	+68	+132	+252	+360	+540	+660	+820	+1000	+1250	+1600	+2100	+2600

The dimensions are in 0.001 mm, except the nominal sizes, which are in millimeters.

Table 8 ISO Standard Fundamental Deviations for Holes

Nominal Size, mm		Grade																				
		01 to 16												6	7	8	≤8	>8	≤8‡	>8	≤8	>8§
		Fundamental (Lower) Deviation EI												Fundamental (Upper) Deviation ES								
Over	To	A*	B*	C	CD	D	E	EF	F	FG	G	H	Js†	J			K††		M††		N††	
...	3	+270	+140	+60	+34	+20	+14	+10	+6	+4	+2	0		+2	+4	+6	0	0	−2	−2	−4	−4
3	6	+270	+140	+70	+46	+30	+20	+14	+10	+6	+4	0		+5	+6	+10	−1+Δ	...	−4+Δ	−4	−8+Δ	0
6	10	+280	+150	+80	+56	+40	+25	+18	+13	+8	+5	0		+5	+8	+12	−1+Δ	...	−6+Δ	−6	−10+Δ	0
10	14	+290	+150	+95	...	+50	+32	...	+16	...	+6	0		+6	+10	+15	−1+Δ	...	−7+Δ	−7	−12+Δ	0
14	18	+290	+150	+95	...	+50	+32	...	+16	...	+6	0		+6	+10	+15	−1+Δ	...	−7+Δ	−7	−12+Δ	0
18	24	+300	+160	+110	...	+65	+40	...	+20	...	+7	0		+8	+12	+20	−2+Δ	...	−8+Δ	−8	−15+Δ	0
24	30	+300	+160	+110	...	+65	+40	...	+20	...	+7	0		+8	+12	+20	−2+Δ	...	−8+Δ	−8	−15+Δ	0
30	40	+310	+170	+120	...	+80	+50	...	+25	...	+9	0		+10	+14	+24	−2+Δ	...	−9+Δ	−9	−17+Δ	0
40	50	+320	+180	+130	...	+80	+50	...	+25	...	+9	0		+10	+14	+24	−2+Δ	...	−9+Δ	−9	−17+Δ	0
50	65	+340	+190	+140	...	+100	+60	...	+30	...	+10	0		+13	+18	+28	−2+Δ	...	−11+Δ	−11	−20+Δ	0
65	80	+360	+200	+150	...	+100	+60	...	+30	...	+10	0		+13	+18	+28	−2+Δ	...	−11+Δ	−11	−20+Δ	0
80	100	+380	+220	+170	...	+120	+72	...	+36	...	+12	0	±IT/2	+16	+22	+34	−3+Δ	...	−13+Δ	−13	−23+Δ	0
100	120	+410	+240	+180	...	+120	+72	...	+36	...	+12	0		+16	+22	+34	−3+Δ	...	−13+Δ	−13	−23+Δ	0
120	140	+460	+260	+200	...	+145	+85	...	+43	...	+14	0		+18	+26	+41	−3+Δ	...	−15+Δ	−15	−27+Δ	0
140	160	+520	+280	+210	...	+145	+85	...	+43	...	+14	0		+18	+26	+41	−3+Δ	...	−15+Δ	−15	−27+Δ	0
160	180	+580	+310	+230	...	+145	+85	...	+43	...	+14	0		+18	+26	+41	−3+Δ	...	−15+Δ	−15	−27+Δ	0
180	200	+660	+340	+240	...	+170	+100	...	+50	...	+15	0		+22	+30	+47	−4+Δ	...	−17+Δ	−17	−31+Δ	0
200	225	+740	+380	+260	...	+170	+100	...	+50	...	+15	0		+22	+30	+47	−4+Δ	...	−17−Δ	−17	−31+Δ	0
225	250	+820	+420	+280	...	+170	+100	...	+50	...	+15	0		+22	+30	+47	−4+Δ	...	−17+Δ	−17	−31+Δ	0
250	280	+920	+480	+300	...	+190	+110	...	+56	...	+17	0		+25	+36	+55	−4+Δ	...	−20+Δ	−20	−34+Δ	0
280	315	+1050	+540	+330	...	+190	+110	...	+56	...	+17	0		+25	+36	+55	−4+Δ	...	−20+Δ	−20	−34+Δ	0
315	355	+1200	+600	+360	...	+210	+125	...	+62	...	+18	0		+29	+39	+60	−4+Δ	...	−21+Δ	−21	−37+Δ	0
355	400	+1350	+680	+400	...	+210	+125	...	+62	...	+18	0		+29	+39	+60	−4+Δ	...	−21+Δ	−21	−37+Δ	0
400	450	+1500	+760	+440	...	+230	+135	...	+68	...	+20	0		+33	+43	+66	−5+Δ	...	−23+Δ	−23	−40+Δ	0
450	500	+1650	+840	+480	...	+230	+135	...	+68	...	+20	0		+33	+43	+66	−5+Δ	...	−23+Δ	−23	−40+Δ	0

The dimensions are given in 0.001 mm, except the nominal sizes which are in millimeters. * Not applicable to sizes up to 1 mm. † In grades 7 to 11, the two symmetrical deviations ±IT/2 should be rounded if the IT value in micro-meters is an odd value, by replacing it with the even value below. For example, if IT = 175, replace it by 174.

†† When calculating deviations for holes K, M, and N with tolerance grades up to and including IT 8, and holes F to ZC with tolerance grades up to and including IT 7, the delta (Δ) values are added to the upper deviation ES. For example, for 25 P7, ES = −0.022 + 0.008 = −0.014 mm.

‡ Special case: for M6, ES = −9 for sizes from 250 to 315 mm, instead of −11. § Not applicable to sizes up to 1 mm.

Table 8 (continued)

Nominal Size, mm		Grade													Values for delta (Δ)††					
		≦7	>7												Grade					
		Fundamental (Upper) Deviation ES																		
Over	To	P to ZC	P	R	S	T	U	V	X	Y	Z	ZA	ZB	ZC	3	4	5	6	7	8
...	3	Same deviation as for grades above 7 increased by Δ	−6	−10	−14	...	−18	...	−20	...	−26	−32	−40	−60	0	0	0	0	0	0
3	6		−12	−15	−19	...	−23	...	−28	...	−35	−42	−50	−80	1	1.5	1	3	4	6
6	10		−15	−19	−23	...	−28	...	−34	...	−42	−52	−67	−97	1	1.5	2	3	6	7
10	14		−18	−23	−28	...	−33	...	−40	...	−50	−64	−90	−130	1	2	3	3	7	9
14	18		−18	−23	−28	...	−33	−39	−45	...	−60	−77	−108	−150	1	2	3	3	7	9
18	24		−22	−28	−35	...	−41	−47	−54	−63	−73	−98	−136	−188	1.5	2	3	4	8	12
24	30		−22	−28	−35	−41	−48	−55	−64	−75	−88	−118	−160	−218	1.5	2	3	4	8	12
30	40		−26	−34	−43	−48	−60	−68	−80	−94	−112	−148	−200	−274	1.5	3	4	5	9	14
40	50		−26	−34	−43	−54	−70	−81	−97	−114	−136	−180	−242	−325	1.5	3	4	5	9	14
50	65		−32	−41	−53	−66	−87	−102	−122	−144	−172	−226	−300	−405	2	3	5	6	11	16
65	80		−32	−43	−59	−75	−102	−120	−146	−174	−210	−274	−360	−480	2	3	5	6	11	16
80	100		−37	−51	−71	−91	−124	−146	−178	−214	−258	−335	−445	−585	2	4	5	7	13	19
100	120		−37	−54	−79	−104	−144	−172	−210	−254	−310	−400	−525	−690	2	4	5	7	13	19
120	140		−43	−63	−92	−122	−170	−202	−248	−300	−365	−470	−620	−800	3	4	6	7	15	23
140	160		−43	−65	−100	−134	−190	−228	−280	−340	−415	−535	−700	−900	3	4	6	7	15	23
160	180		−43	−68	−108	−146	−210	−252	−310	−380	−465	−600	−780	−1000	3	4	6	7	15	23
180	200		−50	−77	−122	−166	−226	−284	−350	−425	−520	−670	−880	−1150	3	4	6	9	17	26
200	225		−50	−80	−130	−180	−258	−310	−385	−470	−575	−740	−960	−1250	3	4	6	9	17	26
225	250		−50	−84	−140	−196	−284	−340	−425	−520	−640	−820	−1050	−1350	3	4	6	9	17	26
250	280		−56	−94	−158	−218	−315	−385	−475	−580	−710	−920	−1200	−1550	4	4	7	9	20	29
280	315		−56	−98	−170	−240	−350	−425	−525	−650	−790	−1000	−1300	−1700	4	4	7	9	20	29
315	355		−62	−108	−190	−268	−390	−475	−590	−730	−900	−1150	−1500	−1800	4	5	7	11	21	32
355	400		−62	−114	−208	−294	−435	−530	−660	−820	−1000	−1300	−1650	−2100	4	5	7	11	21	32
400	450		−68	−126	−232	−330	−490	−595	−740	−920	−1100	−1450	−1850	−2400	5	5	7	13	23	34
450	500		−68	−132	−252	−360	−540	−660	−820	−1000	−1250	−1600	−2100	−2600	5	5	7	13	23	34

The dimensions are given in 0.001 mm, except the nominal sizes which are in millimeters.

†† When calculating deviations for holes K, M, and N with tolerance grades up to and including IT 8, and holes F to ZC with tolerance grades up to and including IT 7, the delta (Δ) values are added to the upper deviation ES. For example, for 25 P7, ES = −0.022 + 0.008 = −0.014 mm.

APPENDIX B

Thread Forms and Threaded Devices Tables

Table 1 Coarse-thread Series—UNC and NC (Bold type indicates unified threads.)

Identification		Screw basic diameters		External threads				Internal threads		Areas of sections	
Size	Threads per in., n	Major diam.	Pitch diam.	Basic minor diam.	Allowances classes 1A and 2A	Class 1A, major diam tolerances	Class 2A and 3A, major diam tolerances	Minor diam.	Minor diam tolerances, classes 1B, 2B, and 3B	Basic min minor diam, sq in.	Stress area, sq in.
1 (0.073)	64	0.0730	0.0629	0.0538	0.0006	......	0.0038	0.0561	0.0062	0.0022	0.0026
2 (0.086)	56	0.0860	0.0744	0.0641	0.0006	......	0.0041	0.0667	0.0070	0.0031	0.0036
3 (0.099)	48	0.0990	0.0855	0.0734	0.0007	......	0.0045	0.0764	0.0081	0.0041	0.0048
4 (0.112)	40	0.1120	0.0958	0.0813	0.0008	......	0.0051	0.0849	0.0090	0.0050	0.0060
5 (0.125)	40	0.1250	0.1088	0.0943	0.0008	......	0.0051	0.0979	0.0083	0.0067	0.0079
6 (0.138)	32	0.1380	0.1177	0.0997	0.0008	......	0.0060	0.1042	0.0098	0.0075	0.0090
8 (0.164)	32	0.1640	0.1437	0.1257	0.0009	......	0.0060	0.1302	0.0087	0.0120	0.0139
10 (0.190)	24	0.1900	0.1629	0.1389	0.0010	......	0.0072	0.1449	0.0106	0.0145	0.0174
12 (0.216)	24	0.2160	0.1889	0.1649	0.0010	......	0.0072	0.1709	0.0098	0.0206	0.0240
1/4	20	0.2500	0.2175	0.1887	0.0011	0.0122	0.0081	0.1959	0.0108	0.0269	0.0317
5/16	18	0.3125	0.2764	0.2443	0.0012	0.0131	0.0087	0.2524	0.0106	0.0454	0.0522
3/8	16	0.3750	0.3344	0.2983	0.0013	0.0142	0.0094	0.3073	0.0109	0.0678	0.0773
7/16	14	0.4375	0.3911	0.3499	0.0014	0.0155	0.0103	0.3602	0.0115	0.0933	0.1060
1/2	13	0.5000	0.4500	0.4056	0.0015	0.0163	0.0109	0.4167	0.0117	0.1257	0.1416
1/2	12	0.5000	0.4459	0.3978	0.0015	0.0172	0.0114	0.4098	0.0125	0.1205	0.1374
9/16	12	0.5625	0.5084	0.4603	0.0016	0.0172	0.0114	0.4723	0.0120	0.1620	0.1816
5/8	11	0.6250	0.5660	0.5135	0.0016	0.0182	0.0121	0.5266	0.0125	0.2018	0.2256
3/4	10	0.7500	0.6850	0.6273	0.0018	0.0194	0.0129	0.6417	0.0128	0.3020	0.3340
7/8	9	0.8750	0.8028	0.7387	0.0019	0.0208	0.0139	0.7547	0.0134	0.4193	0.4612
1	8	1.0000	0.9188	0.8466	0.0020	0.0225	0.0150	0.8647	0.0150	0.5510	0.6051
1 1/8	7	1.1250	1.0322	0.9497	0.0022	0.0246	0.0164	0.9704	0.0171	0.6931	0.7627
1 1/4	7	1.2500	1.1572	1.0747	0.0022	0.0246	0.0164	1.0954	0.0171	0.8898	0.9684
1 3/8	6	1.3750	1.2667	1.1705	0.0024	0.0273	0.0182	1.1946	0.0200	1.0541	1.1538
1 1/2	6	1.5000	1.3917	1.2955	0.0024	0.0273	0.0182	1.3196	0.0200	1.2938	1.4041
1 3/4	5	1.7500	1.6201	1.5046	0.0027	0.0308	0.0205	1.5335	0.0240	1.7441	1.8983
2	4 1/2	2.0000	1.8557	1.7274	0.0029	0.0330	0.0220	1.7594	0.0267	2.3001	2.4971
2 1/4	4 1/2	2.2500	2.1057	1.9774	0.0029	0.0330	0.0220	2.0094	0.0267	3.0212	3.2464
2 1/2	4	2.5000	2.3376	2.1933	0.0031	0.0357	0.0238	2.2294	0.0300	3.7161	3.9976
2 3/4	4	2.7500	2.5876	2.4433	0.0032	0.0357	0.0238	2.4794	0.0300	4.6194	4.9326
3	4	3.0000	2.8376	2.6933	0.0032	0.0357	0.0238	2.7294	0.0300	5.6209	5.9659
3 1/4	4	3.2500	3.0876	2.9433	0.0033	0.0357	0.0238	2.9794	0.0300	6.7205	7.0992
3 1/2	4	3.5000	3.3376	3.1933	0.0033	0.0357	0.0238	3.2294	0.0300	7.9183	8.3268
3 3/4	4	3.7500	3.5876	3.4433	0.0034	0.0357	0.0238	3.4794	0.0300	9.2143	9.6546
4	4	4.0000	3.8376	3.6933	0.0034	0.0357	0.0238	3.7294	0.0300	10.6084	11.0805

Table 2 Fine-thread Series—UNF and NF (Bold type indicates unified threads.)

Identification		Screw basic diameters		External threads				Internal threads		Areas of sections	
Size	Threads per in.	Major diam.	Pitch diam.	Basic minor diam.	Allowances classes 1A and 2A	Class 1A, major diam. tolerances‡	Class 2A and 3A, major diam tolerances	Minor diam.	Minor diam. tolerances, classes 1B, 2B, and 3B	Basic min minor diam. sq in.	Stress area, sq in.
0 (0.060)	80	0.0600	0.0519	0.0447	0.0005	......	0.0032	0.0465	0.0049	0.0015	0.0018
1 (0.073)	72	0.0730	0.0640	0.0560	0.0006	......	0.0035	0.0580	0.0055	0.0024	0.0027
2 (0.086)	64	0.0860	0.0759	0.0668	0.0006	......	0.0038	0.0691	0.0062	0.0034	0.0039
3 (0.099)	56	0.0990	0.0874	0.0771	0.0007	......	0.0041	0.0797	0.0068	0.0045	0.0052
4 (0.112)	48	0.1120	0.0985	0.0864	0.0007	......	0.0045	0.0894	0.0074	0.0057	0.0065
5 (0.125)	44	0.1250	0.1102	0.0971	0.0007	......	0.0048	0.1004	0.0075	0.0072	0.0082
6 (0.138)	40	0.1380	0.1218	0.1073	0.0008	......	0.0051	0.1109	0.0077	0.0087	0.0101
8 (0.164)	36	0.1640	0.1460	0.1299	0.0008	......	0.0055	0.1339	0.0077	0.0128	0.0146
10 (0.190)	**32**	**0.1900**	**0.1697**	**0.1517**	**0.0009**	......	**0.0060**	**0.1562**	**0.0079**	**0.0175**	**0.0199**
12 (0.216)	28	0.2160	0.1928	0.1722	0.0010	......	0.0065	0.1773	0.0084	0.0026	0.0257
1/4	**28**	0.2500	0.2268	0.2062	0.0010	0.0098	0.0065	0.2113	0.0077	0.0326	0.0362
5/16	**24**	0.3125	0.2854	0.2614	0.0011	0.0108	0.0072	0.2674	0.0080	0.0524	0.0579
3/8	**24**	0.3750	0.3479	0.3239	0.0011	0.0108	0.0072	0.3299	0.0073	0.0809	0.0876
7/16	**20**	0.4375	0.4050	0.3762	0.0013	0.0122	0.0081	0.3834	0.0082	0.1090	0.1185
1/2	**20**	0.5000	0.4675	0.4387	0.0013	0.0122	0.0081	0.4459	0.0078	0.1486	0.1597
9/16	**18**	0.5625	0.5264	0.4943	0.0014	0.0131	0.0087	0.5024	0.0082	0.1888	0.2026
5/8	**18**	0.6250	0.5889	0.5568	0.0014	0.0131	0.0087	0.5649	0.0081	0.2400	0.2555
3/4	**16**	0.7500	0.7094	0.6733	0.0015	0.0142	0.0094	0.6823	0.0085	0.3513	0.3724
7/8	**14**	0.8750	0.8286	0.7874	0.0016	0.0155	0.0103	0.7977	0.0091	0.4805	0.5088
1	**12**	1.0000	0.9459	0.8978	0.0018	0.0172	0.0114	0.9098	0.0100	0.6245	0.6624
1 1/8	**12**	1.1250	1.0709	1.0228	0.0018	0.0172	0.0114	1.0348	0.0100	0.8118	0.8549
1 1/4	**12**	1.2500	1.1959	1.1478	0.0018	0.0172	0.0114	1.1598	0.0100	1.0237	1.0721
1 3/8	**12**	1.3750	1.3209	1.2728	0.0019	0.0172	0.0114	1.2848	0.0100	1.2302	1.3137
1 1/2	**12**	1.5000	1.4459	1.3978	0.0019	0.0172	0.0114	1.4098	0.0100	1.5212	1.5799

Table 3 Extra-fine-thread Series—UNEF and NEF (Bold type indicates unified threads.)

Identification		Screw basic diameters		External threads				Internal threads		Areas of sections	
Size	Threads per in.	Major diam	Pitch diam.	Basic minor diam.	Allowances classes 1A and 2A	Class 1A, major diam tolerances	Class 2A and 3A, major diam tolerances	Minor diam.	Minor diam tolerances, classes 1B, 2B, and 3B	Basic min minor diam, sq in.	Stress area, sq in.
12 (0.216)	32	0.2160	0.1957	0.1777	0.0009	......	0.0060	0.1822	0.0073	0.0242	0.2269
1/4	32	0.2500	0.2297	0.2117	0.0010	......	0.0060	0.2162	0.0067	0.0344	0.0377
5/16	32	0.3125	0.2922	0.2742	0.0010	......	0.0060	0.2787	0.0060	0.0581	0.0622
3/8	32	0.3750	0.3547	0.3367	0.0010	......	0.0060	0.3412	0.0057	0.0878	0.0929
7/16	**28**	**0.4375**	**0.4143**	**0.3937**	**0.0011**	......	**0.0065**	**0.3988**	**0.0063**	**0.1201**	**0.1270**
1/2	**28**	**0.5000**	**0.4768**	**0.4562**	**0.0011**	......	**0.0065**	**0.4613**	**0.0063**	**0.1616**	**0.1695**
9/16	24	0.5625	0.5354	0.5114	0.0012	......	0.0072	0.5174	0.0070	0.2030	0.2134
5/8	24	0.6250	0.5979	0.5739	0.0012	......	0.0072	0.5799	0.0070	0.2560	0.2676
11/16	24	0.6875	0.6604	0.6364	0.0012	......	0.0072	0.6424	0.0070	0.3151	0.3280
3/4	**20**	**0.7500**	**0.7175**	**0.6887**	**0.0013**	......	**0.0081**	**0.6959**	**0.0078**	**0.3685**	**0.3855**
13/16	**20**	**0.8125**	**0.7800**	**0.7512**	**0.0013**	......	**0.0081**	**0.7584**	**0.0078**	**0.4388**	**0.4573**
7/8	**20**	**0.8750**	**0.8425**	**0.8137**	**0.0013**	......	**0.0081**	**0.8209**	**0.0078**	**0.5153**	**0.5352**
15/16	**20**	**0.9375**	**0.9050**	**0.8762**	**0.0014**	......	**0.0081**	**0.8834**	**0.0078**	**0.5979**	**0.6194**
1	**20**	**1.0000**	**0.9675**	**0.9387**	**0.0014**	......	**0.0081**	**0.9459**	**0.0078**	**0.6866**	**0.7095**
1 1/16	**18**	**1.0625**	**1.0264**	**0.9943**	**0.0014**	......	0.0087	**1.0024**	**0.0081**	**0.7702**	**0.7973**
1 1/8	18	1.1250	1.0889	1.0568	0.0014	......	0.0087	1.0649	0.0081	0.8705	0.8993
1 3/16	18	1.1875	1.1514	1.1193	0.0015	......	0.0087	1.1274	0.0081	0.9770	1.0074
1 1/4	18	1.2500	1.2139	1.1818	0.0015	......	0.0087	1.1899	0.0081	1.0895	1.1216
1 5/16	18	1.3125	1.2764	1.2443	0.0015	......	0.0087	1.2524	0.0081	1.2082	1.2420
1 3/8	18	1.3750	1.3389	1.3068	0.0015	......	0.0087	1.3149	0.0081	1.3339	1.3684
1 7/16	18	1.4375	1.4014	1.3693	0.0015	......	0.0087	1.3774	0.0081	1.4640	1.5010
1 1/2	18	1.5000	1.4639	1.4318	0.0015	......	0.0087	1.4399	0.0081	1.6011	1.6397
1 9/16	18	1.5625	1.5264	1.4943	0.0015	......	0.0087	1.5024	0.0081	1.7444	1.7846
1 5/8	18	1.6250	1.5889	1.5568	0.0015	......	0.0087	1.5649	0.0081	1.8937	1.9357
1 11/16	18	1.6875	1.6514	1.6193	0.0015	......	0.0087	1.6274	0.0081	2.0493	2.0929
1 3/4	**16**	**1.7500**	**1.7094**	**1.6733**	**0.0016**	......	**0.0094**	**1.6823**	**0.0085**	**2.2873**	**2.2382**
2	**16**	**2.000**	**1.9594**	**1.9233**	**0.0016**	......	**0.0094**	**1.9323**	**0.0085**	**2.8917**	**2.9501**

Table 4 Thread Series—8N

Identification		Screw basic diameters		External threads				Internal threads		Areas of sections	
Size	Threads per in.	Major diam.	Pitch diam.	Basic minor diam.	Allowances classes 1A and 2A	Class 1A, major diam. tolerances	Class 2A and 3A, major diam tolerances	Minor diam.	Minor diam tolerances, classes 1B, 2B, and 3B	Basic min minor diam. sq in.	Stress area, sq in.
1⅛	8	1.1250	1.0438	0.9716	0.0021	......	0.0150	0.9897	0.0150	0.7277	0.7896
1¼	8	1.2500	1.1688	1.0966	0.0021	......	0.0150	1.1147	0.0150	0.9290	0.9985
1⅜	8	1.3750	1.2938	1.2216	0.0022	......	0.0150	1.2397	0.0150	1.1548	1.2319
1½	8	1.5000	1.4188	1.3466	0.0022	......	0.0150	1.3647	0.0150	1.4052	1.4899
1⅝	8	1.6250	1.5438	1.4716	0.0022	......	0.0150	1.4897	0.0150	1.6801	1.7723
1¾	8	1.7500	1.6688	1.5966	0.0023	......	0.0150	1.6147	0.0150	1.9796	2.0792
1⅞	8	1.8750	1.7938	1.7216	0.0023	......	0.0150	1.7397	0.0150	2.3036	2.4107
2	8	2.0000	1.9188	1.8466	0.0023	......	0.0150	1.8647	0.0150	2.6521	2.7665
2⅛	8	2.1250	2.0438	1.9716	0.0024	......	0.0150	1.9897	0.0150	3.0252	3.1469
2¼	8	2.2500	2.1688	2.0966	0.0024	......	0.0150	2.1147	0.0150	3.4228	3.5519
2½	8	2.5000	2.4188	2.3466	0.0024	......	0.0150	2.3647	0.0150	4.2917	4.4352
2¾	8	2.7500	2.6688	2.5966	0.0025	......	0.0150	2.6147	0.0150	5.2588	5.4164
3	8	3.0000	2.9188	2.8466	0.0026	......	0.0150	2.8647	0.0150	6.3240	6.4957
3¼	8	3.2500	3.1688	3.0966	0.0026	......	0.0150	3.1147	0.0150	7.4874	7.6738
3½	8	3.5000	3.4188	3.3466	0.0026	......	0.0150	3.3647	0.0150	8.7490	8.9504
3¾	8	3.7500	3.6688	3.5966	0.0027	......	0.0150	3.6147	0.0150	10.1088	10.3249
4	8	4.0000	3.9188	3.8466	0.0027	......	0.0150	3.8647	0.0150	11.5667	11.7995
4¼	8	4.2500	4.1688	4.0966	0.0028	......	0.0150	4.1147	0.0150	13.1228	13.3683
4½	8	4.5000	4.4188	3.3466	0.0028	......	0.0150	4.3647	0.0150	14.7771	15.0372
4¾	8	4.7500	4.6688	4.5966	0.0029	......	0.0150	4.6147	0.0150	16.5295	16.8042
5	8	5.0000	4.9188	4.8466	0.0029	......	0.0150	4.8647	0.0150	18.3802	18.6694
5¼	8	5.2500	5.1688	5.0966	0.0029	......	0.0150	5.1147	0.0150	20.3290	20.6330
5½	8	5.5000	5.4188	5.3466	0.0030	......	0.0150	5.3647	0.0150	22.3760	22.6945
5¾	8	5.7500	5.6688	5.5966	0.0030	......	0.0150	5.6147	0.0150	24.5211	24.8541
6	8	6.0000	5.9188	5.8466	0.0030	......	0.0150	5.8647	0.0150	26.7645	27.1118

Table 5 12 Thread Series—12UN and 12N (Bold type indicates unified threads.)

Identification		Screw basic diameters		External threads				Internal threads		Areas of sections	
Size	Threads per in.	Major diam.	Pitch diam.	Basic minor diam.	Allowances classes 1A and 2A	Class 1A, major diam. tolerances	Class 2A and 3A, major diam tolerances	Minor diam.	Minor diam tolerances, classes 1B, 2B, and 3B	Basic min minor diam, sq in.	Stress area, sq in.
1/2	12	0.5000	0.4459	0.3978	0.0016	......	0.0114	0.4098	0.0127	0.1205	0.1374
5/8	12	0.6250	0.5709	0.5228	0.0016	......	0.0114	0.5348	0.0115	0.2097	0.2319
11/16	12	0.6875	0.6334	0.5853	0.0016	......	0.0114	0.5973	0.0112	0.2635	0.2883
3/4	12	0.7500	0.6959	0.6478	0.0017	......	0.0114	0.6598	0.0109	0.3234	0.3508
13/16	12	0.8125	0.7584	0.7103	0.0017	......	0.0114	0.7223	0.0106	0.3895	0.4195
7/8	12	0.8750	0.8209	0.7728	0.0017	......	0.0114	0.7848	0.0108	0.4617	0.4943
15/16	12	0.9375	0.8834	0.8353	0.0017	......	0.0114	0.8473	0.0090	0.5000	0.5753
1 1/16	12	1.0625	1.0084	0.9603	0.0017	......	0.0114	0.9723	0.0100	0.7151	0.7556
1 3/16	12	1.1875	1.1334	1.0853	0.0017	......	0.0114	1.0973	0.0100	0.9147	0.9604
1 5/16	12	1.3125	1.2584	1.2103	0.0017	......	0.0114	1.2223	0.0100	1.1389	1.1898
1 7/16	12	1.4375	1.3834	1.3353	0.0018	......	0.0114	1.3473	0.0100	1.3876	1.4438
1 5/8	12	1.6250	1.5709	1.5228	0.0018	......	0.0114	1.5348	0.0100	1.8067	1.8701
1 3/4	12	1.7500	1.6959	1.6418	0.0018	......	0.0114	1.6598	0.0100	2.1168	2.1853
1 7/8	12	1.8750	1.8209	1.7728	0.0018	......	0.0114	1.7848	0.0100	2.4514	2.5250
2	12	2.0000	1.9459	1.8978	0.0018	......	0.0114	1.9098	0.0100	2.8106	2.8892
2 1/8	12	2.1250	2.0709	2.3228	0.0018	......	0.0114	2.0348	0.0100	3.1943	3.2779
2 1/4	12	2.2500	2.1959	2.1478	0.0018	......	0.0114	2.1598	0.0100	3.6025	3.6914
2 3/8	12	2.3750	2.3209	2.2728	0.0019	......	0.0114	2.2848	0.0100	4.0353	4.1291
2 1/2	12	2.5000	2.4459	2.3978	0.0019	......	0.0114	2.4098	0.0100	4.4927	4.5916
2 5/8	12	2.6250	2.5709	2.5228	0.0019	......	0.0114	2.5348	0.0100	4.9745	5.0784
2 3/4	12	2.7500	2.6959	2.6478	0.0019	......	0.0114	2.6598	0.0100	5.4810	5.5900
2 7/8	12	2.8750	2.8209	2.7728	0.0019	......	0.0114	2.7848	0.0100	6.0119	6.1259
3	12	3.0000	2.9459	2.8978	0.0019	......	0.0114	2.9098	0.0100	6.5674	6.6865
3 1/8	12	3.1250	3.0709	3.0228	0.0019	......	0.0114	3.0348	0.0100	7.1475	7.2714
3 1/4	12	3.2500	3.1959	3.1478	0.0019	......	0.0114	3.1598	0.0100	7.7521	7.8812
3 3/8	12	3.3750	3.3209	3.2728	0.0019	......	0.0114	3.2848	0.0100	8.3812	8.5152
3 1/2	12	3.5000	3.4459	3.3978	0.0019	......	0.0114	3.4098	0.0100	9.0349	9.1740
3 5/8	12	3.6250	3.5709	5.5228	0.0019	......	0.0114	3.5348	0.0100	9.7132	9.8570
3 3/4	12	3.7500	3.6959	3.6478	0.0019	......	0.0114	3.6598	0.1000	10.4159	10.4649
3 7/8	12	3.8750	3.8209	3.7728	0.0020	......	0.0114	3.7848	0.0100	11.1433	11.2970
4	12	4.0000	3.9459	3.8978	0.0020	......	0.0114	3.9098	0.0100	11.8951	12.0540
4 1/4	12	4.2500	4.1959	4.1478	0.0020	......	0.0114	4.1598	0.0100	13.4725	13.6411
4 1/2	12	4.5000	4.4459	4.3918	0.0020	......	0.0114	4.4098	0.0100	15.1480	15.3265
4 3/4	12	4.7500	4.6959	4.6478	0.0020	......	0.0114	4.6598	0.0100	16.9217	17.1099
5	12	5.0000	4.9459	4.8978	0.0020	......	0.0114	4.9098	0.0100	18.7936	18.9916
5 1/4	12	5.2500	5.1959	5.1478	0.0020	......	0.0114	5.1598	0.0100	20.7636	20.9717
5 1/2	12	5.5000	5.4459	5.3978	0.0020	......	0.0114	5.4098	0.0100	22.8319	23.0496
5 3/4	12	5.7500	5.6959	5.6478	0.0021	......	0.0114	5.6598	0.0100	24.9983	25.2257
6	12	6.0000	5.9459	5.8978	0.0021	......	0.0014	5.9098	0.0100	27.2628	27.4988

Table 6 16 Thread Series—16UN and 16N (Bold type indicates unified threads.)

Identification		Screw basic diameters		External threads				Internal threads		Areas of sections	
Size	Threads per in.	Major diam.	Pitch diam.	Basic minor diam.	Allowances, classes 1A and 2A	Class 1A, major diam. tolerances	Class 2A and 3A, major diam tolerances	Minor diam.	Minor diam tolerances, classes 1B, 2B, and 3B	Basic min minor diam, sq in.	Stress area, sq in.
13/16	**16**	**0.8125**	**0.7719**	**0.7358**	**0.0015**	......	**0.0094**	**0.7448**	**0.0085**	**0.4200**	**0.4429**
7/8	**16**	**0.8750**	**0.8344**	**0.7983**	**0.0015**	......	**0.0094**	**0.8073**	**0.0085**	**0.4949**	**0.5197**
15/16	**16**	**0.9375**	**0.8969**	**0.8608**	**0.0015**	......	**0.0094**	**0.8698**	**0.0085**	**0.5759**	**0.6025**
1	**16**	**1.0000**	**0.9594**	**0.9233**	**0.0015**	......	**0.0094**	**0.9323**	**0.0085**	**0.6630**	**0.6916**
1 1/16	**16**	**1.0625**	**1.0219**	**0.0958**	**0.0015**	......	**0.0094**	**0.9948**	**0.0085**	**0.7563**	**0.7867**
1 1/8	**16**	**1.1250**	**1.0844**	**1.0483**	**0.0015**	......	**0.0094**	**1.0573**	**0.0085**	**0.8557**	**0.8880**
1 3/16	**16**	**1.1875**	**1.1469**	**1.1108**	**0.0015**	......	**0.0094**	**1.1198**	**0.0085**	**0.9612**	**0.9955**
1 1/4	**16**	**1.2500**	**1.2094**	**1.1733**	**0.0015**	......	**0.0094**	**1.1823**	**0.0085**	**1.0729**	**1.1090**
1 5/16	**16**	**1.3125**	**1.2719**	**1.2358**	**0.0015**	......	**0.0094**	**1.2448**	**0.0085**	**1.1907**	**1.2287**
1 3/8	**16**	**1.3750**	**1.3344**	**1.2983**	**0.0015**	......	**0.0094**	**1.3073**	**0.0085**	**1.3147**	**1.3545**
1 7/16	**16**	**1.4375**	**1.3969**	**1.3608**	**0.0016**	......	**0.0094**	**1.3698**	**0.0085**	**1.4448**	**1.4865**
1 1/2	**16**	**1.5000**	**1.4594**	**1.4233**	**0.0016**	......	**0.0094**	**1.4323**	**0.0085**	**1.5810**	**1.6246**
1 9/16	16	1.5625	1.5219	1.4858	0.0016	......	0.0094	1.4948	0.0085	1.7234	1.7687
1 5/8	16	1.6250	1.5844	1.5483	0.0016	......	0.0094	1.5573	0.0085	1.8719	1.9191
1 11/16	16	1.6875	1.6469	1.6108	0.0016	......	0.0094	1.6198	0.0085	2.0265	2.0757
1 13/16	16	1.8125	1.7719	1.7358	0.0016	......	0.0094	1.7448	0.0085	2.3542	2.4070
1 7/8	16	1.8750	1.8344	1.7983	0.0016	......	0.0094	1.8073	0.0085	2.5272	2.5819
1 15/16	16	1.9375	1.8969	1.8608	0.0016	......	0.0094	1.8698	0.0085	2.7062	2.7269
2 1/16	16	2.0625	2.0219	1.9858	0.0016	......	0.0094	1.9948	0.0085	3.0831	3.1434
2 1/8	16	2.1250	2.0844	2.0483	0.0016	......	0.0094	2.0573	0.0085	3.2807	3.3427
2 3/16	16	2.1875	2.1469	2.1108	0.0016	......	0.0094	2.1198	0.0085	3.4844	3.5483
2 1/4	**16**	**2.2500**	**2.2094**	**2.1733**	**0.0016**	......	**0.0094**	**2.1823**	**0.0085**	**3.6943**	**3.7601**
2 5/16	16	2.3125	2.2719	2.2358	0.0017	......	0.0094	2.2448	0.0085	3.9103	3.9708
2 3/8	16	2.3750	2.3344	2.2983	0.0017	......	0.0094	2.3073	0.0085	4.1324	4.2018
2 7/16	16	2.4375	2.3969	2.3608	0.0017	......	0.0094	2.3696	0.0085	4.3606	4.4319
2 1/2	**16**	**2.5000**	**2.4594**	**2.4233**	**0.0017**	......	**0.0094**	**2.4323**	**0.0085**	**4.4950**	**4.6682**
2 5/8	16	2.6250	2.5844	2.5483	0.0017	......	0.0094	2.5573	0.0085	5.0822	5.1790
2 3/4	**16**	**2.7500**	**2.7094**	**2.6733**	**0.0017**	......	**0.0094**	**2.6823**	**0.0085**	**5.5940**	**5.6745**
2 7/8	16	2.8750	2.8344	2.7983	0.0017	......	0.0094	2.8073	0.0085	6.1303	6.2143
3	**16**	**3.0000**	**2.9594**	**2.9233**	**0.0017**	......	**0.0094**	**2.9323**	**0.0085**	**6.6911**	**6.7789**
3 1/8	16	3.1250	3.0844	3.0483	0.0017	......	0.0094	3.0573	0.0085	7.2765	7.3678
3 1/4	**16**	**3.2500**	**3.2094**	**3.1733**	**0.0017**	......	**0.0094**	**3.1823**	**0.0085**	**7.8864**	**7.9814**
3 3/8	16	3.3750	3.3344	3.2983	0.0017	......	0.0094	3.3073	0.0085	8.5209	8.6194
3 1/2	**16**	**3.5000**	**3.4594**	**3.4233**	**0.0017**	......	**0.0094**	**3.4323**	**0.0085**	**9.1799**	**9.2821**
3 5/8	16	3.6250	3.5844	3.5483	0.0017	......	0.0094	3.5573	0.0085	9.8634	9.9691
3 3/4	**16**	**3.7500**	**3.7094**	**3.6733**	**0.0017**	......	**0.0094**	**3.6823**	**0.0085**	**10.5715**	**10.6809**
3 7/8	16	3.8750	3.8344	3.7983	0.0018	......	0.0094	3.8073	0.0085	11.3042	11.4170
4	**16**	**4.0000**	**3.9594**	**3.9233**	**0.0018**	......	**0.0094**	**3.9323**	**0.0085**	**12.0614**	**12.1779**
4 1/4	**16**	**4.2500**	**4.2094**	**4.1733**	**0.0018**	......	**0.0094**	**4.1823**	**0.0085**	**13.6494**	**13.7730**
4 1/2	**16**	**4.5000**	**4.4594**	**4.4233**	**0.0018**	......	**0.0094**	**4.4323**	**0.0085**	**15.3355**	**15.4662**
4 3/4	**16**	**4.7500**	**4.7094**	**4.6733**	**0.0018**	......	**0.0094**	4.6823	**0.0085**	**17.1199**	**12.2575**
5	**16**	**5.0000**	**4.9594**	**4.9233**	**0.0018**	......	**0.0094**	4.9323	**0.0085**	**19.0024**	**19.1470**
5 1/4	**16**	**5.2500**	**5.2094**	**5.1733**	**0.0018**	......	**0.0094**	5.1823	**0.0085**	**20.9831**	**31.1350**
5 1/2	**16**	**5.5000**	**5.4538**	**5.4233**	**0.0018**	......	**0.0094**	5.4323	**0.0085**	**23.0620**	**23.2208**
5 3/4	**16**	**5.7500**	**5.7094**	**5.6733**	**0.0019**	......	**0.0094**	5.6823	**0.0085**	**25.2390**	**25.4047**
6	**16**	**6.0000**	**5.9594**	**5.9233**	**0.0019**	......	**0.0094**	5.9323	**0.0085**	**27.5142**	**27.6868**

Table 7 Comparison of Bolt and Nut Thread Dimensions for Foreign Thread Forms (All dimensions in mm.)

Nominal Size and Major Bolt Diam.	Pitch	Pitch Diam.	Bolt				Nut				
			Minor Diameter				Major Diameter			Minor Diameter	
			British	French	German	Swiss	British & German	French	Swiss	French, German & Swiss	British
6	1	5.350	4.863	4.59	4.700	4.60	6.000	6.108	6.100	4.700	4.863
7	1	6.350	5.863	5.59	5.700	5.60	7.000	7.108	7.100	5.700	5.863
8	1.25	7.188	6.579	6.24	6.376	6.25	8.000	8.135	8.124	6.376	6.579
9	1.25	8.188	7.579	7.24	7.376	7.25	9.000	9.135	9.124	7.376	7.579
10	1.5	9.026	8.295	7.89	8.052	7.90	10.000	10.162	10.150	8.052	8.295
11	1.5	10.026	9.295	8.89	9.052	8.90	11.000	11.162	11.150	9.052	9.295
12	1.75	10.863	10.011	9.54	9.726	9.55	12.000	12.189	12.174	9.726	10.011
14	2	12.701	11.727	11.19	11.402	11.20	14.000	14.216	14.200	11.402	11.727
16	2	14.701	13.727	13.19	13.402	13.20	16.000	16.216	16.200	13.402	13.727
18	2.5	16.376	15.158	14.48	14.752	14.50	18.000	18.270	18.250	14.752	15.158
20	2.5	18.376	17.158	16.48	16.752	16.50	20.000	20.270	20.250	16.752	17.158
22	2.5	20.376	19.158	18.48	18.752	18.50	22.000	22.270	22.250	18.752	19.158
24	3	22.051	20.590	19.78	20.102	19.80	24.000	24.324	24.300	20.102	20.590
27	3	25.051	23.590	22.78	23.102	22.80	27.000	27.324	27.300	23.102	23.590
30	3.5	27.727	26.022	25.08	25.454	25.10	30.000	30.378	30.350	25.454	26.022
33	3.5	30.727	29.022	28.08	28.454	28.10	33.000	33.378	33.350	28.454	29.022
36	4	33.402	31.453	30.37	30.804	30.40	36.000	36.432	36.400	30.804	31.453
39	4	36.402	34.453	33.37	33.804	33.40	39.000	39.432	39.400	33.804	34.453
42	4.5	39.077	36.885	35.67	36.154	35.70	42.000	42.486	42.450	36.154	36.885
45	4.5	42.077	39.885	38.67	39.154	38.70	45.000	45.486	45.450	39.154	39.885
48	5	44.752	42.316	40.96	41.504	41.00	48.000	48.540	48.500	41.504	42.316
52	5	48.752	46.316	44.96	45.504	45.00	52.000	52.540	52.500	45.504	46.316
56	5.5	52.428	49.748	48.26	48.856	48.30	56.000	56.594	56.550	48.856	49.748
60	5.5	56.428	53.748	52.26	52.856	52.30	60.000	60.594	60.550	52.856	53.748

Table 8 ISO Metric Internal and External Threads (All dimensions in mm.)

THREAD DATA (mm)

Pitch p	$p/8$	$p/4$	H	$H/8$	$H/6$	$H/4$	$\frac{3}{8}H$	$\frac{5}{8}H$	$\frac{17}{24}H$	$\frac{3}{4}H$	$1\frac{1}{4}H$	$1\frac{5}{12}H$
			0·866 03p	0·108 25p	0·144 34p	0·216 51p	0·324 76p	0·541 27p	0·613 44p	0·649 52p	1·082 54p	1·226 88p
0·25	0·031 2	0·062 5	0·216 5	0·027 1	0·036 1	0·054 1	0·081 2	0·135 3	0·153 4	0·162 4	0·270 6	0·306 7
0·35	0·043 8	0·087 5	0·303 1	0·037 9	0·050 5	0·075 8	0·113 7	0·189 4	0·214 7	0·227 3	0·378 9	0·429 4
0·4	0·050 0	0·100 0	0·346 4	0·043 3	0·057 7	0·086 6	0·129 9	0·216 5	0·245 4	0·259 8	0·433 0	0·490 8
0·45	0·056 2	0·112 5	0·389 7	0·048 7	0·065 0	0·097 4	0·146 1	0·243 6	0·276 0	0·292 3	0·487 1	0·552 1
0·5	0·062 5	0·125 0	0·433 0	0·054 1	0·072 2	0·108 3	0·162 4	0·270 6	0·306 7	0·324 8	0·541 3	0·613 4
0·6	0·075 0	0·150 0	0·519 6	0·065 0	0·086 6	0·129 9	0·194 9	0·324 8	0·368 1	0·389 7	0·649 5	0·736 1
0·7	0·087 5	0·175 0	0·606 2	0·075 8	0·101 0	0·151 6	0·227 3	0·378 9	0·429 4	0·454 7	0·757 8	0·858 8
0·75	0·093 8	0·187 5	0·649 5	0·081 2	0·108 3	0·162 4	0·243 6	0·405 9	0·460 1	0·487 1	0·811 9	0·920 2
0·8	0·100 0	0·200 0	0·692 8	0·086 6	0·115 5	0·173 2	0·259 8	0·433 0	0·490 8	0·519 6	0·866 0	0·981 5
1	0·125 0	0·250 0	0·866 0	0·108 2	0·144 3	0·216 5	0·324 8	0·541 3	0·613 4	0·649 5	1·082 5	1·226 9
1·25	0·156 2	0·312 5	1·082 5	0·135 3	0·180 4	0·270 6	0·406 0	0·676 6	0·766 8	0·811 9	1·353 2	1·533 6
1·5	0·187 5	0·375 0	1·299 0	0·162 4	0·216 5	0·324 8	0·487 1	0·811 9	0·920 2	0·974 3	1·623 8	1·840 3
1·75	0·218 8	0·437 5	1·515 6	0·189 4	0·252 6	0·378 9	0·568 3	0·947 2	1·073 5	1·136 7	1·894 4	2·147 0
2	0·250 0	0·500 0	1·732 1	0·216 5	0·288 7	0·433 0	0·649 5	1·082 5	1·226 9	1·299 0	2·165 1	2·453 8
2·5	0·312 5	0·625 0	2·165 1	0·270 6	0·360 8	0·541 3	0·811 9	1·353 2	1·533 6	1·623 8	2·706 4	3·067 2
3	0·375 0	0·750 0	2·598 1	0·324 8	0·433 0	0·649 5	0·974 3	1·623 8	1·840 3	1·948 6	3·247 6	3·680 6
3·5	0·437 5	0·875 0	3·031 1	0·378 9	0·505 2	0·757 8	1·136 7	1·894 4	2·147 0	2·273 3	3·788 9	4·294 1
4	0·500 0	1·000 0	3·464 1	0·433 0	0·577 4	0·866 0	1·299 0	2·165 1	2·453 8	2·598 1	4·330 2	4·907 5
4·5	0·562 5	1·125 0	3·897 1	0·487 1	0·649 5	0·974 3	1·461 4	2·435 7	2·760 5	2·922 8	4·871 4	5·521 0
5	0·625 0	1·250 0	4·330 2	0·541 2	0·721 7	1·082 6	1·623 8	2·706 4	3·067 2	3·247 6	5·412 7	6·134 4
5·5	0·687 5	1·375 0	4·763 2	0·595 4	0·793 9	1·190 8	1·786 2	2·977 0	3·373 9	3·572 4	5·954 0	6·747 8
6	0·750 0	1·500 0	5·196 2	0·649 5	0·866 0	1·299 1	1·948 6	3·247 6	3·680 6	3·897 1	6·495 2	7·361 3

Table 9 Dimensional Specifications for Self-tapping Screw Threads (All dimensions in inches.)

Type AB (Formerly BA)

Nominal Size or Basic Screw Diameter	Threads per inch	Major Diameter		Minor Diameter		Minimum Practical Screw Lengths	
		Max.	Min.	Max.	Min.	90° Heads	Csk. Heads
0 0.0600	48	0.060	0.054	0.036	0.033	1/8	5/32
1 0.0730	42	0.075	0.069	0.049	0.046	5/32	3/16
2 0.0860	**32**	**0.088**	**0.082**	**0.064**	**0.060**	3/16	7/32
3 0.0990	28	0.101	0.095	0.075	0.071	3/16	1/4
4 0.1120	**24**	**0.114**	**0.108**	**0.086**	**0.082**	7/32	9/32
5 0.1250	20	0.130	0.123	0.094	0.090	1/4	5/16
6 0.1380	**20**	**0.139**	**0.132**	**0.104**	**0.099**	9/32	11/32
7 0.1510	19	0.154	0.147	0.115	0.109	5/16	3/8
8 0.1640	**18**	**0.166**	**0.159**	**0.122**	**0.116**	5/16	3/8
10 0.1900	**16**	**0.189**	**0.182**	**0.141**	**0.135**	3/8	7/16
12 0.2160	14	0.215	0.208	0.164	0.157	7/16	21/32
1/4 0.2500	**14**	**0.246**	**0.237**	**0.192**	**0.185**	1/2	19/32
5/16 0.3125	12	0.315	0.306	0.244	0.236	5/8	3/4
3/8 0.3750	12	0.380	0.371	0.309	0.299	3/4	29/32
7/16 0.4375	10	0.440	0.429	0.359	0.349	7/8	1 1/32
1/2 0.5000	10	0.504	0.493	0.423	0.413	1	1 5/32

Type A

Nominal Size or Basic Screw Diameter	Threads per inch	Major Diameter		Minor Diameter		These Lengths or Shorter — Use Type AB	
		Max.	Min.	Max.	Min.	90° Heads	Csk. Heads
0 0.0600	40	0.060	0.057	0.042	0.039	1/8	3/16
1 0.0730	32	0.075	0.072	0.051	0.048	1/8	3/16
2 0.0860	32	0.088	0.084	0.061	0.056	5/32	3/16
3 0.0990	28	0.101	0.097	0.076	0.071	3/16	7/32
4 0.1120	24	0.114	0.110	0.083	0.078	3/16	1/4
5 0.1250	20	0.130	0.126	0.095	0.090	3/16	1/4
6 0.1380	18	0.141	0.136	0.102	0.096	1/4	5/16
7 0.1510	16	0.158	0.152	0.114	0.108	5/16	3/8
8 0.1640	15	0.168	0.162	0.123	0.116	3/8	7/16
10 0.1900	12	0.194	0.188	0.133	0.126	3/8	1/2
12 0.2160	11	0.221	0.215	0.162	0.155	7/16	9/16
14 0.2420	10	0.254	0.248	0.185	0.178	1/2	5/8
16 0.2680	10	0.280	0.274	0.197	0.189	9/16	3/4
18 0.2940	9	0.306	0.300	0.217	0.209	5/8	13/16
20 0.3200	9	0.333	0.327	0.234	0.226	11/16	13/16
24 0.3720	9	0.390	0.383	0.291	0.282	3/4	1

Type U Metallic Drive Screws

Nom. Size	No. of Starts	Out. Dia.		Pilot Dia.		Nom. Size	No. of Starts	Out. Dia.		Pilot Dia.	
		Max.	Min.	Max.	Min.			Max.	Min.	Max.	Min.
00	6	0.060	0.057	0.049	0.046	8	8	0.167	0.162	0.136	0.132
0	6	0.075	0.072	0.063	0.060	10	8	0.182	0.177	0.150	0.146
2	8	0.100	0.097	0.083	0.080	12	8	0.212	0.206	0.177	0.173
4	7	0.116	0.112	0.096	0.092	14	9	0.242	0.236	0.202	0.198
6	7	0.140	0.136	0.116	0.112	5/16	11	0.315	0.309	0.272	0.267
7	8	0.154	0.150	0.126	0.122	3/8	12	0.378	0.371	0.334	0.329

Table 9 (continued)

THREAD FORMING TYPES B AND BP													
Nominal Size or Basic Screw Diameter	Thds per Inch	Major Diameter		Minor Diameter		Point Diameter		Point Taper Length		Minimum Practical Nominal Screw Lengths			
										Type B		Type BP	
		Max	Min	Max	Min	Max	Min	Max	Min	90° Heads	Csk Heads	90° Heads	Csk Heads
0 0.0600	48	0.060	0.054	0.036	0.033	0.031	0.027	0.042	0.031	1/8	1/8	5/32	3/16
1 0.0730	42	0.075	0.069	0.049	0.046	0.044	0.040	0.048	0.036	1/8	5/32	3/16	7/32
2 0.0860	32	0.088	0.082	0.064	0.060	0.058	0.054	0.062	0.047	5/32	3/16	1/4	9/32
3 0.0990	28	0.101	0.095	0.075	0.071	0.068	0.063	0.071	0.054	3/16	7/32	9/32	5/16
4 0.1120	24	0.114	0.108	0.086	0.082	0.079	0.074	0.083	0.063	3/16	1/4	5/16	11/32
5 0.1250	20	0.130	0.123	0.094	0.090	0.087	0.082	0.100	0.075	7/32	9/32	11/32	13/32
6 0.1380	20	0.139	0.132	0.104	0.099	0.095	0.089	0.100	0.075	1/4	9/32	3/8	7/16
7 0.1510	19	0.154	0.147	0.115	0.109	0.105	0.099	0.105	0.079	1/4	5/16	13/32	15/32
8 0.1640	18	0.166	0.159	0.122	0.116	0.112	0.106	0.111	0.083	9/32	11/32	7/16	1/2
10 0.1900	16	0.189	0.182	0.141	0.135	0.130	0.123	0.125	0.094	5/16	3/8	1/2	19/32
12 0.2160	14	0.215	0.208	0.164	0.157	0.152	0.145	0.143	0.107	11/32	7/16	9/16	21/32
1/4 0.2500	14	0.246	0.237	0.192	0.185	0.179	0.171	0.143	0.107	3/8	1/2	21/32	3/4
5/16 0.3125	12	0.315	0.306	0.244	0.236	0.230	0.222	0.167	0.125	15/32	19/32	27/32	31/32
3/8 0.3750	12	0.380	0.371	0.309	0.299	0.293	0.285	0.167	0.125	17/32	11/16	15/16	1 1/8
7/16 0.4375	10	0.440	0.429	0.359	0.349	0.343	0.335	0.200	0.150	5/8	25/32	1 1/8	1 1/4
1/2 0.5000	10	0.504	0.493	0.423	0.413	0.407	0.399	0.200	0.150	11/16	27/32	1 1/4	1 13/32

THREAD CUTTING TYPES BF AND BT[4]											
Nominal Size or Basic Screw Diameter	Thds per Inch	Major Diameter		Minor Diameter		Point Diameter		Point Taper Length		Minimum Practical Nominal Screw Lengths	
		Max	Min	Max	Min	Max	Min	Max	Min	90° Heads	Csk Heads
0 0.0600	48	0.060	0.054	0.036	0.033	0.031	0.027	0.042	0.031	1/8	1/8
1 0.0730	42	0.075	0.069	0.049	0.046	0.044	0.040	0.048	0.036	1/8	5/32
2 0.0860	32	0.088	0.082	0.064	0.060	0.058	0.054	0.062	0.047	5/32	3/16
3 0.0990	28	0.101	0.095	0.075	0.071	0.068	0.063	0.071	0.054	3/16	7/32
4 0.1120	24	0.114	0.108	0.086	0.082	0.079	0.074	0.083	0.063	3/16	1/4
5 0.1250	20	0.130	0.123	0.094	0.090	0.087	0.082	0.100	0.075	7/32	9/32
6 0.1380	20	0.139	0.132	0.104	0.099	0.095	0.089	0.100	0.075	1/4	9/32
7 0.1510	19	0.154	0.147	0.115	0.109	0.105	0.099	0.105	0.079	1/4	5/16
8 0.1640	18	0.166	0.159	0.122	0.116	0.112	0.106	0.111	0.083	9/32	11/32
10 0.1900	16	0.189	0.182	0.141	0.135	0.130	0.123	0.125	0.094	5/16	3/8
12 0.2160	14	0.215	0.208	0.164	0.157	0.152	0.145	0.143	0.107	11/32	7/16
1/4 0.2500	14	0.246	0.237	0.192	0.185	0.179	0.171	0.143	0.107	3/8	1/2
5/16 0.3125	12	0.315	0.306	0.244	0.236	0.230	0.222	0.167	0.125	15/32	19/32
3/8 0.3750	12	0.380	0.371	0.309	0.299	0.293	0.285	0.167	0.125	17/32	11/16
7/16 0.4375	10	0.440	0.429	0.359	0.349	0.343	0.335	0.200	0.150	5/8	25/32
1/2 0.5000	10	0.504	0.493	0.423	0.413	0.407	0.399	0.200	0.150	11/16	27/32

Table 9 (continued)

Nominal Size or Basic Screw Diameter	Threads per inch	Major Diameter		Point Diameter		Point Taper Length: For Short Screws		Point Taper Length: For Long Screws		Determinant Lengths for Point Taper		Minimum Practical Nominal Screw Lengths	
		Max	Min	Max	Min	Max	Min	Max	Min	90° Heads	Csk Heads	90° Heads	Csk Heads
2 0.0860	56	0.0860	0.0813	0.068	0.061	0.062	0.045	0.080	0.062	5/32	3/16	5/32	3/16
2 0.0860	64	0.0860	0.0816	0.070	0.064	0.055	0.039	0.070	0.055	1/8	3/16	1/8	5/32
3 0.0990	48	0.0990	0.0938	0.078	0.070	0.073	0.052	0.094	0.073	3/16	7/32	5/32	7/32
3 0.0990	56	0.0990	0.0942	0.081	0.074	0.062	0.045	0.080	0.062	5/32	3/16	5/32	3/16
4 0.1120	40	0.1120	0.1061	0.087	0.078	0.088	0.062	0.112	0.088	7/32	1/4	3/16	1/4
4 0.1120	48	0.1120	0.1068	0.091	0.083	0.073	0.052	0.094	0.073	3/16	7/32	5/32	7/32
5 0.1250	40	0.1250	0.1191	0.100	0.091	0.088	0.062	0.112	0.088	7/32	9/32	3/16	1/4
5 0.1250	44	0.1250	0.1195	0.102	0.094	0.080	0.057	0.102	0.080	3/16	1/4	3/16	1/4
6 0.1380	32	0.1380	0.1312	0.107	0.096	0.109	0.078	0.141	0.109	1/4	5/16	1/4	5/16
6 0.1380	40	0.1380	0.1321	0.113	0.104	0.088	0.062	0.112	0.088	7/32	9/32	3/16	1/4
8 0.1640	32	0.1640	0.1571	0.132	0.122	0.109	0.078	0.141	0.109	1/4	11/32	1/4	5/16
8 0.1640	36	0.1640	0.1577	0.136	0.126	0.097	0.069	0.125	0.097	7/32	5/16	7/32	9/32
10 0.1900	24	0.1900	0.1818	0.148	0.135	0.146	0.104	0.188	0.146	11/32	7/16	5/16	13/32
10 0.1900	32	0.1900	0.1831	0.158	0.148	0.109	0.078	0.141	0.109	1/4	11/32	1/4	5/16
12 0.2160	24	0.2160	0.2078	0.174	0.161	0.146	0.104	0.188	0.146	11/32	7/16	5/16	13/32
12 0.2160	28	0.2160	0.2085	0.180	0.168	0.125	0.089	0.161	0.125	5/16	13/32	9/32	3/8
1/4 0.2500	20	0.2500	0.2408	0.200	0.184	0.175	0.125	0.225	0.175	13/32	17/32	3/8	1/2
1/4 0.2500	28	0.2500	0.2425	0.214	0.202	0.125	0.089	0.161	0.125	5/16	13/32	9/32	3/8
5/16 0.3125	18	0.3125	0.3026	0.257	0.239	0.194	0.139	0.250	0.194	15/32	19/32	7/16	9/16
5/16 0.3125	24	0.3125	0.3042	0.271	0.257	0.146	0.104	0.188	0.146	11/32	15/32	5/16	15/32
3/8 0.3750	16	0.3750	0.3643	0.312	0.293	0.219	0.156	0.281	0.219	1/2	11/16	15/32	5/8
3/8 0.3750	24	0.3750	0.3667	0.333	0.319	0.146	0.104	0.188	0.146	11/32	1/2	5/16	1/2
7/16 0.4375	14	0.4375	0.4258	0.366	0.344	0.250	0.179	0.321	0.250	19/32	3/4	9/16	23/32
7/16 0.4375	20	0.4375	0.4281	0.387	0.371	0.175	0.125	0.225	0.175	13/32	9/16	3/8	17/32
1/2 0.5000	13	0.5000	0.4876	0.423	0.399	0.269	0.192	0.346	0.269	5/8	25/32	19/32	3/4
1/2 0.5000	20	0.5000	0.4906	0.450	0.433	0.175	0.125	0.225	0.175	13/32	9/16	3/8	17/32

Table 9 (continued)

Nominal Size or Basic Screw Diameter	Threads per inch	D: Major Diameter		P: Point Diameter		S: Point Taper Length				L: Determinant Lengths for Point Taper		L: Minimum Practical Nominal Screw Lengths	
						For Short Screws		For Long Screws					
		Max	Min	Max	Min	Max	Min	Max	Min	90° Heads	Csk Heads	90° Heads	Csk Heads
2 0.0860	56	0.0860	0.0813	0.068	0.061	0.062	0.045	0.080	0.062	5/32	3/16	5/32	3/16
2 0.0860	64	0.0860	0.0816	0.070	0.064	0.055	0.039	0.070	0.055	1/8	3/16	1/8	5/32
3 0.0990	48	0.0990	0.0938	0.078	0.070	0.073	0.052	0.094	0.073	3/16	7/32	5/32	7/32
3 0.0990	56	0.0990	0.0942	0.081	0.074	0.062	0.045	0.080	0.062	5/32	3/16	5/32	3/16
4 0.1120	40	0.1120	0.1061	0.087	0.078	0.088	0.062	0.112	0.088	7/32	1/4	3/16	1/4
4 0.1120	48	0.1120	0.1068	0.091	0.083	0.073	0.052	0.094	0.073	3/16	7/32	5/32	7/32
5 0.1250	40	0.1250	0.1191	0.100	0.091	0.088	0.062	0.112	0.088	7/32	9/32	3/16	1/4
5 0.1250	44	0.1250	0.1195	0.102	0.094	0.080	0.057	0.102	0.080	3/16	1/4	3/16	1/4
6 0.1380	32	0.1380	0.1312	0.107	0.096	0.109	0.078	0.141	0.109	1/4	5/16	1/4	5/16
6 0.1380	40	0.1380	0.1321	0.113	0.104	0.088	0.062	0.112	0.088	7/32	9/32	3/16	1/4
8 0.1640	32	0.1640	0.1571	0.132	0.122	0.109	0.078	0.141	0.109	1/4	11/32	1/4	5/16
8 0.1640	36	0.1640	0.1577	0.136	0.126	0.097	0.069	0.125	0.097	7/32	5/16	7/32	9/32
10 0.1900	24	0.1900	0.1818	0.148	0.135	0.146	0.104	0.188	0.146	11/32	7/16	5/16	13/32
10 0.1900	32	0.1900	0.1831	0.158	0.148	0.109	0.078	0.141	0.109	1/4	11/32	1/4	5/16
12 0.2160	24	0.2160	0.2078	0.174	0.161	0.146	0.104	0.188	0.146	11/32	7/16	5/16	13/32
12 0.2160	28	0.2160	0.2085	0.180	0.168	0.125	0.089	0.161	0.125	5/16	13/32	9/32	3/8
1/4 0.2500	20	0.2500	0.2408	0.200	0.184	0.175	0.125	0.225	0.175	13/32	17/32	3/8	1/2
1/4 0.2500	28	0.2500	0.2425	0.214	0.202	0.125	0.089	0.161	0.125	5/16	13/32	9/32	3/8
5/16 0.3125	18	0.3125	0.3026	0.257	0.239	0.194	0.139	0.250	0.194	15/32	19/32	7/16	9/16
5/16 0.3125	24	0.3125	0.3042	0.271	0.257	0.146	0.104	0.188	0.146	11/32	15/32	5/16	15/32
3/8 0.3750	16	0.3750	0.3643	0.312	0.293	0.219	0.156	0.281	0.219	1/2	11/16	15/32	5/8
3/8 0.3750	24	0.3750	0.3667	0.333	0.319	0.146	0.104	0.188	0.146	11/32	1/2	5/16	1/2
7/16 0.4375	14	0.4375	0.4258	0.366	0.344	0.250	0.179	0.321	0.250	19/32	3/4	9/16	23/32
7/16 0.4375	20	0.4375	0.4281	0.387	0.371	0.175	0.125	0.225	0.175	13/32	9/16	3/8	17/32
1/2 0.5000	13	0.5000	0.4876	0.423	0.399	0.269	0.192	0.346	0.269	5/8	25/32	19/32	3/4
1/2 0.5000	20	0.5000	0.4906	0.450	0.433	0.175	0.125	0.225	0.175	13/32	9/16	3/8	17/32

Table 10 Basic Dimensions for American National Standard Acme Screw Thread Forms (All dimensions in inches.)

Thds. per Inch	Pitch	Height of Thread (Basic)	Total Height of Thread	Thread Thickness (Basic)	Width of Flat	
					Crest of Internal Thread (Basic)	Root of Internal Thread
16	0.06250	0.03125	0.0362	0.03125	0.0232	0.0206
14	0.07143	0.03571	0.0407	0.03571	0.0265	0.0239
12	0.08333	0.04167	0.0467	0.04167	0.0309	0.0283
10	0.10000	0.05000	0.0600	0.05000	0.0371	0.0319
8	0.12500	0.06250	0.0725	0.06250	0.0463	0.0411
6	0.16667	0.08333	0.0933	0.08333	0.0618	0.0566
5	0.20000	0.10000	0.1100	0.10000	0.0741	0.0689
4	0.25000	0.12500	0.1350	0.12500	0.0927	0.0875
3	0.33333	0.16667	0.1767	0.16667	0.1236	0.1184
2½	0.40000	0.20000	0.2100	0.20000	0.1483	0.1431
2	0.50000	0.25000	0.2600	0.25000	0.1853	0.1802
1½	0.66667	0.33333	0.3433	0.33333	0.2471	0.2419
1⅓	0.75000	0.37500	0.3850	0.37500	0.2780	0.2728
1	1.00000	0.50000	0.5100	0.50000	0.3707	0.3655

Thds per Inch	Pitch	Height of Thread (Basic)	Total Height of Thread (All External Threads)	Thread Thick-ness (Basic)	45-Deg Chamfer Crest of External Threads		Max Fillet Radius, Root of Tapped Hole	Fillet Radius at Minor Diameter of Screws	
					Min Depth	Min Width of Chamfer Flat		Min (Classes 5 and 6 Only)	Max (All)
16	0.06250	0.03125	0.0362	0.03125	0.0031	0.0044	0.0040	0.0044	0.0062
14	0.07143	0.03571	0.0407	0.03571	0.0036	0.0050	0.0040	0.0050	0.0071
12	0.08333	0.04167	0.0467	0.04167	0.0042	0.0060	0.0050	0.0058	0.0083
10	0.10000	0.05000	0.0600	0.05000	0.0050	0.0070	0.0060	0.0070	0.0100
8	0.12500	0.06250	0.0725	0.06250	0.0062	0.0090	0.0075	0.0088	0.0125
6	0.16667	0.08333	0.0933	0.08333	0.0083	0.0120	0.0100	0.0117	0.0167
5	0.20000	0.10000	0.1100	0.10000	0.0100	0.0140	0.0120	0.0140	0.0200
4	0.25000	0.12500	0.1350	0.12500	0.0125	0.0180	0.0150	0.0175	0.0250
3	0.33333	0.16667	0.1767	0.16667	0.0167	0.0240	0.0200	0.0233	0.0333
2½	0.40000	0.20000	0.2100	0.20000	0.0200	0.0280	0.0240	0.0280	0.0400
2	0.50000	0.25000	0.2600	0.25000	0.0250	0.0350	0.0300	0.0350	0.0500
1½	0.66667	0.33333	0.3433	0.33333	0.0330	0.0470	0.0400	0.0467	0.0667
1⅓	0.75000	0.37500	0.3850	0.37500	0.0380	0.0530	0.0450	0.0525	0.0750
1	1.00000	0.50000	0.5100	0.50000	0.0500	0.0710	0.0600	0.0700	0.1000

APPENDIX C

Weights and Measures, and Conversion Tables

Table 1 Decimal Equivalents of Common Fractions

8ths	16ths	32nds	64ths	Exact Decimal Values
			1	0.015625
		1	2	0.3125
			3	0.46875
	1	2	4	0.625
			5	0.078125
		3	6	0.09375
			7	0.109375
1	2	4	8	0.125
			9	0.140625
		5	10	0.15625
			11	0.171875
	3	6	12	0.1875
			13	0.203125
		7	14	0.21875
			15	0.234375
2	4	8	16	0.25
			17	0.265625
		9	18	0.28125
			19	0.296875
	5	10	20	0.3125
			21	0.328125
		11	22	0.34375
			23	0.359375
3	6	12	24	0.375
			25	0.390625
		13	26	0.40625
			27	0.421875
	7	14	28	0.4375
			29	0.453125
		15	30	0.46875
			31	0.484375
4	8	16	32	0.50
			33	0.515625
		17	34	0.53125
			35	0.546875
	9	18	36	0.5625
			37	0.578125
		19	38	0.59375
			39	0.609375

Table 1 (continued)

8ths	16ths	32nds	64ths	Exact Decimal Values
5	10	20	40	0.625
			41	0.640625
		21	42	0.65625
			43	0.671875
	11	22	44	0.6875
			45	0.703125
		23	46	0.71875
			47	0.734375
6	12	24	48	0.75
			49	0.765625
		25	50	0.78125
			51	0.796875
	13	26	52	0.8125
			53	0.828125
		27	54	0.84375
			55	0.859375
7	14	28	56	0.875
			57	0.890625
		29	58	0.90625
			59	0.921875
	15	30	60	0.9375
			61	0.953125
		31	62	0.96875
			63	0.984375
8	16	32	64	1

Table 2 U.S. Common Weights and Measures

Linear Measures

1 mile = 1760 yards = 5280 feet
1 yard = 3 feet = 36 inches
1 foot = 12 inches
1 mil = 0.001 inch
1 fathom = 2 yards = 6 feet
1 rod = 5.5 yards = 16.5 feet
1 hand = 4 inches
1 span = 9 inches

Area Measures

1 square mile = 640 acres = 6400 square chains
1 acre = 10 square chains = 480 square yards = 43,560 square feet
1 square chain = 16 square rods = 484 square yards = 4356 square feet
1 square rod = 30.25 square yards = 272.25 square feet = 625 square links
1 square yard = 9 square feet
1 square foot = 144 square inches

Diameter Measures

1 circular inch = area of circle having a diameter of 1 inch = 0.7854 square inch
1 circular inch = 1,000,000 circular mils
1 square inch = 1.2732 circular inches

Cubic Measures

1 cubic yard = 27 cubic feet
1 cubic foot = 1728 cubic inches

Liquid Measures

1 U.S. gallon = 0.1337 cubic foot = 231 cubic inches = 4 quarts = 8 pints
1 quart = 2 pints = 8 gills
1 pint = 4 gills
1 British Imperial Gallon = 1.2009 U.S. gallon = 277.42 cubic inches

Table 3 Metric Weights and Measures

Linear Measures

1 centimeter = 10 millimeter
1 decimeter = 10 centimeters
1 meter = 10 decimeters
1 kilometer = 1000 meters

Area Measures

1 square centimeter = 100 square millimeters
1 square decimeter = 100 square centimeters
1 square meter = 100 square decimeters
1 are = 100 square meters
1 hectare = 100 ares
1 square kilometer = 100 hectares

Cubic Measures

1 cubic centimeter = 1000 cubic millimeters
1 cubic decimeter = 1000 cubic centimeters
1 cubic meter = 1000 cubic decimeters

Liquid and Dry Measures

1 centiliter = 10 milliliters
1 deciliter = 10 centiliters
1 liter = 10 deciliters
1 hectoliter = 100 liters

Weight Measures

1 centigram = 10 milligrams
1 decigram = 10 centigrams
1 gram = 10 decigrams
1 dekagram = 10 grams
1 hectogram = 10 dekagrams
1 kilogram = 10 hectograms
1 metric ton = 1000 kilograms

Dry Measures

1 bushel = 1.2445 cubic foot = 2150.42 cubic inches
1 bushel = 4 pecks = 32 quarts = 64 pints
1 peck = 8 quarts = 16 pints
1 quart = 2 pints
1 heaped bushel = 1.25 struck bushel

Commercial Weights

1 gross or long ton = 2240 pounds
1 net short ton = 2000 pounds
1 pound = 16 ounces = 7000 grains
1 ounce = 16 drahms = 437.5 grains

Table 4 Metric Conversion Factors

Linear Measures

1 kilometer = 0.6214 mile
1 meter = 39.37 inches = 3.2808 feet = 1.0936 yards
1 centimeter = 0.3937 inch
1 millimeter = 0.03937 inch

1 mile = 1.609 kilometers
1 yard = 0.9144 meter
1 foot = 0.3048 meter = 304.8 millimeter
1 inch = 2.54 centimeters = 25.4 millimeters

Area Measures

1 square kilometer = 0.3861 square mile = 247.1 acres
1 hectare = 2.471 acres = 107,639 square feet
1 are = 0.0247 acre = 1076.4 square feet
1 square meter = 10.764 square feet
1 square centimeter = 0.155 square inch
1 square millimeter = 0.00155 square inch

1 square mile = 2.5899 square kilometers
1 acre = 0.4047 hectare, = 40.47 ares
1 square yard = 0.836 square meter
1 square foot = 0.0929 square meter
1 square inch = 6.452 square centimeters

Cubic Measures

1 cubic meter = 35.315 cubic feet = 1.308 cubic yards
1 cubic meter = 264.2 U.S. gallons
1 cubic centimeter = 0.061 cubic inch
1 liter = 0.0353 cubic foot = 61.023 cubic inches
1 liter = 0.2642 U.S. gallons = 1.0567 U.S. quarts

1 cubic yard = 0.7646 cubic meter
1 cubic foot = 0.02832 cubic meter = 28.317 liters
1 cubic inch = 16.38706 cubic centimeters
1 U.S. gallon = 3.785 liters
1 U.S. quart = 0.946 liter

Weight Measures

1 metric ton = 0.9842 ton (long) = 2204.6 pounds
1 kilogram = 2.2046 pounds = 35.274 ounces
1 gram = 0.03527 ounce
1 gram = 15.432 grains

1 long ton = 1.016 metric ton = 1016 kilograms
1 pound = 0.4536 kilogram = 453.6 grams
1 ounce = 28.35 grams
1 grain = 0.0648 gram
1 calorie (kilogram calorie) = 3.968 Btu

Index

S

T

AUDEL®
Over a Century of Excellence
for the Professional
and
Vocational Trades and the Crafts

Contents

Television Service Manual

fifth edition

Robert G. Middleton; revised by Joseph G. Barrile

$5\frac{1}{2} \times 8\frac{1}{4}$ Hardcover 512 pp. 395 illus.

ISBN: 0-672-23395-9 $15.95

Practical up-to-date guide to all aspects of television transmission and reception, for both black and white and color receivers: • step-by-step maintenance and repair • broadcasting • transmission • receivers • antennas and transmission lines • interference • RF tuners • the video channel • circuits • power supplies • alignment • test equipment.

Electrical Course for Apprentices and Journeymen

second edition

Roland E. Palmquist

$5\frac{1}{2} \times 8\frac{1}{4}$ Hardcover 478 pp. 290 illus.

ISBN:0-672-23393-2 $14.95

Practical course on operational theory and applications for training and re-training in school or on the job: • electricity and matter • units and definitions • electrical symbols • magnets and magnetic fields • capacitors • resistance • electromagnetism • instruments and measurements • alternating currents • DC generators • circuits • transformers • motors • grounding and ground testing.

Questions and Answers for Electricians Examinations

eighth edition

Roland E. Palmquist

$5\frac{1}{2} \times 8\frac{1}{4}$ Hardcover 320 pp. 110 illus.

ISBN: 0-672-23399-1 $12.95

Based on the current National Electrical Code®, a review of exams for apprentice, journeyman, and master, with explanations of principles underlying each test subject: • Ohm's Law and other formulas • power and power factors • lighting • branch circuits and feeders • transformer principles and connections • wiring • batteries and rectification • voltage generation • motors • ground and ground testing.

Machine Shop and Mechanical Trades

Machinists Library

fourth edition 3 vols

Rex Miller

$5\frac{1}{2} \times 8\frac{1}{4}$ Hardcover 1,352 pp. 1,120 illus.

ISBN: 0-672-23380-0 $38.95

Indispensable three-volume reference for machinists, tool and die makers, machine operators, metal workers, and those with home workshops.

Volume I, Basic Machine Shop

$5\frac{1}{2} \times 8\frac{1}{4}$ Hardcover 392 pp. 375 illus.

ISBN: 0-672-23381-9 $14.95

• Blueprint reading • benchwork • layout and measurement • sheet-metal hand tools and machines • cutting tools • drills • reamers • taps • threading dies • milling machine cutters, arbors, collets, and adapters.

Volume II, Machine Shop

$5\frac{1}{2} \times 8\frac{1}{4}$ Hardcover 528 pp. 445 illus

ISBN: 0-672-23382-7 $14.95

• Power saws • machine tool operations • drilling machines • boring • lathes • automatic screw machine • milling • metal spinning.

Volume III, Toolmakers Handy Book

$5\frac{1}{2} \times 8\frac{1}{4}$ Hardcover 432 pp. 300 illus.

ISBN: 0-672-23383-5 $14.95

• Layout work • jigs and fixtures • gears and gear cutting • dies and diemaking • toolmaking operations • heat-treating furnaces • induction heating • furnace brazing • cold-treating process.

Mathematics for Mechanical Technicians and Technologists

John D. Bies

$5\frac{1}{2} \times 8\frac{1}{4}$ Hardcover 392 pp. 190 illus.

ISBN: 0-02-510620-1 $17.95

Practical sourcebook of concepts, formulas, and problem solving in industrial and mechanical technology: • basic and complex mechanics • strength of materials • fluidics • cams and gears • machine elements • machining operations • management controls • economics in machining • facility and human resources management.

Millwrights and Mechanics Guide

third edition

Carl A. Nelson

$5\frac{1}{2} \times 8\frac{1}{4}$ Hardcover 1,040 pp. 880 illus.

ISBN: 0-672-23373-8 $22.95

Most comprehensive and authoritative guide available for millwrights and mechanics at all levels of work or supervision: • drawing and sketching • machinery and equipment installation • principles of mechanical power transmission • V-belt drives • flat belts • gears • chain drives • couplings • bearings • structural steel • screw threads • mechanical fasteners • pipe fittings and valves • carpentry • sheet-metal work • blacksmithing • rigging • electricity • welding • pumps • portable power tools • mensuration and mechanical calculations.

Welders Guide

third edition

James E. Brumbaugh

$5\frac{1}{2} \times 8\frac{1}{4}$ Hardcover 960 pp. 615 illus.

ISBN: 0-672-23374-6 $23.95

Practical, concise manual on theory, operation, and maintenance of all welding machines: • gas welding equipment, supplies, and process • arc welding equipment, supplies, and process • TIG and MIG welding • submerged-arc and other shielded-arc welding processes • resistance, thermit, and stud welding • solders and soldering • brazing and braze welding • welding plastics • safety and health measures • symbols and definitions • testing and inspecting welds. Terminology and definitions as standardized by American Welding Society.

Welder/Fitters Guide

John P. Stewart

$8\frac{1}{2} \times 11$ Paperback 160 pp. 195 illus.

ISBN: 0-672-23325-8 $7.95

Step-by-step instruction for welder/fitters during training or on the job: • basic assembly tools and aids • improving blueprint reading skills • marking and alignment techniques • using basic tools • simple work practices • guide to fabricating weldments • avoiding mistakes • exercises in blueprint reading • clamping devices • introduction to using hydraulic jacks • safety in weld fabrication plants • common welding shop terms.

Sheet Metal Work

John D. Bies

$5\frac{1}{2} \times 8\frac{1}{4}$ Hardcover 456 pp. 215 illus.

ISBN: 0-8161-1706-3 $17.95

On-the-job sheet metal guide for manufacturing, construction, and home workshops: • mathematics for sheet metal work • principles of drafting • concepts of sheet metal drawing • sheet metal standards, specifications, and materials • safety practices • layout • shear cutting • holes • bending and folding • forming operations • notching and clipping • metal spinning • mechanical fastening • soldering and brazing • welding • surface preparation and finishes • production processes.

Power Plant Engineers Guide

third edition

Frank D. Graham; revised by Charlie Buffington

5½ x 8¼ Hardcover 960 pp. 530 illus.
ISBN: 0-672-23329-0 $16.95

All-inclusive question-and-answer guide to steam and diesel-power engines: • fuels • heat • combustion • types of boilers • shell or fire-tube boiler construction • strength of boiler materials • boiler calculations • boiler fixtures, fittings, and attachments • boiler feed pumps • condensers • cooling ponds and cooling towers • boiler installation, startup, operation, maintenance and repair • oil, gas, and waste-fuel burners • steam turbines • air compressors • plant safety.

Mechanical Trades Pocket Manual

second edition

Carl A. Nelson

4 × 6 Paperback 364 pp. 255 illus.
ISBN: 0-672-23378-9 $10.95

Comprehensive handbook of essentials, pocket-sized to fit in the tool box: • mechanical and isometric drawing • machinery installation and assembly • belts • drives • gears • couplings • screw threads • mechanical fasteners • packing and seals • bearings • portable power tools • welding • rigging • piping • automatic sprinkler systems • carpentry • stair layout • electricity • shop geometry and trigonometry.

Plumbing

Plumbers and Pipe Fitters Library

third edition 3 vols

Charles N. McConnell; revised by Tom Philbin

5½x8¼ Hardcover 952 pp. 560 illus.
ISBN: 0-672-23384-3 $34.95

Comprehensive three-volume set with up-to-date information for master plumbers, journeymen, apprentices, engineers, and those in building trades.

Volume 1, Materials, Tools, Roughing-In

5½ x 8¼ Hardcover 304 pp. 240 illus.
ISBN: 0-672-23385-1 $12.95

• Materials • tools • pipe fitting • pipe joints • blueprints • fixtures • valves and faucets.

Volume 2, Welding, Heating, Air Conditioning

5½ x 8¼ Hardcover 384 pp. 220 illus.
ISBN: 0-672-23386-x $13.95

• Brazing and welding • planning a heating system • steam heating systems • hot water heating systems • boiler fittings • fuel-oil tank installation • gas piping • air conditioning.

Volume 3, Water Supply, Drainage, Calculations

5½ x 8¼ Hardcover 264 pp. 100 illus.
ISBN: 0-672-23387-8 $12.95

• Drainage and venting • sewage disposal • soldering • lead work • mathematics and physics for plumbers and pipe fitters.

Home Plumbing Handbook

third edition

Charles N. McConnell

8½ x 11 Paperback 200 pp. 100 illus.
ISBN: 0-672-23413-0 $10.95

Clear, concise, up-to-date fully illustrated guide to home plumbing installation and repair: • repairing and replacing faucets • repairing toilet tanks • repairing a trip-lever bath drain • dealing with stopped-up drains • working with copper tubing • measuring and cutting pipe • PVC and CPVC pipe and fittings • installing a garbage disposals • replacing dishwashers • repairing and replacing water heaters • installing or resetting toilets • caulking around plumbing fixtures and tile • water conditioning • working with cast-iron soil pipe • septic tanks and disposal fields • private water systems.

The Plumbers Handbook

seventh edition

Joseph P. Almond, Sr.

4 × 6 Paperback 352 pp. 170 illus.
ISBN: 0-672-23419-x $10.95

Comprehensive, handy guide for plumbers, pipe fitters, and apprentices that fits in the tool box or pocket: • plumbing tools • how to read blueprints • heating systems • water supply • fixtures, valves, and fittings • working drawings • roughing and repair • outside sewage lift station • pipes and pipelines • vents, drain lines, and septic systems • lead work • silver brazing and soft soldering • plumbing systems • abbreviations, definitions, symbols, and formulas.

Questions and Answers for Plumbers Examinations

second edition

Jules Oravetz

5½ x 8¼ Paperback 256 pp. 145 illus.
ISBN: 0-8161-1703-9 $9.95

Practical, fully illustrated study guide to licensing exams for apprentice, journeyman, or master plumber: • definitions, specifications, and regulations set by National Bureau of Standards and by various state codes • basic plumbing installation • drawings and typical plumbing system layout • mathematics • materials and fittings • joints and connections • traps, cleanouts, and backwater valves • fixtures • drainage, vents, and vent piping • water supply and distribution • plastic pipe and fittings • steam and hot water heating.

HVAC

Air Conditioning: Home and Commercial

second edition

Edwin P. Anderson; revised by Rex Miller

5½ x 8¼ Hardcover 528 pp. 180 illus.
ISBN: 0-672-23397-5 $15.95

Complete guide to construction, installation, operation, maintenance, and repair of home, commercial, and industrial air conditioning systems, with troubleshooting charts: • heat leakage • ventilation requirements • room air conditioners • refrigerants • compressors • condensing equipment • evaporators • water-cooling systems • central air conditioning • automobile air conditioning • motors and motor control.

Heating, Ventilating and Air Conditioning Library

second edition 3 vols

James E. Brumbaugh

5½ x 8¼ Hardcover 1,840 pp. 1,275 illus.
ISBN: 0-672-23388-6 $42.95

Authoritative three-volume reference for those who install, operate, maintain, and repair HVAC equipment commercially, industrially, or at home. Each volume fully illustrated with photographs, drawings, tables and charts.

Volume I, Heating Fundamentals, Furnaces, Boilers, Boiler Conversions

5½ x 8¼ Hardcover 656 pp. 405 illus.
ISBN: 0-672-23389-4 $16.95

• Insulation principles • heating calculations • fuels • warm-air, hot water, steam, and electrical heating systems • gas-fired, oil-fired, coal-fired, and electric-fired furnaces • boilers and boiler fittings • boiler and furnace conversion.

Volume II, Oil, Gas and Coal Burners, Controls, Ducts, Piping, Valves

5½ x 8¼ Hardcover 592 pp. 455 illus.
ISBN: 0-672-23390-8 $15.95

• Coal firing methods • thermostats and humidistats • gas and oil controls and other automatic controls •

ducts and duct systems • pipes, pipe fittings, and piping details • valves and valve installation • steam and hot-water line controls.

Volume III, Radiant Heating, Water Heaters, Ventilation, Air Conditioning, Heat Pumps, Air Cleaners
5 1/2 x 8 1/4 Hardcover 592 pp. 415 illus.
ISBN: 0-672-23391-6 $14.95

• Radiators, convectors, and unit heaters • fireplaces, stoves, and chimneys • ventilation principles • fan selection and operation • air conditioning equipment • humidifiers and dehumidifiers • air cleaners and filters.

Oil Burners fourth edition

Edwin M. Field
5 1/2 x 8 1/4 Hardcover 360 pp. 170 illus.
ISBN: 0-672-23394-0 $15.95

Up-to-date sourcebook on the construction, installation, operation, testing, servicing, and repair of all types of oil burners, both industrial and domestic: • general electrical hookup and wiring diagrams of automatic control systems • ignition system • high-voltage transportation • operational sequence of limit controls, thermostats, and various relays • combustion chambers • drafts • chimneys • drive couplings • fans or blowers • burner nozzles • fuel pumps.

Refrigeration: Home and Commercial second edition

Edwin P. Anderson; revised by Rex Miller
5 1/2 x 8 1/4 Hardcover 768 pp. 285 illus.
ISBN: 0-672-23396-7 $17.95

Practical, comprehensive reference for technicians, plant engineers, and homeowners on the installation, operation, servicing, and repair of everything from single refrigeration units to commercial and industrial systems: • refrigerants • compressors • thermoelectric cooling • service equipment and tools • cabinet maintenance and repairs • compressor lubrication systems • brine systems • supermarket and grocery refrigeration • locker plants • fans and blowers • piping • heat leakage • refrigeration-load calculations.

Pneumatics and Hydraulics

Hydraulics for Off-the-Road Equipment second edition

Harry L. Stewart; revised by Tom Philbin
5 1/2 x 8 1/4 Hardcover 256 pp. 175 illus.
ISBN: 0-8161-1701-2 $13.95

Complete reference manual for those who own and operate heavy equipment and for engineers, designers, installation and maintenance technicians, and shop mechanics: • hydraulic pumps, accumulators, and motors • force components • hydraulic control components • filters and filtration, lines and fittings, and fluids • hydrostatic transmissions • maintenance • troubleshooting.

Pneumatics and Hydraulics fourth edition

Harry L. Stewart; revised by Tom Philbin
5 1/2 x 8 1/4 Hardcover 512 pp. 315 illus.
ISBN: 0-672-23412-2 $15.95

Practical guide to the principles and applications of fluid power for engineers, designers, process planners, tool men, shop foremen, and mechanics: • pressure, work and power • general features of machines • hydraulic and pneumatic symbols • pressure boosters • air compressors and accessories • hydraulic power devices • hydraulic fluids • piping • air filters, pressure regulators, and lubricators • flow and pressure controls • pneumatic motors and tools • rotary hydraulic motors and hydraulic transmissions • pneumatic circuits • hydraulic circuits • servo systems.

Pumps fourth edition

Harry L. Stewart; revised by Tom Philbin
5 1/2 x 8 1/4 Hardcover 508 pp. 360 illus.
ISBN: 0-672-23400-9 $15.95

Comprehensive guide for operators, engineers, maintenance workers, inspectors, superintendents, and mechanics on principles and day-to-day operations of pumps: • centrifugal, rotary, reciprocating, and special service pumps • hydraulic accumulators • power transmission • hydraulic power tools • hydraulic cylinders • control valves • hydraulic fluids • fluid lines and fittings.

Carpentry and Construction

Carpenters and Builders Library

fifth edition 4 vols
John E. Ball; revised by Tom Philbin
5 1/2 x 8 1/4 Hardcover 1,224 pp. 1,010 illus.
ISBN: 0-672-23369-x $39.95
Also available in a new boxed set at no extra cost:
ISBN: 0-02-506450-9 $39.95

These profusely illustrated volumes, available in a handsome boxed edition, have set the professional standard for carpenters, joiners, and woodworkers.

Volume 1, Tools, Steel Square, Joinery
5 1/2 x 8 1/4 Hardcover 384 pp. 345 illus.
ISBN: 0-672-23365-7 $10.95

• Woods • nails • screws • bolts • the workbench • tools • using the steel square • joints and joinery • cabinetmaking joints • wood patternmaking • and kitchen cabinet construction.

Volume 2, Builders Math, Plans, Specifications
5 1/2 x 8 1/4 Hardcover 304 pp. 205 illus.
ISBN: 0-672-23366-5 $10.95

• Surveying • strength of timbers • practical drawing • architectural drawing • barn construction • small house construction • and home workshop layout.

Volume 3, Layouts, Foundations, Framing
5 1/2 x 8 1/4 Hardcover 272 pp. 215 illus.
ISBN: 0-672-23367-3 $10.95

• Foundations • concrete forms • concrete block construction • framing, girders and sills • skylights • porches and patios • chimneys, fireplaces, and stoves • insulation • solar energy and paneling.

Volume 4, Millwork, Power Tools, Painting
5 1/2 x 8 1/4 Hardcover 344 pp. 245 illus.
ISBN: 0-672-23368-1 $10.95

• Roofing, miter work • doors • windows, sheathing and siding • stairs • flooring • table saws, band saws, and jigsaws • wood lathes • sanders and combination tools • portable power tools • painting.

Complete Building Construction

second edition
John Phelps; revised by Tom Philbin
5 1/2 x 8 1/4 Hardcover 744 pp. 645 illus.
ISBN: 0-672-23377-0 $19.95

Comprehensive guide to constructing a frame or brick building from the footings to the ridge: • laying out building and excavation lines • making concrete forms and pouring fittings and foundation • making concrete slabs, walks, and driveways • laying concrete block, brick, and tile • building chimneys and fireplaces • framing, siding, and roofing • insulating • finishing the inside • building stairs • installing windows • hanging doors.

Complete Roofing Handbook

James E. Brumbaugh

5½ x 8¼ Hardcover 536 pp. 510 illus.
ISBN: 0-02-517850-4 $29.95

Authoritative text and highly detailed drawings and photographs on all aspects of roofing: • types of roofs • roofing and reroofing • roof and attic insulation and ventilation • skylights and roof openings • dormer construction • roof flashing details • shingles • roll roofing • built-up roofing • roofing with wood shingles and shakes • slate and tile roofing • installing gutters and downspouts • listings of professional and trade associations and roofing manufacturers.

Complete Siding Handbook

James E. Brumbaugh

5½ x 8¼ Hardcover 512 pp. 450 illus.
ISBN: 0-02-517880-6 $23.95

Companion to *Complete Roofing Handbook,* with step-by-step instructions and drawings on every aspect of siding: • sidewalls and siding • wall preparation • wood board siding • plywood panel and lap siding • hardboard panel and lap siding • wood shingle and shake siding • aluminum and steel siding • vinyl siding • exterior paints and stains • refinishing of siding, gutter and downspout systems • listings of professional and trade associations and siding manufacturers.

Masons and Builders Library

second edition 2 vols

Louis M. Dezettel; revised by Tom Philbin

5½ x 8¼ Hardcover 688 pp. 500 illus.
ISBN: 0-672-23401-7 $23.95

Two-volume set on practical instruction in all aspects of materials and methods of bricklaying and masonry: • brick • mortar • tools • bonding • corners, openings, and arches • chimneys and fireplaces • structural clay tile and glass block • brick walks, floors, and terraces • repair and maintenance • plasterboard and plaster • stone and rock masonry • reading blueprints.

Volume 1, Concrete, Block, Tile, Terrazzo
5½ x 8¼ Hardcover 304 pp. 190 illus.
ISBN: 0-672-23402-5 $13.95

Volume 2, Bricklaying, Plastering, Rock Masonry, Clay Tile
5½ x 8¼ Hardcover 384 pp. 310 illus.
ISBN: 0-672-23403-3 $12.95

Woodworking

Woodworking and Cabinetmaking

F. Richard Boller

5½ x 8¼ Hardcover 360 pp. 455 illus.
ISBN: 0-02-512800-0 $16.95

Compact one-volume guide to the essentials of all aspects of woodworking: • properties of softwoods, hardwoods, plywood, and composition wood • design, function, appearance, and structure • project planning • hand tools • machines • portable electric tools • construction • the home workshop • and the projects themselves – stereo cabinet, speaker cabinets, bookcase, desk, platform bed, kitchen cabinets, bathroom vanity.

Wood Furniture: Finishing, Refinishing, Repairing second edition

James E. Brumbaugh

5½ x 8¼ Hardcover 352 pp. 185 illus.
ISBN: 0-672-23409-2 $12.95

Complete, fully illustrated guide to repairing furniture and to finishing and refinishing wood surfaces for professional woodworkers and do-it-yourselfers: • tools and supplies • types of wood • veneering • inlaying • repairing, restoring, and stripping • wood preparation • staining • shellac, varnish, lacquer, paint and enamel, and oil and wax finishes • antiquing • gilding and bronzing • decorating furniture.

Maintenance and Repair

Building Maintenance second edition

Jules Oravetz

5½ x 8¼ Hardcover 384 pp. 210 illus.
ISBN: 0-672-23278-2 $9.95

Complete information on professional maintenance procedures used in office, educational, and commercial buildings: • painting and decorating • plumbing and pipe fitting • concrete and masonry • carpentry • roofing • glazing and caulking • sheet metal • electricity • air conditioning and refrigeration • insect and rodent control • heating • maintenance management • custodial practices.

Gardening, Landscaping and Grounds Maintenance

third edition

Jules Oravetz

5½ x 8¼ Hardcover 424 pp. 340 illus.
ISBN: 0-672-23417-3 $15.95

Practical information for those who maintain lawns, gardens, and industrial, municipal, and estate grounds: • flowers, vegetables, berries, and house plants • greenhouses • lawns • hedges and vines • flowering shrubs and trees • shade, fruit and nut trees • evergreens • bird sanctuaries • fences • insect and rodent control • weed and brush control • roads, walks, and pavements • drainage • maintenance equipment • golf course planning and maintenance.

Home Maintenance and Repair: Walls, Ceilings and Floors

Gary D. Branson

8½ x 11 Paperback 80 pp. 80 illus.
ISBN: 0-672-23281-2 $6.95

Do-it-yourselfer's step-by-step guide to interior remodeling with professional results: • general maintenance • wallboard installation and repair • wallboard taping • plaster repair • texture paints • wallpaper techniques • paneling • sound control • ceiling tile • bath tile • energy conservation.

Painting and Decorating

Rex Miller and Glenn E. Baker

5½ x 8¼ Hardcover 464 pp. 325 illus.
ISBN: 0-672-23405-x $18.95

Practical guide for painters, decorators, and homeowners to the most up-to-date materials and techniques: • job planning • tools and equipment needed • finishing materials • surface preparation • applying paint and stains • decorating with coverings • repairs and maintenance • color and decorating principles.

Tree Care second edition

John M. Haller

8½ x 11 Paperback 224 pp. 305 illus.

ISBN: 0-02-062870-6 $9.95

New edition of a standard in the field, for growers, nursery owners, foresters, landscapers, and homeowners: • planting • pruning • fertilizing • bracing and cabling • wound repair • grafting • spraying • disease and insect management • coping with environmental damage • removal • structure and physiology • recreational use.

Upholstering

updated

James E. Brumbaugh

5½ x 8¼ Hardcover 400 pp. 380 illus.

ISBN: 0-672-23372-x $12.95

Essentials of upholstering for professional, apprentice, and hobbyist: • furniture styles • tools and equipment • stripping • frame construction and repairs • finishing and refinishing wood surfaces • webbing • springs • burlap, stuffing, and muslin • pattern layout • cushions • foam padding • covers • channels and tufts • padded seats and slip seats • fabrics • plastics • furniture care.

Automotive and Engines

Diesel Engine Manual fourth edition

Perry O. Black; revised by William E. Scahill

5½ x 8¼ Hardcover 512 pp. 255 illus.

ISBN: 0-672-23371-1 $15.95

Detailed guide for mechanics, students, and others to all aspects of typical two- and four-cycle engines: • operating principles • fuel oil • diesel injection pumps • basic Mercedes diesels • diesel engine cylinders • lubrication • cooling systems • horsepower • engine-room procedures • diesel engine installation • automotive diesel engine • marine diesel engine • diesel electrical power plant • diesel engine service.

Gas Engine Manual third edition

Edwin P. Anderson; revised by Charles G. Facklam

5½ x 8¼ Hardcover 424 pp. 225 illus.

ISBN: 0-8161-1707-1 $12.95

Indispensable sourcebook for those who operate, maintain, and repair gas engines of all types and sizes: • fundamentals and classifications of engines · engine parts • pistons • crankshafts • valves • lubrication, cooling, fuel, ignition, emission control and electrical systems • engine tune-up • servicing of pistons and piston rings, cylinder blocks, connecting rods and crankshafts, valves and valve gears, carburetors, and electrical systems.

Small Gasoline Engines

Rex Miller and Mark Richard Miller

5 ½ x 8¼ Hardcover 640 pp. 525 illus.

ISBN: 0-672-23414-9 $16.95

Practical information for those who repair, maintain, and overhaul two- and four-cycle engines – with emphasis on one-cylinder motors – including lawn mowers, edgers, grass sweepers, snowblowers, emergency electrical generators, outboard motors, and other equipment up to ten horsepower: • carburetors, emission controls, and ignition systems • starting systems • hand tools • safety • power generation • engine operations • lubrication systems • power drivers • preventive maintenance • step-by-step overhauling procedures • troubleshooting • testing and inspection • cylinder block servicing.

Truck Guide Library 3 vols

James E. Brumbaugh

5½ x 8¼ Hardcover 2,144 pp. 1,715 illus.

ISBN: 0-672-23392-4 $45.95

Three-volume comprehensive and profusely illustrated reference on truck operation and maintenance.

Volume 1, Engines

5½ x 8¼ Hardcover 416 pp. 290 illus.

ISBN: 0-672-23356-8 $16.95

• Basic components · engine operating principles • troubleshooting • cylinder blocks • connecting rods, pistons, and rings • crankshafts, main bearings, and flywheels • camshafts and valve trains • engine valves.

Volume 2, Engine Auxiliary Systems

5½ x 8¼ Hardcover 704 pp. 520 illus.

ISBN: 0-672-23357-6 $16.95

• Battery and electrical systems • spark plugs • ignition systems, charging and starting systems • lubricating, cooling, and fuel systems • carburetors and governors • diesel systems • exhaust and emission-control systems.

Volume 3, Transmissions, Steering, and Brakes

5½ x 8¼ Hardcover 1,024 pp. 905 illus.

ISBN: 0-672-23406-8 $16.95

• Clutches • manual, auxiliary, and automatic transmissions • frame and suspension systems • differentials and axles, manual and power steering • front-end alignment • hydraulic, power, and air brakes • wheels and tires • trailers.

Drafting

Answers on Blueprint Reading

fourth edition

Roland E. Palmquist; revised by Thomas J. Morrisey

5½ x 8¼ Hardcover 320 pp. 275 illus.

ISBN: 0-8161-1704-7 $12.95

Complete question-and-answer instruction manual on blueprints of machines and tools, electrical systems, and architecture: • drafting scale • drafting instruments • conventional lines and representations • pictorial drawings • geometry of drafting • orthographic and working drawings • surfaces • detail drawing • sketching • map and topographical drawings • graphic symbols • architectural drawings • electrical blueprints • computer-aided design and drafting. Also included is an appendix of measurements • metric conversions • screw threads and tap drill sizes • number and letter sizes of drills with decimal equivalents • double depth of threads • tapers and angles.

Hobbies

Complete Course in Stained Glass

Pepe Mendez

8½ x 11 Paperback 80 pp. 50 illus.

ISBN: 0-672-23287-1 $8.95

Guide to the tools, materials, and techniques of the art of stained glass, with ten fully illustrated lessons: • how to cut glass • cartoon and pattern drawing • assembling and cementing • making lamps using various techniques • electrical components for completing lamps • sources of materials • glossary of terminology and techniques of stained glasswork.

Macmillan Practical Arts Library
Books for and by the Craftsman

World Woods in Color

W.A. Lincoln

7 × 10 Hardcover 300 pages
300 photos
ISBN: 0-02-572350-2 $39.95

Large full-color photographs show the natural grain and features of nearly 300 woods: • commercial and botanical names • physical characteristics, mechanical properties, seasoning, working properties, durability, and uses • the height, diameter, bark, and places of distribution of each tree • indexing of botanical, trade, commercial, local, and family names • a full bibliography of publications on timber study and identification.

The Woodworker's Bible

Alf Martensson

8 × 10 Paperback 288 pages 900 illus.
ISBN: 0-02-011940-2 $12.95

For the craftsperson familiar with basic carpentry skills, a guide to creating professional-quality furniture, cabinetry, and objects d'art in the home workshop: • techniques and expert advice on fine craftsmanship whether tooled by hand or machine • joint-making • assembling to ensure fit • finishes. Author, who lives in London and runs a workshop called Woodstock, has also written. *The Book of Furnituremaking.*

Cabinetmaking: The Professional Approach

Alan Peters

$8\frac{1}{2} \times 11$ Hardcover 208 pages 175 illus.
(8 pp. color)
ISBN: 0-02-596200-0 $29.95

A unique guide to all aspects of professional furniture making, from an English master craftsman: • the Cotswold School and the birth of the furniture movement • setting up a professional shop • equipment • finance and business efficiency • furniture design • working to commission • batch production, training, and techniques • plans for nine projects.

The Woodturner's Art: Fundamentals and Projects

Ron Roszkiewicz

8 × 10 Hardcover 256 pages 300 illus.
ISBN: 0-02-605250-4 $24.95

A master woodturner shows how to design and create increasingly difficult projects step-by-step in this book suitable for the beginner and the more advanced student: • spindle and faceplate turning • tools • techniques • classic turnings from various historical periods • more than 30 types of projects including boxes, furniture, vases, and candlesticks • making duplicates • projects using combinations of techniques and more than one kind of wood. Author has also written *The Woodturner's Companion.*

Cabinetmaking and Millwork

John L. Feirer

$7\frac{1}{8} \times 9\frac{1}{2}$ Hardcover 992 pages
2,350 illus. (32 pp. in color)
ISBN: 0-02-537350-1 $47.50

The classic on cabinetmaking that covers in detail all of the materials, tools, machines, and processes used in building cabinets and interiors, the production of furniture, and other work of the finish carpenter and millwright: • fixed installations such as paneling, built-ins, and cabinets • movable wood products such as furniture and fixtures • which woods to use, and why and how to use them in the interiors of homes and commercial buildings • metrics and plastics in furniture construction.

Carpentry and Building Construction

John L. Feirer and Gilbert R. Hutchings

$7\frac{1}{2} \times 9\frac{1}{2}$ hardcover 1,120 pages
2,000 photos (8 pp. in color)
ISBN: 0-02-537360-9 $50.00

A classic by Feirer on each detail of modern construction: • the various machines, tools, and equipment from which the builder can choose • laying of a foundation • building frames for each part of a building • details of interior and exterior work • painting and finishing • reading plans • chimneys and fireplaces • ventilation • assembling prefabricated houses.